T0214260

Lecture Notes in Computer Science 11932

More information about this series at http://www.springer.com/series/7409

Sanjay Madria · Philippe Fournier-Viger ·
Sanjay Chaudhary · P. Krishna Reddy (Eds.)

Big Data Analytics

7th International Conference, BDA 2019
Ahmedabad, India, December 17–20, 2019
Proceedings

 Springer

Editors
Sanjay Madria
Missouri University of Science
and Technology
Rolla, MO, USA

Sanjay Chaudhary 🆔
Ahmedabad University
Ahmedabad, India

Philippe Fournier-Viger 🆔
Harbin Institute of Technology
Shenzhen, China

P. Krishna Reddy 🆔
International Institute of Information
Technology
Hyderabad, India

ISSN 0302-9743   ISSN 1611-3349 (electronic)
Lecture Notes in Computer Science
ISBN 978-3-030-37187-6   ISBN 978-3-030-37188-3 (eBook)
https://doi.org/10.1007/978-3-030-37188-3

LNCS Sublibrary: SL3 – Information Systems and Applications, incl. Internet/Web, and HCI

This Springer imprint is published by the registered company Springer Nature Switzerland AG
The registered company address is: Gewerbestrasse 11, 6330 Cham, Switzerland

Preface

The amount of data stored in computer systems has greatly increased in recent years due in part to advances in networking technologies, storage systems, the adoption of mobile and cloud computing, and the wide deployment of sensors for data collection. To make sense of this big data to support decision-making, the field of Big Data Analytics has emerged as a key research and application area for the industry and other organizations. Numerous applications of big data analytics are found in several important and diverse fields such as e-commerce, finance, healthcare, education, e-governance, finance, media and entertainment, security and surveillance, smart cities, telecommunications, agriculture, astronomy, and transportation.

Analysis of big data raises several challenges such as how to handle massive volumes of data, process data in real-time, and deal with complex, uncertain, heterogeneous, and streaming data, which are often stored in multiple remote repositories. To address these challenges, innovative big data analysis solutions must be designed by drawing expertise from recent advances in several fields such as big data processing, data mining, database systems, statistics, machine learning, and artificial intelligence. There is also an important need to build data analysis systems for emerging applications such as vehicular networks, social media analysis, and to facilitate the deployment of big data analysis techniques.

The 7th International Conference on Big Data Analytics (BDA 2019) was held during December 18–21, 2019, at Ahmedabad University, Ahmedabad, Gujarat, India. This proceedings book includes 13 peer-reviewed research papers and contributions by keynote speakers, invited speakers, and tutorial speakers. This year's program covers a wide range of topics related to big data analytics on themes such as: big data analytics, gesture detection, networking, social media, search, information extraction, image processing and analysis, spatial, text, mobile and graph data analysis, machine learning, and healthcare.

It is expected that research papers, keynote speeches, invited talks, and tutorials presented at the conference will encourage research on big data analytics and stimulate the development of innovative solutions to the problems faced in industry.

The conference received 53 submissions. The Program Committee (PC) consisted of researchers from both academia as well as industry from 13 different countries or territories, namely Australia, Canada, China, Estonia, France, India, Japan, New Caledonia, New Zealand, Norway, Spain, Taiwan, and the USA. Each submission was reviewed by about 3 reviewers in the Program Committee, and was discussed by PC chairs before taking the decision. Based on the above review process, 13 full papers were selected. The overall acceptance rate was about 25%.

We would like to extend our sincere thanks to the reviewers for their time, energy, and expertise in providing support to BDA 2019. Additionally, we would like to thank all the authors who considered BDA 2019 as the forum to publish their research contributions.

The Steering Committee and the Organizing Committee deserve praise for the support they provided. A number of individuals contributed to the success of the conference. We thank Prof. Shashi Shekhar, Prof. Elizabeth Chang, and Prof. R. K. Agrawal for their insightful keynote talks. We also thank all invited speakers and tutorial speakers. We would like to thank the sponsoring organizations, including Space Applications Center of Indian Science Research Organization, Google Research, Gujarat Council of Science and Technology, Meditab Software Pvt. Ltd., Dev Information Technology Ltd., ICICI Bank, and Alumni Association of School of Computer Studies of Ahmedabad University, who deserve praise for the support they provided.

The conference received invaluable support from the management of the Ahmedabad University in hosting and organizing the conference. Moreover, thanks are also extended to the faculty, staff members, and student volunteers of the School of Engineering and Applied Science of the Ahmedabad University for their constant cooperation and support.

November 2019

Sanjay Madria
Philippe Fournier-Viger
Sanjay Chaudhary
P. Krishna Reddy

Organization

BDA 2019 was organized by Ahmedabad University, Ahmedabad, Gujarat, India.

Chief Patrons

Pankaj Chandra	Ahmedabad University, India
Sudhir Jain	IIT Gandhinagar, India

Honorary Chairs

K. S. Dasgupta	DA-IICT Gandhinagar, India
Sunil Kale	Ahmedabad University, India
Rekha Jain	IIM Ahmedabad, India

General Chair

Sanjay Chaudhary	Ahmedabad University, India

Steering Committee Chair

P. Krishna Reddy	IIIT Hyderabad, India

Steering Committee

S. K. Gupta	IIT Delhi, India
Srinath Srinivasa	IIIT Bangalore, India
Krithi Ramamritham	IIT Bombay, India
Sanjay Kumar Madria	Missouri University of Science and Technology, USA
Masaru Kitsuregawa	University of Tokyo, Japan
Raj K. Bhatnagar	University of Cincinnati, USA
Vasudha Bhatnagar	University of Delhi, India
Mukesh Mohania	IBM Research, Australia
H. V. Jagadish	University of Michigan, USA
Ramesh Kumar Agrawal	Jawaharlal Nehru University, India
Divyakant Agrawal	University of California at Santa Barbara, USA
Arun Agarwal	University of Hyderabad, India
Subhash Bhalla	The University of Aizu, Japan
Jaideep Srivastava	University of Minnesota, USA
Anirban Mondal	Ashoka University, India
Sharma Chakravarthy	The University of Texas at Arlington, USA

Program Committee Chairs

Sanjay Madria   Missouri University of Science and Technology, USA
Philippe Fournier-Viger   Harbin Institute of Technology (Shenzhen), China

Organizing Chairs

Kumar Shashi Prabh   Ahmedabad University, India
Mehul Raval   Pandit Deendayal Petroleum University, India

Finance Chairs

Dhaval Patel   Ahmedabad University, India
Amitava Ghosh   Ahmedabad University, India

Sponsorship Chair

Anil Roy   DA-IICT, India

Publication Chair

P. Krishna Reddy   IIIT Hyderabad, India

Workshop Chairs

Srikrishnan Divakaran   Ahmedabad University, India
Punam Bedi   University of Delhi, India

Tutorial Chairs

Mukesh Mohania   IBM, Australia
Takahiro Hara   Osaka University, Japan

Publicity Chairs

Shelly Sachdeva   National Institute of Technology Delhi, India
M. T. Savaliya   Vishwakarma Government Engineering College, India

Web Site Chair

Pratik Trivedi   Ahmedabad University, India

Organizing Committee

C. K. Bhensdadia   Dharmsinh Desai University, India
Amit Ganatra   CHARUSAT, India

Sanjay Sampat	Government Polytechnic College – Surat, India
Aditya Patel	Ahmedabad University, India
George Varughese	Ahmedabad University, India
Kangana Bagani	Ahmedabad University, India
Rahul Bhathiji	Ahmedabad University, India
Barbara Moraswka	Ahmedabad University, India
Anurag Lakhlani	Ahmedabad University, India
Kuntal Patel	Ahmedabad University, India
Shefali Naik	Ahmedabad University, India
Jaydeep Raolji	Ahmedabad University, India
Kunjal Gajjar	Ahmedabad University, India
Hiral Vegda	Ahmedabad University, India
Kinjal Arya	Ahmedabad University, India
Govind Prajapati	Ahmedabad University, India
Amar Gajjar	Ahmedabad University, India
Aman Dave	Ahmedabad University, India
Kruti Yadav	Ahmedabad University, India
Manav Shah	Ahmedabad University, India
Jainam Shah	Ahmedabad University, India
Mohit Vaswani	Ahmedabad University, India
Tirth Jivani	Ahmedabad University, India
Neha Shukla	Ahmedabad University, India
Vinay Vachcharajani	Ahmedabad University, India
Siddhi Shah	Ahmedabad University, India

Program Committee

Amartya Sen	Oakland University, USA
Akhil Kumar	Penn State University, USA
Alok Singh	University of Hyderabad, India
Anirban Mondal	Ashoka University, India
Arkady Zaslavsky	CSIRO, Australia
Engelbert Mephu Nguifo	Université Blaise Pascal – Clermont Ferrand, France
Himanshu Gupta	IBM Research, India
Jerry Chun-Wei Lin	Western Norway University of Applied Sciences, Norway
Ji Zhang	University Southern Queensland, Australia
Jose Marie Luna	University of Cordoba, Spain
Ladjel Bellatreche	ENSMA, France
Lakshmish Ramaswamy	University of Georgia, USA
Lukas Pichl	International Christian University, Japan
Morteza Zihayat	Ryerson University, Canada
Naresh Manwani	IIIT Hyderabad, India
Nazha Selmaoui Folcher	Université de Nouvelle Calédonie, New Caledonia, France
Pradeep Kumar	IIM Lucknow, India

Praveen Rao	University of Missouri – Kansas City, USA
Prem Jayaraman	Swinburne University of Technology, Australia
Roger Nkambou	Université du Québec à Montréal, Canada
Santhanagopalan Rajagopalan	IIIT Bangalore, India
Satish Narayana Srirama	University of Tartu, Estonia
Sebastian Ventura	University of Cordoba, Spain
Srinath Srinivasa	IIIT Bangalore, India
Tin Truong Chi	University of Dalat, Vietnam
Tzung-Pei	Hong National University of Kaohsiung, Taiwan
Uday Kiran	University of Tokyo, Japan
Vasudha Bhatnagar	University of Delhi, India
Vimal Kumar	University of Walkato, New Zealand
Zeyar Aung Masdar	Institute of Science and Technology, UAE

Sponsoring Institutions

Space Applications Center – Indian Science Research Organization
Google Research
Gujarat Council of Science and Technology
Meditab Software Pvt. Ltd.
Dev Information Technology Ltd.
ICICI Bank
Alumni Association of School of Computer Studies of Ahmedabad University

Contents

Machine Learning

Big Data Analytics: Vision and Perspectives

Transforming Sensing Data into Smart Data for Smart Sustainable Cities

Koji Zettsu[✉]

National Institute of Information and Communications Technology,
Tokyo 1848795, Japan
zettsu@nict.go.jp

Abstract. With recent advances in the Internet of Things (IoT), a wide variety of sensing data are disseminated, shared and utilized in smart cities to improve their efficiency and quality of life of citizens. The key is to turn sensing data into actionable information called "smart data", used for planning, monitoring, navigation and intelligent decision making. In order to manipulate smart data, advanced data analytics is indispensable for detecting valuable events from sensing data and discovering and predicting latent associations among different kind of events. Their optimization in collaboration between a variety of observation data and application-specific data collected from users is also a crucial. In NICT Real World Information Analytics Project, an ICT platform called xData (cross-data) platform is constructed for developing smart applications with harnessing the above technologies toward realization of smart and sustainable cities. For example, association discovery from a variety of meteorological and traffic data is performed to create and distribute a map that predicts various transport disturbance risks due to heavy rain, heavy snow and other abnormal weather conditions and to navigate safe, risk-free routes.

Keywords: Smart data · IoT data analytics · xData platform

1 Introduction

Given that an estimated 70% of the world's population will live in cities by 2050, environment surrounding cities is drastically changing and social problems are becoming increasingly complex. A smart sustainable city is an innovative city that uses information and communication technology (ICT) to improve quality of life, efficiency of urban operation and services, while ensuring that it meets the needs of present and future generations with respect to economic, social, environmental aspects [1]. With recent advances of Internet of Things (IoT), cities are getting equipped with many modern devices, employing artificial intelligence (AI) to analyze big data from these devices, and thus higher convergence of physical space and cyber space are promoted. Here, data are the key to produce viable long-living solutions, and to identify the opportunities and challenges of novel best methods and practices on smart and sustainable cities.

"Smart data" refers to the IoT data that has been processed to produce valuable data be turned into actionable information, which can be used for intelligence, planning,

© Springer Nature Switzerland AG 2019
S. Madria et al. (Eds.): BDA 2019, LNCS 11932, pp. 3–19, 2019.
https://doi.org/10.1007/978-3-030-37188-3_1

controlling and decision making efficiently and effectively by governments, industries and citizens. Unprecedentedly large amount and variety of sensory data can be collected to explore how these big data can become smart data and offer intelligence. Advanced data modeling and analytics, as well as data science solutions, are indispensable for transforming big data into smart data. For accelerating the utilization of smart data, NICT Real World Information Analytics Project makes efforts to develop a cross-data analytics technology for utilizing data obtained from a variety of sensing technologies and different kinds of social big data to construct a platform that will help develop and expand smart services with a view towards smart sustainable cities, and will conduct activities for accelerating open innovation using regional IoT infrastructures. In this paper, we introduce and discuss the latest challenges of novel methods and practices for our smart data analytics and utilization.

2   Transforming Smart City Data to Smart Data

A large amount and variety of data from sensors in physical space is accumulated in cyberspace, analyzed by AI, and then the analysis result is delivered to people in physical space in various forms. Cross-sectional data collection and analysis is indispensable for facilitating such innovations as mobility as a service, intelligent wellbeing for everyday life and comprehensive management of city environment (atmosphere, water, waste, etc.). For instance in Society 5.0 [2], through AI analysis of diverse types of sensing data from automobiles, weather, traffic, accommodations, and personal health, it is realized to make travelling and sightseeing comfortable by providing sightseeing routes matching personal preferences and proposing optimal plans taking weather, congestion, health etc. into account or make movement pleasant without congestion and reduce accidents through driving. Understanding associations between weather, traffics, human behavior and/or observation can help to predicting and potential risks to be avoided when traveling. It is achieved by association discovery and predictive modeling for arbitrary factors of cross-sectional sensing data according to the application purpose. Scalable, efficient and robust technologies of association discovery and predictive modeling for cross-sectional sensing data is one of important keys to success for producing smart data from an unprecedentedly large volume & variety of sensing data whose spatial, temporal and thematic information are conserved, analyzed, and reasoned by utilizing AI. The major tasks include:

- Collecting multi-source, multi-modal sensing data and extracting useful event information with faulty data filtering as well as missing/deficient data complementation.
- Discovering high-utility associations from event transactions
- Predictive modeling associative events adaptive to locations and users
- Navigating people using the prediction results in the real space through geographical maps, geo-fencing/alerting, etc.

Here are some motivating examples for utilizing smart data. In a smart mobility service. oar drivers select their routes based on the information obtained about accidents and traffic congestion along the route. In recent years, nowcasting and forecasting of various traffic risk events is being performed by using diverse sensor data. On the roads, drivers face a number of unforeseen hazards, such as sudden weather disasters and hidden accident black spots. To realize safe driving, a map digitizing existence of these risks will be generated by using a variety of sensing data such as extreme weather, traffic congestion, driving recorder, etc., and a car navigation system will be developed that allows drivers to become aware of the risks and helps them make appropriate route decisions. On the other hand in environmental healthcare, exposure to air pollutants have been associated with an increase in health problems (cancers, respiratory disease, depression/fatigue, etc.) as well as in health and social care costs (£2.81 billion and £9.41 billion respectively by 2035 [3]). With the advent of IoT sensors, it is possible to continuously collect environmental data (e.g., pollutant concentrations, weather variables) and personal health data (e.g. heart rate, activity amount, psychophysiological stats) in real-time regularly. Fusion and analytics of these data can be highly helpful for understanding various impacts of an environment on human health at the individual level. Moreover, predictive modeling of environmental health risk in different urban nature can be used for health-optimal route guidance for travel, walk, and relaxation.

3   xData Platform for Smart Data Utilization

3.1   Basic Framework

We are constructing an xData (cross-data) Platform on NICT's Integrated Testbed implementing our data analytics methods (Fig. 1). It provides API and tools for discovering association rules for sensing data from different fields, as well as learning and predicting their association patterns. The platform also enables to collecting and combining sensing data from heterogeneous data sources in a common format. As the a result, it facilitates discovering and predicting associative events spanning between different fields, such as unusual weather → traffic disruptions and transborder air pollution → health impact. We also have developed applications for visualizing the prediction results on a digital map and providing risk-adaptive navigation.

A variety of sensing data has been collected on the xData platform including meteorological observation data (XRAIN, etc.), atmospheric monitoring data (AEROS, etc.), road traffic data in Japan as well as wearable sensor data (personal environment, fitness, activity, lifelog camera) from participatory sensing campaigns. Data consists of several environmental features such as road, park, sightseeing area, coastal way, and woods. The perception of individual participant was also obtained as annotation information. As of March 2019, 11 domains, 24 types, 13.35 billion records of data has been archived since 2016.

In the following sections, fundamental technologies of the xData Platform are introduced.

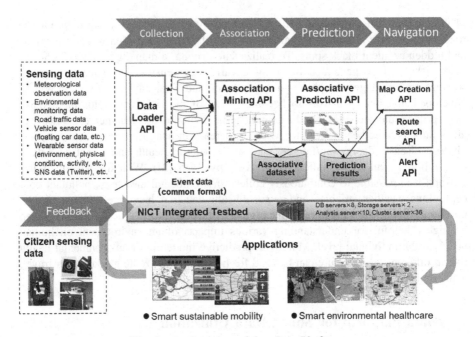

Fig. 1. An Overview of the xData Platform.

3.2 Discovery of High-Utility Association from Sensing Data

Frequent itemset mining is an important data mining model, which aims to discover all itemsets that are occurring frequently in a transactional database. The popular adoption and successful industrial application of this model has been hindered by the following obstacle: "Frequent itemset mining often generates too many patterns, and most of them may be found uninteresting depending on user or application requirements." High utility itemset mining (HUIM) [4] tries to address this problem by taking into account the items' internal utility (such as the number of occurrences of an item within a transaction) and external utility (such as the weight of an item). Given a transactional database, HUIM aims to find itemsets having a utility that is not less than a user-specified minimum utility (minUtil) constraint. HUIs have inspired several important data mining tasks such as high utility sequential pattern mining and high utility periodic pattern mining. We extend conventional HUIM to discover only those HUIs that have items close to one another in a spatiotemporal database, which is called spatial HUI mining (SHUIM) [5]. In a spatiotemporal database (STD), an itemset is said to be an SHUI if its utility is not less than a user-specified minimum utility and the distance between any two of its items is not more than a user-specified maximum distance.

When mining SHUIs, the itemset lattice is a conceptualization of the search space. However, reducing this search space is a challenging task because the generated SHUIs do not satisfy the convertible anti-monotonic, convertible monotonic, or convertible succinct properties. To address this challenge, SHUI-Miner employs two new utility upper bound measures, i.e., Probable Maximum Utility (PMU) and Neighborhood Sub-tree Utility (NSU), to identify itemsets (or items) whose supersets may generate SHUIs. PMU aims to prune items whose supersets in a database (or projected database) cannot be SHUIs. We use a variant of sub-tree utility [6], called Neighborhood sub-tree utility (NSU), to identify these secondary items whose projections (or depth-first search in the itemset lattice) will result in identifying all SHUIs. Both PMU and NSU measures internally utilize the information on the neighborhood of items (or distance between the items) to identify itemsets whose supersets may generate SHUIs. The SHUI-Miner is presented in Fig. 2.

1: **input:** D is a temporal database, SD is a spatial database, UD is an external utility database, $minUtil$ is a user-specified minimum utility constraint, and $maxDist$ is a user-specified maximum distance constraint

2: **output:** A set of SHUIs

3: $N = Neighbors(SD, maxDist)$. Let $N(i_j)$ denote the set of all neighboring items for $i_j \in I$.

4: Let α denote an itemset that needs to be extended. Initially, set $\alpha = \emptyset$;

5: Scan the temporal database to determine the PMU for every item $i_j \in I$.

6: $Secondary(\alpha) = \{i_j | i_j \in I \wedge PMU(i_j) \geq minUtil\}$

7: Let $\succ$ be the total order of PMU in an ascending order of their values in $Secondary(\alpha)$;

8: Scan D to remove each item $i \notin Secondary(\alpha)$ from the transactions, sort items in each transaction according to $\succ$, and delete empty transactions;

9: Calculate *neighborhood sub-tree utility* for all items in $secondary(\alpha)$ by scanning the database D once using utility-bin array;

10: $Primary(\alpha) = \{z \in secondary(\alpha) | NSU(\alpha, z) \geq minUtil\}$;

11: $RecursiveSearch(\alpha, D, Primary(\alpha), Secondary(\alpha), minUtil)$;

Fig. 2. An algorithm for Spatial High Utility Itemset Mining.

One of the applications of SHUIM is improving the road traffic safety in smart cities. To measure and monitor congestion at various road segments, Japan Road Traffic Information Center (JARTIC) has set up a sensor network. The data generated by these sensors represent a spatiotemporal database and an external utility database. An SHUI generated from these databases was observed on the set of roads and provided information regarding the set of neighboring roads where a significant length of congestion was observed. At the time of disasters, the abovementioned information can be very useful to users for various purposes such as diverting the traffic, alerting the pedestrians, and suggesting a patrol path for the police. Japanese Industrial Standard X0410 regional mesh code (or simply, mesh code) [7] represents a portion of the square grid on the Earth's surface. Because a road segment can pass through multiple mesh codes, the congestion length observed by a sensor is equally split among its mesh

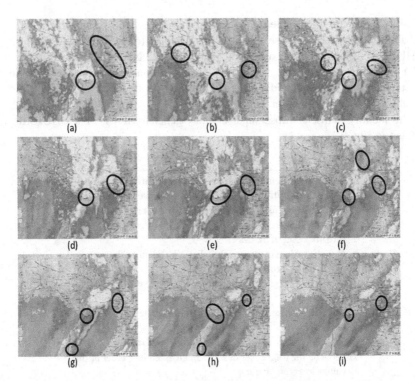

Fig. 3. A SHUIs generated from the hourly congestion data. XRAIN (or precipitation) data is overlaid at hourly intervals. High congested road segments have been shown within the dark ovals. (a) to (i) show some of the SHUIs generated from the hourly congestion data starting from 6 to 14 h (UTC time).

codes. Thus, the internal utility of an item in the temporal database represents the number of occurrences of a road segment with various mesh codes at a particular timestamp. SHUIs in traffic congestion data, we have divided the congestion data into hourly intervals and applied SHUI-Miner algorithm to each hourly congestion data to identify SHUIs (i.e., sets of road segments where a lot of congestion was observed within the traffic network). The minDist = 1 km and minUtil = 10000 m (= 10 km), i.e., a total of at least 10 km of congestion should be observed in a set of neighboring roads in an hour. Figure 3 shown hourly congestion reported from 6:00 h to 14:00 h. The black lines in each of these figures represent the road segments where congestion was observed by a sensor in the entire one hour interval of time. On these black lines, we have overlaid colored lines that represent road segments (or SHUIs) generated in hourly congestion data. Road segments with same color represent a SHUI. Moreover, hourly XRAIN data of the typhoon Nangka was overlaid on the road segments. The following observations can be drawn from these figures: (i) roads with heavy congestion (i.e., SHUIs) vary with time (ii) roads with heavy congestion were observed in areas with high precipitation (Fig. 3(d)–(f)). Thus, the knowledge of SHUIs (or heavy congested roads) can be very useful to the traffic control room in diverting the traffic.

3.3 Complex Event Detection by DNN-Based Raster Data Analysis

Underlaying associations among a variety of sensing data in a smart city (e.g., environmental data, traffic data, and human activity data) can be discovered by exploratory investigation of arbitrary combinations of data according to analysis purposes as well as available data. For instance, conventional approaches tried to discover associations between a fixed number of data based on Structural Equation Modeling [8], Naive Bayes classifier [9], Granger causality test [10] or Apriori algorithm [11]. In order to increase scalability in association discovery for arbitrary combination of sensing data, we propose a brand-new approach for discovering association patterns from an overlayed raster images of sensing data based on deep learning methods in computer vision [12]. A raster image is very popular for geographical representation of sensing data. Different factors of data can be represented by different layers of a raster image, which are combined visually for exploring latent associations among different factors in a scalable manner. Converting different factors of sensing data into spatiotemporal raster images, our approach aims at discovering visual patterns from the overlayed raster images showing spatial, temporal and thematic associations. It consists of (1) collecting and storing sensing data (2) converting sensing data to raster images, (3) detecting events from the raster images, (4) discover common visual patterns showing reasonable associations.

Figure 4 describes the algorithm for creating a raster image from sensing data. Figure 5 also illustrates an example of raster image created for the factors: rain, congestion, and SNS. In this example, the rainfall, congestion, and SNS layers are represented by red, green, and blue channels of the raster image, respectively. For a given time period, a 3D raster image is generated from a time series of the 2D raster images.

Input: the table τ, the time t, the boundary β, the factor ϕ
Output: the raster image $\rho_t^\beta(\phi)$

1: $\tau_t^\beta = filter(\tau, t, \beta)$ #filter table τ so that only events happened at t and inside β are maintained.
2: **if** $(!exist(\rho_t^\beta(\phi)))$ **then**
3:     create $\rho_t^\beta(\phi)$ as empty
4: **end if**
5: i = 1
6: **while** $(!eof(\tau_t^\beta))$ **do**
7:     create a super pixel $\sigma[i]$ of $\rho_t^\beta(\phi)$ by projecting the boundary of the mesh code of $\tau_t^\beta[i]$ to the Google map, and fill this rectangle by the value of ϕ's data of $\tau_t^\beta[i]$
8:     i = i + 1
9: **end while**
10: **return** $\rho_t^\beta(\phi)$

Fig. 4. Algorithm for creating a raster image from a sensing data table with multiple factors.

A deep convolutional neural network, ConvNet is utilized to build a model based on these raster images towards detecting events based on the association between factors. A generic ConvNet architecture consists of convolutional layers and

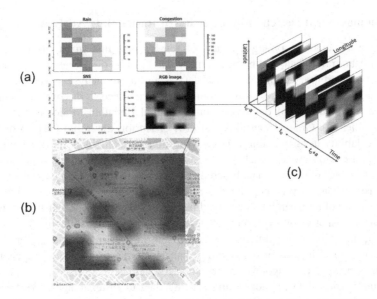

Fig. 5. An example of raster image with three layers: rainfall, congestion and SNS. (Color figure online)

fully-connected layer, each of which is considered as a feature extractor and a classifier, respectively. Figure 6 illustrates our model. Here, we use both transfer learning and fine-tuning approach to build our model due to the small size of dataset. The transfer learning is for using the convolutional layers merely as a feature extractor applying on a new dataset. The last max-pool of VGG16 [13] is used as a feature vector, and feed these feature vectors to a SVM classifier with two classes "event", and "non-event". The fine-tuning is for re-training a part of a pre-trained model on a new dataset with very small weight updates and slow learning rate (e.g., tweak the already trained convolutional layers and/or the full connection layers). Freezing all convolutional layers of VGG16 except the last one, a new fully connected layer is constructed with 1024 nodes followed by a softmax layer with 2 nodes. Thus, the output will show the probability of the event label (e.g., 65% probability the event may happen).

After successfully constructed the model, areas where events of target factor can occur is detected. The event discovery process is analogous to a dynamic scene recognition for video data (i.e., assigning a right scene label to this video for a given an un-labeled video and a model). In our approach, C3D and SVM classifier are used so that the last max-pool of C3D provides a feature vector fed to a SVM classifier with two classes "event", and "non-event". The detected event indicates a strength association between the factors in spatial and temporal dimensions, which lead to a complex event representing co-occurrence or causal relation of the factors. Our proposed approach is evaluated for discovering relative accident risk areas using the datasets on the xData Platform including: weather data (rainfall), traffic data (congestion) and SNS data (twitter) from 2014/05/01 to 2015/10/31 in Kobe area. Figure 7 shows an example. Out of 25,899 car accidents, our approach achieves 86.3%–90.2% of accuracy. The evaluation shows the initial but promising results of our method.

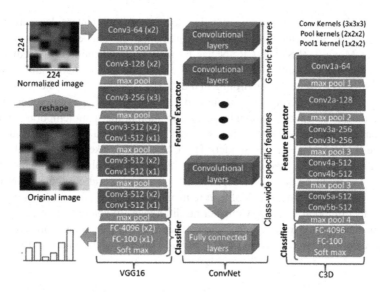

Fig. 6. Proposed Deep CNN architecture based on ConvNet, VGG16 and C3D

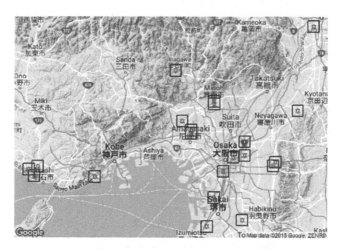

Fig. 7. The example of using CNN and raster images to detect relative accident risk areas in Kobe area at 2015/07/17 17:00 (factors = rainfall, congestion, tweets on rainfall/congestion/accidents, target = traffic accident).

3.4 Prediction Modeling by CRNN for Environmental Data

For predictive modeling, Deep Neural Network (DNN) has already shown promising performance in many recent researches. A convolution recurrent neural network

(CRNN) is class of deep learning which integrates spatial pattern learning by the Convolutional Neural Network (CNN) with temporal pattern learning by the Long Short-Term Memory (LSTM). CRNN has been used classification and prediction of spatiotemporal data in computer vision such as video data. We applied the CRNN to predict associative events for environmental data [14, 15] such as weather observation data, atmospheric monitoring data. In our approach, an optimal integration between the CNN and the LSTM have been sought in consideration of the spatiotemporal continuity of environmental data to achieve high levels of versatility and prediction accuracy.

Figure 8 shows our CRNN model. The CNN model takes a geospatial grid of sensing data as an input, then the convolutional layers and pooling layers will generate a feature map as an output. For a given time series of the grid data, the CNN model yields a corresponding time series of feature maps, then LSTM takes them as inputs for modeling the temporal features. Considering differences in spatiotemporal dispersions of environmental data from local to global areas, individual models are created for different regions of interest (ROIs) and combined for aggregating their outputs.

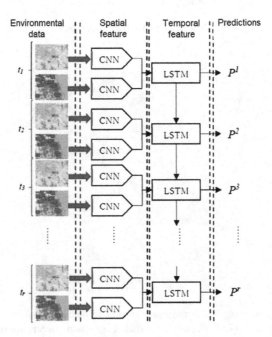

Fig. 8. The architecture of our convolutional recurrent neural network model (CRNN). CRNN processes the spatial environment data with CNNs and the outputs of CNNs as the inputs of a recurrent neural network LSTM.

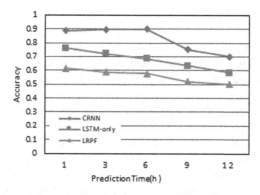

Fig. 9. Results of comparing the CRNN model with LSTM-only and LRPF.

Our CRNN model is applied to short-term prediction of air quality index (AQI) affected by trans-border air pollution. Temporal parameters about the sequence length L, the delay of the transboundary pollution D, and the predict time N are variable in the CRNN model. For training and testing the CRNN model, we collected the atmospheric sensing data of the transboundary air pollution data from 33 coastal cities in China and Fukuoka City in Japan from 2015 to 2017, then 14,401 h of data are labeled by three categories of AQI. Figure 9 show a prediction result for different predict time N (1, 3, 6, 9, 12 h) from the CRNN model with the temporal parameters $L = 24$ and $D = 36$ h, The experimental result shows that our CRNN model using the space-time series is superior to a traditional neural network that only used time series inputs (LSTM-only) and the regression statistics prediction model (LRPF) used by Fukuoka City.

4   Smart Application with Citizen Participation

For the purpose of constructing model cases of smart data, we conducted field experiments of smart applications using the xData Platform with participation of citizens. In this section, we introduce the examples of smart sustainable mobility and smart environmental healthcare.

4.1   Smart Sustainable Mobility

To construct a model case of utilizing the xData Platform, we held a Hackathon event to develop car navigation applications including the display of risk information and customized route guidance (Smart Sustainable Mobility (SSM) Hackathon, 23–24 February 2019, Tokyo) [16]. The xData Platform provides map data of traffic disruption risks for heavy rainfall or heavy snowfall events. These risk maps are generated based on association discovery from the actual recorded data such as extreme weather conditions and traffic congestion, as shown in Fig. 10. According to the risk values in the map, a travel route is calculated using the Dijkstra's algorithm based on the risk values and the movement costs accumulated along the route (see bold paths on the risk maps in Fig. 10).

We have also developed tools for the rapid prototyping of car navigation applications. The prototyping tool is implemented as a Web application using risk map APIs of x Data Platform (see Fig. 11). On the upper part of the Web browser, this tool displays the risk map, travel routes, branch points, and route information, such as the travel time and the amount of risk exposure. On the lower part, alert trigger conditions can be set by using the risk values and the distance from the vehicle. When the alter is triggered, it is possible to display related images and videos on the navigation screen, play audio files, set functional buttons, and change the route (see "Car navig. behavior setting" in Fig. 11). All these settings are saved in a configuration file that can be loaded by the car navigation simulator, which is an Android application developed based on the consumer car navigation application [17].

The hackathon collected twenty-three participants from engineers, researchers, students and professors in IT and ITS fields. The participants were separated into four groups and spent two days for designing and prototyping their car navigation applications. The results are evaluated based on user interface perfection, user experience, innovativeness, feasibility, sustainability, and the fit for purpose. It is worth mentioning that most of all participants took "user mode" into consideration for customizing the navigation application (e.g., travel route guidance). Therefore, it will be considered to collect not only environmental data and traffic data, but also user's information such as driving skill and purpose.

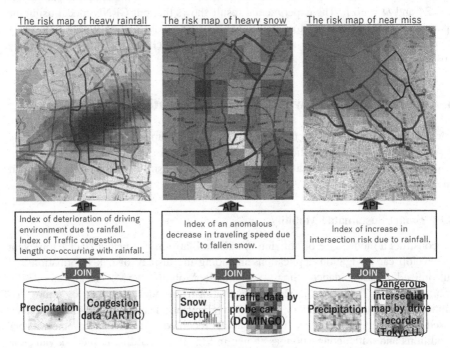

Fig. 10. Examples of risk maps generated by associative prediction of traffic disruption risks for heavy rainfall and heavy snowfall events.

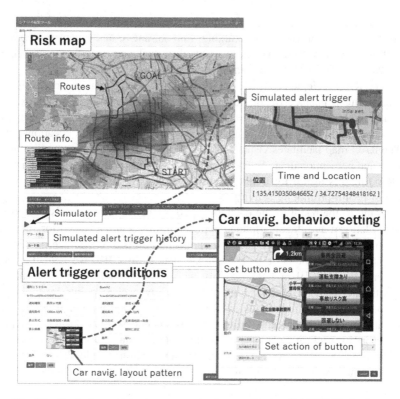

Fig. 11. Tools for prototyping car navigation application for SSM Hackathon.

4.2 Smart Environmental Healthcare

In another model case, the xData Platform provides short-term prediction of air quality for health (AQI) through an associative prediction API using environmental data and health data. We held a user patriating demonstration of the smart environmental healthcare application, called "datathon" [18]. The datathon is a kind of hackathon for collecting and analyzing geolocated sensing data of personal exposure of environment (e.g. temperature, humidity, air pollutants), health condition (e.g., heart rate, amount of activity) and perception (e.g., photos and comments on surrounding urban nature), via wearable sensors, smartphones and lifelog-cameras.

The first datathon was held in Fukuoka City for seven days in March and April, 2018 in Fukuoka City with 133 participants in the total. During the campaign, each participant walked on five routes for about 4–5 km long or 1.5–2.5 h. These routes contain some places showing different urban natures such as arterial roads, parks, sightseeing areas, coastal way, and woods. As shown in Fig. 12, participants brings (1) a mobile atmospheric sensor for collects pollutant concentrations (PM2.5, O3, and NO2) and weather variables (temperature and humidity), and (2) a personal physical condition sensor for monitoring heart rates and nerve activity (calculated from the heart rates). We have developed a smartphone application for collecting these sensing data on the xData Platform. It also collects user perceptions against surrounding environment such as crowdedness, ease of walking, fun, calmness, and quietness based on a five-level questionnaire form. The participants then analyzed underlaying associations between these collected data and created route an air quality (RAQ) map as shown in Fig. 13. The RAQ is depicted by the color on each route segment as well as the radar chart illustrating a balance of air quality index for health, discomfort index, nerve activity (i.e., relaxing level) and user perception levels. User comments and phots characterizing the RAQ is also annotated on the corresponding locations in the map. The RAQ map facilitates visual analytics of associations between environmental, healthcare and perception data. Actually, many participants noticed that visual analytics of the RAQ map is useful for a new discovery of health-optimal locations as well as planning of health-promoting activity.

According to the result of the datathon in 2018, the RQA map is distributed from the xData Platform and the smartphone application has been enhanced for supporting health-optimal route navigation. The new application aims to encourage people to do walking or exercising outdoor in good urban nature. In 2019, we held the second datathon in March, 2019 in Fukuoka City with approximately 40 participants. The participants find a health optimal route along the given checkpoints by looking the RAQ map on the applications provided by CRNN-based short-term AQI prediction the xData Platform. The participants brought environmental sensors, health sensors and lifelog sensors for capturing personal exposure data of urban nature in the same manner as the datathon in 2018. As the reward, the application provided "environmental healthcare points" to each participant converted from the data he/she collected. More

(a)   (b)

(c)

Fig. 12. The participants of datathon (c) carrying personal sensors (a) and smartphone application (b).

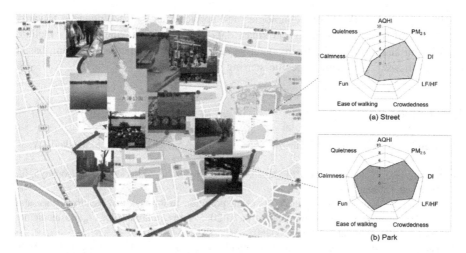

Fig. 13. An example of the road air quality (RAQ) map created in the datathon in 2018. (RAQ colors; blue: very good, green: good, yellow: moderate, brown: bad, red: very bad). (Color figure online)

points are given to a participant when he/she did more activities (measured by foot counts, heart rates, etc.) in better environment (measured by air quality index, discomfort index, perception levels, etc.). The participants also discussed new idea of advanced smart environmental healthcare using the environmental healthcare points.

5 Conclusions

For the purpose of accelerating the use of data in the smart city equipped with IoT, we develop a cross-data analytics technology for utilizing data obtained from a variety of sensing technologies and different kinds of social big data to construct a platform that help develop and expand smart services with a view towards a sustainable society. We are developing the xData Platform on NICT's Integrated Testbed a platform implementing functions of a data loader for data collection, retrieval and conversion from a variety of data sources, association mining for spatiotemporal data integration and discovery of association rules, machine learning for prediction of spatiotemporal association patterns, and creation and distribution of GIS data from prediction results for route search and alert notification. For accelerating open innovation using the xData Platform, we conducted field experiments with participation of citizens. In the smart sustainable motility hackathon, participants developed car navigation applications using risk map data predicting various traffic disturbances with strong association with abnormal weather conditions. In the smart environmental healthcare datathon, the route air quality maps were created by the participants based on environmental data, health data and user perception data collected using personal sensors, and then utilized for the smartphone application of health-optimal route navigation with the short-term AQI prediction.

In our future work, these technologies and applications will be enhanced towards realization of more sustainability in smart data utilization. The major tasks include transferring and customizing the associative discovery and predictive modeling based on location, user, application-specific sensing data.

References

1. Bueti, C.: Shaping Smart Sustainable Cities in Latin America, ITU Green Standards Week (2016)
2. Society 5.0, Cabinet Office of Japan. https://www8.cao.go.jp/cstp/english/society5_0/. Accessed 15 Sept 2019
3. Pimpin, L., et al.: Estimating the costs of air pollution to the National Health Service and social care: An assessment and forecast up to 2035. https://journals.plos.org/plosmedicine/article?id=10.1371/journal.pmed.1002602. Accessed 5 Sept 2019
4. Yao, H., Hamilton, H.J., Butz, C.J.: A foundational approach to mining itemset utilities from databases. In: SIAM, pp. 482–486 (2004)
5. Kiran, R.U., Zettsu, K., Toyoda, M., Kitsuregawa, M., Fournier-Viger, P., Reddy, P.K.: Discovering spatial high utility itemsets in spatiotemporal databases. In: 31st International Conference on Scientific and Statistical Database Management (SSDBM 2019), Santa Cruz, CA, USA, pp. 49–61 (2019)
6. Zida, S., Fournier-Viger, P., Lin, J.C.W., Wu, C.W., Tseng, V.S.: EFIM: a fast and memory efficient algorithm for high-utility itemset mining. Knowl. Inf. Syst. **51**(2), 595–625 (2017)
7. Japanese Industrial Standard X0410. http://www.stat.go.jp/english/data/mesh/02.html. Accessed 28 Feb 2019
8. Lee, J.H., Chae, J.H., Yoon, T.K., Yang, H.J.: Traffic accident severity analysis with rain-related factors using structural equation modeling - a case study of Seoul City. Accid. Anal. Prev. **112**, 1–10 (2018)
9. Li, L., Shrestha, S., Hu, G.: Analysis of road traffic fatal accidents using data mining techniques. In: IEEE 15th International Conference on Software Engineering Research, Management and Applications (SERA), pp. 363–370 (2017)
10. Tse, R., Zhang, L.F., Lei, P., Pau, G.: Social network based crowd sensing for intelligent transportation and climate applications. Mob. Netw. Appl. **23**(1), 177–183 (2018)
11. Tran-The, H., Zettsu, K.: Discovering co-occurrence paterns of heterogeneous events from unevenly-distributed spatiotemporal data. In: 2017 IEEE International Conference on Big Data (BigData 2017), Boston, MA, USA, pp. 1006–1011 (2017)
12. Dao, M.S., Zettsu, K.: Complex event analysis of urban environmental data based on deep CNN of spatiotemporal raster images. In: 2018 IEEE International Conference on Big Data (BigData 2018), Seattle, WA, USA, pp. 2160–2169 (2018)
13. Simonyan, K., Zisserman, A.: Very deep convolutional networks for large-scale image recognition. In: ICLR (2015)
14. Zhao, P., Zettsu, K.: Convolution recurrent neural networks for short-term prediction of atmospheric sensing data. In: 4th IEEE International Conference on Smart Data (SmartData 2018), Halifax, Canada, pp. 815–821 (2018)
15. Zhao, P., Zettsu, K.: Convolution recurrent neural networks based dynamic transboundary air pollution prediction. In: 2019 IEEE Big Data Analytics (ICDBA 2019), Suzhou, China (2019)

16. Itoh, S., Zettsu, K.: Report on a hackathon for car navigation using traffic risk data. In: 3rd International Conference on Intelligent Traffic and Transportation, Amsterdam, The Netherlands (2019)
17. ZENRIN DataCom CO., LTD. https://www.zenrin-datacom.net/en/. Accessed 20 Apr 2019
18. Sato, T., Dao, M.-S., Kuribayashi, K., Zettsu, K.: SEPHLA: challenges and opportunities within environment - personal health archives. In: Kompatsiaris, I., Huet, B., Mezaris, V., Gurrin, C., Cheng, W.-H., Vrochidis, S. (eds.) MMM 2019. LNCS, vol. 11295, pp. 325–337. Springer, Cham (2019). https://doi.org/10.1007/978-3-030-05710-7_27

Deep Learning Models for Medical Image Analysis: Challenges and Future Directions

R. K. Agrawal$^{(\boxtimes)}$ and Akanksha Juneja

Jawaharlal Nehru University, New Delhi 110067, India
rkajnu@gmail.com, akankshajuneja.jnu@gmail.com

Abstract. Artificial neural network (ANN) introduced in the 1950s, is a machine learning framework inspired by the functioning of human neurons. However, for a long time the ANN remained inadequate in solving real problems, because of - the problems of overfitting and vanishing gradient while training a deep architecture, dearth of computation power, and non-availability of enough data for training the framework. This concept has lately re-emerged, in the form of Deep Learning (DL) which initially developed for computer vision and became immensely popular in several other domains. It gained traction in late 2012, when a DL approach i.e. convolutional neural network won in the ImageNet Classification – an acclaimed worldwide computer vision competition. Thereafter, researchers in practically every domain, including medical imaging, started vigorously contributing in the massively progressing field of DL. The success of DL methods can be owed to the availability of data, boosted computation power provided by the existing graphics processing units (GPUs), and ground-breaking training algorithms. In this paper, we have overviewed the area of DL in medical imaging, including (1) machine learning and DL basics, (2) cause of power of DL, (3) common DL models, (4) their applications to medical imaging and (5) challenges and future work in this field.

Keywords: Medical image analysis · Machine learning · Deep Learning · Computer-aided diagnosis

1 Introduction

Machine learning (ML), a subset of "artificial intelligence" (AI), can be described as a class of algorithms which identify patterns within the data automatically and use these for prediction or decision making on future data [1] with minimum interventions by a human. The usage of ML has seen a rapid increase in medical imaging [2–9], and medical image analysis, since the objects in medical images can be quite intricate to be represented correctly by some basic function. Representing such objects requires complex modelling with several parameters. Determining numerous parameters manually from the data is nearly impossible. Therefore, ML plays an essential role in the detection and diagnostics in medical imaging field. Computer-aided diagnosis (CAD) - one of the most prevalent usages of ML [2, 10] - involves classifying objects into predefined classes (e.g., lesions or non-lesions, and benign or malignant) depending upon input data.

© Springer Nature Switzerland AG 2019
S. Madria et al. (Eds.): BDA 2019, LNCS 11932, pp. 20–32, 2019.
https://doi.org/10.1007/978-3-030-37188-3_2

Towards the end of the 1990s, ML techniques, both supervised (that use training data labels to understand patterns in data) and unsupervised (that do not use labelled training data to understand the data patterns), became very popular in medical image analysis. Examples include segmentation of brain regions, extraction of representative features and classification for CAD. ML is immensely popular and happens to be the foundation of several successful commercial medical image analysis products. ML has brought a paradigm shift from manually-defined systems to systems that are computer-trained with the help of historical data by extracting representative features from the data. ML algorithms then derive the optimal decision function. Thus, extracting discriminative features from the input medical images becomes a critical step in designing an ML system. In traditional ML frameworks, feature extraction is defined by human experts. Therefore, such features are called handcrafted features and thus the resultant system performance is subject to the developers' expertise.

Deep learning (DL) is a specific branch of ML. It is based on artificial neural network (ANN) which mimics the multilayered human perception system. DL has further brought a shift from conventional ML systems to a class of self-taught systems that learn the features which capture optimal representation of the data for the given problem thereby eliminating the need for handcrafted features. For decades even after introduction, ANNs found less acceptance as there were severe hinderances in the training of deep architecture to solve real problems. This was mainly because of vanishing gradient and overfitting, lack of powerful computing machines, and the scarcity of enough data for training the system. However, most of these constraints have now been resolved, due to improvements in - the obtainability of data, training algorithms and power of computing using graphics processing units (GPU). DL frameworks have shown promising results in duplicating humans in several arenas, such as medical imaging. The idea of applying DL to medical imaging is a getting increasingly popular however, there are several limitations that impede its progress.

We opine that ultimately, the implementation of ML/DL-based medical image analysis tools in radiology practice at a commercial scale will take place. However, we also believe that this will not completely replace radiologists, although some specific subset of manual tasks will get replaced with complementary modules which will provide an overall augmentation of the entire medical imaging system. In this review, we discuss various state-of-the-art applications of ML and DL in medical image analysis. We also discuss some common DL frameworks, strengths of DL and the challenges to be overcome in future.

2 Machine Learning

2.1 Types of Machine Learning Algorithms

Based on system training, ML algorithms are categorized as supervised and unsupervised.

Supervised Training. Supervised learning determines a function which reconstructs output with inference from the input which is constructed with representative numerical or nominal features vectors, comprising of independent variables and the corresponding output variable also called dependent/target variable. If the output is a numerical variable, the training method is known as regression. When the output is a categorical variable, the method is called classification.

Unsupervised Training. Unsupervised learning performs data processing independent of labels and is trained to describe hidden patterns from unlabeled data and group the data into segments wherein each segment consists of samples which are similar as per some attributes. Dimensionality reduction methods like Principal component analysis and clustering methods like k-means are some examples of unsupervised learning algorithms.

2.2 Applications of Machine Learning in Medical Image Analysis

A large number of research papers have described applications of ML in medical imaging, such as CAD of Parkinson's disease using brain MRI using graph-theory-based spectral features [4] and 3D local binary pattern features [7], CAD of Schizophrenia using functional MRI using non-linear fuzzy kernel-based feature extraction [8] and fuzzy-rough-set feature selection [9]. Some more applications have been - detection of lung nodule in CXR [11–14] and thoracic CT [13–18] diagnosis of lung nodules into malignant or benign in CXR [19], microcalcifications detection in mammography [20–23], masses detection [24] and classification of masses into benign or malignant [25–27] in mammography. Recently, Aggarwal et al. [28] employed surfacelet transform, a 3D multi-resolution transform for capturing complex direction details while extracting features from brain MRI volumes for CAD of Alzheimer's disease. Further, modified and enhanced intuitionistic fuzzy-c-means clustering methods have been suggested for segmenting of human brain MRI scans [29]. A survey of works on ML in CAD has been presented [30]. Some works on ML in CAD and medical image analysis are available in the literature [10, 31].

2.3 Machine Learning Algorithms – Towards Deep Learning

A very prevalent method for classification, support vector machine (SVM), generally shows superior performance in most of the classification problems, owing to its merits of convex optimization and regularization [32, 33]. Lately, ensemble learning, including algorithms leveraging boosting and bagging, is being commonly used for enhanced classification performance [34].

Numerical or nominal representations of input data are called features in ML. Thus, the performance of the ML algorithm largely depends on the quality of features fed to it as input. Determining informative features is an important aspect in ML systems. Many domain experts attempt to learn and create handcrafted features by the help of various estimation techniques, such as performance tests and statistical analysis. In this direction, various feature extraction and selection algorithms have been established for obtaining high-performing features.

Artificial Neural network (ANN) is one of the popular algorithms for ML. ANN models the computational units of human brain neurons and their synapses. ANN comprises of layers of interconnected artificial neurons. Every neuron is an implementation of a basic classifier which generates a decision depending upon the weighted sum of inputs. The weights of the edges in the network are trained by a back-propagation algorithm, wherein tuples of input and expected output are provided, imitating the situation wherein the brain depends on external stimuli for learning about achieving certain goals [35]. ANNs had notable performance in several areas but suffered limits like trapping in the local minima while learning optimized weights of network, and overfitting on the given dataset.

3   Overview of Deep Learning

ANN is a ML framework that derives inspiration from the functioning of human neurons. However, for a long time these did not gain much success. This field of research has recently re-emerged, in the form of DL which was developed in the field of computer vision and is becoming immensely useful in several other fields. It gained immense popularity in 2012, when a convolutional neural network (CNN) won ImageNet Classification [36]. Thereafter, researchers in most domains, including medical imaging, started implementing DL. With fast and increasing improvement in computational power and the availability of massive amounts of data, DL is becoming the most preferred ML technique for implementation since it is able to learn much more high-level, abstract and complex patterns from given data than any other conventional ML techniques. Unlike conventional ML algorithms, DL methods make the feature engineering process significantly simple. Moreover, DL can even be applied directly to raw input data. Hence, this lets more researchers to explore more novel ideas. For instance, DL has been successfully applied to CAD of Alzheimer's [37], Parkinson's [38] and Schizophrenia [39] diseases using MRI, as well as segmentation of brain regions in MRI and ultrasound [40].

Some of the common DL frameworks are - Convolutional Neural Network (CNN), Recurrent Neural Networks (RNNs), Auto-encoders (AEs) and Stacked Auto-encoders (SAEs), among others. CNNs and RNNs are supervised frameworks while AEs and SAEs are unsupervised frameworks. These are explained as follows.

3.1   Convolutional Neural Network

A CNN is a neural network proposed by Fukushima [41] for simulating the human visual system in 1980. A CNN comprises of several layers of connected computation units (neuron-like) with step-wise processing. It has attained substantial advancements in the field of computer vision. A CNN can derive hierarchical information by pre-processing, going from edge-level features to object-representations in images.

The CNN comprises of inter-connected convolutional, pooling and fully connected layers. The input is an image, and the outputs are class categories. The main task of a convolutional layer is detecting characteristic edges, lines, objects and various visual elements. Data propagation in the forward direction is equivalent to a performing shift-invariant convolution. The outcome of convolution is composed into the respective unit in the next layer which further processes it with an activation function and computes subsequent output. The activation function is a non-linear unit like sigmoid or rectified linear unit. For capturing an increasing field of view as we move to deeper layers, feature maps computed at every layer are reduced gradually and spatially by the pooling layer wherein a maximum operation is performed over the pixels within the local region. The convolutional and pooling layers are iterated a few times. After this, fully connected layers are used for integrating the features from the whole of the feature maps. Finally, a softmax layer is generally used as the output layer. Figure 1 depicts the architecture of a CNN.

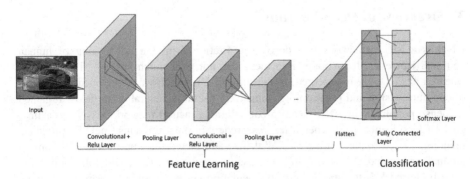

Fig. 1. Convolutional neural network architecture

CNNs are different from simple Deep Neural Networks [42]. In CNNs the network weights are shared so that the network carries out convolution on the input images. Therefore, the network is not required to separately learn detectors for an object that occurs at multiple positions within the image. This makes the network invariant to input translations and considerably decreases the number of parameters to be learnt.

3.2   Recurrent Neural Networks

RNN is a special category of ANN well-suited for temporal data i.e. speech and text for sequence modelling. A conventional ANN assumes that all inputs (and outputs) are independent of each other. This assumption, however, is not applicable for many types of problems. For examples, in machine translation task of natural language processing – prediction of the next word in a sentence is dependent on the words preceding it. RNNs are called "recurrent" since the same computation is performed for each element of the given sequence, with the outcome being dependent on the prior computations. An alternate way to visualize RNNs is that they have a "memory" which saves information of the computations done so far and creates an internal state of the network. This way RNNs yield substantial enhancement in performance in natural language processing, speech recognition and generation, machine translation, handwriting recognition tasks [43, 44].

Although primarily introduced for 1-D input, RNNs are increasingly being used for images as well. In medical applications, RNNs have been employed successfully for segmentation [45] in the MRBrainS challenge.

3.3   Auto-encoders and Stacked Auto-encoders

AEs are networks trained with one hidden layer for reconstructing the input on the output layer. Key aspects of AEs are – (1) bias and weight matrix from input to hidden layer and (2) corresponding bias and weight from the hidden layer to the reconstructed output. A non-linear activation function is employed at the hidden layer. Further, the dimension of the hidden layer is kept smaller than the input layer dimension. This projection of input data onto a lower dimension space captures a dominant latent structure of the data.

The denoising auto-encoder (DAE) [46] is useful in preventing the network from learning a trivial solution. DAEs are trained for reconstructing the input by providing corrupted variant of the input data by adding noise (for example salt-and-pepper-noise). Initially, deep neural networks (DNNs) were difficult to efficiently train. These did not gain popularity until 2006 [47–49] when it was demonstrated that unsupervised training of DNNs in a layer-wise 'greedy' fashion (pre-training), followed by fine-tuning of the whole network in a supervised manner, could achieve better results. Stacked auto-encoder (SAE), also called deep AE, is one such popular architecture. SAEs are constructed by placing layers of AEs on top of each other.

4   Deep Learning in Medical Image Analysis

DL is particularly important for medical imaging analysis because it can take years of manual training to achieve acceptable level of domain expertise for appropriate hand-crafted feature determination required to implement traditional ML algorithm. DL has been extensively implemented in various medical image analysis tasks. The results demonstrate the immense potential of DL in achieving higher accuracies of automated systems in future. Some of these studies, carried out in majorly two application areas, have been discussed as follows:

4.1   Classification

Image classification was initially the area wherein DL made significant contribution to medical imaging. Early reports of DL-based CAD systems for breast cancer [50], lung cancer [51, 52, 58] and Alzheimer's disease (AD) [53–55] demonstrate good performance in detection of the diseases. DL has been implemented in detection, diagnosis and analysis of breast cancer risk [50, 56, 57]. Some works have used DL for AD diagnosis, in which the diseases are detected using multi-modal brain data owing to effective feature computation by DL [53–55]. Numerous studies have surveyed the usage of DL in medicine [59–61].

In such settings, dataset sizes are comparatively smaller (hundreds/thousands) than those in computer vision (millions of samples). Transfer learning, therefore, becomes a natural choice for such applications. Fundamentally, transfer learning is the usage of networks pre-trained on a large dataset of natural images to attempt to work for smaller medical data. Two main transfer learning strategies are: (1) use of pre-trained network for extracting features, (2) fine-tuning a pre-trained network using medical data. Both strategies are prevalent and have been implemented extensively. The first strategy allows plugging the computed features in to image analysis system without the need to train a deep network. However, fine-tuning a pre-trained Google's Inception v3 network on medical data attained a performance close to human expert [62]. CNNs pre-trained using natural images have astoundingly demonstrated promising results in some tasks, matching up to the performance of human experts.

Initially, the medical imaging researchers focused on network architectures like SAEs. Some studies [63–65] have applied SAEs for CAD of Alzheimer's disease using brain MRI. Lately, a move towards CNNs is evident with application such as retinal imaging, brain MRI and lung computed tomography (CT) to name some.

Some researchers also train their networks from scratch rather than utilizing pre-trained networks. Menegola et al. [66] compared training from scratch to fine-tuning of pre-trained networks in their experiments and demonstrated that fine-tuning performed better on a small dataset of about a 1000 skin lesions images.

In a nutshell, DL, along with transfer learning, is capable in improving the performance of existing CAD to the level of commercial use.

4.2 Segmentation

Medical image segmentation, identifying the objects from background medical images to capturing meaningful information about their volumes and shapes, is a major challenge in medical image analysis. The potential of DL approaches has put them as a primary choice for medical image segmentation [67].

Application of CNN in segmentation of medical image has been incorporated in various studies. The overall perception is to carry out segmentation using 2D image as input and to apply 2D filters on it [68]. In an experiment, Bar et al. [69], low-level features are extracted from a model pre-trained on Imagenet. Thereafter, high-level features are extracted from PiCoDes [70] and fused together for segmentation. 2.5D approaches [71–73] are inspired by the fact that these have more spatial information than 2D but less computation costs than 3D. Usually, they encompass extraction of three orthogonal 2D patches in the three planes.

The benefit of a 3D CNN is to compute a powerful representation along all 3 dimensions. The 3D network is trained for predicting the label of a volume element based on the surrounding 3D patch content. The accessibility of 3D medical imaging and vast development in hardware has enabled full utilization of spatial 3D information.

5 Limitations of Deep Learning in Medical Image Analysis

Deep learning applied to medical imaging has the potential to become the most disruptive technology medical imaging has witnessed since the introduction of digital imaging. Researchers recommend that DL systems will take over humans in not only diagnosis but also in prediction of disease, prescribing medicine and guiding in treatments. Even though there are promising results from several research studies, there are a few limitations/challenges yet to be overcome before DL can become a part of mainstream radiological practice. Some of these challenges are discussed below.

5.1 Training Data

DL requires enormous training data as performance of DL classifier depends on the size and quality of the dataset to a large extent. However, scarcity of data is one the main barriers in the success of DL in medical imaging. Further, construction of large medical imaging data is a challenge since annotating the data requires much time and effort from not just single but multiple experts to rule out human error. Furthermore, annotations may not always be feasible due to non-availability of experts. Class imbalance in data is also a major challenge that is common in health sector especially in case of rare diseases wherein the number of samples in disease class are highly insufficient as compared to controls. In addition, due to the variations in prevalence of diseases and imaging machines, protocols and standards used in hospitals across the worlds, it is difficult to build generalized DL systems that can work for varied datasets from different sources.

5.2 Interpretability

The present DL methods are black-box in nature. That is, even when the DL method demonstrates excellent performance, mostly, it is impossible to derive a logical or technical explanation of the functioning of the system. This brings us to a question - whether it is acceptable to use a system which is unable to provide a reasoning for the decision it takes in a critical domain like healthcare?

5.3 Legal and Ethical Aspects

Issues may arise concerning the usage of clinical imaging data for development of DL systems at a commercial scale as subjects' private and sensitive information is captured in their medical image data. With the growth of healthcare data, medical image analysis practitioners also face a challenge of how to anonymize patient information to avert its disclosure. Additional legal issues would arise if we implement a DL system in clinical practice, independent from intervention of a clinician. This raises a question - who would be accountable for any error that causes harm to patients?

6 Discussion and Conclusion

Currently, clinicians are experiencing a growing number of complex imaging standards in radiology. This adds to their difficulty to complete analysis on time and compose accurate reports. However, Deep Learning is showing impressive performance outcomes in a wide range of medical imaging applications. In many of the applications DL methods have mostly outperformed traditional ML methods. In future, DL is expected to support clinicians and radiologists in providing diagnosis with more accuracy at a commercial scale. However, penetration of DL in medicine is not as fast as in other real-world problems such as automated speech recognition, machine translation, object recognition etc.

Innovation and development in DL has led to disruption in ubiquitous technology space including virtual digital assistants and autonomous driving, to name some. With these ground-breaking technological advancements, it is not unreasonable to foresee that there will be critical changes in radiology in the coming years owing to DL.

In this review, we have discussed DL from the perspective of medical image analysis. The usage of DL in medical image analysis is currently at very initial stage. With the recent technological innovations, larger fully annotated databases are required to progress with DL-based progress in medical imaging. It will be essential to train the DL network, and to evaluate its performance. There are several other problems and challenges to resolve and overcome. Such as, legal and ethical issues arising due to the presence of identification related patient information present in the patient data used for developing commercial DL systems.

As we contemplate the use of DL in medical imaging, we see this revolution to be more of a collective medium in reducing the load from several repetitive tasks and increasing objective computations in presently subjective processes, rather than replacing clinical experts. The active involvement of experts is indispensable in this critical and sensitive area of medical imaging.

References

1. Murphy, K.P.: Machine Learning: A Probabilistic Perspective, 1st edn., p. 25. The MIT Press, Cambridge (2012)
2. Wang, F., Shen, D., Yan, P., Suzuki, K. (eds.): MLMI 2012. LNCS, vol. 7588. Springer, Heidelberg (2012). https://doi.org/10.1007/978-3-642-35428-1
3. Shen, D., Wu, G., Zhang, D., Suzuki, K., Wang, F., Yan, P.: Machine learning in medical imaging. Comput. Med. Imaging Graph. 41, 1–2 (2015)
4. Rana, B., et al.: Graph-theory-based spectral feature selection for computer aided diagnosis of Parkinson's disease using T 1-weighted MRI. Int. J. Imaging Syst. Technol. 25(3), 245–255 (2015)
5. Suzuki, K., Zhou, L., Wang, Q.: Machine learning in medical imaging. Pattern Recognit. 63, 465–467 (2017)
6. El-Baz, A., Gimel'farb, G., Suzuki, K.: Machine learning applications in medical image analysis. Comput. Math. Methods Med. 2017, 2361061 (2017)
7. Rana, B., et al.: Relevant 3D local binary pattern based features from fused feature descriptor for differential diagnosis of Parkinson's disease using structural MRI. Biomed. Signal Process. Control 34, 134–143 (2017)
8. Juneja, A., Rana, B., Agrawal, R.K.: fMRI based computer aided diagnosis of schizophrenia using fuzzy kernel feature extraction and hybrid feature selection. Multimed. Tools Appl. 77 (3), 3963–3989 (2018)
9. Juneja, A., Rana, B., Agrawal, R.K.: A novel fuzzy rough selection of non-linearly extracted features for schizophrenia diagnosis using fMRI. Comput. Methods Programs Biomed. 155, 139–152 (2018)
10. Suzuki, K.: Computational Intelligence in Biomedical Imaging. Springer, New York (2014). https://doi.org/10.1007/978-1-4614-7245-2

11. Shiraishi, J., Li, Q., Suzuki, K., Engelmann, R., Doi, K.: Computer aided diagnostic scheme for the detection of lung nodules on chest radiographs: localized search method based on anatomical classification. Med. Phys. **33**(7), 2642–2653 (2006)
12. Coppini, G., Diciotti, S., Falchini, M., Villari, N., Valli, G.: Neural networks for computer-aided diagnosis: detection of lung nodules in chest radiograms. IEEE Trans. Inf. Technol. Biomed. **7**(4), 344–357 (2003)
13. Hardie, R.C., Rogers, S.K., Wilson, T., Rogers, A.: Performance analysis of a new computer aided detection system for identifying lung nodules on chest radiographs. Med. Image Anal. **12**(3), 240–258 (2008)
14. Chen, S., Suzuki, K., MacMahon, H.: A computer-aided diagnostic scheme for lung nodule detection in chest radiographs by means of two-stage nodule-enhancement with support vector classification. Med. Phys. **38**, 1844–1858 (2011)
15. Arimura, H., Katsuragawa, S., Suzuki, K., et al.: Computerized scheme for automated detection of lung nodules in low-dose computed tomography images for lung cancer screening. Acad. Radiol. **11**(6), 617–629 (2004)
16. Armato III, S.G., Giger, M.L., MacMahon, H.: Automated detection of lung nodules in CT scans: preliminary results. Med. Phys. **28**(8), 1552–1561 (2001)
17. Ye, X., Lin, X., Dehmeshki, J., Slabaugh, G., Beddoe, G.: Shape-based computer-aided detection of lung nodules in thoracic CT images. IEEE Trans. Biomed. Eng. **56**(7), 1810–1820 (2009)
18. Way, T.W., Sahiner, B., Chan, H.P., et al.: Computer-aided diagnosis of pulmonary nodules on CT scans: improvement of classification performance with nodule surface features. Med. Phys. **36**(7), 3086–3098 (2009)
19. Aoyama, M., Li, Q., Katsuragawa, S., MacMahon, H., Doi, K.: Automated computerized scheme for distinction between benign and malignant solitary pulmonary nodules on chest images. Med. Phys. **29**(5), 701–708 (2002)
20. Wu, Y., Doi, K., Giger, M.L., Nishikawa, R.M.: Computerized detection of clustered microcalcifications in digital mammograms: applications of artificial neural networks. Med. Phys. **19**(3), 555–560 (1992)
21. El-Naqa, I., Yang, Y., Wernick, M.N., Galatsanos, N.P., Nishikawa, R.M.: A support vector machine approach for detection of microcalcifications. IEEE Trans. Med. Imaging **21**(12), 1552–1563 (2002)
22. Yu, S.N., Li, K.Y., Huang, Y.K.: Detection of microcalcifications in digital mammograms using wavelet filter and Markov random field model. Comput. Med. Imaging Graph. **30**(3), 163–173 (2006)
23. Ge, J., Sahiner, B., Hadjiiski, L.M., et al.: Computer aided detection of clusters of microcalcifications on full field digital mammograms. Med. Phys. **33**(8), 2975–2988 (2006)
24. Wu, Y.T., Wei, J., Hadjiiski, L.M., et al.: Bilateral analysis based false positive reduction for computer-aided mass detection. Med. Phys. **34**(8), 3334–3344 (2007)
25. Huo, Z., Giger, M.L., Vyborny, C.J., Wolverton, D.E., Schmidt, R.A., Doi, K.: Automated computerized classification of malignant and benign masses on digitized mammograms. Acad. Radiol. **5**(3), 155–168 (1998)
26. Delogu, P., Evelina Fantacci, M., Kasae, P., Retico, A.: Characterization of mammographic masses using a gradient-based segmentation algorithm and a neural classifier. Comput. Biol. Med. **37**(10), 1479–1491 (2007)
27. Shi, J., Sahiner, B., Chan, H.P., et al.: Characterization of mammographic masses based on level set segmentation with new image features and patient information. Med. Phys. **35**(1), 280–290 (2008)
28. Aggarwal, N., Rana, B., Agrawal, R.K.: Role of surfacelet transform in diagnosing Alzheimer's disease. Multidimens. Syst. Signal Process. **30**(4), 1839–1858 (2019)

29. Verma, H., Agrawal, R.K., Sharan, A.: An improved intuitionistic fuzzy c-means clustering algorithm incorporating local information for brain image segmentation. Appl. Soft Comput. **46**, 543–557 (2016)
30. Suzuki, K.: Machine learning in computer-aided diagnosis of the thorax and colon in CT: a survey. IEICE Trans. Inf. Syst. **E96-D**(4), 772–783 (2013)
31. Suzuki, K.: Machine Learning in Computer-Aided Diagnosis: Medical Imaging Intelligence and Analysis. IGI Global, Hershey (2012)
32. Byvatov, E., Fechner, U., Sadowski, J., Schneider, G.: Comparison of support vector machine and artificial neural network systems for drug/nondrug classification. J. Chem. Inf. Comput. Sci. **43**, 1882–1889 (2003)
33. Tong, S., Chang, E.: Support vector machine active learning for image retrieval. In: Proceedings of the 9th ACM International Conference on Multimedia, Ottawa, Canada, 5 October–30 September 2001, p. 107. ACM, New York (2001)
34. Arbib, M.A.: The Handbook of Brain Theory and Neural Networks, 2nd edn. The MIT Press, Boston (2003)
35. Haykin, S.S.: Neural Networks: A Comprehensive Foundation, pp. 107–116. Macmillan College Publishing, New York (1994)
36. Krizhevsky, A., Sutskever, I., Hinton, G.E.: ImageNet classification with deep convolutional neural networks. In: Advances in Neural Information Processing Systems, pp. 1097–1105 (2012)
37. Basaia, S., et al.: Automated classification of Alzheimer's disease and mild cognitive impairment using a single MRI and deep neural networks. NeuroImage Clin. **21**, 101645 (2019). Alzheimer's disease neuroimaging initiative
38. Chen, L., Shi, J., Peng, B., Dai, Y.: Computer-aided diagnosis of Parkinson's disease based on the stacked deep polynomial networks ensemble learning framework. Sheng wu yi xue gong cheng xue za zhi = J. Biomed. Eng. = Shengwu yixue gongchengxue zazhi **35**(6), 928–934 (2018)
39. Zeng, L.L., et al.: Multi-site diagnostic classification of schizophrenia using discriminant deep learning with functional connectivity MRI. EBioMedicine **30**, 74–85 (2018)
40. Milletari, F., et al.: Hough-CNN: deep learning for segmentation of deep brain regions in MRI and ultrasound. Comput. Vis. Image Understand. **1**(164), 92–102 (2017)
41. Fukushima, K.: Neocognitron: a self organizing neural network model for a mechanism of pattern recognition unaffected by shift in position. Biol. Cybern. **36**(4), 193–202 (1980)
42. Suzuki, K.: Overview of deep learning in medical imaging. Radiol. Phys. Technol. **10**(3), 257–273 (2017)
43. Mikolov, T., Karafiát, M., Burget, L., Cernocký, J., Khudanpur, S.: Recurrent neural network based language model. In: Proceedings of the 11th Annual Conference of the International Speech Communication Association (INTERSPEECH 2010), Makuhari, Japan, 26–30 September 2010. International Speech Communication Association, pp. 1045–1048 (2010)
44. Gregor, K., Danihelka, I., Graves, A., Rezende, D.J., Wierstra, D.: DRAW: a recurrent neural network for image generation. In: Proceedings of the 32nd International Conference on Machine Learning (ICML 2015), Lille, France, 6–11 July 2015, pp. 1462–1471. JMLR (2015)
45. Stollenga, M.F., Byeon, W., Liwicki, M., Schmidhuber, J.: Parallel multi-dimensional LSTM, with application to fast biomedical volumetric image segmentation. In: Advances in Neural Information Processing Systems, pp. 2998–3006 (2015)
46. Vincent, P., Larochelle, H., Lajoie, I., Bengio, Y., Manzagol, P.A.: Stacked denoising autoencoders: learning useful representations in a deep network with a local denoising criterion. J. Mach. Learn. Res. **11**, 3371–3408 (2010)

47. Bengio, Y., Lamblin, P., Popovici, D., Larochelle, H.: Greedy layer-wise training of deep networks. In: Advances in Neural Information Processing Systems, pp. 153–160 (2007)
48. Hinton, G.E., Osindero, S., Teh, Y.W.: A fast learning algorithm for deep belief nets. Neural Comput. **18**(7), 1527–1554 (2006)
49. Hinton, G.E., Salakhutdinov, R.R.: Reducing the dimensionality of data with neural networks. Science **313**(5786), 504–507 (2006)
50. Wang, D., Khosla, A., Gargeya, R., Irshad, H., Beck, A.H.: Deep learning for identifying metastatic breast cancer. arXiv preprint: arXiv:1606.05718, 18 June 2016
51. Hua, K.L., Hsu, C.H., Hidayati, S.C., Cheng, W.H., Chen, Y.J.: Computer-aided classification of lung nodules on computed tomography images via deep learning technique. OncoTargets Ther. **8**, 2015–2022 (2015)
52. Kumar, D., Wong, A., Clausi, D.A.: Lung nodule classification using deep features in CT images. In: Proceedings of the 2015 12th Conference on Computer and Robot Vision, Halifax, Canada, 3–5 June 2015, pp. 133–138. IEEE (2015)
53. Suk, H.I., Lee, S.W., Shen, D.: Hierarchical feature representation and multimodal fusion with deep learning for AD/MCI diagnosis. Neuroimage **101**, 569–582 (2014). Alzheimer's disease neuroimaging initiative
54. Suk, H.I., Shen, D.: Deep learning-based feature representation for AD/MCI classification. Med. Image Comput. Comput. Assist. Interv. **16**(Pt 2), 583–590 (2013)
55. Liu, S., Lis, S., Cai, W., Pujol, S., Kikinis, R., Feng, D.: Early diagnosis of Alzheimer's disease with deep learning. In: Proceedings of the IEEE 11th International Symposium on Biodmedical Imaging, Beijing, China, 29 April–2 May 2014, pp. 1015–1018. IEEE (2014)
56. Cheng, J.Z., Ni, D., Chou, Y.H., Qin, J., Tiu, C.M., Chang, Y.C., et al.: Computer-aided diagnosis with deep learning architecture: applications to breast lesions in US images and pulmonary nodules in CT scans. Sci. Rep. **6**, 24454 (2016)
57. Kallenberg, M., et al.: Unsupervised deep learning applied to breast density segmentation and mammographic risk scoring. IEEE Trans. Med. Imaging **35**, 1322–1331 (2016)
58. Chen, J., Chen, J., Ding, H.Y., Pan, Q.S., Hong, W.D., Xu, G., et al.: Use of an artificial neural network to construct a model of predicting deep fungal infection in lung cancer patients. Asian Pac. J. Cancer Prev. **16**, 5095–5099 (2015)
59. Liu, Y., Wang, J.: PACS and Digital Medicine: Essential Principles and Modern Practice, 1st edn. CRC Press, Boca Raton (2010)
60. Collins, F.S., Varmus, H.: A new initiative on precision medicine. N. Engl. J. Med. **372**, 793–795 (2015)
61. Aerts, H.J., Velazquez, E.R., Leijenaar, R.T., Parmar, C., Grossmann, P., Carvalho, S., et al.: Decoding tumour phenotype by noninvasive imaging using a quantitative radiomics approach. Nat. Commun. **5**, 4006 (2014)
62. Esteva, A., et al.: Dermatologist-level classification of skin cancer with deep neural networks. Nature **542**(7639), 115 (2017)
63. Suk, H.-I., Shen, D.: Deep learning-based feature representation for AD/MCI classification. In: Mori, K., Sakuma, I., Sato, Y., Barillot, C., Navab, N. (eds.) MICCAI 2013. LNCS, vol. 8150, pp. 583–590. Springer, Heidelberg (2013). https://doi.org/10.1007/978-3-642-40763-5_72
64. Suk, H.I., Lee, S.W., Shen, D.: Latent feature representation with stacked auto-encoder for AD/MCI diagnosis. Brain Struct. Funct. **220**(2), 841–859 (2015). Alzheimer's disease neuroimaging initiative
65. Suk, H.I., Wee, C.Y., Lee, S.W., Shen, D.: State-space model with deep learning for functional dynamics estimation in resting-state fMRI. NeuroImage **129**, 292–307 (2016)

66. Menegola, A., Fornaciali, M., Pires, R., Avila, S., Valle, E.: Towards automated melanoma screening: exploring transfer learning schemes. arXiv preprint: arXiv:1609.01228, 5 September 2016
67. Hesamian, M.H., Jia, W., He, X., Kennedy, P.: Deep learning techniques for medical image segmentation: achievements and challenges. J. Digit. Imaging **32**(4), 582–596 (2019)
68. Zhang, W., et al.: Deep convolutional neural networks for multi-modality isointense infant brain image segmentation. NeuroImage **108**, 214–224 (2015)
69. Bar, Y., Diamant, I., Wolf, L., Greenspan, H.: Deep learning with non-medical training used for chest pathology identification. In: Medical Imaging 2015: Computer-Aided Diagnosis, vol. 9414, p. 94140V. International Society for Optics and Photonics, 20 March 2015
70. Bergamo, A., Torresani, L., Fitzgibbon, A.W.: PiCoDes: learning a compact code for novel-category recognition. In: Advances in Neural Information Processing Systems, pp. 2088–2096 (2011)
71. Moeskops, P., et al.: Deep learning for multi-task medical image segmentation in multiple modalities. In: Ourselin, S., Joskowicz, L., Sabuncu, M.R., Unal, G., Wells, W. (eds.) MICCAI 2016, Part II. LNCS, vol. 9901, pp. 478–486. Springer, Cham (2016). https://doi.org/10.1007/978-3-319-46723-8_55
72. Prasoon, A., Petersen, K., Igel, C., Lauze, F., Dam, E., Nielsen, M.: Deep feature learning for knee cartilage segmentation using a triplanar convolutional neural network. In: Mori, K., Sakuma, I., Sato, Y., Barillot, C., Navab, N. (eds.) MICCAI 2013, Part II. LNCS, vol. 8150, pp. 246–253. Springer, Heidelberg (2013). https://doi.org/10.1007/978-3-642-40763-5_31
73. Roth, H.R., Lu, L., Farag, A., Sohn, A., Summers, R.M.: Spatial aggregation of holistically-nested networks for automated pancreas segmentation. In: Ourselin, S., Joskowicz, L., Sabuncu, M.R., Unal, G., Wells, W. (eds.) MICCAI 2016, Part II. LNCS, vol. 9901, pp. 451–459. Springer, Cham (2016). https://doi.org/10.1007/978-3-319-46723-8_52

Recent Advances and Challenges in Design of Non-goal-Oriented Dialogue Systems

Akanksha Mehndiratta[(✉)] and Krishna Asawa

Computer Science and Engineering,
Jaypee Institute of Information Technology, Noida, India
mehndiratta.akanksha@gmail.com,
krishna.asawa@jiit.ac.in

Abstract. Human being is pervasively surrounded by smart devices that provide numerous services to them. These devices are equipped with interfaces that are natural and intuitive with the aim of providing user an effortless and organic interaction with the devices. A dialogue agent is one such interface that interacts with the user in natural language. Recent paradigm classifies them as goal-oriented and non-goal-oriented dialogue systems. The aim of goal-oriented dialogue systems is to assist the user in completing a task. Evidently the design of goal-oriented-dialogue agents has made a lot of progress. But the interactions with non-goal-oriented dialogue systems are reasonable, open- domain and more applicable to real-world applications. This paper reviews the state-of-the-art in Non-goal-oriented dialogue systems. The design of such systems has advanced due to Big Data hence most of the recent models are data-driven. This paper is a comprehensive study of data driven systems - advances in learning models and recent frontiers. Also provide an insight on datasets and evaluation methods and the limitations that the data-driven methods, datasets and evaluation methods present.

Keywords: Natural Language Processing · Dialogue management systems · Machine learning · Deep learning · Dialogue agent

1 Introduction

The idea of intelligent agents has transitioned from a fictional character in a sci-fi movie to reality. They have paved their path from our office space to personal space. They provide numerous services ranging from complex tasks to everyday household tasks. These agents are smart, adaptive and provide human-machine interaction that is intuitive and organic. Speech-driven interface provide a platform that stimulates smart and natural human-machine interaction that is effective and effortless. Speech-driven interfaces have garnered a lot of attention in recent years the resultant of which is Dialogue Agents. Recent advances in the field of big data and learning models, especially deep models, have resulted in variants of dialog agents that provide user with an experience that is natural and that adapt dynamically the context of the interaction. Broadly the dialogue systems can be classified as goal-oriented and non-goal-oriented dialogue systems.

© Springer Nature Switzerland AG 2019
S. Madria et al. (Eds.): BDA 2019, LNCS 11932, pp. 33–43, 2019.
https://doi.org/10.1007/978-3-030-37188-3_3

Goal-oriented Dialog Systems are designed as assistants to a user for rendering certain service. Their aim is to help user in completing a task. They perform short conversation and extract certain information from a session. The design structure adopted by modern-day goal-oriented dialogue agents for Natural Language Understanding is based on a predefined ontology, example frame-based, for knowledge representation. Recent work in the direction of goal-oriented dialogue agents focuses on Spoken Language Understanding. The objective is to perform the following tasks in order to provide the user with the service demanded.

Identify the Domain - Extract the frame(s) that the systems need to consider. They are generally used for systems that perform multifaceted tasks.
Determine Intent - Identify the goal that the user is trying to accomplish.
Slot Filling - Extract the slots and the value by evaluating user's utterance that maps user's interpretation.

The frame-based dialog system are generally inclined to a control structure based on *(a) finite state automata (b) production rule system* and typically employs a template based generation. Also, such systems tend to be domain specific, hence the conversational session are short consisting of 2–3 utterances.

Non-goal-oriented Dialogue Agent is designed to perform extended conversations that exhibit characteristics of chit-chat. Although such conversation has no contribution in business but help provide user a satisfactory and natural experience [61]. One of the reasons for not retaining the ontology-based framework for non-goal-oriented dialogue systems is that they tend to be open-domain and so the no of ontology's needed to define is incalculable. Also, structured frameworks are unable to maintain dialogue history as it entails additional storage and processing overhead too.

The objective of a dialogue agent is to model a mapping between the input and output utterance in a dialogue. Goal-oriented-systems were designed to learn such mappings by focusing on feature engineering for a specific domain. Non-goal-oriented dialogue agent allows a researcher to design dialogue agents for tasks that are open-domain, reasonable and may not have any domain knowledge available.

This paper surveys the current state of the art for non-goal-oriented dialogue systems. It presents the models used for Natural language interpreter/understanding and generation, though the models discussed have also proven useful for goal-oriented agents. Most of the current empirical work focuses on data-driven models for dialogue generation.

2 Non-goal-Oriented Dialogue Agent

Non-goal-oriented dialog systems are an extension of speech-driven interfaces that process and produce responses that mimic the characteristics of human-human interaction. They can be classified as rule-based and data-driven.

In rule-based systems, typically user provides some input to a knowledge base. The knowledge base processes the input to extract a collection of facts or in other words patterns. A rule base contains the collection of rules, based on predicate logic, to be used by the system. The pattern action rules represent the action/goal the system will perform. An inference engine is designed to find the rule that matches closest to the pattern provided by the knowledge base. ELIZA [55], based on Rogerian psychology, and PARRY [9], to study patients suffering from schizophrenia, are examples of rule based conversational agent. The agents designed were able to generate responses efficiently but lacked the capability of performing natural language understanding that imitates human.

Recently data-driven systems have gained a lot of attention. With the abundance of data available it is easy to design models that learn to extract relevant information. Unlike rule base systems, these systems may not have an established set of actions that corresponds to the predetermined set of rules. Rather such system learns and adapts itself by mining largely available human-human and human-machine corpora. Data driven systems are further classified as generative and retrieval based models. A generative model generates a variable length response while retrieval based models select from a set of candidate responses. They can further be classified based on techniques employed as shown in Fig. 1.

This study presents a review of the recent learning methods, dataset and evaluation strategies adopted to fashion a non-goal-oriented dialogue agent and also discuss limitations data-driven methods, dataset and evaluation method exhibit. Section 3 presents the architectures and approaches used in the design of data-driven non-goal-oriented dialogue agent. Section 4 describes the challenges that the dataset, evaluation methods and architectures used in data-driven models present followed by discussion and conclusion in Sect. 5.

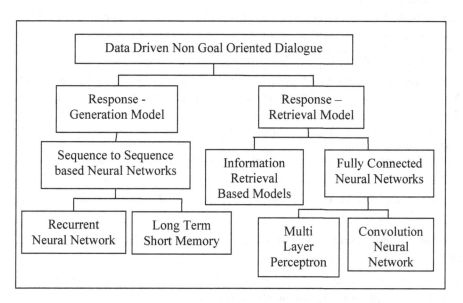

Fig. 1. The classification of data driven goal oriented dialogue agents

3 Data-Driven Dialogue Agents: Architectures and Approaches

Data driven systems can further be classified based on techniques employed:

3.1 Information Retrieval (IR) Based Model

The idea behind retrieval based strategy is to mine a corpus of input-output utterance pair, to determine a set of candidate responses from a set of responses. Generally the model outputs a response that is most similar to the current user input utterance. Although it is also relevant to generate response by performing matching between current user input utterance and available input utterances. Theoretically the first approach is intuitive but empirically the second approach outperformed the former. Researchers have applied various IR techniques exemplar nearest neighbor [16, 50], TF-IDF [28, 29], to determine similarity between input–output utterance pair. The main drawback of retrieval-based techniques is that it's inability to capture strong structural relationship between adjacent utterances.

3.2 Neural Network (NN) Based Model

Neural Networks are designed to learn complex representations to represent all suitable responses by exploiting back propagation.

Fully Connected Networks
A Statistical Machine Translation [21] based model is another perspective of modeling response. It learns a Probability Distribution Function (PDF), by employing Multilayer Perceptron (MLP), over a corpus of input-output utterance pair. Ritter [38] proposed a phrase based SMT, as phrase-level translations are more efficient than word-level that learns PDF that is conditioned on input-output pair in both directions. Although SMT based approaches perform better than IR based approaches but a status-response pair does not have the same relationship as that of a translation pair. Also SMT based models are unable to extract or update context or focus of attention in multi-turn conversation [38].

Ji [17] proposed a following retrieval-matching-ranking framework for selecting an appropriate response as output.

Retrieval: Searches for relevant responses from a set of output utterances using IR based techniques.

Matching: Generates a feature representation by performing Translation based model, deep matching model and topic word model.

Ranking: Assigns scores to each candidate response and select the highest scoring response as output using Linear ranking SVM [13].

Lowe [28, 29] also proposed a retrieval model using a probabilistic neural network over the input-output representations generated by Recurrent Neural Network based

Language Model (RNN-LM) [33] and Long Short Term Memory Based Language Model [48].

Convolution Neural Networks [22] became very popular amongst researchers for retrieval based models as they perform well in classification tasks in Natural Language Processing (NLP) [10, 15, 18, 36] proposed a convolution deep structured semantic retrieval model to perform semantic matching. Attention based model were found to be effective in machine translation [2, 23, 32]. An Attention gate is introduced in the model that signals certain parts of information from the input. Wu [57] introduced an attention gate at the pooling layer to propose a retrieval model over the input-output representations generated by the recent character-level CNN [20, 63]. Yin [62] proposed a model that determines semantic relevance between input-output pair in early stages of the model rather than at later stages. He presented three variations of attention model with bi-directional CNN by applying attention gate after the input layer, after the convolution layer and after both input and pooling layer.

Sequence to Sequence Models

Fully connected networks work effectively for retrieval based models as length of input and output is fixed while sequence to sequence models work well for generative model as they are capable of generating an arbitrary length sequence vector. Non-goal-oriented dialogue system generative models are generally fashioned in an encoder-decoder framework [5]. For a given input-output utterance pair, an encoder generates a representation, known as context vector, by processing the input utterance as a sequence of words. The decoder learns a probability distribution of the output utterance over the context vector. Recurrent neural network is the preferred sequence to sequence model for generating response. It has the ability to model a variable length input to a fixed length representation and learn a stochastic function that maps the fixed length representation to a variable length output [5, 33, 49]. Shang [45] proposed a RNN for both encoder and decoder with beam search. He also proposed attention mechanism in the encoder-decoder structure. Mei [31] proposed an extension of Shang's model by infusing attention dynamically in the same structure. Given a pair of context-input-output utterance pair, [47] generated context sensitive representation of the utterance pair using an RNNLM over 1-hot encoding in the encoder–decoder structure. Empirically a RNN can model context to a certain length due to which it is unable to model semantic dependencies for longer conversations. The problem can be addressed by using LSTM in place of RNN in the encoder-decoder framework [30, 35, 53, 56]. Due to limitation of variability in input-output pair in the corpus the beam search may result in generation of redundant responses. To increase diversity in the decoder [30] uses Latent Dirichlet Allocation (LDA) while Shao [46] and Choudhary [6] suggested a stochastic beam search Algorithm. Also, Mei [31] uses LDA based re-ranker to generate topic aware responses.

Hierarchical Models

Retrieval based models tend to be incoherent yet precise whereas generative models are coherent yet becomes stuck in infinite loop of repetitive and predictable dull responses. Few researchers have developed hierarchical models, by combining generative and retrieval models in order to generate diverse outputs.

The retrieval-matching-ranking framework determine effectively cohesion between a given input-output utterance pair, but lacks the ability of engineering semantic relevance in conversations that are not limited to 2-turn. For modeling semantic coherence over utterances and response over multi-turn conversations [58, 64] employs a hierarchical model that employs Gated Recurrent Unit (GRU) [7] over CNN for mapping utterance sequences. LSTM [14] and GRU are also sequence to sequence based models and perform equally comparable in various NLP tasks [7]. Zhou [66, 67] proposed encoder-diverter-decoder model using GRU to retrieve response that are coherent with the mechanism of user style of writing. Even the encoder-decoder framework for generative model follows a flat architecture as they learn probability distribution function only at the decoder level. Pierre [35] and Serban [43] suggested a hierarchical encoder-decoder structure. The first-level maps the sequence of words in an utterance while the second maps the sequence of utterance in the conversation. Serban [39] suggested a variant of hierarchical model that encodes and decodes the utterance at word and phrase level. Another, hierarchical generative model was suggested by Xing [59] using Twitter LDA [65] to jointly lean context and topic, Xing [60] also suggested use of attention gate over sequence of utterances while Serban [44] introduces a set of latent variable [8] in the hierarchical encoder model.

4 Research Challenges

4.1 Dataset

Serban [40, 41] presented a study of various corpora available for data-driven dialogue agents. But as our domain is restricted to non-goal-oriented dialogue agents hence corpus collected from Social networking sites, troubleshooting forums and dialogues from movie and TV shows are considered for performing a study.

Most researchers analyze the efficiency of their model on a corpus collected from various Social Networking Sites exemplar Twitter [47], Sina Weibo [54], Tencent Weibo [66] and Baidu Tieba [59]. Social networking Sites provide a platform where users interact with each other in a public domain. Although the conversations on such platform are short and comments are repetitive [29] hence the model is unable to learn how context is updated over long conversations and maintain coherence in conversations that involve many turns.

The dialogue for movies and TV shows, identified as human-human scripted corpora, has notable contribution in this domain. The dataset contains dialogue exchange between characters in movie and TV show. The SubTleCorpus [1], OpenSubtitles [52] and MovieTriples [43] are examples of such dataset. But the dataset comprises of abstract and non-representational data.

Another source of dataset in this domain is the written dataset available on various forums regarding trouble shoot. Although the topics are constrained to a limited set but a conversation session has multiple turns and are unstructured and unlabeled. Ubuntu Dialog Corpus [28], largest dataset available in this domain, is a trouble shooting dataset for ubuntu related problems. But the conversational sessions available in the dataset are few or have limited number of turns per session.

The abundant multi-turn human-human chat dataset is unavailable due to proprietary issues and the dataset available pose limitations in modeling response for multi-turn conversations. Neural networks contribute largely in retrieval based and generative models for response modeling and they tend to perform well with large noisy dataset [35].

4.2   Evaluation Strategies

Most of the empirical work in the domain of non-goal-oriented dialogue agents uses statistical evaluation methods like- Recall, Precision, Perplexity (PPL), Word error rate (WER), Mean Average Precision (MAP), Mean Reciprocal Rank (MRR). Statistical techniques perform well in evaluating a language model but are unable to evaluate the outputted response on semantic and syntactic relevance [45]. Some studies may have used translation based evaluation techniques BLEU [34] and METEOR [3] for evaluating the generated response. But compared to judgments' of humans does not correlate well with BLEU and METEOR judgment [26]. Hence for evaluating such systems most researchers employ a team of human annotators to label the response generated from various models. Pierre [35] suggested an evaluation technique based on specific discourse markers [11]. Kannan [19], Lowe [27] and Li [25] proposed adversarial evaluation for evaluating such dialogue systems.

4.3   Data-Driven Models

The advent of stochastic models provided tools to researcher to develop models that output response utterance that are coherent syntactically and semantically with the input utterances. The tradeoff with such stochastic models is between generating a good response and generating an interesting user engaging response. Most of the empirical work tends to focus more on former and sacrifice the latter. As recent research has shown models aiming at balancing these two are susceptible to considerable computational cost, though such models accomplish a unified view of language generation, thus making them compatible with generation of formal and informal language.

5   Discussion and Conclusion

The study presented various models that perform Natural Language Understanding and Generation employed for generation and retrieval of responses. Most of the work highlighted the role of deep learning models specially the models that are based on sequence to sequence framework. Sequential models are capable of learning contextual dependencies using a probabilistic distribution function that are word based, Phrase based or utterance based. But the stochastic models perform narrow interpretation due to – Restricted Dataset, Models that generate or retrieve dull and repetitive outputs and evaluation models are shallow.

Tian [51] performed a empirical study on various neural conversational models and concluded that hierarchical model perform well in an encoder-decoder framework, Attention mechanism based models are able to model context for conversation session that are longer than 2 turns and that semantic coherence between input – output utterance pair is captured efficiently by context aware models.

Li [24] and Serban [42] proposed a model based on reinforcement learning. Although the field of goal-oriented dialogue systems is flourishing due to reinforcement learning techniques developed recently but non-goal-oriented systems lack a concrete policy learning mechanism and reward function. Inspired from the work of Goodfellow [12], Bowman [4], Li [25] and Rajeswar [37] proposed a generative adversarial model (GAN) for natural language. Although generative adversarial networks cannot be effectively applied to natural language as they are not continuous and differentiable.

References

1. Ameixa, D., Coheur, L., Redol, R.A.: From subtitles to human interactions: introducing the subtle corpus. Technical report, INESC-ID, November 2014 (2013)
2. Bahdanau, D., Cho, K., Bengio, Y.: Neural machine translation by jointly learning to align and translate. arXiv preprint arXiv:1409.0473 (2014)
3. Banerjee, S., Lavie, A.: METEOR: an automatic metric for MT evaluation with improved correlation with human judgments. In: Proceedings of the ACL Workshop on Intrinsic and Extrinsic Evaluation Measures for Machine Translation and/or Summarization (2005)
4. Bowman, S.R., et al.: Generating sentences from a continuous space. arXiv preprint arXiv:1511.06349 (2015)
5. Cho, K., et al.: Learning phrase representations using RNN encoder-decoder for statistical machine translation. arXiv preprint arXiv:1406.1078 (2014)
6. Choudhary, S., et al.: Domain aware neural dialog system. arXiv preprint arXiv:1708.00897 (2017)
7. Chung, J., et al.: Empirical evaluation of gated recurrent neural networks on sequence modeling. arXiv preprint arXiv:1412.3555 (2014)
8. Chung, J., et al.: A recurrent latent variable model for sequential data. Advances in Neural Information Processing Systems (2015)
9. Colby, K.M., Weber, S., Hilf, F.D.: Artificial paranoia. Artif. Intell. 2(1), 1–25 (1971)
10. Collobert, R., et al.: Natural language processing (almost) from scratch. J. Mach. Learn. Res. 12(Aug), 2493–2537 (2011)
11. Fraser, B.: What are discourse markers? J. Pragmat. 31(7), 931–952 (1999)
12. Goodfellow, I., et al.: Generative adversarial nets. Advances in Neural Information Processing Systems (2014)
13. Herbrich, R.: Large margin rank boundaries for ordinal regression. Advances in Large Margin Classifiers, pp. 115–132 (2000)
14. Hochreiter, S., Schmidhuber, J.: Long short-term memory. Neural Comput. 9(8), 1735–1780 (1997)
15. Hu, B., et al.: Convolutional neural network architectures for matching natural language sentences. Advances in Neural Information Processing Systems (2014)
16. Jafarpour, S., Burges, C.J., Ritter, A.: Filter, rank, and transfer the knowledge: learning to chat. Adv. Rank. 10, 2329–9290 (2010)
17. Ji, Z., Lu, Z., Li, H.: An information retrieval approach to short text conversation. arXiv preprint arXiv:1408.6988 (2014)
18. Kalchbrenner, N., Grefenstette, E., Blunsom, P.: A convolutional neural network for modelling sentences. arXiv preprint arXiv:1404.2188 (2014)

19. Kannan, A., Vinyals, O.: Adversarial evaluation of dialogue models. arXiv preprint arXiv: 1701.08198 (2017)
20. Kim, Y., et al.: Character-aware neural language models. In: Thirtieth AAAI Conference on Artificial Intelligence (2016)
21. Koehn, P.: Statistical Machine Translation. Cambridge University Press, Cambridge (2009)
22. LeCun, Y., Bengio, Y.: Convolutional networks for images, speech, and time series. Handb. Brain Theory Neural Netw. **3361**(10), 1995 (1995)
23. Li, J., Luong, M.-T., Jurafsky, D.: A hierarchical neural autoencoder for paragraphs and documents. arXiv preprint arXiv:1506.01057 (2015)
24. Li, J., et al.: Deep reinforcement learning for dialogue generation. arXiv preprint arXiv:1606. 01541 (2016)
25. Li, J., et al.: Adversarial learning for neural dialogue generation. arXiv preprint arXiv:1701. 06547 (2017)
26. Liu, C.-W., et al.: How not to evaluate your dialogue system: an empirical study of unsupervised evaluation metrics for dialogue response generation. arXiv preprint arXiv: 1603.08023 (2016)
27. Lowe, R., et al.: Towards an automatic turing test: learning to evaluate dialogue responses. arXiv preprint arXiv:1708.07149 (2017)
28. Lowe, R., et al.: The Ubuntu dialogue corpus: a large dataset for research in unstructured multi-turn dialogue systems. arXiv preprint arXiv:1506.08909 (2015)
29. Lowe, R.T., et al.: Training end-to-end dialogue systems with the Ubuntu dialogue corpus. Dialogue Discourse **8**(1), 31–65 (2017)
30. Luan, Y., Ji, Y., Ostendorf, M.: LSTM based conversation models. arXiv preprint arXiv: 1603.09457 (2016)
31. Mei, H., Bansal, M., Walter, M.R.: Coherent dialogue with attention-based language models. In: Thirty-First AAAI Conference on Artificial Intelligence (2017)
32. Meng, F., et al.: Encoding source language with convolutional neural network for machine translation. arXiv preprint arXiv:1503.01838 (2015)
33. Mikolov, T., et al.: Recurrent neural network based language model. In: Eleventh Annual Conference of the International Speech Communication Association (2010)
34. Papineni, K., et al.: BLEU: a method for automatic evaluation of machine translation. In: Proceedings of the 40th Annual Meeting on Association for Computational Linguistics. Association for Computational Linguistics (2002)
35. Pierre, J.M., et al.: Neural discourse modeling of conversations. arXiv preprint arXiv:1607. 04576 (2016)
36. Prakash, A., Brockett, C., Agrawal, P.: Emulating human conversations using convolutional neural network-based IR. arXiv preprint arXiv:1606.07056 (2016)
37. Rajeswar, S., et al.: Adversarial generation of natural language. arXiv preprint arXiv:1705. 10929 (2017)
38. Ritter, A., Cherry, C., Dolan, W.B.: Data-driven response generation in social media. In: Proceedings of the Conference on Empirical Methods in Natural Language Processing. Association for Computational Linguistics (2011)
39. Serban, I.V., et al.: Multiresolution recurrent neural networks: an application to dialogue response generation. In: Thirty-First AAAI Conference on Artificial Intelligence (2017)
40. Serban, I.V., et al.: A survey of available corpora for building data-driven dialogue systems. arXiv preprint arXiv:1512.05742 (2015)
41. Serban, I.V., et al.: A survey of available corpora for building data-driven dialogue systems: the journal version. Dialogue Discourse **9**(1), 1–49 (2018)
42. Serban, I.V., et al.: A deep reinforcement learning chatbot. arXiv preprint arXiv:1709.02349 (2017)

43. Serban, I.V., et al. Building end-to-end dialogue systems using generative hierarchical neural network models. In: Thirtieth AAAI Conference on Artificial Intelligence (2016)
44. Serban, I.V., et al.: A hierarchical latent variable encoder-decoder model for generating dialogues. In: Thirty-First AAAI Conference on Artificial Intelligence (2017)
45. Shang, L., Lu, Z., Li, H.: Neural responding machine for short-text conversation. arXiv preprint arXiv:1503.02364 (2015)
46. Shao, L., et al.: Generating high-quality and informative conversation responses with sequence-to-sequence models. arXiv preprint arXiv:1701.03185 (2017)
47. Sordoni, A., et al.: A neural network approach to context-sensitive generation of conversational responses. arXiv preprint arXiv:1506.06714 (2015)
48. Sundermeyer, M., Schlüter, R., Ney, H.: LSTM neural networks for language modeling. In: Thirteenth Annual Conference of the International Speech Communication Association (2012)
49. Sutskever, I., Vinyals, O., Le, Q.V.: Sequence to sequence learning with neural networks. Advances in Neural Information Processing Systems (2014)
50. Swanson, R., Gordon, A.S.: Say anything: a massively collaborative open domain story writing companion. In: Spierling, U., Szilas, N. (eds.) ICIDS 2008. LNCS, vol. 5334, pp. 32–40. Springer, Heidelberg (2008). https://doi.org/10.1007/978-3-540-89454-4_5
51. Tian, Z., et al.: How to make context more useful? An empirical study on context-aware neural conversational models. In: Proceedings of the 55th Annual Meeting of the Association for Computational Linguistics, vol. 2, Short Papers (2017)
52. Tiedemann, J.: News from OPUS-A collection of multilingual parallel corpora with tools and interfaces. Recent Adv. Nat. Lang. Process. **5**, 237–248 (2009)
53. Vinyals, O., Le, Q.: A neural conversational model. arXiv preprint arXiv:1506.05869 (2015)
54. Wang, H., et al.: A dataset for research on short-text conversations. In: Proceedings of the 2013 Conference on Empirical Methods in Natural Language Processing (2013)
55. Weizenbaum, J.: ELIZA—a computer program for the study of natural language communication between man and machine. Commun. ACM **9**(1), 36–45 (1966)
56. Wen, T.-H., et al.: Semantically conditioned LSTM-based natural language generation for spoken dialogue systems. arXiv preprint arXiv:1508.01745 (2015)
57. Wu, B., Wang, B., Xue, H.: Ranking responses oriented to conversational relevance in chatbots. In: Proceedings of COLING 2016, the 26th International Conference on Computational Linguistics: Technical Papers (2016)
58. Wu, Y., et al.: Sequential matching network: a new architecture for multi-turn response selection in retrieval-based chatbots. arXiv preprint arXiv:1612.01627 (2016)
59. Xing, C., et al.: Topic aware neural response generation. In: Thirty-First AAAI Conference on Artificial Intelligence (2017)
60. Xing, C., et al.: Hierarchical recurrent attention network for response generation. In: Thirty-Second AAAI Conference on Artificial Intelligence (2018)
61. Yan, Z., et al.: Building task-oriented dialogue systems for online shopping. In: Thirty-First AAAI Conference on Artificial Intelligence (2017)
62. Yin, W., et al.: ABCNN: attention-based convolutional neural network for modeling sentence pairs. Trans. Assoc. Comput. Linguist. **4**, 259–272 (2016)
63. Zhang, X., LeCun, Y.: Text understanding from scratch. arXiv preprint arXiv:1502.01710 (2015)
64. Zhang, Z., et al.: Modeling multi-turn conversation with deep utterance aggregation. arXiv preprint arXiv:1806.09102 (2018)
65. Zhao, W.X., et al.: Comparing Twitter and traditional media using topic models. In: Clough, P., et al. (eds.) ECIR 2011. LNCS, vol. 6611, pp. 338–349. Springer, Heidelberg (2011). https://doi.org/10.1007/978-3-642-20161-5_34

66. Zhou, G., et al.: Mechanism-aware neural machine for dialogue response generation. In: Thirty-First AAAI Conference on Artificial Intelligence (2017)
67. Zhou, G., et al.: Elastic responding machine for dialog generation with dynamically mechanism selecting. In: Thirty-Second AAAI Conference on Artificial Intelligence (2018)

Data Cube Is Dead, Long Life to Data Cube in the Age of Web Data

Selma Khouri[1], Nabila Berkani[1], Ladjel Bellatreche[2(✉)], and Dihia Lanasri[1]

[1] Ecole nationale Supérieure d'Informatique, Algiers, Algeria
{s_khouri,n_berkani,ad_lanasri}@esi.dz
[2] LIAS/ISAE-ENSMA – Poitiers University, Futuroscope, France
bellatreche@ensma.fr

Abstract. In a short time, the data warehouse (DW) technology took an important place in the academic and industrial landscapes. This place materialized in the large majority of engineering and management schools that adopted it in their curriculum and in the small, medium-size and large companies that enhanced their decision making capabilities thanks to it. The 1990s saw the advent of conferences such as DaWaK and DOLAP that carried the acronyms DW and OLAP in their titles. Then, all of a sudden, this technology has been upset by the arrival of Big Data. Consequently, those actors have replaced DW and OLAP by Big Data Analytics. We are well placed to assert that this brutal move may have a negative impact on schools, academia, and industry. This technology is not dead, today's context, with the connected world and Web of Data, is more favorable than when building DW merely stemmed from company internal sources. In this invited paper, we attempt to answer the following question: how does DW technology interact with Linked Open Data (LOD)? To answer the question, we provide a complete vision to augment the traditional DW with LOD, to capture and quantify the added value generated through this interaction. This vision covers the main steps of the DW life-cycle. This value is estimated through two different perspectives: **(i)** a source-oriented vision, by calculating the rate of the DW augmentation in terms of multidimensional concepts and instances, and **(ii)** a goal-oriented vision where the value is calculated according to the ability of the DW to estimate the performance levels of defined goals that reflect the strategy of a company, using the defined DW of the case study of a leading Algerian company.

Keywords: DW augmentation · LOD · Value · KPI

1 Introduction

For more than three decades, DW technology has been leading technologies for data analytics solutions. Despite the evolution of the IT Global Market, this technology remains at the forefront of many data-driven solutions. Based on a

© Springer Nature Switzerland AG 2019
S. Madria et al. (Eds.): BDA 2019, LNCS 11932, pp. 44–64, 2019.
https://doi.org/10.1007/978-3-030-37188-3_4

report published Mordor Intelligence[1], the global active DW market was valued at USD 5.67 billion in 2018 and is expected to reach a value of USD 10.75 billion by 2024. Certainly, several studies and reports mentioned the hardness of computing the real **R**eturn **O**f **I**nvestment (ROI) of a DW project, but nobody can deny that this technology offers direct and indirect benefits for companies. Direct benefits are usually materialized by the *time savings* in extracting, transforming and integrating data from various sources into one unified repository. The main indirect benefits concern[2]: (i) the flexibility of BI tools that offer decision-makers the possibility to decide on the fly what type of analysis they have to perform on their stored data and (ii) the productivity gain usually considered as a measure of innovation [9] allowing decision-makers switching from maintaining rigid legacy systems to productive and creative tasks thanks to accurate and timely reports generated by BI tools.

The advent of Big Data technologies has brought new and exciting opportunities for data-centric systems in general, and for DW technology in particular. As there are many companies owing and providing solutions for DWs, engineering and management schools offering curriculum covering this technology, and researchers, they are then obliged to get the best of the Big Data environment (its different V's, computation platforms and technologies, programming paradigms, etc.). This passes through the process of augmentation of their DWs. This process is defined as *the combination of traditional and big data dimensions in order to increase the effectiveness of existing and add more value to DWs*[3].

The first track to augment a DW is based on the following finding widely commented in literature [6,7]: in traditional DWs, a large amount of requirements is not fully satisfied by internal sources. This situation penalizes companies looking for value and contradicts with the initial goal of DW technology. Certainly, the obtained DWs create value, but it remains proper to the context of these companies (closed value) which does not bring surprises [9]. Since the beginning of the 2000s, the term "open" was used everywhere [11]. This gives rise to several concepts such as open data, open innovation, open medical records system, open science, open knowledge, and open education, etc. Open value of a DW is a new term opening up any DW beyond company boundaries in order to increase the closed value through integrating external sources related to activities of these companies.

In the Big Data landscape, Linked Open Data (LOD) are typical external data sources that can bring value to DW systems. The definition provided by Open Data provided by European Data Portal[4] follows our philosophy: *"data that anyone can access, use, and share. Governments, businesses, and individuals can use open data to bring about social, economic, and environmental benefits"*.

---

[1] https://www.mordorintelligence.com/industry-reports/global-active-data-warehous ing-market-industry.

[2] https://datavirtuality.com/blog-calculating-the-return-on-investment-roi-of-busine ss-intelligence-projects/.

[3] https://developer.ibm.com/tutorials/ba-augment-data-warehouse1/.

[4] https://www.europeandataportal.eu/elearning/en/module1/#/id/co-01.

YAGO and DBpedia[5] are examples of popular LOD portals. LOD is characterized by two main aspects that are relevant for DWs: (i) they are open and freely available for companies, that can find valuable information completing their internal sources. (ii) They are based on Semantic Web standards like *RDF* definition language based on a graph-oriented formalism and *Sparql* query language. Semantic web technologies have been largely used to eliminate data conflicts and to automatize the data integration process for DW systems. In our study, we focus on *DW augmentation using LOD data sources*. We claim that this augmentation must be strongly connected to the value augmentation for the DW.

The presence of LOD sources can rapidly be seen as an opportunity for DWs and for analysts in companies. But their integration in the DW represents a crucial issue. It has to be driven by the business requirements to be fully valuable for the company's strategies. The evaluation of a company strategy is crucial for measuring its success. For DWs, this strategy is commonly carried out by analyzing metrics such as key performance indicators (KPIs) that must be identified from the beginning of the design process [35]. KPIs are defined according to the set of goals defined by the company's managers and analysts for the DW. To give a brief definition, goals represent the desired state-of-affairs, defined during strategic planning and pursued during business operation [3]. KPIs present metrics that evaluate the performance of the defined goals.

In this paper, we propose a complete approach that spans the DW life-cycle from the definition of goals requiring internal and external LOD sources to the integration of the set of sources that meet the defined goals. Since the demand for LOD sources is motivated by the augmentation of the value for the company, our approach estimates the added-value of LOD sources, using quantitative and qualitative metrics, based on KPI values. This value is estimated according to two visions: (i) a source-oriented vision, where the rate of the DW augmentation is calculated in terms of multidimensional concepts and instances, and (ii) a goal-oriented vision, where the value is calculated according to the ability of the DW to estimate the performance levels of defined strategic goals.

The rest of the paper is organized as follows: Sect. 2 presents a motivating example that illustrates our objectives. Section 3 reviews the related work. Section 4 details our proposal. Section 5 presents a case study inspired by a real scenario of an Algerian company. The last section concludes the paper and sketches some future works.

2   Motivating Example

In this section, we present a motivating example that will be used in our case study. The case study is inspired by an application of a leading Algerian private company in the field of Home appliances industry, that commercializes its products in the national and international markets competing for big brands. In order to enhance its activity, enforce its presence on the market and win the

---

[5] http://wiki.dbpedia.org/.

satisfaction of its clients, the managers of this company defined a set of goals
and their related KPIs in their dashboard.

The set of goals and their related KPIs are organized in a tree hierarchy as
illustrated in Fig. 1, where a strategic goal is satisfied by the satisfaction of its
sub-goals using AND/OR relationships (Eg. goal "Stand out in the market" is
satisfied by the satisfaction of goals "Satisfy the clients" OR "Enhance innovation"). Influence relationships are also present in the tree, indicating that a goal
G1 is positively or negatively influence by another goal G2 (+ or ++ according
to the strength of the influence) (Eg. goal "Stand out in the market" influence
positively (++) the goal "Increase the sales revenue"). The set of goals at the
bottom of the tree (the leaves) are atomic goals and they indicate actions to
achieve and fulfill goals of higher level.

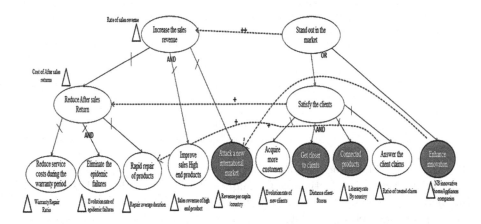

Fig. 1. Tree of goals and their KPIs basing on internal & external sources

For designing a DW, each goal allows identifying the fragment of external
and/or internal source required for identifying the performance of the goal. We
considered a relational dataset containing information about sales and after-sales
as an internal source. The source schema is illustrated in Fig. 2. We also considered Dbpedia, Linked Geo Data, and DataGov as external LOD sources. Figure 1
illustrates the goals that require external sources; these goals are represented by
gray nodes. In our approach, goals are considered incrementally to illustrate a
realistic scenario.

The performance of each goal is calculated through the value of its related
KPI. A Key Performance Indicator (KPI), is an industry term for a measure or
metric that evaluates performance with respect to some objective [3]. Each KPI
is used to measure the fulfillment of a goal. The set of KPIs forms consequently
a hierarchy tree similar to the hierarchy of goals.

Different approaches are proposed in the literature in order to calculate KPI
values [3]. Atomic KPIs provide formulas that are directly calculated to the set
of sources, they present the component that relies on the sources to the goals of

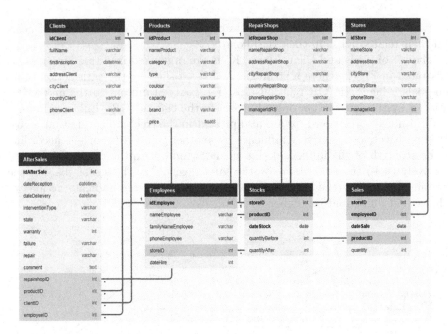

Fig. 2. Relational schema of the internal database

the target DW. The values of KPIs of higher level are calculated by propagation using qualitative and quantitative techniques. The value of the highest KPI in the hierarchy indicates the performance level of its related goal, which provides a management vision to decision-makers. Knowing this information is the value that the DW (and consequently the set of sources that aliment the DW) provides to decision-makers. Our approach allows to incrementally identifying the calculation of performance level of goals requiring external LOD resources and provides insights on the relevance of incorporating LOD sources into the target DW.

One main issue that we had to manage for augmenting the DW by new external sources, is to allow the integration of internal and external sources, using semantic techniques in order to unify the sources syntactically and semantically to ensure the flow of data from the sources to the target DW. This semantic unification also allowed the calculation of KPIs values on a semantic schema, which makes the approach independent of the various physical formats of sources.

3   Related Work

In this section, we first present the main generations of augmented-DWs. Then, we present the main studies dealing with value in DW projects.

3.1   Augmentation of DW with New Sources

DW technology has faced different views of design and usage since its appearance
in the late 80's. Different generations of augmented-DWs have emerged through
the years according to the emergence of new data sources.

The B1.0 era has largely studied the design of relational DWs exclusively
from relational sources. Many issues have been addressed from optimization to
modeling issues [16]. Since the emergence of the web, research studies pivoted
their interest into the augmentation of DW with external data. The first works
were on integrating HTML data from web sites [14] then XML data from DTD
[17] or XML schema [37].

The B2.0 era started with the explosion of Web2.0 and Web3.0, where different studies proposed integrating semantic data [30], social data [31,39], NoSQL
data [13], data from data lakes [27], and LOD [15,32].

The LOD augmented DWs with semantic web data freely available. Some
studies focused on the definition of multidimensional patterns from LOD datasets
[1] or LOD query logs [23]. Other studies managed the integration process of
LOD using conventional ETL process [12] or an adaptation of ETL consisting
in extracting transforming and querying (ETQ) the required data on demand
[2]. Our approach provides a new vision by estimating the value of a LOD
augmented-DW, following a goal-driven approach.

3.2   Value in DWs

Existing works on the value creation process, all interpret value differently, which
has a negative impact on the semantic quality of the value modeling and instanti-
ation. To resolve this issue, some works attempt to design it at different levels. In
the case of value-based requirements engineering, authors of [36] have reviewed
different issues related to poor understanding of stakeholder values and proposed a taxonomy of stakeholders, motivations, and emotions (VME). Other
works have investigated the ontological analysis of the conceptualization of value
and employ a number of concepts and theories that serve as ontological analysis related to competition [33], value proposition [34], and value ascription [19].
Other works have proposed some value-modeling methods such as: e^3value [18],
SEAM [38] that aim to identify the exchange of value objects between actors.
However, all these approaches deal with value in a general context and do not
provide a clear conceptualization on how insights from DW are transformed to
added value.

Recently, review of literature emphasizes the ability of organizations to create
value through the use of DW technology and analytic [8,10,25,26,28,29]. They
try to analyze customer's behavior and opinions, and the ability of an organization to create value. They have recognized the role of DWs in augmenting
the value such as new business opportunities and increases the ROI through the
acquisition of valuable information about customer interests and requirements.
On the one hand, existing literature does not provide a process for the estimation
of the added value brought by external sources w.r.t to internal sources. On the

other hand, and despite increased interest, the process of value ascription and its measurement from the insights triggered by DW remains vague. Therefore, we call for a deeper analysis of what can enable creating value and measure it from the use of DW.

4 Augmentation of DW by LOD

From the beginning, we emphasize on the maturity of DW technology. Based on this maturity, we attempt to model different phases of DW design. This modeling will contribute to augmenting these phases by external sources. These modeling efforts are then presented in the following sections.

4.1 Modeling of DW Phases

Before detailing the approach for designing a LOD augmented DW, we first provide the meta-model underlying the DW construction. The proposed metamodel follows a design-cycle view where we distinguish four main phases: requirements definition, conceptual modeling, ETL modeling and the implementation phase (logical and physical modeling). We have largely studied the DW design cycle in previous studies [24] which allows us to have a whole view of the different design artifacts composing each design phase and their various correlations present inside each phase and between the phases.

The proposed meta-model details the requirements phase using a goal-oriented view, we distinguish three main classes (Fig. 3): (i) *Goal* class defining each goal provided by the decision-makers and managers, (ii) *Relationship* class defining the relationships between goals that can be of and main types: AND/OR and influence relationship. (iii) *KPI* class, evaluating the satisfaction of each goal. Figure 1 illustrates the goal tree and the relationships between goals. Each KPI is evaluated either by propagation or from other KPIs or from the set of sources that may be internal and/or external (LOD). Each KPI has a current value which is evaluated, and the performance level of each KPI is defined by comparing the current value calculated with other values defined by the business managers, which are: the worst value, the threshold value and the target value of the KPI.

The set of goals are defined by the analysts and projected on the set of sources for identifying the set of relevant fragments to integrate. The main issue to solve in DW projects is the integration of various heterogeneous data sources. When the DW considers external LOD sources, this variety is intensified and is solved by choosing the appropriate pivot model that can unify sources at the two levels: at the schema level (semantic unification) and at the format level.

On the one hand, LOD sources are semantic data that can be provided with their ontological schema. Ontologies have been largely used for unifying the semantics of data sources in DW projects. On the other hand, LOD sources follow a graph-based representation, which is usually used in literature for unifying the format of sources. The LOD environment represents thus a nice opportunity to be the elected model, which allows designers to simultaneously manage the levels of variety present in the vocabulary of the sources (using ontologies) and in their formalisms. In this vision, the LOD schema is defined as the pivot semantic schema and mappings between the internal sources and this schema are defined for unifying the semantics of sources. For unifying the formalisms, the graph format of LOD is chosen and mappings between the format of internal sources (for instance, the relational format) and the graph formalism are identified so that all sources can be considered as graphs.

In this paper, we adopt this solution that impacts significantly the ETL environment in which the ETL process is defined at the semantic conceptual level (*CM_Element* class), where the ontological schema is seen as a pivot model to unify the set of sources and load the data to the target DW schema. The ETL environment includes activities, workflows, and operators. An ETL workflow (*ETLWorkflow* Class) is the global collection of ETL activities (*ETLActivity* Class) and transitions between them. A transition determines the execution sequence of activities to generate a data flow from sources to the target DW. Each ETLActivity is an expression for extracting data from sources, transforming them and loading data to the target data store. This process is achieved using a set of defined ETL operators (*ETLOperator* class) that are classified into five categories: source, store, transform, flow and query operators. Note that the ETL process is defined at the conceptual (ontological) level so that it can be implemented using any physical schema (class PhysicalFormat). This implementation only requires translating the required ETL operators on the physical format (source schema or target DW schema). Note also that for calculating the current values of KPI, each KPI that has to be evaluated on the sources is defined using a formula that can be translated to a query on the semantic schema integrating the sources, using the adapted semantic query language (Sparql language).

4.2 Augmenting a DW by LOD

Our approach assumes that a set of goals (and associated KPIs) are considered incrementally. If a given goal is not fully supported by internal sources, the call for external LOD sources is required. By analyzing different actions of building a DW from internal and external sources, we have identified common patterns used for all scenarios. Our process considers the given goal through the whole design cycle of the DW using the metamodel described in the previous section, as follows:

- The incorporation of the new goal (resp. KPI) in the tree hierarchy of goals (resp. hierarchy of KPIs).
- The identification of the fragments of internal and external sources required by the goal.
- The integration of the sources into the DW using different scenarios that will be detailed in what follows.
- The quantification of the augmentation of the DW through the rate of multidimensional concepts and instances that augment the DW.

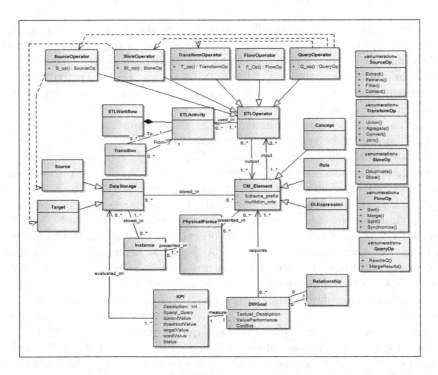

Fig. 3. Metamodel of design artefacts used for designing an augmented DW from internal and external sources.

– The estimation of the added value of the external concepts integrated through the performance of KPIs and goals in the goal tree.

1. Identification of the Required Sources. Once the new goals (resp. related KPIs) are inserted in the tree hierarchy of goals (resp. hierarchy of KPIs), they are considered in the DW design. The first step consists in identifying the required fragments of the sources that allow answering the goal.

The set of goals considered are first projected on the internal sources, or directly on the target DW (in case it is operational). In this last case, goals are represented by queries posed on the DW. If the goals (or their related queries) are not fully satisfied, they are projected on the LOD. Note that the LOD is very large in terms of concepts. Therefore, it has to be customized to the application domain. This customization is performed by the usage of *context* operator introduced in [20] (method *Context()* in class *SourceOp* in the meta-model). This operator generates a fragment of the LOD satisfying the set of defined goals. This process allows identifying relevant concepts and properties augmenting the target DW schema. Consequently, concepts of the DW schema are identified from the schemas of data sources, enriched with concepts and roles from LOD following the set of goals. Each new concept (internal or external) enriching the DW schema has to be annotated with multidimensional labels (*Measure, Fact,*

Dimension and *Dimension attribute*). The algorithm we proposed in [5] can be used for the multidimensional annotation.

2. Integration of the Sources into the DW. By analyzing the literature related to building a DW from internal and external sources, we have identified two main policies of companies for incorporating the LOD:

- The LOD and internal sources are integrated and physically materialized in the DW.
- Both LOD and DW query results are merged, where the integration between internal and external data is not made at the DW level, but during the querying process.

We have identified three main scenarios following these policies illustrated in Fig. 4. The first two scenarios (a) and (b) are identified for the first policy, whereas scenario (c) reflects the second policy. We detail in what follows the three scenarios.

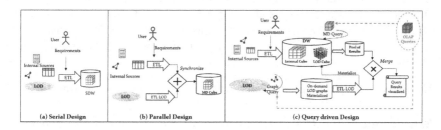

Fig. 4. Defined scenarios for integrating LOD sources in the DW

Scenario (a): Serial Design. It is feasible when a company decides to build its DW by considering both internal and external sources from scratch. This scenario follows a *conventional* DW design. The LOD is considered as a new semantic source to manage in addition to internal sources. An ETL process is defined considering all the sources. This scenario is realistic for constructing a DW from scratch, it may be not realistic if it requires a redefinition of the whole ETL process each time new requirements are needed for extracting value from the LOD. We have proposed in [5] a detailed ETL algorithm that manages this scenario.

Scenario (b): Parallel Design. This scenario assumes that the target DW is operational and keeps integrating data from internal sources and from the LOD. The ETL process from the LOD is generated and then synchronized with the initial ETL process from internal sources before the loading step (Fig. 4-b). This proposal describes the *reaction* of the parts of the ETL process affected by a flow change. It requires the consolidation of two parallel ETL flows (internal and external) which is needed to keep the target warehouse up to date. We formalize the problem of consolidation, by introducing the main design operation *synchronize*. It corresponds to synchronization between: (i) *Current flow*: existing ETL flows satisfying the n current information requirements at t time and (ii) *New flow*: ETL flow satisfying the upcoming requirement at $t + 1$ time. Because all sources can be considered as graphs (pivot format), the *synchronize* operation has to be defined using the graph format, as follows:

- $Synchronize(G, G_i, G_j, CS)$: Synchronize two sub-graphs G_i and G_j based on some criteria CS (AND-JOIN/OR-JOIN).

Scenario (c): Query-Driven Design. This scenario corresponds to the on-demand ETL to feed the target DW. Here, data are incrementally fetched from existing DW and the LOD (in the case where it is necessary), then loaded into the DW only when they are needed to answer some OLAP queries (Fig. 4-c). This scenario requires rewriting the OLAP queries on the LOD, extracting required fragments of LOD (using *Context* operator), applying the transformations required (using the *Class Transform-Operator*) by mean on an ETL process dedicated to LOD and then materializing using *Store* operator, the resulting graphs in case they are required later. The results are first integrated into a data cube reserved for LOD data analysis, and final results of queries (on LOD) are *merged* with the results of OLAP queries executed on the internal DW in order to display the query result to the end-user. Merge operator is defined using the graph format as follows:

- $Merge(G, G_i, G_j)$: merges two sub-graphs G_i and G_j into one graph G.

3. Estimation of the DW Augmentation. We have defined two mechanisms for identifying the added value of internal and external sources in the DW. The first mechanism is source-driven. It provides quantitative hints on the augmentation of the DW in terms of multidimensional concepts and instances, as it is detailed in this section. The second mechanism that will be detailed in the next section is goal-driven. It estimates the value of external LOD sources according to their capacity to evaluate the performance of the goals that the company identified for the defined DW.

We detail in what follows the first mechanism. The estimation of the DW augmentation is formalized as follows, given: (i) a set of internal sources $S_I = \{Si_1, Si_2, \dots Si_m\}$, (ii) a set of external LOD sources: $S_E = \{Se_1, Se_2, \dots Se_m\}$. (iii) a set of goals G to be satisfied. (vi) A DW (to be defined or operational). The augmentation rate of the target DW (Aug_Rate) regarding a given goal can be calculated as follows:

$$Augmentation = \sum_{S_i \in S_I \cup S_E} Weight(S_i) * Augmentation(S_i) \qquad (1)$$

where $weight(S_i)$ describes the weight of each source as it can be estimated for a given organizational sector.

In our work, the augmentation of the DW is strongly related to multidimensional concepts and instances provided by sources. This information provides some hints to the designers about the value that can be provided by external sources. We defined two augmentation metrics associated to each source S_i: *AugMD* and *AugInst* that are defined as follows. Note that these equations measure the percentage of the augmentation from external sources in terms of MD concepts to be met:

$$AugMD(S_i) = \frac{Number\_Concepts(S_i)}{TotalNumber\_Concepts(DW)} \qquad (2)$$

where $Number\_Concepts(S_i)$ is the number of multidimensional concepts of DW schema by integrating the source $_i$ and

$TotalNumber\_Concepts(DW))$ describes the total number of multidimensional concepts of DW.

$$AugInst(S_i) = \frac{NumberInstancesInt(S_i)}{TotalIns(DW)} \qquad (3)$$

where $NumberInstancesInt(S_i)$ and $TotalIns(DW)$ represent respectively the number of instances of DW by integrating the source S_i and the total number of instances of the DW.

4. Estimation of the Added-Value for the DW. The last step of our approach estimates the added value of external sources required by a given goal G, according to the ability of the source to estimate the performance of this goal G and all its associated goals until the highest goal of the hierarchy indicating the vision of the company's managers.

The performance of goals is estimated according to the performance level of their related indicators (KPIs). Considering a hierarchy of indicators (Fig. 1), the current values of leaf indicators (called also atomic or component indicators) is extracted from data sources, and the values of non-leaf indicators (called composite indicators) is calculated by propagation using a quantitative approach (when metric expressions are available) or estimated following a qualitative approach. The ideal case would be that all relations between atomic and composite indicators are described using mathematical equations, in this case, a fully quantitative approach can be applied.

Different studies detail two well-known techniques for the quantitative approach: conversion factors and range normalization [21]. The first category of techniques (conversion factors) estimates the current value of composite indicators using metric expressions containing current values for component indicators and conversion factors. These metric expressions are defined by the designer relying on historical data and/or her domain knowledge. If the indicator is measured using metrics measured in different units, then the conversion factor must be added to the formula.

The second category of techniques (range normalization) derives composite indicators by considering values spanning a specific range and representing them in another range. This category also relies on a defined metric expression for performance levels, where the performance level of an indicator is defined using performance levels of other indicators in the hierarchy.

In both techniques, the effect of goals influencing the goal which KPI is evaluated can refine the metric expressions of indicators by adding weights showing the importance of some parts of the formula.

In case where mathematical formula are not available, the qualitative approach can be followed [21]. This approach relies on two variables: positive performance (per−) and negative performance (per−) that are assigned to each indicator, and that are calculated from the set of sources for atomic indicators. According to the values of (per+) and (per−), the performance level of the indicator is assigned that ranges from fully non-performant to fully performant. The performance levels of each indicator (*ind*) of higher levels are defined using some defined mapping rules defined in [21], that considers the indicators of the lower level in the hierarchy and the relationships between these indicators and the indicator (*ind*) (AND/OR and influence relationships).

If only some business formula is available for a subset of indicators, it is possible to combine the qualitative and the quantitative approaches, in order to follow a hybrid approach, which is the approach we followed in our case study.

Following either a quantitative or qualitative approach, we notice that the values of atomic KPIs are the key component that links the data sources to the goals and thus the strategy of the company for its DW. The calculation of KPIs values indicates their performance level and consequently the performance of each goal. Because the calculation of composite KPIs values is achieved on the set of sources (internal and/or external), we have to define the process that allows this calculation.

As discussed in Sect. 4.2 (step 2), we defined three main scenarios that reflect the policies of a company that requires external LOD sources in its DW. The calculation of KPIs values is discussed according to these scenarios.

For scenario (a) and (b), the integration process allows the definition of an integrated schema. Recall that we have elected the LOD (with its semantic model and graph format) as the pivot model to unify the set of sources. The integrated schema is thus a semantic schema that unifies the instances of internal and external sources. The calculation of atomic KPIs values is calculated on this semantic schema. Note that the formulas of atomic KPIs can be translated as queries on the set of sources. Since the integrated schema is semantic, the KPIs formulas require a translation to a semantic query language. We opted for Sparql, which is the semantic web standard language for querying ontologies. The translation process is based on the algorithm we defined in [22].

For scenario (c), the integration process is achieved at the querying level and LOD fragments are not materialized. The calculation of the result of the atomic KPIs is achieved similarly to the integration process. The query of the indicator is exposed on the LOD and also on the DW, and the final results of the queries (on LOD) are *merged* with the results of the query executed on the internal DW in order to calculate the value of the corresponding KPI.

5   Case Study

As explained in Sect. 2, our case study is inspired from a real company, it is based on the set of goals illustrated in Fig. 1. The list of KPIs formulas is given in Table 2. In order to test our approach and to appreciate the added value of LOD sources in the DW, the goal hierarchy requiring only an internal source is illustrated in Fig. 5.

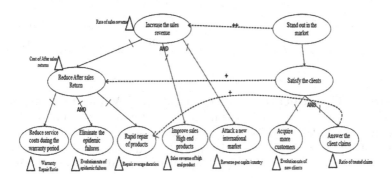

Fig. 5. Goals based on internal source

Source-Driven Quantification of the Augmented Value. As described in our approach, we can provide insights on the DW augmentation in terms of multidimensional concepts and instances enriching the DW. In our experiments, we first calculated the added value considering the integrated data sources as a whole (internal and external sources) then per goal. In this latter, we have integrated fragments of internal and external sources per goal where each LOD fragment corresponds to a goal, for example, the fragment of LOD Dbpedia (DBpedia: company) corresponds to the goal Enhance innovation.

Table 1 summarizes the results obtained and demonstrates the value added by considering LOD in the design of DW. The results show the added value of the integration of external sources compared to internal sources according to the given formulas in the previous section. Note that we have considered the third scenario in our experiments. These results clearly indicate that taking into account LOD data provides added value in terms of the size of the target DW: concepts (number of dimensions and measures) and instances, compared to internal data sources integration. Thereafter, we have considered the integration of external data sources per goals. We can observe the results on Table 1 that indicates, for every goal the integration of external data sources (LOD) enriches the concepts and instances of the DW which provides new dimensions and measures for a more in-depth OLAP analysis and will help to meet the company's changing requirements.

Table 1. Added value by integrating internal and external sources

Scenarios	Goals	Sources	Value(Si)$_{MD}$	Value(Si)$_I$
Whole data sources	–	Internal	62%	47%
	–	Internal & LOD	82%	74%
Data sources per goal	Attack a new international market	LOD	9%	19%
	Connected products	LOD	7%	11%
	Get closer to clients	LOD	17%	26%
	Enhance innovation	LOD	26%	31%

Goal-Driven Quantification of the Augmented Value. Here, we start by calculating the performance of goals using only internal data sources.

The results of the analysis and performance according to the period January 2018–December 2018 of the different KPIs and goals using internal sources are given in Table 2. In this table, each goal is associated to its KPI with its formula that is translated on the internal source. To evaluate the performance of goals, each KPI has a target value (best value to reach), worst values (to avoid) and a threshold of accepted values indicating good progress. The real value column (i.e. current value of the KPI) represents the calculated value using internal data source (sales & after sales) for the year 2018. The last column evaluates the performance of goals according to these KPIs. As resumed in the table, the KPIs (**EREF, SRHEP, ERNCC, TCR**) reached a very performant values and they accede the target value, that is the reason why their associated goals (respectively **Eliminate the epidemic failures, Improve sales High end products, Acquire more customers, Answer the client claims**) are fully satisfied. While the goal **Reduce service costs during the warranty period** is fully denied and **Rapid repair of products** is partially denied because their associated KPIs are non-performant in this period. Based on the hybrid reasoning, the evaluation of goals of the level L-1 is based on the goals of level L (leaf goals). **Reduce After-sales Return** is associated with **CASR** KPI which is calculated according to the KPIs of level L, this goal is partially denied because the real value of its KPI is not performant. While the goal **Satisfy the clients** is fully satisfied according to the goals of level L. The highest goals **Increase the sales revenue** and **Stand out in the Market** are respectively partially satisfied (quantitative value is acceptable) and fully satisfied (qualitative evaluation) according to the previous goals.

Because the approach considers each goal incrementally, we will augment the goal hierarchy with the new goals requiring external LOD sources, represented by gray nodes in Fig. 1. We followed the hybrid reasoning approach (mixing quantitative and qualitative reasoning) for calculating the values of KPIs. The new goals using external sources and the new results obtained based on this reasoning are given in Table 3.

Table 2. Results of KPIs and performance goals basing on internal source

Goals (using only internal source)	KPI (needing internal sources)	Formula	Target values	Threshold values	Worst values	Real values	Evaluation (Before integrating external sources)
Reduce service costs during the warranty period	Warranty Repair Ratio(WRR)	nb warranty repairs/ nb total repairs	WRR=5%	5%<WRR<15%	WRR>15%	90%	Fully denied
Eliminate the epidemic failures	Evolution rate of epidemic failures(EREF)	Nb epidemic failures(n)-Nb epidemic failures(n-1)/Nb epidemic failures(n-1)	EREF=0,2%	0,2%<EREF<0.5%	EREF>0,5%	0,01%	Fully satisfied
Rapid repair of products	Repair average duration(RAD)	AVG(product date delivery-product date reception)	RAD=2d	2d<RAD<5d	RAD>5d	10d	Partially denied
Improve sales High end products	Sales revenue of high end product(SRHEP)	revenue of high end product	SRHEP=35M	25M<SRHEP<35M	SRHEP<25M	32M	Fully satisfied
Acquire more customers	Evolution rate of new clients(ERNCC)	(nb clients(n)-nb clients(n-1)/ nb client n-1)	ERNCC=7%	3%<ERNCC<7%	ERNCC<3%	14%	Fully satisfied
Answer the client claims	Ratio of treated claims(TCR)	Nb treated calls/ total calls	TCR=90%	80%<TCR<90%	TCR<80%	97%	Fully satisfied
Reduce After sales Return	Cost of After sales returns (CASR)	(WWR+EREF)*TotalRepairs* (RAD*CostEmployee+PriceRepair)	CASR=5M	4M<CASR<5M	CASR<4M	7,5M	Partially denied
Satisfy the clients	No KPI						Fully Satisfied
Increase the sales revenue	Rate of sales revenue (RSR	SR(nonHighProduct)+ SRHEP+ CASR/ [SR(nonHighProduct)+ SRHEP]	RSR=80%	60%<RSR<80%	RSR<60%	82%	Partially satisfied
Stand out in the Market	No KPI						Fully Satisfied

Table 3. Results of KPIs and performance goals basing on internal & external sources

Goal (basing on internal & external sources)	KPI (needing internal & external sources)	Target values	Threshold values	Worst values	Real values of KPIs using internal & external sources	Evaluation (After integrating external sources)
Attack a new international market	AVG(Revenue per capita /country) (RPCC)	RPCC= 5000$	4000$<RPCC< 5000$	RPCC< 4000$	6500$ (DBpedia)	Fully satisfied
Get closer to clients	Literacy rate By country (LR)	LR=80%	60%<LR<80%	LR<60%	72% (DataGov)	Partially satisfied
Connected products	Distance between clients and Stores / RepairShop (DC)	DC=90%	60%<DC<90%	DC<60%	83% (Linked Geo data)	Partially satisfied
Enhance innovation	Home appliances innovative companies (HC)	HC=100	HC<100	HC>100	(75) (DBpedia)	Fully satisfied
Reduce After sales Return	Cost of After sales returns (CASR)	CASR=5M	4M<CASR<5M	CASR<4M	7,5M	Partially denied
Satisfy the client	No KPI					Partially satisfied
Increase the sales revenue	Rate of sales revenue (RSR)= SR(nonHighProduct)+ SRHEP+RPCC*3000 - CASR/ |SR(nonHighProduct) +SRHEP|	RSR=80%	60%<RSR<80%	RSR<60%	87%	Fully satisfied
Stand out in the Market	No KPI					Partially satisfied

The new goals associated to their KPIs calculated from the external sources (*DBpedia*, *Liked geo data*, *DataGov*) are included in this table (**Attack a new international market, Get closer to clients, Connected products, Enhance innovation**). These new KPIs are used to evaluate the parent goals (**Satisfy the clients, Increase the sales revenue** and **Stand out in the Market**) as illustrated in the hierarchy of goals. The atomic KPIs are evaluated using the third integration scenario described in the proposed approach (cf. Sect. 4.2, scenario "Query-driven Design"), where some parts of the KPI formula are translated on the LOD and the results are merged with the results of the parts of the formula related to the internal source.

As expected, according to these new goals, and the previous goals (of Table 2) the performance of parent goals has changed. The goal **Satisfy the client** that was Fully Satisfied is now Partially satisfied (based on a qualitative reasoning). **Increase the sales revenue** that was Partially satisfied (82%) is now Fully satisfied (87%) (based on a quantitative reasoning). **Stand out in the Market** that was Fully Satisfied is now Partially satisfied (based on a qualitative reasoning).

In the decision-making process of this company, the managers should control periodically their KPIs in order to evaluate the progress of their activity and measure their goals reaching. KPIs are powerful means that help in the management process and allow readjusting the actions and reaching the defined strategic goals. Calculating KPIs and evaluating the satisfaction of goals using just internal sources is very important for top management but as we can conclude from our case study, the augmentation of the DW with external data give more information and help the managers to have an eye on the market and their competitors that will allow them to adjust their strategic plans. This information provide the Value that our approach pursues.

6 Conclusion

In this paper, we attempt to have a controversial discussion about the future of Data Warehouse Technology in the Big Data and Machine Learning Era. The long life that we are expecting for this technology is motivated by the externalization phenomenon widely adopted by companies that already concerned storage, computation, and recently data. We proposed then to exploit the external data to augment/enrich company' DWs. Another aspect that got less attention from the DW community is related to the computation of the added value of the constructed DW and its ability to satisfy strategic requirements. We arrived to this point after our presentation given in Dagstuhl Seminar [4], where the Return of Investment question was largely mentioned by the participants. With these notions in mind (externalization and strategic requirements), we proposed to revisit traditional methodology of designing DW, by proposing a new one in the context of Web of Data offering decision-makers measurements to evaluate the value of the constructed DWs.

As a perspective, we plan to investigate the impact of LOD evolution on the DW and the impact of its credibility on the estimated value.

References

1. Abelló Gamazo, A., Gallinucci, E., Golfarelli, M., Rizzi Bach, S., Romero Moral, O.: Towards exploratory OLAP on linked data. In: SEBD, pp. 86–93 (2016)
2. Baldacci, L., Golfarelli, M., Graziani, S., Rizzi, S.: QETL: an approach to on-demand ETL from non-owned data sources. DKE **112**, 17–37 (2017)
3. Barone, D., Jiang, L., Amyot, D., Mylopoulos, J.: Reasoning with key performance indicators. In: Johannesson, P., Krogstie, J., Opdahl, A.L. (eds.) PoEM 2011. LNBIP, vol. 92, pp. 82–96. Springer, Heidelberg (2011). https://doi.org/10.1007/978-3-642-24849-8_7
4. Bellatreche, L.: Value-driven approach for BI application design. Dagstuhl Reports: Next Generation Domain Specific Conceptual Modeling: Principles and Methods (Dagstuhl Seminar 18471), vol. 8, no. 11, p. 69 (2019)
5. Berkani, N., Bellatreche, L., Benatallah, B.: A value-added approach to design BI applications. In: Madria, S., Hara, T. (eds.) DaWaK 2016. LNCS, vol. 9829, pp. 361–375. Springer, Cham (2016). https://doi.org/10.1007/978-3-319-43946-4_24
6. Berkani, N., Bellatreche, L., Guittet, L.: ETL processes in the era of variety. In: Hameurlain, A., Wagner, R., Benslimane, D., Damiani, E., Grosky, W.I. (eds.) Transactions on Large-Scale Data- and Knowledge-Centered Systems XXXIX. LNCS, vol. 11310, pp. 98–129. Springer, Heidelberg (2018). https://doi.org/10.1007/978-3-662-58415-6_4
7. Berkani, N., Bellatreche, L., Khouri, S., Ordonez, C.: Value-driven approach for designing extended data warehouses. In: DOLAP (2019)
8. Božič, K., Dimovski, V.: Business intelligence and analytics for value creation: the role of absorptive capacity. IJIM **46**, 93–103 (2019)
9. Chakrabarti, S., Sarawagi, S., Dom, B.: Mining surprising patterns using temporal description length. In: VLDB, pp. 606–617 (1998)
10. Chen, H., Chiang, R.H., Storey, V.C.: Business intelligence and analytics: from big data to big impact. MIS Q. **36**(4), 1165–1188 (2012)
11. Corrales-Garay, D., Mora-Valentín, E., Ortiz-de-Urbina-Criado, M.: Open data for open innovation: an analysis of literature characteristics. Futur. Internet **11**(3), 77–102 (2019)
12. Deb Nath, R.P., Hose, K., Pedersen, T.B.: Towards a programmable semantic extract-transform-load framework for semantic data warehouses. In: DOLAP, pp. 15–24 (2015)
13. Dehdouh, K.: Building OLAP cubes from columnar NoSQL data warehouses. In: Bellatreche, L., Pastor, Ó., Almendros Jiménez, J.M., Aït-Ameur, Y. (eds.) MEDI 2016. LNCS, vol. 9893, pp. 166–179. Springer, Cham (2016). https://doi.org/10.1007/978-3-319-45547-1_14
14. Domingues, M.A., Jorge, A.M., Soares, C., Leal, J.P., Machado, P.: A data warehouse for web intelligence. In: 13th Portuguese Conference on Artificial Intelligence (EPIA), pp. 487–499 (2007)
15. Gallinucci, E., Golfarelli, M., Rizzi, S., Abelló, A., Romero, O.: Interactive multidimensional modeling of linked data for exploratory OLAP. Inf. Syst. **77**, 86–104 (2018)
16. Golfarelli, M., Maio, D., Rizzi, S.: The dimensional fact model: a conceptual model for data warehouses. Int. J. Cooper. Inf. Syst. **7**(02n03), 215–247 (1998)
17. Golfarelli, M., Rizzi, S., Vrdoljak, B.: Data warehouse design from XML sources. In: ACM OLAP, pp. 40–47 (2001)

18. Gordijn, J., Akkermans, J.: Value-based requirements engineering: exploring innovative e-commerce ideas. Requir. Eng. **8**(2), 114–134 (2003)
19. Guarino, N., Andersson, B., Johannesson, P., Livieri, B.: Towards an ontology of value ascription. In: FOIS, pp. 331–344 (2016)
20. Hoffart, J., et al.: YAGO2: exploring and querying world knowledge in time, space, context, and many languages. In: WWW, pp. 229–232 (2011)
21. Horkoff, J., et al.: Strategic business modeling: representation and reasoning. SSM **13**(3), 1015–1041 (2014)
22. Khouri, S., Ghomari, A.R., Aouimer, Y.: Thinking the incorporation of LOD in semantic cubes as a strategic decision. In: Schewe, K.-D., Singh, N.K. (eds.) MEDI 2019. LNCS, vol. 11815, pp. 287–302. Springer, Cham (2019). https://doi.org/10.1007/978-3-030-32065-2_20
23. Khouri, S., Lanasri, D., Saidoune, R., Boudoukha, K., Bellatreche, L.: LogLInc: LoG queries of linked open data investigator for cube design. In: Hartmann, S., Küng, J., Chakravarthy, S., Anderst-Kotsis, G., Tjoa, A.M., Khalil, I. (eds.) DEXA 2019. LNCS, vol. 11706, pp. 352–367. Springer, Cham (2019). https://doi.org/10.1007/978-3-030-27615-7_27
24. Khouri, S., Semassel, K., Bellatreche, L.: Managing data warehouse traceability: a life-cycle driven approach. In: Zdravkovic, J., Kirikova, M., Johannesson, P. (eds.) CAiSE 2015. LNCS, vol. 9097, pp. 199–213. Springer, Cham (2015). https://doi.org/10.1007/978-3-319-19069-3_13
25. Larson, D., Chang, V.: A review and future direction of agile, business intelligence, analytics and data science. Int. J. Inf. Manag. **36**(5), 700–710 (2016)
26. LaValle, S., Lesser, E., Shockley, R., Hopkins, M.S., Kruschwitz, N.: Big data, analytics and the path from insights to value. MIT Sloan Manag. Rev. **52**(2), 21 (2011)
27. Llave, M.R.: Data lakes in business intelligence: reporting from the trenches. Proc. Comput. Sci. **138**, 516–524 (2018)
28. McAfee, A., Brynjolfsson, E., Davenport, T.H., Patil, D., Barton, D.: Big data: the management revolution. Harv. Bus. Rev. **90**(10), 60–68 (2012)
29. Mithas, S., Lee, M.R., Earley, S., Murugesan, S., Djavanshir, R.: Leveraging big data and business analytics. IT Prof. **15**(6), 18–20 (2013)
30. Nebot, V., Berlanga, R.: Building data warehouses with semantic data. In: Proceedings of the 2010 EDBT/ICDT Workshops, p. 9. ACM (2010)
31. Rehman, N.U., Weiler, A., Scholl, M.H.: OLAPing social media: the case of Twitter. In: IEEE/ACM International Conference on Advances in Social Networks Analysis and Mining, pp. 1139–1146 (2013)
32. Rizzi, S., Gallinucci, E., Golfarelli, M., Abelló, A., Romero, O.: Towards exploratory OLAP on linked data. In: SEBD, pp. 86–93 (2016)
33. Sales, T.P., Baião, F., Guizzardi, G., Almeida, J.P.A., Guarino, N., Mylopoulos, J.: The common ontology of value and risk. In: Trujillo, J.C., et al. (eds.) ER 2018. LNCS, vol. 11157, pp. 121–135. Springer, Cham (2018). https://doi.org/10.1007/978-3-030-00847-5_11
34. Sales, T.P., Guarino, N., Guizzardi, G., Mylopoulos, J.: An ontological analysis of value propositions. In: EDOC, pp. 184–193 (2017)
35. Silva Souza, V.E., Mazon, J.N., Garrigos, I., Trujillo, J., Mylopoulos, J.: Monitoring strategic goals in data warehouses with awareness requirements. In: ACM SAC, pp. 10–75 (2012)
36. Thew, S., Sutcliffe, A.: Value-based requirements engineering: method and experience. Requir. Eng. **23**(4), 443–464 (2018)

37. Vrdoljak, B., Banek, M., Rizzi, S.: Designing web warehouses from XML schemas. In: Kambayashi, Y., Mohania, M., Wöß, W. (eds.) DaWaK 2003. LNCS, vol. 2737, pp. 89–98. Springer, Heidelberg (2003). https://doi.org/10.1007/978-3-540-45228-7_10
38. Wegmann, A.: On the systemic enterprise architecture methodology (SEAM). In: ICEIS, pp. 483–490 (2003)
39. Yangui, R., Nabli, A., Gargouri, F.: Towards data warehouse schema design from social networks-dynamic discovery of multidimensional concepts. In: ICEIS, no. 1, pp. 338–345 (2015)

Search and Information Extraction

Improving Result Diversity Using Query Term Proximity in Exploratory Search

Vikram Singh$^{(\boxtimes)}$ and Mayank Dave

Computer Engineering Department, National Institute of Technology,
Kurukshetra, Kurukshetra 136119, Haryana, India
{viks,mdave}@nitkkr.ac.in

Abstract. In the information retrieval system, relevance manifestation is pivotal and regularly based on document-term statistics, i.e. term frequency (tf), inverse document frequency (idf), etc. Query term proximity within matched documents is mostly under-explored. In this paper, a novel information retrieval framework is proposed, to promote the documents among all relevant retrieved ones. The relevance estimation is a weighted combination of document statistics and query term statistics, and term-term proximity is a simply aggregates of diverse user preferences aspects in query formation, thus adapted into the framework with conventional relevance measures. Intuitively, QTP is exploited to promote the documents for balanced exploitation-exploration, and eventually navigate a search towards goals. The evaluation asserts the usability of QTP measures to balance several seeking tradeoffs, e.g. relevance, novelty, result diversity (Coverage and Topicality), and overall retrieval. The assessment of user search trails indicates significant growth in a learning outcome.

Keywords: Exploratory analytics · Information retrieval · Query term proximity · Relevance · Retrieval strategy

1 Introduction

Information-seeking is a fundament endeavor of human being and several information search systems has been designed to assist a user to pose queries and retrieves informative data to accomplish search goals. The traditional systems strongly trust user's capability of phrasing precise request and perform better if requests are short and navigational. A potential obstacle to such systems is an astonishing rate of information overload that makes difficult to a user for identifying useful information. Therefore nowadays, search focus is shifting from finding to understanding information [1], especially in discovery-oriented search aka exploratory search [2, 3]. When a user wants information for learning purpose, decision making or other cognitive activity, the conventional search methodologies are not capable to assist, though data exploration is helpful.

The retrieval of relevant data requires either formal awareness of complex schema or familiarity with content to formulate the retrieval request 'query' [4, 5]. An ideal information system employs implicit relevance estimate to outline the data objects and explicit measures to navigate the search. Most existing retrieval models score a

© Springer Nature Switzerland AG 2019
S. Madria et al. (Eds.): BDA 2019, LNCS 11932, pp. 67–87, 2019.
https://doi.org/10.1007/978-3-030-37188-3_5

document predominantly on documents-terms statistics, i.e. term-frequency, inverse document frequencies, etc. [5, 6]. Intuitively, the query terms proximities (QTPs) within pre-fetched result set/documents could be exploited for re-position/re-raking of the results in which the matched query terms are close to each other. For example, a user search considering the query 'big data' on documents,

$$Doc_1 : \{\ldots big\ data\ldots\}$$
$$Doc_2 : \{\ldots big\ldots data\ldots\}$$

Intuitively, doc_1 should be ranked higher, as occurrences of both query terms are closest to each other. In compare to the doc_2, where both query terms are far apart and their combination does not necessarily imply the meaning of 'big data'.

The term-term affinity within matched results has role to play during the retrieval and eventually to offers diverse result [5–7]. As a user specify data request in more than one term with an anticipated inherent closeness. The proximity in query terms characterizes the constraints the importance between two matched results during the information search. The query term proximity (QTP) is, however, has been principally under-explored in retrieval framework and models designed to support exploratory search or analytics; primarily due to intrinsic design concerns (*how we can model proximity*) and its overall usability (*what it serve*) into a retrieval model.

In this paper, we systematically deliberate the QTP heuristic for the relevance manifestation with document-terms statistics. Following two inherent research questions (RQs) are the focus of the overall work conducted in this paper:

RQ1: How can diverse QTP are formalized? How to design a retrieval framework that, account user' search task while rewarding/penalizing the QTP statistics.
RQ2: Finally, how can QTP be adapted with document-terms relevance to optimize information exploitation and eventually exploration efforts?

The QTP measures proposed in our previous work in [5] are revised by adapting the normalization factor. The normalization of each query term proximity measure significantly improves overall result diversity. The significance of proposed measures is conducted using defined search trails (STs), and over huge *scientific publication* records (approximates 50 thousands).

1.1 Contribution and Outline

The key contribution is a novel retrieval framework for the discovery-oriented information-seeking that, demonstrate the role of query-term statistics with document-term statistics. Additional contributions of work apart from the previous researches:

(1) We investigated diverse relevance aspects of a data retrieval strategy, e.g., user search efforts, relevance types, relevance attributes, exploitation vs. exploration balance, potential assistance for learning, etc. The aim is to find relevant results and browse through influential ones alongside, by exploiting the QTP within matched results.

(2) We observe that, contextual factors of traditional relevance measures (e.g. tf & idf) are common with term-term proximity and correlation among contextual parameters and generalization may play a key role. Hence, both (similarity and proximity) measures are amalgamate in retrieval framework.

(3) The experimental analysis validate usability of the intuitive information retrieval framework based on implicit and explicit relevance factors, and its feasibility for enhanced tradeoffs among relevance, novelty, and diversity (coverage and topicality), and overall retrieval (precision, recall, f-measure). The superiority of proposed framework over the baseline system is evident to balance the inherent exploitation-exploration tradeoffs.

The organization of paper as follows: Sect. 2 discussed the related research efforts. Section 3 elaborates, the components of proposed work, including, definition of proposed proximity measures, and algorithms. The experimental analysis and assessment is presented in Sect. 4. Section 5 briefly summaries the finding of the proposed QTP based retrieval framework for a search task, and last conclusion.

1.2   Materials and Methods

The paper investigates the following questions and co-related issues of relevance measures in the context of exploratory information-seeking:

(i) When QTP relevance measure is useful?
   (a) When the closeness among query terms is required.
   (b) When the matched search results are not ordered on search preferences.
(ii) In what ways does term-term proximity affect the overall relevance?
   (a) It promotes the matched document with higher proximity for a user query;
   (b) It guides to a list of cohesive results over a matched results for search need.

A dataset of over 50 thousands scientific publication records extracted from DBLP, ACM, and other sources. The dataset contains *title, abstract, keywords, author names, and publication year*. The aim of the work is to exploit the query term proximity to model search intent and result re-ranking. The relevance-based measure, i.e. algorithm relevance, topical relevance and affective relevance, is complemented by QTP measures. The performance is validated by feasibility assessments on potential search trails (extracted from TREC QA 2004 Database) and comparative analysis with potential exploratory search system, such as YmalDB [27], AIDE [28], IntentRadar [29], and uRank [30] on retrieval tradeoffs.

2   Related Work

The primary focus of information retrieval (IR) systems has been to optimize for relevance, as existing retrieval approaches used to rank documents or evaluate IR systems do not account for *"user effort"*. Most of the existing IR strategies rely on the term statistics estimated from the document to query terms, e.g. *document length (length), term frequency (tf), inverse document frequency (idf)*, etc., [5, 7–9]. These *document- term* (DTs) measures essentially are used to position/rank a document in the relevance order for user search [9, 10]. Though, DT measures rarely undertook the coherence aspects of the *user queries*, for example the *proximity of query terms* (QTs) within a matched result set or document. Intuitively, the proximity of query-terms (QTs) can be exploitation addition to traditional DT measures for the manifestation of overall relevance score. The overall relevance scheme will produces semantically improved results, as QT measures promotes the scores of documents (or results set), in which the matched query terms are proximate to each other [11–14]. Though, proximity heuristics has been largely under-explored in traditional models; primarily due to uncertainty on *how we can model proximity* and *how much is significant* into an existing retrieval model [6, 7, 15–17].

Interestingly, the proximity measure is conceptually appealing in the user information-seeking behaviors. Indeed, several existing studies have covers the proximity aspects into seeking behavior [18–20] and the outcomes are not conclusive. The studies assert the significance of proximity in query terms within matched result, but unable to establishes the clear directions for modeling a proximity information retrieval framework. The conceptualization of proximity aspects and its consequence on the overall *relevance* scheme are the two main concerns raised in the studies. The *proximity heuristic* has also been *indirectly* captured in some retrieval models through using larger indexing units than words that are derived based on term proximity (e.g., [5], but these models can only exploit proximity to a limited extent since they do not measure the proximity of terms [21, 22].

With the above limitations, the proposed work explores the possibility to accommodate both relevance measures (*document-terms* and *query-terms*) into information retrieval framework. The proposed framework steers both the key drivers, i.e. *focused search* based on *document-term* similarity measures and *exploratory browsing* based on *query-term* proximity measure within matched results, eventually to personalize the support in information-seeking tasks.

3 Promoting Result Relevance

The retrieval of relevant information requires either formal awareness understanding of complex schema and content to formulate a data retrieval request or assistance from information system [7, 8, 23, 24]. Most existing retrieval models score a document predominantly on documents-terms statistics, i.e. term-frequency, inverse document frequencies, etc. [5, 6]. Intuitively, the query terms proximities (QTPs) within pre-fetched result set/documents could be exploited for re-position/re-raking of the results in which the matched query terms are close to each other. Figure 1 illustrates the various relevance aspects involve in the information-seeking and correlated notions.

A relevance measure is primarily responsible to outline relevant data objects, and eventually navigate user search towards a *region-of interest*. For a search task (W), the user has cognitive perceptions (CW) and its equivalent abstraction as search needs (N). Information needs (N) are metamorphosed into data request 'Query' (q) and often transformations $(q-q^n)$ fetch respective results $(r-r^n)$, the retrieval of $r-r^n$ is purely based on the system/algorithmic relevance between information objects and query terms. Additionally, topical relevance (T) and Pertinent relevance (P) is derived from user relevance feedback, to define a co-relation among information objects ($O-O^n$) and needs ($N-N^n$), and cognitive information need (CW) respectively [5].

The proposed *QTP* capsulate the topical relevance (T) and pertinence relevance (P) of user's search context. Here, the term-term proximity of query plays a pivotal role to promote the document relevance. The proposed strategy for relevance manifestation achieves user-centric data exploitation and eventually improves exploration by offering the diverse results [5].

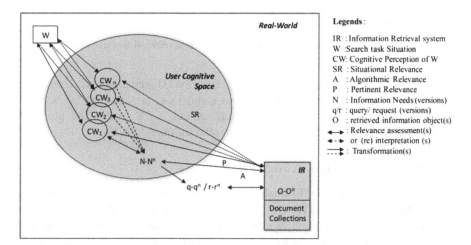

Fig. 1. Illustrate the overall search contexts, User Information Needs (N), Relevance estimate types (A, P, SR), and, evolution in Cognitive Perception (CW) of Search task (W).

3.1 Proposed Framework

The traditional relevance measure of information retrieval are solely based on the occurrence of the query terms within a result, and implies the term-frequency (*tf*) to a term in a document to emphasis the relevance of the document for the user query. Similarly, inverse document frequency (*idf*) enhances the search context bubbles and extracts larger set of query results containing query terms [25], contrary to *tf* that, implies smaller set of query search results and from a *'local context'*. Eventually, *tf* offers results for *focused search* and *idf* for the *exploratory browsing* during the search [26]. Both, traditional measures unable to capture contextual preference of user information-search query, i.e. *query term proximity*, *query term semantics*, etc. and eventually insulate into the relevance estimation for the retrieval.

Query term proximity (QTP) is one potential measures that incorporates the con-textual relevance of query in information retrieval model [5]. The query term proximity is characterized either *implicitly* or *explicitly*, a *explicit* proximity deal with the distance between the positions of a pair of query terms within a matched document, whereas *implicit* proximity measured based on the length of a text segment covering all the query terms. Figure 2 illustrates the schematic point of imposing *QTP* in retrieval [5].

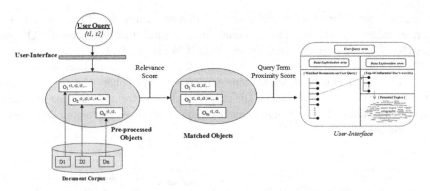

Fig. 2. Schematic view of proposed information retrieval framework.

Here, initial query results are extracted by employing the DT_score (evaluated based on tf-idf values of user query terms on document corpus) and at next level QT_score is adapted to introduce the user-eccentric results among top-k ranks. The proposed retrieval framework, delivers diverse yet more relevant query results.

3.2   Proposed Query Term Proximity (QTP)

The query term proximity (QTP) implies user preferences in relevance manifestation, characteristically proposed QTP are grouped as *implicit and explicit*. The implicit proximity is based on the length of a text segment covering all the query terms and explicit deals with the distance between the pair of query term positions within matched results. The adapted definition re described below with example document:

$$\text{Document (D)} = \{t_1^1, t_2^2, t_1^3, t_3^4, t_5^5, t_4^6, t_2^7, t_3^8, t_4^9\}.$$

Definition 1: *Span* is estimated as *the length of the shortest segment in result/document that covers all user query terms occurrences, including repeated occurrences.* In the short document d, the *Span* value is 7 for the user query $\{t_1, t_2\}$.

Definition 2: *Minimum coverage* is estimated as *the shortest segment of the result/document which covers all the user query terms which are present in that document,* and formalized as,

$$\text{Minimum Coverage} = \frac{QL}{\text{Min\_len}} * \frac{1}{(\text{Diff} + 1)} \tag{1}$$

where, *QL* is the *length of user input query*, *Min_len* is the minimum coverage of query terms, and *Diff* is *number of query terms* which are not present in documents. In document closet the closest distance for each pair is 1for a user query $\{t_1, t_2, t_3\}$.

Definition 3: *Minimum pair distance* is estimated as *the smallest pair distance value of all query terms pairs*, and formalized as,

$$\text{Minimum Distance} = \text{Min}(q_1, q_2 \in D_{q_1 = q_2}\{\text{Dist}(q_1, q_2; D)\}) \tag{2}$$

Definition 4: *Average distance* is estimated *as the average distance between every path of query terms for all position combinations within document*, and formalized as,

$$\text{Average Distance} = \sum\nolimits_{1 \leq i \leq j \leq n} \frac{\text{Freq}_i * \text{Freq}_j}{\sum \|\text{position}_k - \text{position}_l\|} \tag{3}$$

Here, *position$_k$* and *position$_l$* indicates the kth position of ith and j^{th} query terms respectively.

Definition 5: *Match distance* is estimated *the smallest distance achievable when each co-occurrence of a query terms is uniquely matched to another occurrence of a term*, and formalized as,

$$\text{Match Distance} = \text{Max}(q_1, q_2 \in Q \cap D_{q_1 = q_2}\{\text{Dist}(q_1, q_2; D)\}) \tag{4}$$

Definition 6: *Different average position* is estimated as sum of difference of average position of query term$_1$ and average position of query term $_1$, for each query term pair within a matched document, and formalized as,

$$\text{Difference Position Average} = \sum\nolimits_{K \in QP} \left(\sum\nolimits_{i \in M_1} \frac{\text{Pos}_i^{Qt1}}{m} - \sum\nolimits_{j \in M_2} \frac{\text{Pos}_j^{Qt2}}{m} \right) \tag{5}$$

Each proximity measure implies different perspective of user relevance via term-term of user query in the retrieval of results; therefore all the 06 proximity values are aggregated to eventually enhance the diverse results.

3.3 Relevance Manifestation and QTP Heuristic

The proposed information retrieval framework model relevance score based on two statistics: DT and QT. DT statistics are estimated by traditional notions, whereas the estimation of QT statistics are based on adapted defections. A user search begins, by submitting the initial user query (Q_i), e.g. single term query (Q_i^w), text query (Q_i^{text}), phrase query (Q_t^{phrase}). The text-processing measures (e.g. stemming, lemmatization, etc.) are applied to both Q_i and *document corpus*. The validated query is utilized for the estimation of both statistics, here DT_score extracts initial list of m matched documents ($D_1, D_2, ..., D_m$) for Q_i^t.

The proposed strategy extract initial query-results based on implicit relevance (DT_score) and subsequently adapt explicit relevance (QT_score) measures for the re-ranking. This re-ranking eventually promotes the relevance of the extracted result and diversity and novelty tradeoffs. The traditional tf-idf measure extracted initial list of m matched documents (*MatchedDocList* $(D_1, D_2, ... D_m)$) for Q_i^t. Here, *tf* (i, j) is the simplest count of term i in j^{th} document and denoted by $n_{i, j}$ to simply implies $tf(i, j) = n_{i,j}$. Now average *tf-idf* of the k documents is defined as $DT\_score(avg) = n_{i,j} / \sum_{k \in j} n_{k,j}$, where $n_{k, j}$ is the *tf-idf* of term k in document j. At the end the aggregated QT_score is implies to re-introduces diverse result into top-k ranks.

Algorithm 1 : *User Information Search*

Input: *User Initial Query* Q_i (Q_i^{t1} or $Q_i^{t1,t2,t,...,tk}$ or Q_i^{phrase}) & *Document Corpus*$(D_1, D_2, ..., D_n)$
Output: *Matched DocumentList* $(D_1, D_2, ..., D_m)$

Initialize
Submit Initial User Query Q_i ($Q_i^{t1}/Q_i^{t1,t2,t3...tn}/Q_i^{phrase}$),   // Q_i alternates//
          Begin   //Pre-processing of Q_i//
          if Q_i is a one_word_query, thenccontinue
          else Qi is text or phrase query
Compute DT_score (TF-IDF)   // Evaluation of Pre-search relevance score//
          for each Document (di) in Document Corpus
              MatchedDocList= **Extract** (D₁,D₂,..., Dₙ); //for each User Query (Qi)//
Rank the *Matched DocList* on descending Relevance measure(i.e. *DT_score*)
          for each documents in *MatchedDocList* (D₁,D₂,..., Dₘ) // if m documents are matched//
          Apply Query-Term score (QT_score)   // QT score is evaluated in *Algorithm 2*//
          Evaluate Final Relevance score
              Final_score_{doci}= (w₁*DT_scores)+ (w₂*QT_score);
          end
Visualize *Top-K* ranked *Documents* $(D_1, D_2, ..., D_k)$

The contribution of each score in overall relevance is quantified with the help of strength factors (SFs). Hence, corresponding strength factors to *DT_score* and *QT_score* are defined as SF_1 and SF_2. SF_1 controls term- frequency and defined as:

$$SF_1 = \frac{n_{i,j}}{\sum_k n_{i,j} * \log(|D|/(1 + \{j \in D : i \in j\}))} \tag{6}$$

where, D is total number of documents in the corpus and $n_{i,j}$ is the term frequency $tf(i, j)$ of ith term in j^{th} document. Next, SF_2 controls query term proximity and defined,

$$SF_2 = \frac{(QT\_score)}{6} \tag{7}$$

where, QT_score is accumulated proximity for six different proximity characteristics (discussed in Sect. 3.2). Therefore, the resultant weights will be computed as,

$$w_p = \frac{SF_p}{\sum_{i=1}^2 SF_i} \tag{8}$$

where, p has two values 1 and 2 and explored as follows in Eqs. 9 and 10,

$$w_1 = \frac{SF_1}{\sum_{i=1}^2 SF_i} \tag{9}$$

and

$$w_2 = \frac{SF_2}{\sum_{i=1}^{2} SF_i} \tag{10}$$

Algorithm 2 demonstrates that how evaluated *QTP* is aggregated into QT_score with a suitable weight factors. Intuitively, we expect the distance measures to delivers relevant result than non-relevant, since the query terms are expected to be close. The counter-intuition indicates that we may have missed important factors in relating proximity to document relevance. We notice that not all query terms appear in every result, and also some terms appear more frequently in one result than in another. When a document has more query terms, those terms would tend to span widely. To correct this bias, normalization factor proposed. Specifically, we adapt term-frequency (tf) as the most credible approach to normalize the bias in the QTP. Each proximity definition is revised, e.g. the *Span* is normalized by dividing it by the total number of occurrences of query terms in the span segment. The *Minimum Coverage* is normalized and revised as,

$$\text{Minimum Coverage} = \frac{QL}{SS} * \frac{(f - \text{Min\_len})}{(f - tf)} \tag{11}$$

The *Minimum pair distance* is revised as,

$$\text{Minimum Distance} = \sum_{i=1}^{k} tf(q_{t1}q_{t2}) * \text{Min}(q_{t1}, q_{t2} \in D_{q_{t1}=q_2}\{\text{Dist}(q_1, q_2; D)\}$$
$$\text{for } K \cong \sum_{i \in k} tf(q_{i1}, q_{i2}) \tag{12}$$

Average distance is revised as,

$$\text{Average Distance} = \sum_{1 \le i \le j \le n} \frac{\text{Freq}_i * \text{Freq}_j}{\sum \|\text{position}_k - \text{position}_l\|} \tag{13}$$

Match distance is revised as,

$$\text{Match Distance} = \text{Max} \sum (q_{t1}, q_{t2} \leftarrow Q \cap D_{q_{t1} \ne q_{t2}})\{\text{Dist}(q_{t1}, q_{t2}, D)\} * \frac{idf}{tf} \tag{14}$$

and *Different average position* is revised as

$$\text{Difference Position Average} = \sum_{K \in QP} \left(\sum_{i \in M_i} \frac{\text{Pos}_i^{Qt1}}{m} - \sum_{j \in M_2} \frac{\text{Pos}_j^{Qt2}}{m} \right) / T \tag{15}$$

The evaluated values of each *QTp* measure will contribute to the derivation for query-term score (QT_score) for the query. Hence, estimation of contribution is pivotal. The strength factors (SF) are adapted in it fundamental definitions and formalized to evaluate the contribution of each QTP in the computation of QT_score as follows:

Algorithm 3.2: *QT_score evaluation*

Input: *Query Terms* $Q_i^{(t1,t2,\ldots,tn)}$ and *Matched document List* $(D_1, D_2, \ldots, D_m)$

Output: *QT_score*$(D_{1\,(Score)}, D_{2\,(Score)}, \ldots, D_{m(Score)})$

Begin

      for each document D_i in *MatchedDocList*   // *documents* based on DT_score//

        $QT\_score_i = 0;$

        for each QTP measure$_p$ //p equal to 11 to 15, for different QTP adapted //

          document_score=QT_score_evalution_j (document_i);

            QT_score(i)+= c_p * document_score; // as per definition in //

      end

      return *QT_score*

      end

end

Factor c_p evaluated for six characteristics and there is need to control the contribution of individual in the overall term proximity for a specific result. Six different strength factors are defined for each characteristic and denoted by sf_i.

The contribution of first QTP measure, *span* is mainly affected by the shortest segment (SS) of text within a matched document, thus strength factor is defined as:

$$sf_1 = \left(\frac{1}{ss}\right) * QL \tag{16}$$

where is *QL* is *user query length (number of query terms)*. Similarly, *minimum coverage* strength-factor is SF_2 defined as,

$$sf_2 = \frac{(tf - f)}{tf} * \frac{QL}{SS} \tag{17}$$

The distance based proximity measures are primarily driven by the evaluated distance values and total document length (DL). The contribution of the *minimum distance* measure is computed as,

$$sf_3 = \frac{Min\_Distance}{DL} \tag{18}$$

match distance measures as,

$$sf_4 = \frac{Match\_Distance}{DL} \tag{19}$$

average distance measure as,

$$sf_5 = \frac{Avg\_Distance}{DL} \tag{20}$$

and for *difference average position distance*, as

$$\mathrm{sf}_6 = \frac{\mathrm{Normalized}_{\mathrm{AvgPositionsalDistance}}}{\mathrm{QL}} \tag{21}$$

The proposed query term score accumulates the contributions of each QTP measures, via iterative computation individual scores. Here, *factor c_k* drives the encapsulation of various QTP values with dynamic contributions into single score for each result, as:

$$c_p = \frac{\mathrm{sf}_p}{\sum_{p=1}^{6} \mathrm{sf}_i} \tag{22}$$

and simply incorporates the contribution of each QTP measure via its strength factors.

4 Experimental Setup and Performance Assessment

A dataset of over 50 thousands scientific publication records extracted from *DBLP, ACM*, and other sources related to *computer science* domain. The dataset contains meta-information, e.g. *title, abstract, keywords, author names, and publication year* of each paper. The aim of the work is to exploit the document-based similarity and query term-based proximity to model search intent and result extraction. The relevance-based measure, i.e. *algorithm relevance, topical relevance* and *affective relevance*, is complemented by proximity measures (shown in Fig. 1). The performance analysis focuses on the evaluation of the following goals in line with research questions (RQs),

(i) Feasibility analysis of proposed relevance measures, e.g. *query-term statistics* (QTs) into novel *information retrieval framework*.

(ii) Assert the impact of proposed framework for the document *promotion*, on information *novelty, search result diversification* (*coverage* and *topicality*) and overall *retrieval* indicators (*precision, recall* and *f-measure*).

In an exploratory information-seeking, a user performs searches in trails. A *search trail* (ST) describes the user's search behaviours in a context and characterizes information-seeking coverage. To investigate the overall performance, 05 most frequent STs are extracted from *TREC 2004 QA* database. For the simplicity, STs are listed as sequence of keyword queries, in Table 1.

Table 1. List of Search Trails (STs).

	Search Trails (STs) description (as query chain: $Q_i^{term}, Q_{i+1}^{term} \ldots\ldots Q_{i+k}^{term}$)
ST_1	{machine learning, image processing, supervised learning}
ST_2	{computer vision, operating system, centralized network}
ST_3	{interactive modeling, user intention, interactive interface}
ST_4	{radio networks, cognitive radio, mobile networks, measurement of radio networks}
ST_5	{database management system, structured database, transaction, DB normalization}

4.1   Overall Relevance Manifestation

The first observation is aimed to establish the significance of each *Query term proximity* scores for the retrieval of relevant and no-relevant documents. For this evaluation, the 2000 records are extracted based on *tf-idf* based weight scheme and labeled as relevant and non-relevant. Next, both *implicit QTP measures (Span* and *MinCoverage)* are employed in 2000 matched documents to assess the significance of each score (average of *normalized relevance score*), listed in Tables 2 and 3. The Query terms *Span* measure shows less effective growth for *relevance*, as result under *relevant* column are lesser on most of the search trails, though *MinCover* measure is now indeed slightly smaller on relevant documents than on non-relevant documents in most cases, suggesting the existence of weak signals.

Similar, exercise is done for explicit measures, listed in Table 3. The results are clearly indicative of the fact that *MaxDist* results are still non-preferable, both *AvgDist* and *MinDist* are consistent; particularly *MinDist* delivers better than among *explicit* query term proximity measures. A consolidated information retrieval performance on the user search trails on the *average precision* with aggregated *query proximity scores*, shown in Fig. 3. The evaluation of *precision* is as '*the fraction of relevant instances among the retrieved instances*'. Figure 3 depicts the *average precision* delivers by a relevance measures, among the top-2000 document/results.

Table 2. Performance of Implicit QTP measures.

Search Tail (STs)	Span		MinCoverage	
	Relevant	Non-relevant	Relevant	Non-relevant
ST_1	46.43	50.78	27.63	30.9
ST_2	150.9	104.13	127.93	127.93
ST_3	57.67	56.25	20.13	20.13
ST_4	153.48	103.38	56.88	56.88
ST_5	156.398	108.38	9.857	5.857

Table 3. Performance of Explicit QTP measures.

STs	MinDist		MaxDist		AvgDist		MatchDist		Diff_avg_pos	
	Rel	N-rel	Rel	N-rel	Rel	N-rel	Rel	N-rel	Rel	N-rel
ST_1	16.18	30.64	89.41	82.78	43.3	52.25	43.3	52.25	43.3	52.25
ST_2	39.35	39.83	415.02	133.62	148.82	72.07	148.82	72.07	148.82	72.07
ST_3	19.15	31.77	49.92	48.52	32.25	39.33	32.25	39.33	32.25	39.33
ST_4	61.15	67.91	146.92	100.42	96.65	82.73	96.65	82.73	96.65	82.73
ST_5	7.66	11.68	13.97	15.31	10.57	13.38	10.57	13.38	10.57	13.38

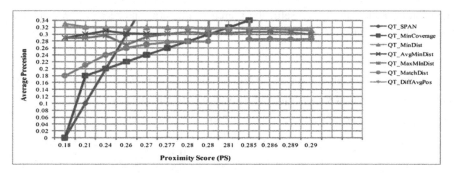

Fig. 3. Average precision score delivered by QTP measures.

4.2   Information Retrieval Performance

To evaluate the overall information novelty, three aspects of search results are considers, for search trails (STs) listed in Fig. 4. Traditionally, the novelty described by three main factors among extracted results: number of unique results, number of re-retrieved results, and number of useful results. For the simplicity of the evaluation, numbers of result are clustered into three sets from top-2000 relevant result for each ST. A noteworthy point is any unique pattern is not identified from result although, high degree of results common and task ST2 lines are higher than ST1 task, which shows that the is performing better even for un-cleaned explorations. The *information novelty* hints at the total results for each search trails during exploratory information-seeking, for a search task with more opportunity for feedback offers higher number of results (in ST_5) and lesser intermediate relevance feedback leads to higher number of re-retrieved results (in ST_3).

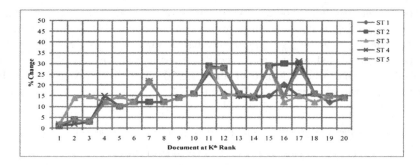

Fig. 4. Novelty introduced in Query results among top-20 ranks/positions

Similarly, Table 4 lists the retrieval performance of proposed framework. For the evaluation, traditional definition are adapted, e.g. precision is a ratio of 'relevant objects to total relevant retrieved', recall (sensitivity) is ratio of 'relevant objects with total retrieved', and f-measure is a harmonic mean of precision and recall values. The precision characterizes 'how useful the search results are' and recall 'how complete results are' of an information search.

Table 4. Overall information retrieval performance of proposed framework.

User Search Tails (STs)	Precision	Recall	F-Score
ST_1	0.280	0.549	0.371
ST_2	0.120	0.594	0.200
ST_3	0.150	0.415	0.147
ST_4	0.173	0.569	0.265
ST_5	0.047	0.251	0.079

For the proposed framework *exploitation-exploration* balance is validated though a detailed analysis on concerned parameters: *MAP*, P_{10}, P_{20}, P_{100} and B_{pref} for user effort in *exploitation* and *MAR*, R_{10}, R_{20}, R_{100} and E_{pref} for *Exploration-efforts* over search tails listed in Table 5.

The evaluation indicators *MAP*, P_{10}, P_{20}, P_{100} and B_{pref} are accepted for the assessment of *focused search* efforts in an IR. *MAP* characterizes the mean of the precision scores obtained after each relevant document is retrieval; B_{pref} is a preference-based IR measure that considers whether relevant documents are ranked above irrelevant ones [5, 26]. There are generally 10 search results in one page in most of IR systems, and P_{10} indicate the precision of 1st page (*as all users prefer to view page 1*); similarly P_{20} is *precision* in page 1 and page 2 (most users will click next page at least once). P_{100} means the precision in pages 1–10 (*most users will not see the pages after page 11*). The precision scores clearly indicate the significant enhanced exploitation and resultant into reduced user efforts [5, 24, 26].

Similarly, *MAR*, R_{10}, R_{20}, R_{100} and E_{pref} adapted to evaluate exploration effects [5, 7]. The *Recall* measure emphasis towards retrieval of potentially relevant results additions to precisely matched result for user's current search, *MAR* is the mean recall scores after retrievals and E_{pref} is a preference-based indicator of whether how relevant documents are predicted. R_{10} means the recall score on in page 1 results; R_{20} means the recall in page 1 and page 2. Moreover, there are 20 results in one page in some IR systems; R_{100} means the precision in pages 1–10.

Table 5. Performance on the Exploration-Exploitation aspects.

STs	Exploitation efforts					Exploration efforts				
	MAP	P_{10}	P_{20}	P_{100}	B_{pref}	MAR	R_{10}	R_{20}	R_{100}	E_{pref}
ST_1	0.871	0.956	0.756	0.613	0.513	0.896	0.809	0.889	0.687	0.587
ST_2	0.891	0.885	0.701	0.673	0.673	0.928	0.789	0.870	0.784	0.684
ST_3	0.658	0.833	0.722	0.341	0.344	0.732	0.933	0.780	0.483	0.430
ST_4	0.779	0.660	0.862	0.475	0.389	0.897	0.760	0.964	0.729	0.799
ST_5	0.800	0.556	0.97 0	0.432	0.532	0.858	0.892	0.970	0.606	0.605

Search Result Diversity (% Information Coverage within Top-k results): Most of the *state-of-art* information system offer result re-ranking in order to deliver diverse results. For this often adapts added relevance measures, we employed QTP, to promote the results relevance. The analyses affirm that, QTP steers the retrieval process by promoted results. The result diversity is undertaken in two aspects: *Information Coverage* and *Information Topicality*.

Information coverage mainly affected by the % of changes (*re-ranking*) in top-k results set that implies the re-positioning of results into top-k and personalized results. A user often reviews results places on *top-10* or *top-20* positions, due to which diversity among top query results is significant. The results diversification (on information coverage among top-k searched result is formalized as:

$$\% \, of \, change = \frac{1}{i} \sum_{1 \leq j \leq i} \left| \left(pos_{tf} - \frac{1}{ws} \right) - \frac{j-1}{ws} \right| + 1/\left(x^2/ws \right) \tag{23}$$

Where, *ws* represent *number of result per page*, pos_{tf} indicates the *result rank via tf-idf* based measure and x is number of relevant results. In Eq. 8, the % change characterizes the re-positioning of results during the search, largely due to QTP.

Figure 5 illustrates the % change occurred due re-ranked positions of results among top-k results. The result established the feasibility of both *QTP-score* into retrieval framework, as steers to significant diverse results (improved coverage within top-k results). The *QTP* revise the search intents and introduces result to delivers higher information coverage. The peaks indicate the point of search-interactions.

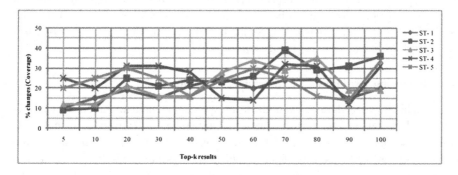

Fig. 5. Result diversity within Top-k results (% change among top-k rank/positions)

Other aspect of result diversity is information topicality. *Topicality* of retrieved result with information needs is pivotal. *Information topicality* often characterizes the pertinence of user intents or the material's degree of information provided; and utility, or the item's usefulness in fulfilling the information need. For the evaluation of the

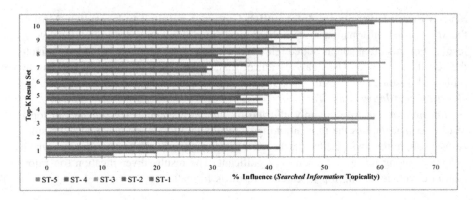

Fig. 6. Information topicality delivered (% of influence among Top-k results).

proposed framework, *topicality* is defined as *influence* of extracted information that implies the presence of search terms within *top-k* results, formalized as:

$$\% \text{ of influence} = \frac{1}{i} \sum_{1 \leq j \leq i} \left| \left(\left(\text{newRelS} - \frac{1}{\text{WS}} \right) - \frac{j-1}{\text{WS}} \right) \right| + 1 / \left(\frac{x^2}{\text{WS}} \right) \quad (24)$$

Where, *WS* is *number of results per page*, *NewRelS* is the relevance score via document-terms and *x* is number of relevant results., Hence, Eq. 24 characterizes the proximity the *topicality* in search, e.g. in ST_1 the *influence* improves 35% and 12% among *top-5* and *top-10* results respectively (in Fig. 6). This asserts that, higher search-interactions opportunity leads to better *topicality*, e.g. in ST_3 relevance opportunities are high leads to better *topicalities*. Similarly, inferred that higher *number of query terms*, such as in ST_5, delivers higher *topicality*.

4.3   Balancing Exploitation-Exploration Tradeoffs

In an exploratory search task, the tradeoff in exploitation and exploration are vital and contextually contradicting, as steers the overall growth in user knowledge-state. For the ease of assessment, let, $a = \sum_{i=1}^{n} \left((\text{tf} - \text{idf})_i \right) / n$, where n is the total number of relevant documents in the list and $(\text{tf\_idf})_i$ indicate the tf_idf of ith result, b is the number of terms in user query and c is the number of user interactions $d = (\frac{1}{t} * \left(\sum_{i=1}^{t} (\text{tf} - \text{idf}_i) \right) / \text{tf} - \text{idf}_{D_i}$ where $\text{tf} - \text{idf}_{D_i}$ is tf-idf of the 1st document in relevance list. If $\left(\frac{1}{t} * \left(\sum_{i=1}^{t} (\text{tf} - \text{idf}_i) \right) - a \right)$ is significantly positive, then

$$\beta = \left(\left(\frac{d * b}{5} \right) + ((1 - d) * c) \right) * PRF/2, \quad (25)$$

otherwise,

$$\beta = \left(\left(\frac{(1-d)*b}{5} \right) + (d*c) \right) * PRF/2, \qquad (26)$$

where a, b, c, d, and β are the variables. Another variable x and y are defined as

$$x = (tf - idf + \beta * URF + \beta * PRF)/(tf - idf + URF + PRF) \qquad (27)$$

$$y = (N_{QT} + (1 - \beta * URF) + (1 - \beta) * URF))/(N_{QT} + URF + PRF) \qquad (28)$$

a derived variable α is defined as, $\alpha = (1-y)/(1-x)$.

Finally, the tradeoff is defined as $(x + \alpha y)/(x + y)$. Here, N_i^{qt} number of query terms in i^{th} query, URF is number of user-interactions on i^{th} query results. The exploitation-exploration tradeoffs described as a ratio of two different expressions and variables x, y, and α. Initially, both search behaviours (exploitation and exploration) are modelled in terms of variables x and y and defined in the range of 0 to 1. The ratio of the complementary behaviour of exploitation and exploration is defined by the variable y that eventually defines tradeoffs, which consider the nature of the data along with user feedback and query term proximity. The tradeoffs among both key search behaviour keep in the range of 0 to 1 with a tendency to achieve a balance at the mean score.

The behaviour of the proposed framework in terms of exploitation can also be analyses using performance indicators, such as MAP and MAR. MAP increases with the search progression, and MAR reduces. It clearly shows that both behaviours are reciprocal and share similar parameters to balance (as shown in Eqs. 27 and 28). Figures 7 and 8 illustrates the MAP and MAR score for the five ESS, such as YmalDB [27], AIDE [28], IntentRadar [29], and uRank [30]. Contextually, lesser re-retrieved result produces the desired MAR level and gradually reduces the user uncertainty on search needs and goals.

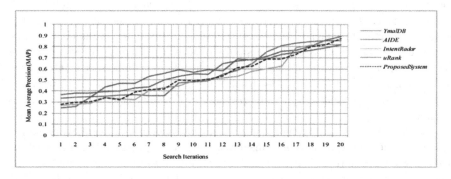

Fig. 7. Evaluation of exploitation via MAP

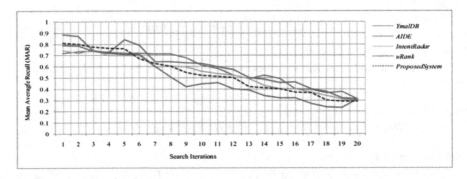

Fig. 8. Evaluating exploration via MAR

The analysis clearly indicate that the proposed system perform close to YmalDB and IntentRadar, as both model intermediate search-interactions into relevance. In contrary, both AIDE and uRank systems are based on result classification. It is clear that the relevance of the document increased and leads to enhanced diverse results.

4.4 Discussion and Analysis

The key objective of work discussed is to incorporate a proximity measures with existing term-weighting scheme in retrieval framework. This also ensures that these *N documents* have an ample supply of *query-terms*. Consequently, performance analysis among the top 2000 documents from retrieval run and examines the correlation between the measures outlined. The proposed framework balance both search behaviors, and assigns higher weights initially on exploration and emphasis exploitation on later. This assertion confirms the solution to *RQ1*, as user's *uncertainty* at initial phase improves with a search progression with expanded precision among retrieved results and enhanced exploitation. Though, achieving the shift change of the search focus is complex and requires adaptive decision-making. Further, additional relevance measures are adapted to enhance relevance of results in line with evolution of search intents, experiment result assert the feasibility of relevance factors, this also improves the overall retrieval performance and result diversification on both aspects: information coverage and topicality among top-k result set. The adaption of additional relevance factors also plays pivotal role, for the promotion of results objects among matched documents.

Similarly, for *RQ 2* capturing the user search interactions and intent evolution is a fruitful direction in exploratory information-seeking. The opportunity of search related interactions, e.g. *relevance feedback, query reformulations, new query insertion*, significantly affects entire search evolution and personalization of results. Further, adaption of additional relevance measures navigate the entire exploration process to real region-of-interest, as *query term proximity* (QTP) promotes the document within matched list to characterize the importance of document. The experimental analysis confirms the usability for the proximity to enhance the overall exploration and controlling the focal shift.

5   Conclusion

In this paper, document-terms (DTs) and query-terms (QTs) statistics are amalgamated to estimate the overall relevance to steers the balanced informational search. To navigate the proposed strategy, particularly for the *relevance manifestation*, six contextually diverse proximity measures are formalized. The QTP measures are designed to enhanced semantic and contextual relevance of user query into relevance estimate, as each measure captures different aspects of query terms proximity. The computed QTP scores are adapted to re-rank the initially matched document set, with an objective to introduce the more relevant and diverse result document. The QT based statistics/scores significantly promote the document and incorporate user-preferences context in retrieval framework. The experimental assessment validates the significant growth on several tradeoffs of information search, e.g. *Information novelty,* overall *relevance, search result diversification* (on both aspects *Coverage* and *Topicality*), and overall *information retrieval* (on indicators *precision, recall,* and *f-measure*). The assessment framework evaluates the usability of proposed strategy with potential equivalent system, e.g. YmalDB, AIDE, IntentRadar and uRank. The comprehensive assessment affirms the intuitive role of proposed measures in steering the retrieval and data analysis over big database.

The *future scope* of current work may include an *adaptive query completion* approach based on term proximities (inherited from DTs and QTs) within matched documents. The query terms prediction for prospective user search will emulate a term/word level intents for improved information-search.

References

1. White, R.W., Roth, R.A.: Exploratory search: beyond the query-response paradigm. Synthesis Lect. Inform. Concepts Retrieval Serv. **1**(1), 1–98 (2009)
2. Idreos, S., Papaemmanouil, O., Chaudhuri, S.: Overview of data exploration techniques. In: Proceedings of the 2015 ACM SIGMOD International Conference on Management of Data, pp. 277–281. ACM, May 2015
3. Marchionini, G.: Exploratory search: from finding to understanding. Commun. ACM **49**(4), 41–46 (2006)
4. Kersten, M.L., Idreos, S., Manegold, S., Liarou, E.: The researcher's guide to the data deluge: querying a scientific database in just a few seconds. PVLDB Chall. Vis. **3**(3) (2011)
5. Singh, V.: Predicting search intent based on in-search context for exploratory search. Int. J. Adv. Pervasive Ubiquit. Comput. (IJAPUC) **11**(3), 53–75 (2019)
6. Van Rijsbergen, C.J.: A theoretical basis for the use of co-occurrence data in information retrieval. J. Doc. **33**(2), 106–119 (1977)
7. Salton, G., Buckley, C.: Term-weighting approaches in automatic text retrieval. Inf. Process. Manage. **24**(5), 513–523 (1988)
8. Cosijn, E., Ingwersen, P.: Dimensions of relevance. Inf. Process. Manage. **36**(4), 533–550 (2000)
9. Barry, C.L.: User-defined relevance criteria: an exploratory study. J. Am. Soc. Inform. Sci. **45**(3), 149–159 (1994)

10. Rasolofo, Y., Savoy, J.: Term proximity scoring for keyword-based retrieval systems. In: Sebastiani, F. (ed.) ECIR 2003. LNCS, vol. 2633, pp. 207–218. Springer, Heidelberg (2003). https://doi.org/10.1007/3-540-36618-0_15

11. Qiao, Y.N., Du, Q., Wan, D.F.: A study on query terms proximity embedding for information retrieval. Int. J. Distrib. Sens. Netw. 13(2), 1550147717694891 (2017)

12. Keen, E.M.: Some aspects of proximity searching in text retrieval systems. J. Inform. Sci. 18 (2), 89–98 (1992)

13. Beigbeder, M., Mercier, A.: An information retrieval model using the fuzzy proximity degree of term occurrences. In: Proceedings of the 2005 ACM Symposium on Applied Computing, pp. 1018–1022. ACM, March 2005

14. Büttcher, S., Clarke, C.L., Lushman, B.: Term proximity scoring for ad-hoc retrieval on very large text collections. In: Proceedings of the 29th Annual International ACM SIGIR Conference on Research and Development in Information Retrieval, pp. 621–622. ACM, August 2006

15. Schenkel, R., Broschart, A., Hwang, S., Theobald, M., Weikum, G.: Efficient text proximity search. In: Ziviani, N., Baeza-Yates, R. (eds.) SPIRE 2007. LNCS, vol. 4726, pp. 287–299. Springer, Heidelberg (2007). https://doi.org/10.1007/978-3-540-75530-2_26

16. Svore, K.M., Kanani, P.H., Khan, N.:. How good is a span of terms?: exploiting proximity to improve web retrieval. In: Proceedings of the 33rd International ACM SIGIR Conference on Research and Development in Information Retrieval, pp. 154–161. ACM, July 2010

17. Zhao, J., Yun, Y.: A proximity language model for information retrieval. In: Proceedings of the 32nd International ACM SIGIR Conference on Research and Development in Information Retrieval, pp. 291–298. ACM, July 2009

18. He, B., Huang, J.X., Zhou, X.: Modeling term proximity for probabilistic information retrieval models. Inf. Sci. 181(14), 3017–3031 (2011)

19. Sadakane, K., Imai, H.: Text retrieval by using k-word proximity search. In: Proceedings of 1999 International Symposium on Database Applications in Non-Traditional Environments (DANTE 1999) (Cat. No. PR00496), pp. 183–188. IEEE (1999)

20. Borlund, P.: The IIR evaluation model: a framework for evaluation of interactive information retrieval systems. Inform. Res. Int. Electron. J. 8(3) (2003)

21. Song, R., Taylor, M.J., Wen, J.-R., Hon, H.-W., Yu, Y.: Viewing term proximity from a different perspective. In: Macdonald, C., Ounis, I., Plachouras, V., Ruthven, I., White, R.W. (eds.) ECIR 2008. LNCS, vol. 4956, pp. 346–357. Springer, Heidelberg (2008). https://doi.org/10.1007/978-3-540-78646-7_32

22. Miao, J., Huang, J.X., Ye, Z.: Proximity-based rocchio's model for pseudo relevance. In: Proceedings of the 35th International ACM SIGIR Conference on Research and Development in Information Retrieval, pp. 535–544. ACM, August 2012

23. Zhao, J., Huang, J.X., Ye, Z.: Modeling term associations for probabilistic information retrieval. ACM Trans. Inform. Syst. (TOIS) 32(2), 7 (2014)

24. Ye, Z., He, B., Wang, L., Luo, T.: Utilizing term proximity for blog post retrieval. J. Am. Soc. Inform. Sci. Technol. 64(11), 2278–2298 (2013)

25. Saracevic, T.: The notion of relevance in information science: everybody knows what relevance is. But, what is it really? Synthesis Lect. Inform. Concepts Retrieval Serv. 8(3), i-109 (2016)

26. Borlund, P.: The concept of relevance in IR. J. Am. Soc. Inform. Sci. Technol. 54(10), 913–925 (2003)

27. Drosou, M., Pitoura, E.: YmaLDB: exploring relational databases via result-driven recommendations. VLDB J.—Int. J. Very Large Data Bases 22(6), 849–874 (2013)

28. Dimitriadou, K., Papaemmanouil, O., Diao, Y.: Explore-by-example: an automatic query steering framework for interactive data exploration. In: Proceedings of the 2014 ACM SIGMOD International Conference on Management of Data, pp. 517–528. ACM, June 2014
29. Ruotsalo, T., et al.: IntentRadar: search user interface that anticipates user's search intents. In: CHI 2014 Extended Abstracts on Human Factors in Computing Systems, pp. 455–458. ACM, April 2014
30. di Sciascio, C., Sabol, V., Veas, E.E.: Rank as you go: user-driven exploration of search results. In: Proceedings of the 21st International Conference on Intelligent User Interfaces, pp. 118–129. ACM, March 2016

Segment-Search vs Knowledge Graphs: Making a Key-Word Search Engine for Web Documents

Rashmi P. Sarode[1], Shelly Sachdeva[2], Wanming Chu[1], and Subhash Bhalla[1(✉)]

[1] University of Aizu, Aizu-Wakamatsu, Fukushima, Japan
{d8202102,w-chu,bhalla}@u-aizu.ac.jp
[2] National Institute of Technology (NITD), Delhi, India
shellysachdeva@nitdelhi.ac.in

Abstract. It is becoming increasingly popular to publish data on the web in the form of documents. Segment-search is a semantic search engine for web documents. It presents a query language. It is suitable for skilled and semi-skilled domain experts, who are adept at the use of a specific collection of documents. It returns suitable documents selected by using document fragments, that satisfy user's query. In contrast to knowledge graph approach, the technique is based on performing web page segmentation as per user perceived objects. Thus, it allows users' to query without the knowledge of complex query languages or learning about the data organization schemes. The proposed system is scalable and can cater to large scale web document sources.

Keywords: Data analytics · Cloud-based databases · Heterogeneous data · Distributed data · Polystore data management · Query language support · Scalability

1 Introduction

The World Wide Web is a vast and rapidly growing repository of information. Various kinds of valuable semantic information exist within Web pages. The information is spread out in its simplest (document) form. For example, the earlier paper-based specialized documents in various domains such as healthcare are now available on the Web in form of specialized documents. These repositories provide consumer health information for skilled and semi-skilled users in everyday clinical activities. Currently, information on the Web is discovered primarily by two mechanisms: browsers and search engines. The novice users in specialized domains query the Web resources in a top-down manner. They are not aware of the underlying schema. They use a keyword to find the relevant results. In addition to the top-down approach of keyword based searching, the domain experts have specific and complex queries for the underlying resources. These users are experts and refer to records to search content related to their queries. These users can query the database using query languages such as SQL and

© Springer Nature Switzerland AG 2019
S. Madria et al. (Eds.): BDA 2019, LNCS 11932, pp. 88–107, 2019.
https://doi.org/10.1007/978-3-030-37188-3_6

XQuery. The high-level graphical query interfaces (such as the XQBE [4] and QBE [18]), can simplify the life of the skilled and semi-skilled users.

Most professionals and researchers need information from reputable sources to accomplish their work [42]. For them (skilled and semi-skilled users) the outcome of the existing Web search engines poses a significant barrier to get usable information. The medical repositories exist in form of wiki-like large number of HTML document collections. These are in the form of governmental and commercial medical encyclopedias and dictionaries. For example, MedlinePlus and NHS Direct Online, information is checked for accuracy by licensed medical professionals prior to being published [39]. The patients are advised to utilize these resources for searching health information. Each of their articles has descriptive, sections on disease overview, diagnosis, prognosis, treatment, procedures, common medical conditions and other medical terms. The availability of such distributed repositories of quality medical information on the Internet has placed access methods at the center of research [43].

Most information access systems consider Web pages as the smallest and undividable units. Existing search engines such as Yahoo, Google, service millions of searches a day using Web pages/documents as a unit. The problems of information overload and retrieval are prevalent across many disciplines. The ability to accurately search for, access, and process information is particularly pressing in the field of medicine [43]. Hence, the guideline of our study is to target the domain-experts and help them with intelligent querying methods. We assume that the unskilled users have trivial and less frequent query needs which can be served by existing raw searches such as keyword search or alphabetical search. Our methodology enables the experts to perform in-depth querying on medical repositories. In this study, we present a Web page segmentation technique based on the layout analysis of the Web page and medical domain knowledge repository [38]. We propose to create a database, considering the user's view (semantic objects) of the Web document as attributes of the database and then allow the users to query this semantics-based database. A basic understanding of the structure and the semantics of Web pages could significantly improve people's browsing and searching experience over specialized domains [28]. The hierarchy within a Web page creates defines the relationships among the contents. Moreover, the entire page may be divided into regions for different contents using layout properties. Thus, each of the sub-regions represents coherent semantics.

Rest of the paper is organized as follows. Section 2 presents the background and motivation for the study. Section 3 defines the problem statement for the study. Section 4 presents the details on the semantic structure of a Web document. Sections 5 and 6, explain offline and online components of the model. Section 7 considers related studies. In Sect. 8, present performance related topics. Section 9 summarizes the study and presents the conclusions.

Fig. 1. Semantic groups within the MedlinePlus medical encyclopedia

2 Background and Motivation

Recently, many studies seek to find a general solution to the search and query diffi-culties. A semantic search engine for XML (XSEarch) has been described by Cohen et al. [46]. In the field of text analytics, a search engine has been described by Kan-dogan et al. [47]. Jagadish et al. have proposed Schema free SQL approaches [48, 49]. Most recently, related work has appeared [50, 51]. These have been discussed in the related work section later, in this report. Considering specialized domains such as Astronomy or healthcare, current medical search engines such as OmniMedicaSearch medical Web [29], HealthLine [30], Medical World Search [31] and PogoFrog [32] for physicians and healthcare professionals, are unsupervised and untrained and can be used by the novice users. The user of such system has to decide the relevance of results in his or her terms from all the results returned from these engines. Few efforts have been made for providing interfaces for the querying needs of the domain experts. Such users need precise results and do not prefer browsing through complete Web pages. A Web page is a document or a collection of documents. A collection of contents can form semantically coherent groups. These may allow the users to query them individually. In this manuscript, we wish to utilize the concept of Web page segmentation for the purpose of performing in-depth querying on the medical repositories. Web page segmentation is utilized in applications such as Web page dynamics, information retrieval, phishing detection and generation of result snippets. The earlier works in Web page segmentation are based on the DOM structure of a Web page [33–35] or rely on the visual heuristics [5, 6, 8, 36]. Some approaches make use of the image analysis on the Web pages [16] or perform graph based segmentation [37]. Among these the VIPS (Vision-based Page Segmentation) algorithm [6] is very similar to our proposal. In VIPS algorithm, a tree structure is used to model the page. Each node corresponds to a visual block in the page, and has a Degree of Coherence (DoC) associated with it.

The medical or healthcare articles are structured in a similar way. A clear understanding of the layout features of these Web pages allows querying on the transformed page segments database schema. Our segmentation algorithm avoids many ad hoc rules, and adopts a clear model, which makes the querying more efficient and reliable.

2.1   Nature of Web Documents

The Web resources often comprise of the electronic form of simple document collections. This is to improve their reach for novice users. In healthcare domain, these are referred by the clinicians for in-patient diagnosis and also referred by general users for preliminary symptom recognition. As in the case of a text document, headings are employed to organize the content of the document. Headings are located at the top of the sections and subsections. These preview and succinctly summarize upcoming content and show subordination. These lead to a hierarchical structure for a document. It plays a role in understanding the relationships between its contents. Thus, the logical structure becomes useful for segmenting a Web document and to perform passage retrieval. The Web pages of the most commonly cited medical Web resources such as the encyclopedias [9, 11–13] are similarly structured and static in nature. Thus, the Web collection holds information about diseases, medical conditions, tests, symptoms, injuries, and surgeries. Their contents do not change (much) over a period of time. The schema and these resources have evolved from common practice terms over 100s of years. Typically, a Web page represents a single theme or a topic. Elements for a medical concept are semantically grouped in a Web page form. These elements are arranged and populated for the enriching the content and efficient browsing of a Website.

The header of any Web page represents the general contents of the Website it belongs to. And the Web page footer is common to all the pages of a Website. It represents the author contents and other metadata. The free text may comprise of paragraphs and lists. Figure 1 shows a sample Web document describing the various components given above which form a complete Web page. The main content of the similarly organized Web pages can be logically divided into several information segments such as causes, symptoms, home care, alternatives and references.

A Web document, W can be mathematically defined as:

$$W = \sum s_i, \text{ where each } s_i = (1,c)_i \text{ and } (L, c) = \sum(1,c)$$

The pair (l, c) represents a segment s of the Web page, where l is the label of the content c enclosed within W.

A Web document can be visualized in form of an ordered tree where the leaves correspond to text elements and internal nodes correspond to labels of the semantic groups within the Web document. The root represents the label of the Web document. Thus, for subsequent queries, the ordered tree follows a pre-order traversal. The root and the internal nodes of the tree expose the labels (or attributes) within which a user may query. The content flow through a medical encyclopedia page. It plays a major role in the transformation of the content blocks into a hierarchical structure of the page. Each of the content groups can be segmented internally into semantic groups (which in turn may be free text, hyperlink text or lists). The depth of the tree depicts the nesting of content groups. Figure 1 shows a snapshot of a MedlinePlus (Encyclopedia) Web page about the topic of "heart attack". Each rectangular box represents a coherent semantic group such as symptom, cases, post-care and so-on.

2.2   Case of Web-Based Medical Resources

MedlinePlus is a free access Web site maintained by U.S. National Library of Medicine [13]. It provides consumers health information for patients, families, doctors and Health care providers. The site brings together information from the United States National Library of Medicine, the National Institutes of Health (NIH), other U.S. government agencies, and other health organizations. Over 150 million people from around the world use MedlinePlus each year. The existing keyword search and menu based (interface) searches on MedlinePlus often give redundant results that are of not much use to the medical domain experts.

The documents on the Web are not well-structured, such that a program can reliably extract the routine information that a human wishes to discover. In deciding which disease does a patient has, both the presence and absence of certain symptoms provide clues (e.g. whether sputum is accompanied by coughing). For example, Web pages describing the diseases that do not have the symptom "Sy", can either mention "without Sy" explicitly in many different ways or do not mention "Sy" at all. For a Web page P describing a disease D, the presence or absence of the keywords for "Sy" on P cannot be directly used as the criterion for deciding whether D has "Sy". Hence, it is difficult for a medical information searcher to obtain useful search results purely through traditional keyword search. Moreover, considering only one of the attributes like the symptoms is not enough, the physician or medical expert may have some considerations w.r.t the symptoms. For example, a physician examining a patient for headache, after observing the patient may have a query such as, "patient not having eye pain but swelling under the eye". The web documents returned as result, cannot be considered for further consideration if such a documents has one symptom is absolutely missing from it.

In the light of these existing issues with the searches we define "In-depth Querying" as enabling the user to specify the contents he (or she) wishes to query. For instance, in case of the traditional keyword search, a disease name in the user's search criteria will return all the results "disease name" along with the Web pages. In our model for page segmentation, a Web user is able to perform DB-style queries on an originally static source of information. For example, if he or she wishes to query "cases where *helicobacter pylon bacteria* causes *peptic ulcer*". The query is formulated in terms of occurrence of the "peptic ulcer" within a segment "symptom" within the medical resource say, the encyclopedia. The user is taken into a more semantic and more granular level. Its aims are to provide him or her specific set of results (say, where the match occurs as a topic or where it occurs within a sub-topic).

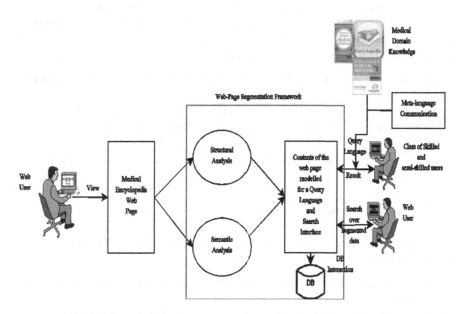

Fig. 2. The web document segmentation and query language interface

2.3   Skilled and Semi-skilled Users

According to White et al. [22], the most common task on the Web today is searching for healthcare information. Medical domain experts have complex query requirements over the data and have a well-evolved meta-language. They often use well-formed query expression to seek the answer set by expressing several terms using medical terminology. Several studies of domain expertise have highlighted the differences between experts and novices, including: vocabulary and search expression [22]. The amount of knowledge the domain experts have about a domain is an important determinant of their search behavior.

Medical experts carry out depth-first searches, following deep-trails of information and evaluate the information based on the most sophisticated and varied criteria. Whereas, the novice users concentrate on breadth-first searches and evaluate through overview knowledge [42]. For example, consider medical personnel accessing the MedlinePlus medical encyclopedia. E.g. A nurse understands if systolic is high - (i) Systolic is a part of blood pressure ii) or when it is a 'symptom' or a 'cause'). He (or she) can query using meta-language, for example, with such option, as, symptom = 'systolic'.

We consider the example from the cited work [22], the expert users have queries such as, *find the categories of people who should or should not get flu shot*? For such a query the normal path of querying that is followed by an end-user (in-case of medical domain, a health expert) is that he first explores the government Website such as the MedlinePlus and then access the related resources. This Website is his first choice because of the existence of domain-specific search knowledge in healthcare and his expertise. Websites they choose to visit in order to find the solution for the above query are described in [22]. The target of this study is to remove the difficulty of acquiring such knowledge from just using general-purpose search engines and to move to query directly over the Web pages (with segments) of the medical resources. For the skilled and semi-skilled users the following assumptions can be stated:

1. The user is well-versed with the data (user view of the database), and the associated meta-language.
2. The users have highly focused search and query needs that can only be supported
3. Moreover, the skilled users expect precise and complete answers.

The challenge is to simplify their querying experience by providing the expert user tools.

3   Problem Statement

This study aims to support structured queries for domain experts. The challenges for the goal are listed as below:

i. To understand the structure of the documents within a medical document repository.
ii. To understand the semantics represented by these documents.
iii. Mapping the structural elements and the corresponding semantics.
iv. Transform the Web page semantics to an enriched database the content-labels in the Web document are mapped to attributes in the database.
v. Adaption of easy to use query interface to support general purpose queries.

3.1   Web Page Segmentation vs. Searching and Querying

The positions of tags containing main content differ in a variety of Websites. Finding the main content in such an unstructured Web documents needs consideration and special algorithms. The earlier work in the area of block level searches, segment a Webpage into semantic blocks, and compute the importance values of the blocks using a block importance model [21]. These semantic blocks, along with their importance values, can be used to build block-based Web Search engines [7]. This approach improves the relevance of search results for record-based or data intensive Websites, such as, yahoo.com, and amazon.com. In the case of queries on medical repositories and Web documents (for example MedlinePlus [13]), such a search may not be useful. In *Object-Level Vertical Search* [28] all the Web information about a real world object or entity is extracted and integrated to generate a pseudo page for the object and is indexed to answer user queries. This search technology has been used to build Microsoft Academic Search (http://libra.msra.cn) and Windows Live Product Search (http://products.live.com). Another type of search, the *Entity Relationship Search* [28], deploys an Entity Relationship Search Engine in the China search market called Renlifang (http://renlifang.msra.cn). The disadvantage of the approach is that if the search terms are scattered across various segments with different semantics, it leads to low-level of precision. None of the mentioned studies capture the querying requirements of the Web documents.

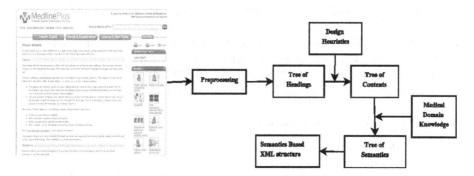

Fig. 3. The web page segmentation model (Offline process)

Document Object Model (http://www.w3.org/DOM/) provides each Web page a fine-grained structure, illustrating the content and the presentation of the page. In case of using DOM directly sometimes simple text lines correspond to complex DOM trees [Zhou]. Moreover, even the visually similar pages can have completely different DOM trees. This is a major shortcoming when schema for a whole document repository such as, an encyclopedia is to be generated. Some algorithms consider the content or link information besides the tag tree. Some consider the heuristic rules to discover record boundaries within a page [26]. In [19] and [17], some visual cues are used along with DOM analysis. These try to identify the logical relationships within the Web content, based on visual layout information. These approaches rely mostly on the DOM structure. Since a normal Web page may have hundreds of basic elements. Such algorithms are time-consuming and inflexible. Moreover, not all the DOM nodes contain some relevant content. They are incapable to deal with a large number of Web pages.

The existing approaches as above are mostly suitable for keyword search. The lack of a well-formed query interface for the Web documents, prompted us to segment the Web document using the underlying structure and pre-existing domain knowledge for successful application of a useful query language interface. Web page segmentation expands a user's querying and searching horizons by enabling him or her query within headings and subheadings of the page.

4 Semantic Structure of Web Documents

In the medical domain the medical dictionary, thesaurus, encyclopedia and the database of common medical concerns are the resources for the user's knowledge base. Figure 3, displays the outline of Web page segmentation based query interaction between the various Web users. The left side of the Fig. 2 displays an arbitrary Web user browsing a medical encyclopedia page. Whereas the right side of the Fig. 2 distinguishes the class of domain experts (medical or healthcare experts) and the novice (Web) users' w.r.t the way, the users query the medical encyclopedia. The domain knowledge of the experts aids them in the query formulation with exact medical vocabulary. And the novice or the general Web users use keywords related to their search questions. The center part describes the Web page segmentation framework to map the HTML documents into a semantics enriched database that can be queried.

The proposed approach aims at understanding a medical Web page and extracting the semantic structure as per the user's view of data. The proposed approach has two components, an offline component to generate a semantic schema for the Web documents and second is an online process which allows querying over the generated schema instead of the original HTML resources. The DOM-tree is utilized along with the layout heuristics for the semantic Web page segmentation (and construction of its hierarchical structure).

Assumptions for the segmentation framework:

i. The headings and the subheadings within each of the pages are extracted by scanning using the HTML parser.
ii. The related content of the Web pages is not of prime concern to the end-user. This focus is on the main content of the Web documents.

Figures 3, 4, 5, and 6 depicts the complete transformation of a static HTML information source to a query-able instance. In this study, we use the Web documents of a medical encyclopedia, to demonstrate the proposed algorithm.

Preprocessing Stage: The HTML pages are preprocessed. The headings are extracted and the hierarchy is captured (termed as *Tree of Headings*).

Tree of Contents: To generate the tree of contents the layout (design) heuristics are used and the nodes of the *tree of headings* are associated with the corresponding content from the Web document.

Tree of Semantics: The nodes of *tree of contents* are labeled at this step using the medical domain knowledge to generate a tree with semantics.

This hierarchical structure is then transformed into a corresponding XML structure which has elements representing medical concepts and sub-concepts (Fig. 6). Table 1 shows the outline of proposed segmentation algorithm highlighting the two major phases, construction of (i) structural tree, (ii) semantic tree.

We propose the following 8 layout (design) rules that guide the identification of segments. These have been designed by combining the syntactic structure of the contents with their visual structure on a rendered Web page. These rules are the most common observations made from the study of various similar to medical resources of Web documents such as [11–14].

Rule 0:   Web pages are mostly organized top-down and left-right.

Rule 1:   Contents within Web documents are organized in semantic units of information, best suited to user's understanding.

Rule 2:   Headings or subheadings within the Web document denote the semantics of the content enclosed within them.

Rule 3:   Headings and subheadings of the same group have same text font, size and background color.

Rule 4:   Each subheading has different text style (size, font) than the heading within which it is enclosed.

Rule 5:   All headings and subheadings have the same orientation.

Rule 6:   The components in the Web document may be text/anchors/forms/lists/ tables or images which are recognizable units.

Rule 7:   Not all the sub-topic labels are present in all of the Web pages of a repository.

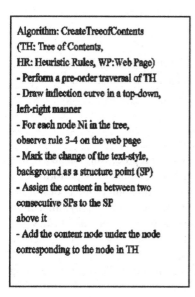
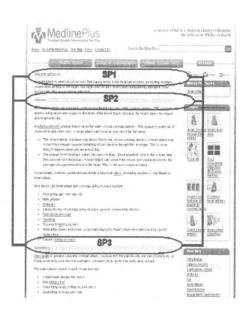

Fig. 4. The structural curve construction algorithm around a MedlinePlus web page

5   The Offline Process

In this section, we present the structural analysis and the semantic analysis for constructing the tree-schema. Our aim is to enable querying using the pre-established medical processes that follow a time tested workflow cycle for data quality enhancement procedures.

5.1   Generation of Schema Tree

We aim to create a hierarchical structure of Web documents based on the semantic and structural characteristics of its contents on the Web page. The grouping of contents is such that it is non- overlapping and can be placed back, to recreate the entire page. Such a structure clearly explains the "is-a", "belongs-to" relationship among the contents of the Web documents. The domain knowledge is used to generate possible set of candidate labels for the headings and subheadings. The algorithm takes into account the complete description of each of the subtopics by maintaining their subheadings and headings as a part of the resultant schema structure.

Table 1. Semantic schema tree generation algorithm

CreateSemanticSchemaTree (TH: Tree of headings, CS: Candidate Labelsets) CreateTreeOfContents(TH) CreateTreeOfSemantics(TC, CS)

First the method *CreateTreeOfContents* is invoked. It associates the nodes of the *tree of headings* to the contents enclosed within each of the topics and sub-topics. Secondly the *CreateTreeOfSemantics* method is invoked. It semantically labels the tree of contents in the previous step (Table 1).

5.1.1   The Structural Analysis

The tree of headings and subheadings generated at the preprocessing stage are the place-holders for the semantic labels. The content group association is performed by associating the text under the heading or subheading corresponding to the nodes of the tree of headings. This is important as the rendered Web page contents may differ from how the tags are organized in the HTML DOM of the Web page. The layout rules 1–2 (above) are used to identify content organization on the Web page whereas 3–5 help in heading recognition.

We define the term, *structural curve* as a curve drawn across the page segregating the content groups. As shown in Fig. 7, any change observed with respect to rules 3–5 is marked as a *structure point* (SP). On any Web page the content groups are contained between these structure points. The algorithm captures any change in the font, text size and color of text distinguishes a heading from the rest of the text. The formal algorithm is given in Fig. 7. Mathematically, it can be defined as,

Definition 5.1.1 (Tree of Contents): Given a Web document (W), the tree of contents (TC) is defined as:

$$TC: \sum (hs, c)_i, \, i = \{\# \text{ of segments s of the Web page W}\}$$

Where hs_i is node of the DOM tree (placeholder for the label l_i), representing a heading or subheading in the Web page and c_i refers to the contents under i^{th} label l_i.

The result of this step is a tree where each node acts as a place-holder for a semantic label. And each leaf node is associated with the content segment.

The domain knowledge of the medical encyclopedias is enriched by the medical dictionary, thesaurus and the medical concerns database. This domain knowledge helps in formulating a candidate label set for headings (topics) and the sub-headings (sub-topics) within any Web page belonging to a medical Web-resource. These candidate label sets are the basis of queries by the skilled and semi-skilled users. These form the meta-language components and attributes for queries over contents. At this step of the model, the candidate labels are mapped to the corresponding nodes of the structural tree. The result is a semantic tree which can be directly mapped to a schema for the document repository.

The candidate label set for the root level is represented as, $CL_{root} = \{$Medical concepts$\}$. Here, the medical concepts may refer to name of diseases, tests, surgery and other topics. Candidate labels for the intermediate nodes may refer to the terminologies that may describe a sub-concept of a medical concept. For instance, a disease (fever) is a medical concept which may have several sub-concepts describing it such as, $CL_{fever} = \{$symptoms, causes, exams and tests and treatment$\}$. Mathematically, the *Tree of Semantics* can be defined as,

Definition 5.1.2 (Tree of Semantics): Given the tree of contents (TC) and set of labels (concepts) CL, from the domain. We define the tree of semantics TS as:

$$TS = \sum (n_i, L)$$

Where n_i is a node in TC (Definition 5.1.1), and L: TC $\rightarrow$ l maps each node of n_i to a label l where l $\subseteq$ CL, based on the medical concepts.

Table 2. Algorithm for constructing tree of semantics

CreateTreeOfSemantics (TC: Tree of Contents, CL_{node})
- Perform pre-order traversal of TC
- For each node $n_i \in$ TC
- determine_level(n_i)
- determine label set corresponding to the level
- Assign_label (*l*)

A pre-order traversal of the *tree of contents* is performed to create the semantic schema tree. The procedure *determine_level*, determines whether the node is a root node (page title) or is one of the intermediate nodes (subheadings or subtitles). As given in, Table 2, each node N_i is picked while performing the pre-order traversal and its level is determined. The procedure *Assign_label* assigns the labels to the nodes after determining the category of the page (disease, treatment or laboratory test) and then picking the labels from the candidate label-set for that category.

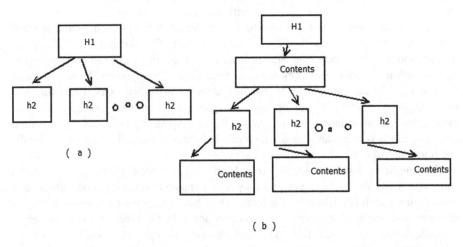

Fig. 5. (a) Tree of HTML tags in a DOM tree (b) Tree of Contents similar to VIPS [6].

Once the level is determined the corresponding candidate label set is used for label assignment. Figure 5 represents the transition of "tree of headings" to "semantic schema tree" with an intermediate form of "tree of contents". The semantic schema tree for the MedlinePlus Web pages, is utilized this structure for enhancing the searching and querying. For this purpose, we generate an enriched schema by converting each of the nodes of the semantic schema tree into the XML tags. The contents become the contents of these XML elements. These tags represent semantics such as, the causes, symptoms. Figures 4, 5 and 6, show tags of a Medline Plus article on "Aarskog syndrome" as the XML database tags generated by the above algorithm.

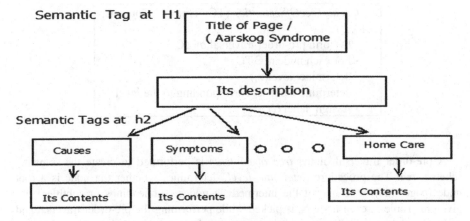

Fig. 6. Converted semantic tags in XML database

6 The Online Process

In this section, we explain the online component of improving the searching and querying experience of the user. It utilizes the semantic schema generated by the offline component for answering user queries.

6.1 Improving Querying

The complex task of in-depth querying the medical Web resources is reduced to a simpler task of utilizing XQuery, query language (or XQBE query language interface) [4] over these. Each of the segments of the Web pages of the medical document repository may be viewed as a query-able entity. The schema tree may be visualized as a data graph [23], where the edges represent the attributes and the nodes represent the data. The notion of "path-expression" may be associated with it. Any query can be mapped as a sub-graph or sub-tree of this data graph. One of the main distinguishing features of semi-structured query language is its ability to reach arbitrary depths in the data graph. The semi-structured data can be viewed as an edge-labeled graph with a function

$$F_E : E \rightarrow L_E,$$

Where L_E represents the candidate label set which form the domain of edge labels.

Figure 7, represents the data graph for the medical encyclopedia documents, where the database (db) is the root of the graph and its children are components of the documents that are segmented into various categories (disease, test and surgery). The child nodes show the segments within these Web pages. A user query follows a path

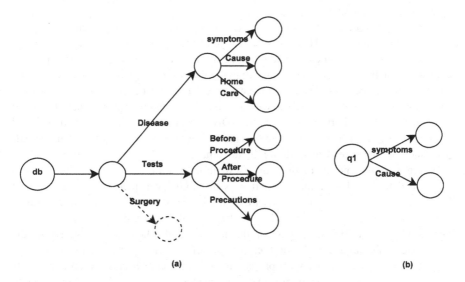

(a)   (b)

Fig. 7. Query is a sub-tree of the schema tree

from leaf to root i.e. it is a sub-tree of the larger tree called as the schema tree. It helps the user to visualize his query path within a graph. The hierarchy of the node provides the user the in-depth information as per the query.

7   Related Studies

Recent study by Yang et al., proposes to consider knowledge graphs (such as Wikipedia) for performing keyword search. Our proposal is similar to the knowledge graph approach. In case of Wikipedia, a user may gather pages of interest. For example, people, movies, places, etc. Over such a collection of documents, the proposed approach to search and query using segments can be applied. Over the documents of one kind, segments of user interest can be found. Similar other approaches have been proposed earlier, for XML documents [46], annotation based [47], Schema-free SQL approaches [48], Interactive browsing [50]. In ETable is a easy to use interface users can interactively browse and navigate through databases on ER level without writing SQL. Users having knowledge of entity relationship (ER) level can easily query the databases. Its proposed transformation from ER level to SQL level is achieved through Typed Graph Model described in paper. Table 3 presents an analysis of recent approaches of querying databases.

8   Usability, Acceptance and Further Studies

Preliminary results indicate that medical experts are skilled and semi-skilled users and need powerful query interfaces. They do not trust searches, such as the advanced keyword search. They are not confident about the outcome of computation or ranking method from which the results are fetched. Nor they can be sure about the procedures implemented for fetching the results.

The expert medical practitioner (skilled or semi-skilled user) can create a query using the XQBE interface. He (or she) is aware of the relationships among the various medical terminologies and can relate the nodes and their corresponding output to make the best use of the interface. Previous studies point out that QBE is suited for simple (single relation) queries [24]. Most of the medical expert's queries usually concern a single disease. Secondly, the user is well aware of the domain contents, meta-language and content linkages. Further, the document DB is as per the user's view of the DB and not as per the DB normalization theory. This facilitates simplified query using XQBE (QBE such as interface) for MedlinePlus documents.

Figure 8, shows the interactions of a domain user and a novice user with the Web document database. The domain expert uses the high level query language interface like the XQBE, whereas the novice user uses the keyword search. We aim to generalize the proposed segmentation approach over the general purpose domains which have a large number static HTML pages (or documents) and where the end-users of the domain are skilled and face the problem of lack of interfaces for querying.

Our approach transforms Web pages which form a collection of Web documents into a knowledge resource, where it makes a transition from an HTML Web pages to

Table 3. Analysis of approaches of querying databases

	Main Point	Overheads/Difficulties	Mechanism	Query type	User skill
SQL	General purpose query language for structured data	Schema knowledge is required; programming skill required	Based on Relational algebra	Ad-hoc; Predictable	For high Skilled users only
XSEarch [46]	Semantic search engine for XML documents -	Ranking through extended information retrieval techniques	Finds related fragments, ranked by estimated relevance	Uses node labels., Simple query language	Skilled, semi-skilled and naive users
Avatar Semantic Search [47]	A prototype search engine for semi-structured data	Often Keywords is insufficient to build semantics; Same keyword with more than one meaning may exist	Facts, concepts and relation-ships are extracted from the text, in an offline manner	Operates on a annotations; advanced keyword search	Skilled, semi-skilled and naive users
Schema-Free SQL [48]	Partial schema knowledge is sufficient to query	Requires prior Information requirement elicitation (IRE) steps	Simpler than SQL with no part schema; no join path	Quasi-relational	Skilled, semi-skilled users
ETable [50]	Users can interactively browse and navigate through databases at ER level without writing SQL	E-tables are made for each action fired by the users after interactive browsing	Relational schema is converted into Typed Graph schema	Familiarity with Entities and relationships is utilized	Skilled, Semi-Skilled users
Wikisearch [51]	Search on Wikipedia, Freebase and Yogo using knowledge graphs	Knowledge graphs grow in size at rapid pace	Based on Keyword search approach	Users express advanced keyword search query	Skilled and novice users
Segment-search (proposed)	Searches within segments in documents; provides exact matches	Prior segment-keywords catlogue is generated from web documents	Makes segment as unit of search query over web documents	Users express advanced keyword queries	Skilled and novice users

104   R. P. Sarode et al.

semantics based XML database. Further it enables the user to perform queries on it. Figure 9 shows the transformations that are performed for this process. The Web objects are identified from the HTML Web document from the user's perspective and then the user can query them using the query languages such as the SQL and XQuery. As we can see in the figure, the XML and the DTD enable the high level query language interface such as the XQBE, or the XQuery query language, to query the tags within the XML and DTD defines the query schema. Hence, the Web page can now be queried within the segments (in the form of semantic objects).

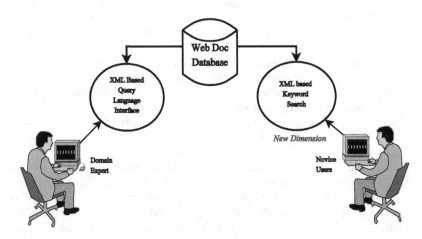

Fig. 8. Query and search interfaces (according to user category)

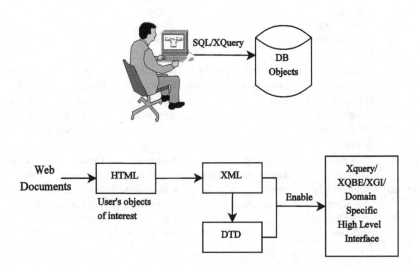

Fig. 9. Transformation of the web page to a query-able Entity

9 Summary and Conclusions

Our attempt is to provide exact results to these users, with minimal effort. In this study, we propose semantics based page segmentation technique for the Web-based document repositories. The proposed schema generation scheme has been implemented to generate user-level objects database view (virtual database). The proposed approach is divided into two phases: construction of the semantic database and further querying using XML base. This technique is designed especially for the Web pages which contain documents with structured contents as per the user's view. We believe that this approach can be improved to make it applicable to a large set of HTML pages on the Web which cannot be easily queried or searched by the expert-users. Identifying and understanding the domain experts view, makes it possible to provide an expert query language interface and tune the system based on search and query results of these users. It will also help in improving the search interface for the non-experts.

References

1. Spink, A., et al.: Health Inform. Libr. J. Health Libr. Group (2004)
2. Chen, J., Zhou, B., Shi, J., Zhang, H., Wu, Q.: Function-based object model towards website adaptation. In: The 10th International World Wide Conference (2001)
3. Kohlschütter, C., Nejdl, W.: A densitometric approach to web page segmentation. In: Proceedings of CIKM 2008, 26–30 October (2008)
4. Braga, D., Campi, A., Ceri, S.: XQBE (XQuery By Example): a visual interface to the standard XML query language. ACM Trans. Database Syst. TODS **30**(2), 398–444 (2005). https://doi.org/10.1145/1071610.1071613
5. Fernandes, D., de Moura, E.S., da Silva, A.S., Ribeiro-Neto, B., Braga, E.: A site oriented method for segmenting web pages. In: Proceedings of SIGIR 2011, 24–28 July 2011
6. Cai, D., Yu, S., Wen, J.-R., Ma, W.-Y.: Extracting hierarchical structure for web pages based on visual representation. In: Proceedings of 5th Asia-Pacific Web Conference, APWeb 2003, Xian, China, 23–25 April 2003, pp. 596–596 (2003)
7. Cai, D., He, X., Wen, J.-R., Ma, W.-Y.: Block-based web search. In: Proceedings of SIGIR (2004)
8. Gu, X., Chen, J., Ma, W., Chen, G.: Visual based content understanding towards web adaptation. In: Proceedings of 2nd International Conference on Adaptive Hypermedia and Adaptive Web-based Systems (AH2002), Spain, pp. 29–31 (2002)
9. http://adam.about.net/encyclopedia/
10. http://dbgroup.elet.polimi.it/xquery/tool/
11. http://www.drugs.com/medical_encyclopedia.html
12. http://www.mgo.md/encyclopedia.cfm
13. http://www.nlm.nih.gov/medlineplus/encyclopedia.html
14. http://www.umm.edu/ency/
15. http://www.w3schools.com/htmldom/default.asp
16. Cao, J., Mao, B., Luo, J.: A segmentation method for web page analysis using shrinking and dividing. Int. J. Parallel Emergent Distrib. Syst. **25**(2), 93–104 (2010)
17. Ramaswamy, L., Iyengar, A., Liu, L., Douglis, F.: Automatic detection of fragments in dynamically generated web pages. In: Proceedings of the 13th International Conference on World Wide Web (2004)

18. Zloof, M.M.: Query-By-Example: a data base language. IBM Syst. J. **16**(4), 324–343 (1977)
19. El-Shayeb, M.A., El-Beltagy, S.R., Rafea, A.: Extracting the latent hierarchical structure of web documents. In: Proceedings of SITIS (2006)
20. Asfia, M., Pedram, M.M., Rahmani, A.M.: Main content extraction from detailed web pages. Int. J. Comput. Appl. (IJCA) **4**(11), 18–21 (2010)
21. Song, R., Liu, H., Wen, J.-R., Ma, W.-Y.: Learning block importance models for webpages. In: Proceedings of WWW (2004)
22. White, R.W., Dumais, S., Teevan, J.: How medical expertise influences web search interaction. In: Proceedings of SIGIR 2008, 20–24 July 2008, Singapore (2008)
23. Abiteboul, S., Buneman, P., Suciu, D.: Data on the web: from relations to semistructured data and XML (2000)
24. Saito, T.L., Morishita, S.: Relational-style XML query. In: SIGMOD, pp. 303–314 (2008)
25. Hong, T.W., Clark, K.L.: Towards a universal web wrapper. In: Proceedings of FLAIRS Conference (2004)
26. Liu, W., Meng, X., Meng, W.: ViDE: a vision-based approach for deep WebData extraction. IEEE Trans. Knowl. Data Eng. **22**, 447–460 (2010). Member, IEEE
27. Diao, Y., Lu, H., Chen, S., Tian, Z.: Toward learning based web query processing. In: Proceedings of the 26th International Conference on Very Large Databases, Cairo, Egypt (2000)
28. Nie, Z., Wen, J.-R., Ma, W.-Y.: Webpage understanding: beyond page-level search. Sigmod Rec. **37**(4), 48–54 (2008)
29. http://www.omnimedicalsearch.com/forumsearch.html
30. http://www.healthline.com/
31. http://www.mwsearch.com/mwsframetemplate.htm?
32. http://www.pogofrog.com/
33. Chung, C.Y., Gertz, M., Sundaresan, N.: Reverse engineering for web data: from visual to semantic structures. In: Proceedings of the 18th International Conference on Data Engineering (ICDE 2002)
34. Juan, H., Zhiqiang, G., Hui, X., Yuzhong, Q.: DeSeA: a page segmentation based algorithm for information extraction. In: Proceedings of the First International Conference on Semantics, Knowledge, and Grid, SKG 2005
35. Yang, Y., Zhang, H.J.: HTML Page Analysis Based on Visual Cues. IEEE (2001)
36. Pnueli, A., Bergman, R., Schein, S., Barkol, O.: Web page layout via visual segmentation. HP Laboratories (2009)
37. Chakrabarti, D., Kumar, R., Punera, K.: A graph-theoretic approach to webpage segmentation. In: Proceedings of WWW 2008, Refereed Track: Search-corpus Characterization and Search Performance, Beijing, China (2008)
38. Zou, J., Le, D., Thoma, G.R.: Combining DOM tree and geometric layout analysis for online medical journal article segmentation. In: JCDL 2006, Chapel Hill, North Carolina, USA, 11–15 June 2006
39. Zhang, C.: Medical students, and healthcare professionals use Wikipedia? UBCMJ, **3**(2) (2012)
40. http://www.cad.zju.edu.cn/home/dengcai/VIPS/VIPS.html
41. Cai, D., He, X., Wen, J.-R., Ma, W.-Y.: Block-level link analysis. In: SIGIR 04, Sheffield, South Yorkshire, UK, July 2004
42. Jenkins, C., Corritore, C.L., Wiedenbeck, S.: Patterns of information seeking on the web: a qualitative study of domain expertise and web expertise. IT & Soc. **1**(3), 64–89 (2003)
43. Chen, H., Lally, A.M., Zhu, B., Chau, M.: HelpfulMed: intelligent searching for medical information over the internet. J. Am. Soc. Inf. Sci. Technol. **54**(7), 683–694 (2003)
44. http://saxon.sourceforge.net/dtdgen.html

45. http://www.altova.com/
46. Cohen, S., Mamou, J., Kanza, Y., Sagiv, Y.: XSEarch: a semantic search engine for XML. In: Proceedings of the 2003 VLDB Conference, Berlin, Germany (2003)
47. Kandogan, E., Krishnamurthy, R., Raghavan, S., Vaithyanathan, S., Zhu, H.: Avatar semantic search: a database approach to information retrieval. In: SIGMOD 2006, 27–29 June 2006, Chicago, Illinois, USA (2016)
48. Li, F., Pan, T., Jagadish, H.V.: Schema-free SQL. In: SIGMOD 2014, Snowbird, UT, USA (2014)
49. Jagadish, H.V., Nandi, A., Qiun, L.: Organic databases. In: DNIS 2014 Workshop, pp. 49–63 (2014)
50. Kahng, M., Navathe, S.B., Stasko, J.T., Chau, D.H.: Interactive browsing and navigation in relational databases. In: 2016 Proceedings Of VLDB, vol. 9, no. 12, pp. 1017–1028 (2016)
51. Yang, Y., Agrawal, D., Jagadishy, H.V., Tung, A.K.H., Wu, S.: An efficient parallel keyword search engine on knowledge graphs. In: ICDE, pp. 338–349 (2019)

Pairing Users in Social Media
via Processing Meta-data
from Conversational Files

Meghna Chaudhary, Ravi Sharma, and Sriram Chellappan[(✉)]

Department of Computer Science and Engineering, University of South Florida,
Tampa 33620, USA
{meghna1,ravis}@mail.usf.edu, sriramc@usf.edu

Abstract. Massive amounts of data today are being generated from users engaging on social media. Despite knowing that whatever they post on social media can be viewed, downloaded and analyzed by unauthorized entities, a large number of people are still willing to compromise their privacy today. On the other hand though, this trend may change. Improved awareness on protecting content on social media, coupled with governments creating and enforcing data protection laws, mean that in the near future, users may become increasingly protective of what they share. Furthermore, new laws could limit what data social media companies can use without explicit consent from users. In this paper, we present and address a relatively new problem in privacy-preserved mining of social media logs. Specifically, the problem here is the feasibility of deriving the topology of network communications (i.e., match senders and receivers in a social network), but with only meta-data of conversational files that are shared by users, after anonymizing all identities and content. More explicitly, if users are willing to share only (a) whether a message was sent or received, (b) the temporal ordering of messages and (c) the length of each message (after anonymizing everything else, including usernames from their social media logs), how can the underlying topology of sender-receiver patterns be generated. To address this problem, we present a Dynamic Time Warping based solution that models the meta-data as a time series sequence. We present a formal algorithm and interesting results in multiple scenarios wherein users may or may not delete content arbitrarily before sharing. Our performance results are very favorable when applied in the context of Twitter. Towards the end of the paper, we also present interesting practical applications of our problem and solutions. To the best of our knowledge, the problem we address and the solution we propose are unique, and could provide important future perspectives on learning from privacy-preserving mining of social media logs.

Keywords: Social media · Privacy · Big-data · Meta-data · Dynamic Time Warping

S. Madria et al. (Eds.): BDA 2019, LNCS 11932, pp. 108–123, 2019.
https://doi.org/10.1007/978-3-030-37188-3_7

1   Introduction

As of today, social media is a major platform that citizens across the globe choose to communicate over. Such communications span many forms including bi-directional one-on-one messaging among peers; posting content to be viewed by a larger group; following posts from favorite personalities; advertising products to customers; reviewing products and services; and so much more. It is a fact that mining such data is a billion dollar business with so many players in the market now. One of the downsides of this scenario is compromising the privacy of common citizens.

As of today, a vast majority of users on social media do not care about privacy, and even if they do, they still actively communicate over many social media platforms. In fact, this is true even amongst the more educated citizens. However, this trend may change in the future. There are numerous reports now wherein common citizens are becoming victims because the content they post/share on social media has been accessed by third-party entities beyond the scope/knowledge of the victim. These include people losing their jobs, denied admissions to universities, being charged with crimes and fines, losing custody of children, and so much more. With increasing awareness of such reports, citizens of the future are likely to be increasingly aware of privacy violations. In parallel, governments across the world are also monitoring un-restricted access of social media content by data mining agencies, and newer laws and regulations are being generated today, one prime example being The EU General Data Protection Regulation (GDPR), that came into force in 2018.

Given these developments, there are now urgent efforts in the academia and the industry to investigate the paradigm of privacy-preserved learning from big-data. Essentially, the issue at hand is how can we learn meaningful information from data (whether generated from social media or not), which still preserving the privacy of the data as intended by users. There are number of studies currently looking at this paradigm from perspectives of AI [1–3], Crypto [4,5], and Social Sciences [6,7]. In this paper, we make contributions towards this paradigm. Specifically, the scenario we address is one where users wish to protect identities (i.e., their own username and those of peers) and content of what they share on social media (texts, images, multi-media etc.), but are willing to share meta-data of such content - for example, whether a message was sent or received; the temporal ordering of messages; and the size of the message (e.g., number of characters in the text). In this scenario, our specific problem is identifying the network/communication topology - that is pairing the sender and receiver among multiple conversational files from multiple users containing only meta-data.

We propose a Dynamic Time Warping (DTW) based approach to our problem. Essentially, our solution models the meta-data sequences from each conversational file as a temporal sequence, and then uses DTW technique to find the best similarity match among multiple conversational files. We evaluate the technique using a limited sample of Twitter users/logs (47 users and 128 conversational files), and performance results are very favorable, and our technique is also scalable. What is important is that our proposed method gracefully degrades when users can choose to delete content (and hence the underlying meta-data) within conversational files before sharing. To the best of our knowledge, our problem in this paper is unique and has not been addressed before. There are important practical ramifications of this problem from multiple perspectives including one related to cyber-abuse, which we elaborate on towards the end of the paper.

The rest of the paper is organized as follows. Section 2 discusses about important related work. Section 3 presents information about our data source and the formal problem statement. Section 4 elucidates the methodology followed for matching of conversational files. In Sect. 5, we discuss performance evaluations. Section 6 discusses about the practical relevance of our contributions in this paper. Finally, we conclude the paper in Sect. 7.

2 Related Work

We now present a brief overview of important work related to this paper.

2.1 Identifying Potential Friends in Social Networks

In [8], a study is conducted in order to determine effectiveness of textual features and network trends to make recommendations for friends on the Twitter platform. In this study, 200 most recent tweets from 100, 000 users are collected and analyzed. Also, another 200 recent tweets from 10 friends and 10 followers for each user is also collected. Different information sources include posts by a user, posts by user's mentions, and friends and followers are processed, and a model integrating Bag of Words and Principal Component Analysis is designed to identify potential friends. Another model that considers network-level metrics is also used, wherein two users are considered to be friends if their social connections (i.e., those they follow and those that follow them) share similar content. Based on this network structure also, friends are predicted for each user. The paper argues that network level structures are better suited for friends recommendation compared to purely textual based features. While using pure network level properties does preserve privacy to a certain degree, the fast that user names are shared and processed raises privacy issues.

In [9], another study is conducted wherein pictures and contact tags of 10, 000 Flickr users are collected, wherein the 10, 000 users belong to 2000 social groups. Features from these picture and contact tags are processed to see how similarities as computed from the features match the actual underlying topology. The paper demonstrates that picture and contact tag features can model the underlying social topology very well. Based on these results, a recommendation system is also proposed for friends matching. But processing images, can have serious privacy implications.

2.2   Computational Techniques to Preserve and Compromise Privacy

There are a number of studies now that design novel encryption techniques to search from encrypted data [4,5]. The basic idea is to ensure that legitimate users have a degree of access to meaning of the data even under encryption, while adversaries do not derive any meaningful information. The limitations though are that even with state-of-the-art encryption today, the quality and quantity of information gleaned from encrypted data is still minimal, and not enough for many applications. There are also other recent papers in the realm of [1–3] that investigate how much information do machine learning models remember after training, to the point where information about ground truth data used to train the models can be recovered. These avenues of research demonstrate serious privacy breaches from the perspective of exposing data used to train algorithms today. There are also a number of studies also in the realm of using the privacy-preserving and spatially compact Bloom Filters to store and retrieve records based on similarities. Most of these studies [10,11] and [12], are in medical contexts though, and only for data searching operations.

3   Data Source and Formal Problem Formulation

The source of our dataset is Twitter. For this study, we extracted tweets from a group of 47 socially connected users and obtained 3200 most recent tweets per user. Conversational connections in terms of user names and their mentions were collected, and tweets which do not represent a conversation were discarded. The period of data collection was from January 2019 to March 2019.

The process of anonymization is a little more complex and is elaborated below. First, for each of the 47 users whose data we collect, we anonymize their identities as $User_1$ to $User_{47}$. For example, consider $User_1$. Let $User_1$ engage in communications with 10 other contacts in the timeframe of our data collection. Now when $User_1$ wishes to share his/her history, with privacy expectations, $User_1$ will not reveal the identities of these contacts. What $User_1$ will instead do is anonymize their identities arbitrarily before sharing. As such in our data collection process, every contact of $User_1$ is anonymized as $User_1^{c1}, User_1^{c2}, \ldots User_1^{c10}$ for ten such contacts. Naturally, to protect the content shared between $User_1$ and any of its contacts, the records of corresponding communications shared are only whether a message was sent or received; the temporal ordering; and the overall length of the message/content. The same process is followed for every other user, and for every contact of that user in our study.

One thing is very important to note here, and this stems from the way each user's data is independently anonymized before processing. Let us for now assume that two users anonymized as $User_2$ and $User_3$ are indeed conversing with each other. Due to the way we independently anonymize the contact for each user in our study, it could be the case that $User_3$ is identified as $User_2^{c1}$ in the dataset corresponding to $User_2$; and $User_2$ is identified as $User_3^{c5}$ in the dataset corresponding to User$_3$. That is, there is no linkage among these contacts due to independence of anonymization under our data collection process. The problem we address here is how to still derive the social/network topology by identifying that User$_2$ and User$_3$ are indeed communicating with each other using only the metadata collected from each user independently, and especially when certain logs of data could be deleted arbitrarily by each user, irrespective of whether or not the other party deletes them.

Figure 1 presents a snapshot of the dataset collected in the above manner. In the dataset illustrated, User$_1$ has 10 connections labeled from User$_1^{c1}$ to User$_1^{c10}$; User$_2$ has 7 connections labeled from User$_2^{c1}$ to User$_2^{c7}$; User$_3$ has 5 connections labeled from User$_3^{c1}$ to User$_3^{c5}$; User$_5$ has 8 connections labeled from User$_5^{c1}$ to User$_5^{c8}$; and so on. Finally, User$_{46}$ has 9 connections labeled from User$_{46}^{c1}$ to User$_{46}^{c9}$ and User$_{47}$ has 8 connections labeled from User$_{47}^{c1}$ to User$_{47}^{c8}$. The corresponding sent/received patterns and length of messages for these users are also presented in Fig. 1. Given this dataset, our problem is to determine the overall social/network topology. Essentially, our goal is to identify (based on alignment of sent/received patterns and lengths in the dataset) that User$_1$ is a social contact of User$_3$; User$_2$ is a social contact of User$_3$; User$_3$ is a social contact of User$_5$; User$_5$ is a social contact of User$_{46}$; User$_{46}$ is a social contact of User$_{47}$ and so on. While this is easier to do if all messages among all contacts are retained and shared, this will not be the case in reality. Users can arbitrarily choose to delete messages in their conversations with one or more or contacts, irrespective of whether or not the same message is deleted by the other party. Under such situations, which is common in practice, the problem of establishing social connections becomes much harder with potential for false negatives and false positives as well.

4   Our Dynamic Time Warping Framework

In this paper, we employ a Dynamic Time Warping (DTW) approach to solve our problem. DTW is an ideal technique for our problem, since, the calculation of the distance (or similarity) metric between time-series sequences of datasets using DTW can overcome problems related to comparing short patterns of data and class imbalances. The technique is also independent of certain non-linear variations in the time dimension [13–16].

To apply DTW for our problem, we need to encode our dataset as a time-series sequence. This is doable in our case because the sent/received patterns over time for any user are essentially temporal orderings, to which the length of a corresponding message can be easily appended. With this temporal ordering in place across all conversational files for all users, the problem now is find the best alignment between encoded files in order to generate the connection between users, and hence the overall social topology. As we present below, the DTW technique involves determination of a warping path between temporal sequences, from which the path that optimally minimizes the corresponding distance is the ideal one.

Note that, there are certain critical steps that need to be executed to compute the warping path, and hence the distance metric. These are presented below.

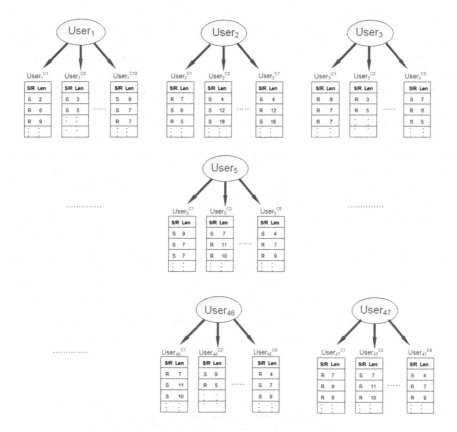

Fig. 1. A Snapshot of our dataset for analysis. Our problem is to determine social connections even in the presence of arbitrary message drops

1. **The boundary case**: We ensure that the starting points and the ending points of each pair of sequences are identified and matched as such. This ensures that all data points in the entire sequences are compared in determining the similarity metric, and hence prevents data loss.
2. **Ensuring monotonicity**: When comparing sequences of data points, any set of points in time in one sequence that are already aligned with points in another sequence are not used to evaluate for matching with later points in time. In order words, comparison of similarities between points in multiple sequences is monotonically increasing.
3. **The step size**: In computing the distance metric across a warping path, we ensure that every point within the neighborhood of a data point is considered for distance measurement, and as such jumps across data points are not allowed [17].

Algorithm Description: We now present our DTW algorithm for the problem above. But first, we present the encoding technique for our ground truth data to facilitate discussions. Consider a sequence of messages from any arbitrary User (say User$_x$) to another arbitrary user (say User$_y$) as $[S, R, S, S, R, ..., S, R]$. This sequence will be encoded as $[1, 0, 1, 1, 0, ..., 1, 0]$ during processing. If the corresponding message lengths are $[10, 12, 4, 5, 6, ..., 12, 23]$ are considered, the encoded sequence now becomes $[101, 120, 41, 51, 60, ..., 121, 230]$ that integrates both message lengths and sent/received status.

We now present a simple example of how to execute the DTW technique to find similarities for our problem statement. We present the example for the case of considering sent/received patterns only, without considering message lengths, but the technique is straightforward to integrate lengths also. Consider a sent/received sequence denoted as $T_1 = [1, 0, 1, 0]$ (read from bottom to top in the left in green in Fig. 2). Consider another sequence $T_2 = [0, 0, 1]$ (in red in the bottom in Fig. 2). Note that two sequences are not the same even in the number of entries they have. This can happen in our case, since users are allowed to arbitrarily delete content before sharing, and our technique will accommodate this case. In the DTW technique, for this example, we first compute the distance matrix $Dist_{Matrix}$ of dimensions 4×3, where each entry in the matrix is computed using the following equation (where ED stands for Euclidean Distance):

$$DTW\_Dist[i, j] = ED[T_1[i], T_2[j]] + min(DTW\_Dist[i - 1, j],$$
$$DTW\_Dist[i - 1, j - 1], DTW\_Dist[i, j - 1]) \tag{1}$$

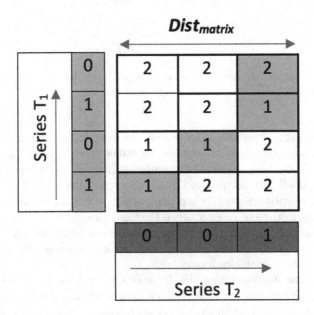

Fig. 2. Distance matrix computation for two series using DTW (Color figure online)

Finally, we traverse the optimal warping path from the end point to the starting point between the two paths (which in the distance matrix is from top-right cell to the bottom-left cell) by choosing the adjacent cell with the minimum distance. In Fig. 2, this is shown in blue shaded color. The sum of the entries in these cells is the overall distance computed, and denotes the similarity between the two time sequences, which in this case is 5.

It is easy to infer that the above procedure can be expanded to include sequences encoded with message lengths. Furthermore, for two sequences that are exactly the same, the distance computed will be 0. Finally, we point out that in order to compare a sequence of messages shared by one user with a sequence of messages shared by another user for similarity matching, we invert the sequence of messages in the other user prior to determining the warping path. This ensures that a message sent by the first user is compared with a message received message by the other user; and that a message received by the first user is matched with a message sent by the other user. Algorithm 1 presents the formal sequence of steps in our solution to compute similarity across message sequences. The complexity here is $O(f^2 \times m^2)$ where f denotes the number of conversation files to compare, and m denotes the maximum length among sequences to compare. Our technique to implement DTW was built using Python. Note that Python provides open-source Just-in-Time (JIT) compiler Numba to generate faster machine code which helps accelerate computation.

5   Results

5.1   Overview

We present results of our technique to match users that are communicating with each other, based on processing only meta-data logs of conversational files. Before that we present some preliminaries. First, of all we point out that the total number of Twitter users in our experiment was 47. The total number of conversational files was 128. The 3200 most recent tweets per user were obtained, and the period of data collection was from January to March 2019. On an average, each conversational file contained anywhere from $1,500$ to $3,000$ messages. We only processed textual messages between users. As mentioned above, all user identities were anonymized. Only sent/received status of messages, their temporal ordering and lengths were processed. No actual textual content was processed. We considered two classes of features for determining the connections between users. The first one only included the sent/received patterns (without considering the lengths). The second includes the lengths of the corresponding messages along with the sent/received patterns.

Note that in our problem formulation, users can choose to arbitrarily delete messages from one or more conversational files before sharing. As such in our dataset, we provision for that. Specifically, we define two new parameters denoted as α and β. Here α denotes the percentage of files whose content a user opts to delete before sharing. In other words, when $\alpha = 0$, the user does not modify any file; and when $\alpha = 100\%$, the user chooses to modify every file shared. The next parameter β denotes the percentage of messages within a file that a user chooses to delete. Here again, when $\beta = 0$ for a particular file, the user does

---

Algorithm 1. DTW to Compute Similarity

User Files = U_{files}
Source User File Series = X
Receiver User File Series = Y
Length of Source User File Series = m
Length of Receiver User File Series = n
Array $DTW_{Dist}[0\ldots,0\ldots n]$
Euclidean Distance = ED
Similarity score = $similarity_{score}$
Input: List X and Y of patterns in two conversation files
Output: The similarity score (measure of distance)

1: **for** $p \leftarrow 1, number\ of\ user\ files - 1$ **do**
2:      **for** $q \leftarrow p + 1, number\ of\ user\ files$ **do**
3:          $X = U_{files}[p]$
4:          $Y = U_{files}[q]$
5:          cost=0
6:          **function** DTW$((X[1\ldots m], Y[1,\ldots.n]))$
7:              **for** $i \leftarrow 1, m$ **do**
8:                  **for** $j \leftarrow 1, n$ **do**
9:                      $DTW\_Dist[i,j]$ $=$ $ED[X[i], Y[j]]$ $+$ $min(DTW\_Dist[i - 1, j], DTW\_Dist[i - 1, j - 1], DTW\_Dist[i, j - 1]$
10:                 **end for**
11:             **end for**
12:             $path \leftarrow [m, n]$
13:             $y \leftarrow m$
14:             $z \leftarrow n$
15:             **while** $z >= 0\ and\ y >= 0$ **do**
16:                 **if** $z = 0$ **then**
17:                     $z \leftarrow z - 1$
18:                 **else if** $y = 0$ **then**
19:                     $y \leftarrow y - 1$
20:                 **else**
21:                     **if** $DTW\_Dist[y, z - 1]$ $=$ $min(DTW\_Dist[y, z - 1], DTW\_Dist[y - 1, z], DTW\_Dist[y - 1, z - 1])$ **then**
22:                         $z \leftarrow z - 1$
23:                     **else if** $DTW\_Dist[y - 1, z] = min(DTW\_Dist[y, z - 1], DTW\_Dist[y - 1, z], DTW\_Dist[y - 1, z - 1])$ **then**
24:                         $y \leftarrow y - 1$
25:                     **else**
26:                         $y \leftarrow y - 1$
27:                         $z \leftarrow z - 1$
28:                     **end if**
29:                 **end if**
30:             **end while**
31:             $Add\ path[z, y]\ to\ path\ array$
32:             **for** $[z, y] \in path$ **do**
33:                 $cost \leftarrow cost + ED[y, z]$
34:                 return cost
35:             **end for**
36:         **end function**
37:         $similarity_{score} = DTW(X[1\ldots m], Y[1,\ldots.n])$
38:         **print** $similarity_{score}$
39:     **end for**
40: **end for**

---

not delete any conversation in that file; and when $\beta = 100\%$ for a file, the user chooses to delete every conversation in that file. In presentation of results below, we vary α and β from 0 to 90%. Note that the files chosen for deletion and the messages chosen to be deleted within a file for each user are random. Also, note that content is deleted independently for each user, irrespective of whether or not the corresponding content is deleted in the communication files of the other party. This is the most practical scenario, and hence we evaluate our system under this scenario.

5.2 Metrics and Results

We employ three standard measures to evaluate our algorithm for our problem. These are Precision, Recall and the ROC curve.

Precision and Recall: In Fig. 3(a) and (b), we plot the Precision of our system for varying values of α and β. Here, Fig. 3(a) is the plot for the case where only sent/received patterns are considered (without considering message lengths). Figure 3(b) is the plot for the case where sent/received patterns and message lengths are both considered. Matching of conversational files across users is done for those pair of files that have the lowest similarity scores. It is straightforward to calculate True Positives, False Positives and False Negatives from this, which will directly yield the computation of Precision and Recall.

First off, we see from that when α and β are low, the Precision and Recall are high for both feature sets. This is straightforward, since matches among communicating users are more reliable when there are very few message drops. As α and β increase, we see lower precision and recall values. We also see that Precision is more sensitive to β than to α. This is because, even when more conversational files are chosen to be deleted (i.e., higher values of α), the fact that only few messages within each file are deleted (i.e., lower values of β) enables our system to match correct conversational files better. On the other hand, even when α is lower, the situation when β is higher, means that more messages within files are deleted, which makes correct matches harder. This is why our performance metrics are more sensitive to β than to α. Finally, we see that Precision and Recall are better in Figs. 3(b) and 4(b) than in Figs. 3(a) and 4(a), since the inclusion of message lengths along with sent/received patterns in Figs. 3(b) and 4(b) results in better performance than the case with only considering sent/received patterns as done in Figs. 3(a) and 4(a).

ROC Curves: In Figs. 5 and 6, we plot the ROC curves for a range of α and β values. We once again get very good ROC curves when α and β are low, and they get progressively poor when α and β increase. We also see superior ROC curves when message lengths are integrated with sent/received patterns (Fig. 6) compared to considering only sent/received patterns (Fig. 5).

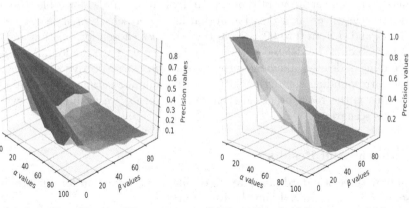

(a) Only Sent/Received Patterns

(b) Sent/Received Patterns and Message Lengths

Fig. 3. Precision plots for two classes of features, and for varying α and β values

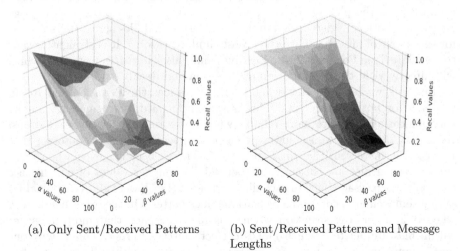

(a) Only Sent/Received Patterns

(b) Sent/Received Patterns and Message Lengths

Fig. 4. Recall plots for two classes of features, and for varying α and β values

5.3   Summary

To summarize, we believe that our technique is satisfactory despite limited users and limited datasets. The fact that our technique is sensitive to message drops is reasonable, but for relatively smaller drops, the performance is still acceptable. We believe if system deployers have a fair idea of their user profile, and some knowledge of message drop parameters, the right thresholds can be chosen for better accuracies. Furthermore, with larger scale datasets, more advanced machine learning techniques could also be developed and performance improved.

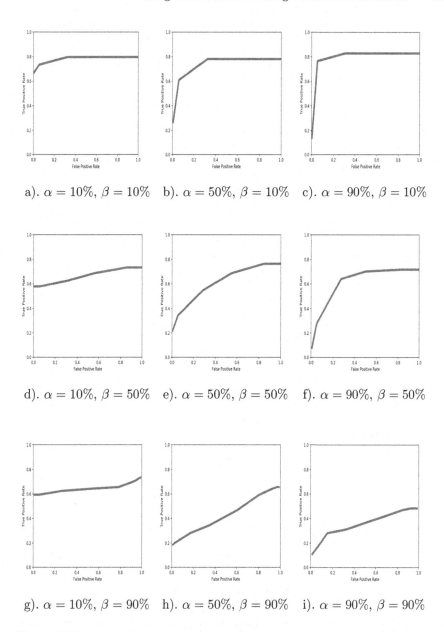

a). $\alpha = 10\%, \beta = 10\%$ b). $\alpha = 50\%, \beta = 10\%$ c). $\alpha = 90\%, \beta = 10\%$

d). $\alpha = 10\%, \beta = 50\%$ e). $\alpha = 50\%, \beta = 50\%$ f). $\alpha = 90\%, \beta = 50\%$

g). $\alpha = 10\%, \beta = 90\%$ h). $\alpha = 50\%, \beta = 90\%$ i). $\alpha = 90\%, \beta = 90\%$

Fig. 5. ROC curves with only sent/received patterns for various α and β values

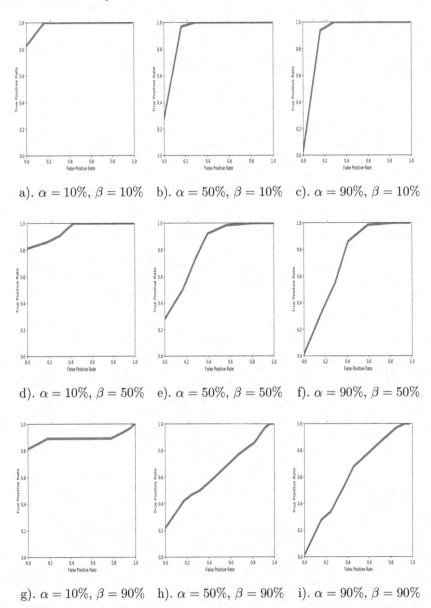

a). $\alpha = 10\%$, $\beta = 10\%$ b). $\alpha = 50\%$, $\beta = 10\%$ c). $\alpha = 90\%$, $\beta = 10\%$

d). $\alpha = 10\%$, $\beta = 50\%$ e). $\alpha = 50\%$, $\beta = 50\%$ f). $\alpha = 90\%$, $\beta = 50\%$

g). $\alpha = 10\%$, $\beta = 90\%$ h). $\alpha = 50\%$, $\beta = 90\%$ i). $\alpha = 90\%$, $\beta = 90\%$

Fig. 6. ROC curves with sent/received patterns and message lengths for various α and β values

6   Discussions on Practical Relevance of This Paper

We now present some important perspectives on the practical impact of our work in this paper. First off, with massive scale participation of citizens on social media platforms, there are billion dollar industries that focus on mining social media data for profit. Unfortunately, the policies of such companies in terms of what they process, what they share, who they sell results to are not transparent at all. However, as mentioned earlier in the paper, these trends are likely to change with increasing privacy awareness among multiple governmental agencies, civil liberties organizations, and the educated public. As such, in the near future, organizations will have to contend with people imposing limits on what companies can and cannot use about data from the public. Now, of course, different users may have different privacy expectations, and naturally, there will be a notion of adaptive privacy requirements and data sharing in such cases. The dataset we generate in this paper (and, in our opinion) is highly privacy preserving, since all user identities are anonymized, and only meta-data of content is processed (only sent/received patterns and lengths of messages).

Now a question may arise regarding the practical utility of our problem statement - which is to determine entities that communicate with each other based on processing the meta-data, and hence the overall social topology. We present perspectives below. The first application relates to cyber-abuse, especially as it pertains to young people. Currently, the process of research on cyber-abuse primarily involves adult researchers looking at content of messages and then indicating whether or not a message (or maybe an image) constitutes abuse. This is fundamentally flawed, since the perspectives of the actual victim - in this case children are not solicited. It is common knowledge that unique slang that children use, context of communications (as relates to events in schools or playgrounds), code-mixing etc. make it very hard for adults (that are absent from the social contexts of younger victims) to decipher the emotional impact of messages. This issue is actually well studied in area called digital divide [18–20].

On the other hand though, requesting content directly from victims (again, children in this case) is also problematic because of IRB regulations, and the high risk it entails when sensitive data from an already sensitive population is analyzed. This issue significantly impedes the possibility of getting robust ground truth data, which is important for research with realistic outcomes. We believe that a system like ours can mitigate these shortcomings. We are currently designing smart-phone apps where users (of any age), can willingly assent to share meta-data of their communication logs, wherein the meta-data will only be lengths of messages, sizes of files, results after performing sentiment analysis on the messages (within the device). All user identities will be anonymized. Subsequently, and if there is larger scale adoption of our system, we could use results from this paper to derive the social/network topologies of young people. In addition, if the user is also willing to mark certain messages as abusive, then the meta-data of those messages, along with the preceding and succeeding messages can be analyzed to derive signatures of abuse. If these results can be mapped back to the derived network topologies, we could perform research with significant impact to answer questions like (a) feasibility of early warning of cyber-abuse from meta-data alone; (b) identify victims and abusers in the topology, and apply graph theoretic results to understand topology evolution; (c) use metrics like graph centrality to see which nodes are more significant;

(d) model how the graph enables the dissemination of abusive content across various nodes and so much more. We strongly believe that results from this paper, coupled with the insights mentioned above have significant impact to cyber-abuse research in the near future. Naturally, the impact of this research can also extend to other domains where users are willing to share meta-data for critical applications like - modeling the efficiency of an office environment as it pertains to social communications among employees; model and analyze privacy-preserving topologies in the realm of doctor-patient or nurse-patient interactions for better dissemination of health related information etc. Designing systems for these applications, and furthering research in analyzing meta-data in these unique contexts is part of our future work.

7   Conclusions

In this paper, we presented a unique problem in the realm of privacy-preserved mining of social media logs to derive network topologies. The source of data for our study was Twitter, and included 47 users. The novelty of our study is the significant privacy accorded to the dataset during analysis, wherein all user identities were anonymized, and only message lengths of content was processed, and not the actual content. We presented a Dynamic Time Warping based algorithm for our problem, and presented interesting results on the accuracy of deriving network topology from meta-data alone. Towards, the end of the paper, we presented some practical impact of work accomplished in this paper. With increasing privacy awareness across the globe, coupled with newer privacy laws coming into effect, we believe our work in this paper is timely and relevant, and can create new societal scale and privacy-preserving big-data applications.

Acknowledgment. This work was supported in part by US National Science Foundation (Grant # 1718071). Any opinions, findings and conclusions are those of the authors alone, and do not reflect views of the funding agency.

References

1. Melis, L., Song, C., De Cristofaro, E., Shmatikov, V.: Exploiting unintended feature leakage in collaborative learning. arXiv preprint arXiv:1805.04049 (2018)
2. Hunt, T., Song, C., Shokri, R., Shmatikov, V., Witchel, E.: Chiron: privacy-preserving machine learning as a service. arXiv preprint arXiv:1803.05961 (2018)
3. Song, C., Ristenpart, T., Shmatikov, V.: Machine learning models that remember too much. In: Proceedings of the 2017 ACM SIGSAC Conference on Computer and Communications Security, pp. 587–601. ACM (2017)
4. Bost, R., Minaud, B., Ohrimenko, O.: Forward and backward private searchable encryption from constrained cryptographic primitives. In: Proceedings of the 2017 ACM SIGSAC Conference on Computer and Communications Security, pp. 1465–1482. ACM (2017)
5. Demertzis, I., Papamanthou, C.: Fast searchable encryption with tunable locality. In: Proceedings of the 2017 ACM International Conference on Management of Data, pp. 1053–1067. ACM (2017)

6. Jung, A.R.: The influence of perceived Ad relevance on social media advertising: an empirical examination of a mediating role of privacy concern. Comput. Hum. Behav. **70**, 303–309 (2017)
7. Tsay-Vogel, M., Shanahan, J., Signorielli, N.: Social media cultivating perceptions of privacy: a 5-year analysis of privacy attitudes and self-disclosure behaviors among facebook users. New Media Soc. **20**(1), 141–161 (2018)
8. Benton, A., Arora, R., Dredze, M.: Learning multiview embeddings of twitter users. In: Proceedings of the 54th Annual Meeting of the Association for Computational Linguistics (Volume 2: Short Papers), pp. 14–19 (2016)
9. Huang, S., Zhang, J., Wang, L., Hua, X.: Social friend recommendation based on multiple network correlation. IEEE Trans. Multimedia **18**(2), 287–299 (2016). https://doi.org/10.1109/TMM.2015.2510333
10. Vatsalan, D., Christen, P.: Privacy-preserving matching of similar patients. J. Biomed. Inform. **59**, 285–298 (2016). https://doi.org/10.1016/j.jbi.2015.12.004. http://www.sciencedirect.com/science/article/pii/S1532046415002841
11. Randall, S.M., Ferrante, A.M., Boyd, J.H., Bauer, J.K., Semmens, J.B.: Privacy-preserving record linkage on large real world datasets. J. Biomed. Inform. **50**, 205–212 (2014). https://doi.org/10.1016/j.jbi.2013.12.003. http://www.science direct.com/science/article/pii/S1532046413001949. Special Issue on Informatics Methods in Medical Privacy
12. Chi, Y., Hong, J., Jurek, A., Liu, W., O'Reilly, D.: Privacy preserving record linkage in the presence of missing values. Inf. Syst. **71**, 199–210 (2017). https://doi.org/10.1016/j.is.2017.07.001. http://www.sciencedirect.com/science/article/pii/S030643791630504X
13. Fulcher, B.D., Jones, N.S.: Highly comparative feature-based time-series classification. IEEE Trans. Knowl. Data Eng. **26**(12), 3026–3037 (2014)
14. SerrÃ, J., Arcos, J.L.: An empirical evaluation of similarity measures for time series classification. Knowl.-Based Syst. **67**, 305–314 (2014). https://doi.org/10.1016/j.knosys.2014.04.035. http://www.sciencedirect.com/science/article/pii/S0950705114001658
15. Bellman, R., Kalaba, R.: On adaptive control processes. IRE Trans. Autom. Control. **4**(2), 1–9 (1959)
16. Myers, C., Rabiner, L., Rosenberg, A.: Performance tradeoffs in dynamic time warping algorithms for isolated word recognition. IEEE Trans. Acoust. Speech Signal Process. **28**(6), 623–635 (1980)
17. Senin, P.: Dynamic time warping algorithm review. Inf. Comput. Sci. **855**(1–23), 40 (2008). Department University of Hawaii at Manoa Honolulu, USA
18. Chassiakos, Y.L.R., Radesky, J., Christakis, D., Moreno, M.A., Cross, C., et al.: Children and adolescents and digital media. Pediatrics **138**(5), e20162593 (2016)
19. Ballano, S., Uribe, A.C., Munté-Ramos, R.À.: Young users and the digital divide: readers, participants or creators on internet? (2014)
20. Miller, J.L., Paciga, K.A., Danby, S., Beaudoin-Ryan, L., Kaldor, T.: Looking beyond swiping and tapping: review of design and methodologies for researching young children's use of digital technologies. Cyberpsychology: J. Psychosoc. Res. Cyberspace **11**(3), 6 (2017)

Large-Scale Information Extraction from Emails with Data Constraints

Rajeev Gupta$^{(\boxtimes)}$, Ranganath Kondapally, and Siddharth Guha

Microsoft R&D, Hyderabad, India
{rajeev.gupta, rakondap, sidguha}@microsoft.com

Abstract. Email is the most frequently used web application for communication and collaboration due to its easy access, fast interactions, and convenient management. More than 60% of the email traffic constitutes business to consumer (B2C) emails (e.g., flight reservations, payment reminder, order confirmations, etc.). Most of these emails are generated by filling a *template* with user or transaction specific values from databases. In this paper we describe various algorithms related to extracting important information from these emails.

Unlike web pages, emails are personal and due to privacy and legal considerations, no other human except the receiver can view them. Thus, adapting extraction techniques used for web pages, such as HTML wrapper-based techniques, have privacy and scalability challenges. We describe end-to-end information extraction system for emails—data collection, anonymization, classification, building the information extraction models, deployment, and monitoring. To handle the privacy and scalability issues, we focus on algorithms which can work with minimum human annotated samples for building classifier and extraction techniques. Similarly, we present algorithms to minimize samples for human inspection to detect precision and recall gaps in the extraction pipeline.

Keywords: Emails · Information extraction · Machine learning · Learning by examples · Anonymization

1 Introduction

Email is the most frequently used web application. It is estimated that around 270 billion emails are sent and received per day with the total number of emails expected to reach 320 billion in 2021 [2]. More than 60% of the email traffic constitutes business to consumer (B2C) emails [1]. Most B2C emails are generated by filling an HTML template (*form*) with user or transaction specific values from databases. Typically, users only look for these details, while other static or visual content helps in beautifying, locating and explaining these. Information extracted from the emails can be presented in a more structured format to the user or can be used for further inference by computer algorithms to assist users [7]. For example, digital assistants such as Cortana, Alexa, and Siri help users with *flight reminders*. In this paper we describe an end-to-end email information extraction system and the practical considerations based on our experience

© Springer Nature Switzerland AG 2019
S. Madria et al. (Eds.): BDA 2019, LNCS 11932, pp. 124–139, 2019.
https://doi.org/10.1007/978-3-030-37188-3_8

in building such a system. Our system processes billions of emails every day extracting several billions of structured records from them.

1.1 Information Extraction from Emails

Information extraction from webpages has been a popular area of research for last several years [8]. Theoretically, one can adapt techniques used for webpages to extract information from emails. In most of these techniques one needs several human annotated examples identifying various parts or attributes of the entity/record to be extracted. However, emails contain personal information and, for legal and privacy considerations, access to emails is restricted [23]. Therefore, one must either work with limited amount of donated emails or use anonymized data. We describe various scalable algorithms and techniques using which one can develop extractors in such a data-constrained environment. Although we focus our attention on extracting information from emails, this paper is also applicable to extractions from other data sources with similar data constraints such as *consumer chat-bot* interactions.

A typical information extraction system from emails involves identifying emails from which the information should be extracted. A set of classifiers are used to perform this task. Further, since email characteristics change regularly, we need to monitor the system ensuring that it is extracting the relevant information while maintaining very high precision and high recall. We present various components of the information extraction workflow so that we continue to extract high quality information while dealing with the privacy and scalability issues. Here is the outline of the paper.

1.2 Outline of the Paper

We provide an overview of the information extraction workflow in Sect. 2. Specifically, we motivate the needs for email classification, extraction model building, and continuous monitoring. In Sect. 3, we outline the constraints under which we need to extract information from emails. Specifically, we compare the email information extraction with web information extraction. In both the cases, we need to extract information from HTML page but there are different privacy and scalability challenges in extracting information from emails. In Sect. 4, we describe algorithms to develop email classifiers in the data constraint environments. In Sect. 5, we present information extraction techniques which are adapted from the wrapper-based algorithms, i.e., techniques which use HTML DOM structure (*template*) of the email to identify and extract the information. As wrapper-based techniques have scalability challenges, we present techniques in Sect. 6 which scale better for diverse kinds of emails – their structures and contents. We need to continuously monitor the extraction pipeline to ensure the quality and quantity of extractions. But, due to privacy and scalability considerations humans cannot validate a large number of emails. In Sect. 7, we present efficient monitoring strategies towards the same. The paper concludes in Sect. 8.

Fig. 1. Building information extraction system

2 Information Extraction Workflow

Figure 1 shows different components of our information extraction system. For extracting information from any B2C document first we sample the data and anonymize it. It should be noted that these documents being private, humans cannot read raw documents of the customers. Anonymization is used to scrub all personally identifiable information (PII) [23, 25–30]. This is followed by classification, extraction model building, and monitoring, as described next.

2.1 Classify Data

We classify the documents to the scenario they belong to. This gives type of the document—whether it is *flight confirmation, hotel cancellation* or *purchase order*. The classification helps in segregating documents based on different types [6, 21] while building the extraction models (i.e., different models for different classes) whereas in the production pipeline (run-time) the classifier helps in narrowing down the extraction model to be used for the information extraction. The classification also helps decide the output schema for the extraction. We describe a number of algorithms in Sect. 4 which can be used for building classifier in a data constraint environment.

2.2 Information Extraction Model

As shown in Fig. 1, the extraction model is built using the classified data. In the process of building the model, we need to identify location of each information required from the document and extract that. There are number of ways using which the information can be extracted. Specifically, one or more of HTML DOM structure, text in the document, format of the text, etc., are used to extract information. After testing the model over a small sample of classified emails, the model is deployed in production.

Typically, emails are generated by filling the user and transaction specific details in an HTML form. By using the structure of the HTML form an extraction model can be developed for all the emails generated using the same *form* or *template*. We present such models in Sect. 5. As there are way too many templates (in one scenario we have more than 10000 templates just for English language), we need techniques which work across the templates. Such techniques are described in Sect. 6.

2.3 Deploy and Monitor

After preparing the extraction model, it is deployed to production. Deployment ensures that the run time of the whole pipeline—classification and extraction—is within a prescribed limit; so that the extracted data can be shown to the user as an *event card* along with the email (Fig. 2). The deployed model is consciously monitored to ensure that precision and recall of the extraction are maintained. Since we need to ensure a very high precision and a high recall for the extraction, we need to continuously sample the documents and see that the model is extracting all the output attributes correctly. There may be various reasons why a model may not extract all the attributes or extract them accurately. The sender of the document might change any of the signals the model uses to identify the required information (e.g., DOM structure, text, or format of the text) which may lead to classifier identifying a model to be used for extraction but the model is not extracting or extracting wrong information. If, during monitoring, it is detected that the extracted model has extracted incorrect and/or incomplete information, then we need to re-create and re-deploy the extraction model. It should be noted that all the deployed models are thoroughly tested before being deployed. Hence, there is very small probability that those models will misfire for the *vendor* or *template* for which they are created. In Sect. 7 we describe how do we handle different types of errors using monitoring with the minimal human verifications.

Fig. 2. An email and its corresponding event card

3 Data Constraints

Before describing different blocks of the information extraction workflow, in this section, we describe the additional constraints email extraction systems need to work on compared to that from webpages. The traditional approach for extracting data from web pages is to write specialized programs called *wrappers* that identify data of interest and map them to a suitable structured format. In template-based techniques, metadata such as the underlying DOM structure provides multiple presentational and structural signals that are exploited algorithmically to extract relevant information [8]. There are a number of issues we need to resolve before we can adapt those technique for emails:

1. As emails are meant for individual users, the variety it introduces is more than what we have in webpages where each viewer sees the same content of the webpage. This introduces scalability issues.

2. For developing *wrappers*, humans need to see several annotated examples for each template. Emails, being personal data, are not accessible in an open fashion the way web pages are [23]. Thus, it is not easy to get emails and create the corresponding models.

3. Web content can be uniquely identified by a URL using which one can select the model to be used for a webpage. But for emails, the same sender can use different templates for conveying the similar information. Hence, one needs to come-up with technique to construct *signatures* which can be used to identify the appropriate extraction model for an email.

4. If an extraction model is not functioning correctly for a webpage we can see the differences compared to previous version of web page (where the model was working) to identify the precision and recall issues. The same is not applicable in data constrained environment where we do not store older data.

5. One can use anonymized data for developing wrappers. But unlike the typical privileged information extraction scenarios [25], emails need to take care of a wide variety of personal data. Generating superior quality anonymized data is another challenge—one which srubs PII information in the training data but keeps the required characteristics of the input intact.

6. One needs to deal with a very large number of templates even for emails for the same scenario (e.g., hotel reservation) or the provider (e.g., Amazon). For example, in a popular scenario, we are dealing with tens of thousands of templates. Further, we need to deal with billions of emails daily which is much larger in volume compared to any web-page extraction

Next, we present classification and model building algorithms which work with the above-mentioned constraints.

4   Classification Algorithms

In this section, we describe how to build classifier while adhering to the data constraints mentioned in Sect. 3. Classification is a supervised technique, which learns a function using a set of labeled data known as *training data* and predicts the labels of *unlabeled data*. Classifiers can be developed using various machine learning techniques such as SVM, logistic regression, deep learning, etc. All these techniques need a lot of training data to predict correct labels. In some classification problems, training data is prepared using heuristics, e.g., for spam detection, activities like person viewing, time spent on the email, etc., can give the indication whether the email is a spam or not. However, there are other classification problems where the available data needs to be human labeled. Manual labeling is an expensive task, both in terms of time and money. Further, emails being private data, makes it costlier. Since classification of emails is a precursor for any information extraction task, it is important to have a scalable way of training classification models with minimal manual annotation while achieving the high quality required for production systems. Next, we describe a semi supervised labelling technique to reduce manual labelling significantly [16–18].

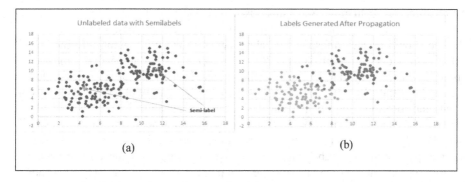

Fig. 3. (a) Unlabeled data with semi-labels (b) Propagated labels

4.1 Scalable Label Propagation

Our approach employs small amount of manually labelled data and huge unlabelled data to generate large amount of labelled data [40]. Specifically, the labelled data sample was used and its label was propagated to *similar* emails with the confidence depending on the similarity between the labelled sample and unlabelled samples [19, 20, 22]. We then segregate the high confidence data points and iterate on the low-confidence data to further label them [18]. Using the labelled data thus generated, we train a classifier. We find that within a few iterations, we get enough data so that the classifiers trained on them have high accuracy (precision and recall). Though we use this approach on the dataset consisting of anonymized B2C emails, the approach can still be used for any classification problem. We provide details of the same next.

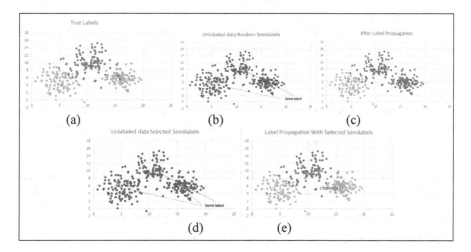

Fig. 4. (a) True labels (b) Unlabeled data with randomly selected semi-labels (c) Propagated labels with random semi-labels (d) Chosen semi labels by our techniques (e) Propagated labels with chosen semi-labels

Semi-supervise Learning. Semi-supervised machine learning algorithms [41] work on smoothness assumption, i.e., if two points $\{x_1, x_2\}$ in a *high-density region* are close, then so should be the corresponding labels $\{y_1, y_2\}$. We use a graph inference algorithm, which uses the RBF kernel [39] to project data into alternative dimensional spaces in such a way that the same labelled points in the semi-label set come closer. Figure 3 illustrate the way this technique works. In Fig. 3(a) we observe two clusters of points and corresponding labelled data. Using label propagation, we get labels of all the data points. There is one drawback in the label propagation approach which we explain with the help of Fig. 4. Suppose the true data looks like one shown in Fig. 4(a). On this sample space, if the semi-labels are selected randomly (Fig. 4(b)), the label propagation algorithm propagates wrong labels to the data point as shown in Fig. 4(c). This shows that the label propagation technique [18] is dependent on the selection of the labelled data. This technique works well when the semi-labels are chosen using some sample space heuristics. Specifically, we use DBscan [42] to recognize data points for human annotations (Fig. 4(d)) and use them as semi-label for the classification problem (Fig. 4 (e)). Our proposed approach is cost-effective as it reduces the amount of data to be annotated significantly. Specifically, in one case, we used this technique to develop a classifier for 400K emails. In general, we need 5000 human annotated emails so that the classifier gives high precision and recall (95/90). By using our algorithm, number of human labeled data for the same precision and recall reduces to 250 emails, i.e., a reduction by 20 times.

4.2   Expansion Across Languages

It is important for a global product such as Outlook to be available across languages since users are spread across continents and speak multiple languages. This requires classification of emails to be supported across multiple languages. As mentioned before, manual annotation for building classification models is expensive and having experts for different languages is even more challenging. Hence, it is desirable to have models built for one language such as English but leveraged to develop classifiers across other languages with little additional annotations [38].

For developing classifiers for any non-English language, we may need to use a translator. A typical production system must ensure that a user receives the email as quickly as possible. Hence, the production system has a very tight temporal budget (e.g., classification needs to perform in under 5 ms). The translation is performed using translation APIs which are usually hosted remotely [43]. Hence, in our production system, using language translation at runtime on hundreds of millions of emails daily is not feasible. Another restriction which comes due to tight time budget is the size of the model. If the model is very bulky, it is likely to take more time and memory resources to execute the model. Higher resource requirements lead to higher costs for executing the model on millions of emails. Hence, n-grams based logistic regression models are preferred. We considered a number of methods for generating labeled data in non-English languages which can be used to develop such models:

- **Get emails in non-English languages, translate them into English, and running English language classifier to get its label:** This approach is not likely to work in production as this requires use of translator for the incoming email leading to unacceptable delays.
- **Translating labeled English data to non-English languages:** No translation of emails in production is required for this algorithm. We translated labeled English emails into non-English emails and built a classifier using that data. Performance of this classifier was very poor compared to an English language classifier. We got a precision of 69 with a recall of 84. Though promising, with low precision we cannot use this classifier in production. Structure of different languages is different. Simple translation doesn't give us representative annotated data.
- **Using neural network to generate labeled examples:** In this scheme, we used the English labeled data to train a CNN-based classifier for English emails classification. Then we use unlabeled non-English email, translated them into English and used English language embedding to represent sequence of words in the email. We use convolution with varying filter sizes. Output of these filters is fed to max-pooling layers. Output of max-pooling layers is concatenated to create features for a fully connected layer. Finally, a soft-max layer is used to generate the classification label. After getting labeled emails in non-English language, we used them to create a logistic regression model for non-English languages. We got precision and recall of 95/90 using CNN based methods with language embeddings.

Using these studies, we show that it is possible to extend the information extraction pipeline across multiple languages without incurring the linear cost.

5   Information Extraction: Wrapper Based Models

The most popular technique of information extraction from webpages has been wrappers [9]. In databases literature, a wrapper was originally defined as a component in an information integration system to provide a single uniform query interface to access multiple information sources. In an information extraction (IE) system, a wrapper is generally a program that "wraps" an information source (e.g. a Web server, emails) such that a single desired output can be extracted from multiple sources. HTML wrappers process documents that are semi-structured and usually generated automatically by a server-side application program. They typically use the pattern mining techniques to exploit the syntactical structure or layout (template) of the documents [13–15, 35].

5.1   Handwritten Wrappers

In this technique, we collect a number of donated emails of popular scenarios (e.g., *flights confirmation*) and sender domains (e.g., *American airlines*). A set of documents having the same template or DOM structure are observed by human annotators and the information to be identified is highlighted. Common patterns for different information across documents are discovered and used for information extraction. Metadata such as underlying DOM provides multiple presentational and structural signals. There are

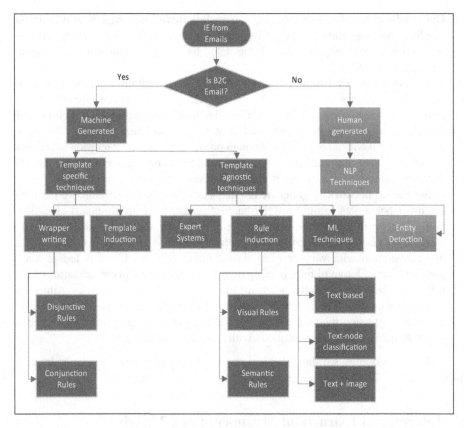

Fig. 5. Information extraction modelling

several tools available to better address the issue of wrapper generation [8]—using declarative language, HTML structure extraction, data modelling, etc. TSIMMIS was the first framework for manually building wrappers using these. Similarly, there are tools where one can provide annotated examples and the system learns the DOM structure to extract the useful information [11, 12]. Figure 5 gives an overview of email information extraction models. We use a classifier to detect whether an incoming email is a B2C email. Features for such classifier include presence of HTML tags, HTML tables, reference links, etc. Developing wrappers manually has well-known scalability issues mainly due to writing and maintaining them. In various techniques presented in Fig. 5, these scalability issues are handled differently.

5.2   Wrapper Induction

Template induction [4], the technique of generating a skeleton of repeated content based on previously seen examples, has seen substantial success for structured content such as HTML structured emails. These techniques use similarity of structures across webpages to write disjunctive and conjunctive rules for information extraction. Despite using the same templates, XPaths for various information to be extracted have some minor and major variations (e.g., advertisement, banner, different terms, payment methods, etc.). Further, providers keep changing the templates and the extractors need to adjust accordingly. Thus, instead of fixed XPaths, a sequence of features is used to extract the information. Various supervised and unsupervised techniques are available for wrapper induction.

Despite all the tooling, wrappers still require annotated data for all the templates. Instead techniques which can work with limited number of email examples are preferable. In the next section we present such techniques. As shown in Fig. 5, such techniques include expert systems, machine learning techniques, etc. These techniques typically use number of other features besides DOM structure as we explain next.

6 Information Extraction: Generic Models

While template-based extraction guarantees high precision and using automation tools even eases development of the template-based models, but it has scalability issues which are difficult to overcome. One need to develop different models for different templates, and we need to monitor those models. Some of the template cover very small fraction of emails making those template models uneconomical. In this section, we describe a number of techniques using which non-template or template agnostic models can be developed.

6.1 Different Template Agnostic Signals

Figure 5 depicts a number of template agnostic techniques. In these techniques, instead of HTML DOM tree, one uses a number of visual and semantic signals. We explain these signals with the help of Fig. 6. It shows important portions of three B2C emails. Using visual signals, we can find that "*date* appearing in front of *Check-in* is the *guest arrival date*" (Fig. 6(a)); where the "date" appearing below *Salida* is the *departure date* (Fig. 6(c)). There can be various types of visual signals:

- Presence of a word or phrase (e.g., *Llegada, Nuemro de vuela*, etc.)
- Presence of special characters and/or regular expressions (e.g., *string:string*, etc.)
- Different formats and colors of texts
- Location and closeness (e.g., address below a *hotel name* is the *hotel-address*).

Similarly, one can use semantics associated with the text and its presentation to identify the required information. Here are examples of various semantic clues which can be used:

- Entities of specified types (e.g., *dates, phone numbers, person names, zip-codes*).
- Keyword dictionaries (e.g., all the words/phrases in a dictionary having the same semantic meaning—*checking-in, checkin date, arrival*, etc.).

These signals can be used to develop different types of generic, non-template, models as described next.

Fig. 6. Different layouts for important information

6.2 Rule-Based Models

A human can go through a set of emails and using the visual and semantic signals she can write a number of rules to extract the information. In these expert systems, experts look at examples from a scenario (e.g., *flight confirmation, hotel reservation*, etc.) and specify all the steps required to extract the relevant information. These steps can be encoded in the form of *if-then-else* rules. Thus, the expert specifies all the steps, the basis of taking those steps, and how to handle exceptions. As these signals are common for a scenario across a number of sender domains, such extractors can be written for a number of templates.

But, do we really need experts to come-up with these rules? We explore this possibility using rule induction. In this technique one applies the principles outlined in [10, 11] for extracting relevant information from emails. These *learning by examples* techniques, rely on an explicit input-output examples specification from the user to automatically synthesize programs. Authors of [11], describe a *predictive* program synthesis algorithm that infers programs in a general form of extraction domain specific languages (DSLs) given input-only examples. We define DSLs to identify various non-template signals and use them to develop extraction models automatically.

6.3 Machine Learning Models

Prior work has shown how the conditional random fields (CRF) and expectation maximization (EM) can be used to extract relevant information from emails [3]. In different neural network techniques [24], one can use various features outlined in Sect. 6.1 and detect whether a particular information is there in certain email or part of the email. In these techniques an email can be represented as an image (with character recognition), running text with horizontal and vertical correlations, an array of text nodes, etc. Depending on the representation, we can use CNN, CRF, LSTM, DNN or a combination of them to extract the relevant information [31–34, 36]. It is very difficult to generalize a single ML model across a large variety of diverse templates; hence, an ensemble of models is used to generalize the technique. Authors of [2], describe such an approach. In this approach emails are clustered so that all the emails belonging to the same or similar templates are clustered together. For each of these cluster, different ML models are created to extract different information from emails.

7 Efficient Monitoring

We need different monitoring strategies for different types of extraction models. As explained in the previous sections, we can have template-based wrappers, expert systems, or ML based models. For template-based models, we need different monitoring sets for different templates. As template-based models and rule-based techniques are hand coded by humans, we deploy them after getting precision and recall of 100%. In comparison, machine learning algorithms are not expected to provide 100% precision and recall. Hence, monitoring algorithms for machine learning models are different compared to other techniques. We start by providing various sampling techniques we used to identify the samples for human verification.

7.1 Sampling for Monitoring

As mentioned previously, monitoring is done to identify issues with model building. Specifically, we need to find if the model deployed in the production pipeline is extracting the information correctly. Thus, one way is to uniformly sample the documents belonging to the sender domains for which a given model is applicable. This way of sampling is called as *influence sampling*. But, as model is deployed after thorough testing and we design the model for a very high precision and a high recall (e.g., 99/95), we require a large number of samples to catch any mistake.

Another option is to do the sampling such that we sample more from emails which have higher likelihood of mistakes. We use *outlier sampling* to identify such samples as described next.

7.2 Outlier Sampling

In this technique we sample more instances which can contribute more to the model improvement. Specifically, we identify document extractions which have higher probability of being wrong. We model various features of the extracted attributes and create a histogram of the same. Using that we identify anomalous attribute values. If any extraction leads to anomalous attribute value, we sample such emails with higher probability. There are various algorithms for outlier detection. Only the documents selected using these algorithms are sent to human annotators. We calculate the fraction of documents having the wrong extraction as the performance measure for outlier detection. As a baseline, performance measure of *influence sampling* is 0.02, i.e., whatever sample we give to human annotators only 2% of them have any issues related to extraction.

Table 1 shows the various rules we use for extractions from *flight emails* using outlier detection. First three rules use a common method for outlier detection. Specifically, they derive some numeric values from extracted attributes. A large number of these numeric values are fitted in a Gaussian distribution. We can calculate lower and upper bounds of these derived features using mean (μ) and standard deviation (σ) of the Gaussian distribution as ($\mu - 2\sigma$) and ($\mu + 2\sigma$) respectively. In the first rule we use number of characters in the *flight reservation id*. If, in any flight ticket, the number of characters in the *flight reservation id* is less than ($\mu - 2\sigma$) this rule is fired. This rule identifies incorrect extractions with an accuracy of 33%, saving human annotation efforts by more than 16 times compared to the influence sampling.

Table 1. Outlier detection rules and their performance

S. No.	Outlier detection rule	Accuracy
1	FlightReservation_Id_LessThan_MinChar	0.33
2	Duration_More_Than_Threshold	0.15
3	Duration_Less_Than_Threshold	0.02
4	Classifier_&_Status_Mismatch	0.16
5	Flight_AMPM_InstancesLesser_Than_Threshold	0.11

Second and third rules are fired when extracted values indicate that *flight duration* in the ticket is above or below certain threshold, respectively. We get much better performance from the upper threshold compared to the lower threshold probably indicating that there should not be any lower threshold on flight durations. The fourth rule follows the philosophy that if we can extract the same information from multiple methods/sources, then, any mismatch in the extraction may denote the anomaly. Specifically, using classifier we get that a document is of particular type and sometimes, that class of the document is also written in the document itself (e.g., status field of a *flight ticket* indicates that the ticket is *confirmed*, which can be obtained in the classification stage also as *flight confirmation*). The rule fires when there is mismatch in these values. This rule identifies the incorrect extraction with precision of 16% saving human annotation effort by 8 times compared to the influence sampling.

Last rule depicted in Table 1 is a statistical rule, applied over a set of documents rather than a single document. Specifically, this rule assumes that the extraction issues affect a number of documents at one go. For example, when extracting part of the day (AM/PM), the rule is fired when statistical distribution of values differs widely compared to the *expected distribution*. We have more than 20 other such rules. This work can be extended such that the system learns rules itself using correct and incorrect extractions (e.g., using a classifier with diverse kinds of features).

7.3 Discussion on Monitoring Algorithms

In practice, we use a combination of *uniform* and *outlier* sampling to ensure that we don't ignore popular sender domains while identifying all the issues with minimum human annotation cost. Specifically, we use outlier sampling with small period and influence sampling with large period to get best features of both types of sampling. Further, we can use machine learnt model for efficient monitoring as well. In this scheme we train machine learning models using output of other models, possibly, developed using template-based and rule-based techniques. These machine learning models are also deployed along with the other models. We pick documents for human validation only if output of the machine learnt model is different compared to the other deployed models. This technique is likely to reduce the monitoring requirements by orders of magnitude.

8 Conclusions and Future Work

In this paper we described various algorithms for extracting information from emails. We outlined the privacy and scalability challenges in the extraction. A number of our algorithms are targeted to reduce requirements of human annotations at various stages of the extraction pipeline. Using label propagation, we reduce the annotation cost by 20 times while developing classifiers. Similarly, using outlier sampling we reduce the number of samples we need to verify to identify mistakes in the extraction pipeline. We presented a number of template and non-template techniques for extracting information from emails. Using template agnostic techniques, we can overcome the scalability challenges. Current research areas in this field include machine learning techniques which can cover wider diversity of emails, new anonymization techniques which preserve user's privacy without compromising much on the email structure and content, entity extraction from human to human emails [5, 37], etc.

Acknowledgements. We would like to acknowledge the efforts of various members of our team for this paper. Specifically, we would like to thank Richa Bhagat for her work on reducing human involvement in the extraction pipeline. We would also like to thank Pankaj Khanzode and Chakrapani Ravi for their helpful comments.

References

1. Ailon, N., Karnin, Z.S., Liberty, E., Maarek, Y.: Threading machine generated email. In: Proceedings of the Sixth ACM International Conference on Web Search and Data Mining, WSDM 2013 (2013)
2. Sheng, Y., Tata, S., Wendt, J.B., Xie, J., Zhao, Q., Najork, M.: Anatomy of a privacy-safe large-scale information extraction system over email. In: Proceedings of the 24th ACM SIGKDD International Conference on Knowledge Discovery & Data Mining, KDD 2018 (2018)
3. Zhang, W., Ahmed, A., Yang, J., Josifovski, V., Smola, A.J.: Annotating needles in the haystack without looking: product information extraction from emails. In: Proceedings of the 21st ACM International Conference on Knowledge Discovery and Data Mining, KDD 2015 (2015)
4. Proskurnia, J., Cartright, M.-A., Garcia-Pueyo, L., Krka, I.: Template induction over unstructured email corpora. In: Proceedings of the International Conference on World Wide Web, WWW 2017, Perth, Australia (2017)

5. Hua, W., Wang, Z., Wang, H., Zheng, K., Zhou, X.: Short text understanding through lexical-semantic analysis. In: International Conference on Data Engineering, ICDE 2015 (2015)
6. Grbovic, M., Halawi, G., Karnin, Z., Maarek, Y.: How many folders do you really need? classifying email into a handful of categories. In: Proceedings of the 23rd ACM International Conference on Conference on Information and Knowledge Management, CIKM 2014 (2014)
7. Guha, R.V., Brickley, D., Macbeth, S.: Schema.org: evolution of structured data on the web. Commun. ACM 59(2), 44–51 (2016)
8. Chang, C.-H., Kayed, M., Girgis, M.R., Shaalan, K.F.: A survey of web information extraction systems. IEEE Trans. Knowl. Data Eng. 18(10), 1411–1428 (2006). TKDE 2006
9. Zheng, S., Song, R., Wen, J.-R., Giles, C.L.: Efficient record-level wrapper induction. In: Proceedings of the 18th ACM Conference on Information and Knowledge Management, CIKM 2009 (2009)
10. Polozov, O., Gulawani, S.: LaSEWeb: automating search strategies over semi-structured web data. In: Proceedings of the 20th ACM International Conference on Knowledge Discovery and Data Mining. KDD 2014 (2014)
11. Gulwani, S., Jain, P.: Programming by examples: PL meets ML. In: Asian Symposium on Programming Languages and Systems, November 2017
12. Microsoft PROSE SDK Tutorial. https://microsoft.github.io/prose/documentation/prose/tutorial/
13. Penna, G.D., Magazzeni, D., Orefice, S.: Visual extraction of information from web pages. J. Vis. Lang. Comput. 21(1), 23–32 (2010)
14. Penna, G.D., Magazzeni, D., Orefice, S.: A spatial relation-based framework to perform visual information extraction. Knowl. Inform. Syst. 30(3), 667–692 (2012)
15. Chiticariu, L., Li, Y., Reiss, F.R.: Rule-based information extraction is dead! long live rule-based information extraction systems. In: Proceedings of the Conference on Empirical Methods in Natural Language Processing, EMNLP 2013 (2013)
16. Wendt, J.B., Bendersky, M., Garcia-Pueyo, L., et al.: Hierarchical label propagation and discovery for machine generated email. In: Proceedings of the 9th ACM International Conference on Web Search and Data Mining, WSDM 2016 (2016)
17. Ratner, A., Bach, S.H., Ehrenberg, H., Fries, J., Wu, S., Re, C.: Snorkel: rapid training data creation with weak supervision. Proc. VLDB Endowment 11(3). VLDB 2017
18. Bengio, Y., Delalleau, O., Le RouxForman, N.: Label propagation and quadratic criterion, pp. 193–216. MIT Press (2006)
19. Wang, F., Tan, C., König, A.C., Li, P.: Efficient document clustering via online nonnegative matrix factorizations. In: 11th SIAM International Conference on Data Mining Society for Industrial and Applied Mathematics, 28 April 2011
20. Dhillon, I.S., Guan, Y., Kulis, B.: Kernel k-means: spectral clustering and normalized cuts. In: Proceedings of the Tenth ACM International Conference on Knowledge Discovery and Data Mining, KDD 2004 (2004)
21. Prabhu, Y., Verma, M.: FastXML: a fast accurate and stable tree-classifier for eXtreme multi-label learning. In: Proceedings of the 20th ACM International Conference on Knowledge Discovery and Data Mining, KDD 2014 (2014)
22. Wang, F., Li, P., König, A.C., Wan, M.: Improving clustering by learning a bi-stochastic data similarity matrix. Knowledge and Information Systems (KAIS), August 2011
23. Safari, B.A.: Intangible privacy rights: how europe's GDPR will set a new global standard for personal data protection. 47 Seton hall l. Rev. 809, 820–822 (2017)
24. Graepel, T., Lauter, K., Naehrig, M.: ML confidential: machine learning on encrypted data. In: International Conference on Information Security and Cryptology, ICISC 2012 (2012)

25. Bayardo, R.J., Agrawal, R.: Data privacy through optimal k-Anonymization. In: Proceedings of the International Conference on Data Engineering, ICDE 2005 (2005)
26. Inan, A., Kantarcioglu, M., Bertino, E.: Using anonymized data for classification. In: Proceedings of the International Conference on Data Engineering, ICDE 2009 (2009)
27. Benjamin, C.M., Fung, K.W., Yu, P.S.: Anonymizing classification data for privacy preservation. IEEE Trans. Knowl. Data Eng. **19**(5), 711–725 (2007). TKDE 2007
28. Brickell, J., Shmatikov, V.: The cost of privacy: destruction of data-mining utility in anonymized data publishing. In: Proceedings of the 14th ACM International Conference on Knowledge Discovery and Data Mining, KDD 2008 (2008)
29. Dwork, C.: Differential privacy: a survey of results. In: Theory and Applications of Models of Computation—TAMC, April 2008
30. Gkountouna, O., Terrovitis, M.: Anonymizing collections of tree-structured data. IEEE Trans. Knowl. Data Eng. TKDE **27**(8), 2034–2048 (2015)
31. Gogar, T., Hubacek, O., Sedivy, J.: Deep neural networks for web page information extraction. Artificial Intelligence Applications and Innovations. September 2016
32. Wojna, Z., et al.: Attention based extraction of structured information from street view imagery (2017). http://arxiv.org/abs/1704.03549
33. Raffel, C., Ellis, D.P.W.: Feed forward networks with attention can solve some long-term memory problems, June 2017. http://arxiv.org/abs/1512.08756
34. Zhu, J., Nie, Z., Zhang, B., Wen, J., et al.: 2D conditional random fields for web information extraction. In: Proceedings of the 22nd International Conference on Machine Learning, ICML 2005 (2005)
35. Zhu, J., Nie, Z., Wen, J., Zhang, B., et al.: Simultaneous record detection and attribute labeling in web data extraction. In: Proceedings of the 12th ACM SIGKDD International Conference on Knowledge Discovery and Data Mining, KDD 2006 (2006)
36. Huang, Z., Xu, W., Kai, Yu.: Bidirectional LSTM-CRF Models for Sequence Tagging. http://arxiv.org/abs/1508.01991
37. Fader, A., Soderland, S., Etzioni, O.: Identifying relations for open information extraction. In: Proceedings of the Conference on Empirical Methods in Natural Language Processing, EMNLP 2011 (2011)
38. Joty, S., Nakov, P., Màrquez, L., Jaradat, I.: Cross-language learning with adversarial neural networks: application to community question answering. In: The SIGNLL Conference on Computational Natural Language Learning; Cross-Language Adversarial Neural Network (CLANN) Model, CoNLL 2017 (2017)
39. Kuo, B.-C., Ho, H.-H., Li, C.-H., Hung, C.-C., Taur, J.-S.: A kernel-based feature selection method for SVM with RBF kernel for hyperspectral image classification. IEEE 2013, pp. 317–326
40. De Bie, T., Maia, T.T., Braga, A.P.: Machine learning with labeled and unlabeled data. In: European Symposium on Artificial Neural Networks - Advances in Computational Intelligence and Learning. Bruges (Belgium), 22–24 April 2009
41. Chapelle, O., Sholkopf, B., Zien, A. (eds.): Semi-Supervised Learning, MIT Press, London (2006)
42. Ester, M., Kriegel, H.-P., Sander, J., Xu, X.: A density-based algorithm for discovering clusters in large spatial databases with noise. In: Simoudis, E., Han, J., Fayyad, U.M. (eds.) Proceedings of the Second International Conference on Knowledge Discovery and Data Mining (KDD-96), pp. 226–231. AAAI Press (1996)
43. Microsoft Translator Text API. https://www.microsoft.com/en-us/translator/business/translator-api/

Comparative Analysis of Rule-Based, Dictionary-Based and Hybrid Stemmers for Gujarati Language

Nakul R. Dave[1]([✉]) [iD] and Mayuri A. Mehta[2] [iD]

[1] Vishwakarma Government Engineering College, Gujarat Technological University,
Ahmedabad, Gujarat, India
davenakulr@gmail.com
[2] Sarvajanik College of Engineering and Technology, Surat, Gujarat, India
mayuri.mehta@scet.ac.in

Abstract. Gujarati is an Indo-Aryan language spoken substantially by people of Gujarat state of India. It is highly and actively used for communication in Gujarat government's educational institutes and offices, local industries, businesses as well as in media such as newspapers, magazines, radio and television programs. In all these areas, Internet is the keen requirement today. Its utilization will be increased if contents are provided on web in regional language using the notion of Natural Language Processing (NLP). In NLP, stemming plays a vital role in retrieving accurate contents and producing effective results for web search query. It identifies the root word from morphological variants of respective word. There are three typical approaches to perform stemming: rule-based approach, dictionary-based approach and hybrid approach. In this paper, we present a comparative empirical study of these three approaches for Gujarati language. The aim of the study is to evaluate the effectiveness of different types of stemmers for Gujarati language. Firstly, we discuss the rule-based algorithm and present its evaluation with 152 different suffix stripping rules. Next, we illustrate stemming mechanism developed using Gujarati dictionary that contains around 20000 root words. Lastly, we discuss the hybrid approach that is a combination of rule-based and dictionary-based approaches. Experimental results reveal that hybrid approach retrieves more accurate stemmed words compared to rule-based and dictionary-based approaches.

Keywords: Gujarati · Indian language · Natural Language Processing · Rule-based stemmer · Dictionary-based stemmer · Hybrid stemmer

1 Introduction

Gujarati language is morphologically very rich language. It is widely spoken by more than 55 million people around the world. Due to globalization, people are

© Springer Nature Switzerland AG 2019
S. Madria et al. (Eds.): BDA 2019, LNCS 11932, pp. 140–155, 2019.
https://doi.org/10.1007/978-3-030-37188-3_9

connecting and communicating with each other through the internet. Typically, they are using the web content for knowledge sharing and gaining, reading news and articles, watching videos and listening audios and for their business marketing and promotion. It would be more beneficial, relevant and convenient for the users, if online contents are available in their respective regional language. The amount of content being pushed on the internet is growing exponentially and Gujarati content is now no longer behind as compared to dominant languages such as English and Chinese. Since last decade, numerous Gujarati literature, videos, documents, news and articles have been published online. Even digital copy of few ancient Gujarati resources are also available on the internet. Therefore, to leverage the maximum benefit from these resources, search engine's query result should be optimized and should be relevant to the query. However, web search results for Gujarati scripts are very poor. The results can be improved with the efficient utilization of natural language processing concepts. Specifically, stemming plays a vital role in NLP and is extremely useful to produce relevant results for the keyword based Gujarati search query.

Stemming is useful in information retrieval (IR) system to reduce the inflectional and derivational words to their root form. It is capable to convert various grammatical forms of word such as noun, adjective, verb, adverb and pronoun to its base form [1]. The correctness of the resultant root form decides how accurate the stemming algorithm is! There are three key approaches to evaluate and develop stemming algorithm: rule-based approach, Dictionary-based approach and hybrid approach. Rule-based stemmer uses suffix stripping, prefix stripping and/or substitution techniques. It reduces the total number of terms in IR system and thereby reduces the size and complexity of data in the system which is always advantageous [2,3]. Further, the rule-based stemmer can be either inflectional stemmer or derivational stemmer. In Dictionary-based approach, language based dictionary is created and is periodically updated. Subsequently, it is used to reduce the actual word into its corresponding root word. Hybrid approach is the combination of rule-based and Dictionary-based approaches. It gives more advance and accurate stemmed results.

Numerous stemmers have been proposed in literature for various vernacular languages. In 1968, Lovins proposed a rule-based stemmer for English language which includes two major principals to derive the root word: iteration and longest match [2]. The rule-based suffix stripping technique that uses knowledge as well as combination of vowels and consonants to derive the root word was introduced in [3]. Context free Nepali stemmer using string similarity and hybrid approach has been presented in [4]. In [5], Arabic light stemmer that enhances the effect of monolingual IR using extended suffixes was proposed. Stemmers for Indian languages are also available in literature. Rule-based Hindi lemmatizer [6] and Hindi stemmer [7] have been designed to find out precise lemma from input word. Subsiquently, additional rules were integrated in lemmatizer and stemmer to extract more precise word from lemma word. Rule-based stemmers for Bengali [8,9], Tamil [10] and Urdu [11] languages are also available. An unsupervised rule-based stemmer for Marathi language was proposed in [9].

Apart from Hindi, Bengali, Tamil, Urdu and Marathi, stemmers have also been created for Gujarati language in large number. Rule-based Gujarati stemmers using suffix stripping, prefix stripping and/or substitution techniques are available in [12–15]. Dictionary-based Gujarati stemmer has been introduced in [16]. Some rule-based and take-all-splits inflectional stemmers as well as derivational stemmers have been presented in [17–19]. The paradigm based stemmer, knowledge based stemmer and statistical stemmer have been discussed in [15,20]. Gujarati stemmer to deal with errors such as overstemming and understemming has been presented in [21]. Stemmers including calculation of stemming error rate for the Gujarati text have been illustrated in [15,19,22]. Part-of-Speech (POS) method based Gujarati stemmers have been introduced in [12,20,22–28]. They work on noun, pronoun, adjective and verb based paradigm technique.

Though numerous Gujarati stemmers are available in literature, a less amount of work is carried out to compare the different types of Gujarati stemmers to know their effectiveness. Hence, in this paper, we present the theoretical as well as experimental study of rule-based, dictionary-based and hybrid Gujarati stemmers. Such study discloses strengths and weaknesses of different types of stemmers and open research problems in the area.

The rest of the paper is organized as follows: In Sect. 2, we discuss necessary background of Gujarati stemmers. Rule-based Gujarati stemmer, dictionary-based Gujarati stemmer and hybrid Gujarati stemmer are presented in Sects. 3, 4 and 5 respectively alongwith their comparative evaluation based on identified parameters. There experimental evaluation considering dataset from Hindi-Gujarati parallel chunked text corpus ILCI-II and considering application domains entertainment, tourism, health and agriculture is presented in Sect. 6. Finally conclusion is specified in Sect. 7.

2   Background of Gujarati Stemmers

Gujarati stemmer falls in one of two categories: inflectional stemmer and derivational stemmer [17,18]. Inflectional stemmer processes the words by producing several inflections without changing the word class [29]. It considers word variants related to language specific syntactic deviations like plural, gender, case and so on [6]. For example, if the word is "ગુજરાત", it is converted into its various inflectional forms such as "ગુજરાતી, ગુજરાતીઓ, ગુજરાતનું and ગુજરાતમાથી". In derivational stemmer, the word variants related to POS of a sentence where the word occurs are considered [29]. For example, "singer" is the derived word of "to sing" and the "ગાયક" is the derived word of "ગાવું". Inflectional and derivational stemmers generally uses rule-based approach, dictionary-based approach and hybrid approach.

By applying specific rules from the rule set, rule-based stemmer removes the inflectional or derived part from the word and generates the root word. To find the applicability of a specific rule on the input word, the rule-based stemmer uses different techniques such as suffix stripping, prefix stripping and substitution. Usually, suffix is attached after root word, whereas prefix is attached before it.

Suffix stripping is the process to chop of the suffixes from the word and derive the root word. Prefix stripping is used in the same way to derive the base word by removing unwanted prefixes from the test word. Substitution is applied in case of inefficient information after applying suffix stripping or prefix stripping.

Dictionary-based approach uses dictionary (database) of Gujarati language. It searches input word in the dictionary to find out actual stem word. It works efficiently if the dictionary in specific domain is available and designed adequately. However, the dictionary of word-stem pairs is not readily available for many languages and hence, it is difficult to produce the stem words [30].

Hybrid approach combines the functioning of rule-based and dictionary-based approaches. Hence, it is capable of handling both known words as well unknown words by deriving the accurate stem words from its inflectional variant or derivational variant [17, 18].

Some Gujarati stemmers also deal with stemming errors such as overstemming and understemming [15, 19, 21, 22]. In overstemming, two separate inflected words are stemmed to the same root word which reduces the precision of IR [30]. Overstemming also occurs when two or more words which are not morphological variants conflate to a single stem word. For example, "વહેતું" (flowing) and "વહેમ" (superstition) may result to the same stem word "વહે" (to flow); however they are not morphological variants. This phenomenon is also known as false positive. Under-stemming refers to words that are supposed to be grouped together by stemming; however that are not grouped together. It occurs when words that are indeed morphological variants are not conflated. Such errors tend to decrease the recall value in IR search [30] and thereby produce false negative values. An example of understemming in English language as follows: "compile" is stemmed to "comp" and "compiling" is stemmed to "compil". Example in Gujarati language is as follows: "સાબરકાંઠાની" is stemmed to "સાબરકાંઠાન" and "સાબરકાંઠાનું" is stemmed into "સાબરકાંઠા".

3   Rule-Based Gujarati Stemmer

In this section, first we describe the various rule-based Gujarati stemmers. Then we present parametric evaluation of rule-based stemmer based on identified parameters. Subsequently, we present a rule-based Gujarati stemmer.

Affix removal technique that uses suffix removal and prefix removal methods to generate the stem word from Gujarati text has been presented in [15]. It includes dictionary lookup and rule-based approach for inflectional words and also handles stemming errors. GUJSTER: a rule-based stemmer that uses 144 suffix rules, 14 prefix rules and 13 various substitution rules has been proposed in [19]. These rules are applied on 5116 words of EMILLE corpus by removing 206 stop words from the dataset and has achieved higher accuracy using online Gujarati dictionary. Rule-based lightweight stemmer has been described in [23]. Authors have used 167 suffix rules, executed them on 3000 words of text data and have achieved higher accuracy. This stemmer focuses on morphology of Gujarati language such as noun formation, verb formation, adverb formation and adjective formation of Gujarati contents. Improvised lightweight stemmer using linguistic resources has been presented in [16]. It uses 15 prefixes rules, 152 suffixes rules, 95 stop words and 736 dictionary words as root words. It has been tested on 1099 derived words from divyabhaskar e-newspaper. DHIYA stemmer [22] that uses EMILLE corpus [31] deals with morphological features of Gujarati language. For example, છોકરી , છોકરીઓ , વેલી, વેલીઓ belong to noun and લુગડું, લુગડાં represents verb inflection. It is implemented considering 52 stemming rules. It deals with stemming errors. It has been utilized to build Saaraansh: Gujarati text summarization system [24]. In [14], authors have described the naive approach for Gujarati stemmer and presented the study on various components of stemmers such as 1. GUJ Stem-suffix Frequency Table (GSFT), 2. Gujarati Language Rule Set (GLRS) and 3. GUJ_Strip function.

Preprocessing steps for Gujarati text summarization have been discussed in [32]. In [21], authors have shown the importance of morphology, stemming learning methods, stemming algorithms and stemming errors for Gujarati language. In [13], morphological analyzers for South Asian languages have been presented. They have been tested with Hindi and Gujarati languages creating dataset of 2000 words. They have shown the most frequent suffix list for the Gujarati language with their base grammatical category. Morphological analysis tool based on knowledge oriented approach and statistical approach has been illustrated in [20]. It has been evaluated considering 81000 words and considering noun, pronoun, verb and adjective morphology. In [33], authors have discussed basics of Gujarati language morphology and have designed the rule sets for noun, pronoun, verb and adjective morphology. Moreover, they have built lexical Gujarati dictionary using Gujarati grammar book [34] and Gujarati dictionary from Gujarati Lexicon [37]. Conditional Random Fields (CRF) model has been prepared and analyzed with 26 different POS tags for Gujarati language [27]. Authors have created two different models: tagged data of 600 sentences and untagged data of 5000 sentences. 5000 test words on the 10000 training words were tested. Moreover, they have presented error analysis with respect to different tags.

Based on our study on various rule-based Gujarati stemmers, we have identified the following parameters to evaluate them: dataset used, number of words considered for experiments, number of rules applied, stemming approach, Is POS tagging used?, stemming method, accuracy and stemming errors. Table 1 shows the comparative analysis of existing Gujarati stemmers.

It has been observed from Table 1 that the majority of stemmers are rule-based and use inflectional stemming method. The bulk of existing stemmers have used EMILLE dataset for experiments and some of them have measured stemming errors. To analyze the performance of rule-based, dictionary-based and hybrid stemmers, we have created Gujarati dictionary that consists of around 20000 words and 152 suffix stripping rules that vary in size from length 1 to length 8. Few examples of these rules are listed in Fig. 1.

Table 1. Comparative analysis of Gujarati stemmers

Stemmer	Dataset used	Number of words Considered for experiments	Number of Rules applied	Stemming Approach	Is POS tagging used?	Stemming Method	Accuracy	Stemming Errors
Affix Removal Stemmer [15]	EMILLE	6530	156 suffix rules, 57 substitution rules and 14 prefix rules	Rule based and Dictionary based	No	Inflectional	High	Overstemming errors and Understemming errors
Lightweight Stemmer [23]	Own Dataset	3000	167 suffix rules	Rule Based	Yes	Inflectional	High	Overstemming errors and Understemming errors
Improved Lightweight Stemmer [16]	e-newspaper (Divyabhaskar)	1099	152 suffix rules and 15 prefix rules	Rule based	No	Inflectional	-	-
DHIYA stemmer [22]	EMILLE	3935	-	Rule based	No	Inflectional	High	Overstemming errors and Understemming errors
GUJSTER [19]	EMILLE	5116	144 suffix rules, 14 prefix rules and 13 substitution rules	Rule based	Yes	Inflectional	High	Overstemming errors and Understemming errors
Morphological Analyzer [20]	Own dataset	81000	-	Hybrid	Yes	Inflectional	Low in Rule based, High in Knowledge based and High Statistical	-
Hybrid Stemmer [18]	EMILLE	8525649	59 handcrafted suffix rules	Hybrid	No	Inflectional	Low	-
Hybrid Inflectional Stemmer and Rule-based Derivational Stemmer [17]	EMILLE	-	-	Rule based and Hybrid	Yes	Inflectional and Derivational	High in Inflectional and Low in Derivational	-
Saaraansh [24]	EMILLE	78260	-	Hybrid	No	Inflectional and Derivational	-	-

Suffix of size 8: {'વાળાઓનું','વડાવવું}   Suffix of size 7: {'નારાઓની', 'વાલીનું"}

Suffix of size 6: {'ોમાંથી', 'વેરાને'}   Suffix of size 5: {'માંથી', 'વાનું' }

Suffix of size 4: {'ાઓથી','ીમાં'}   Suffix of size 3: {'ઓથી','ેની"}

Suffix of size 2: {'ના', 'થી"}   Suffix of size 1: {'ી','ો}

Fig. 1. Suffix stripping rules

In Table 2, we describe the notations that are used throughout the paper. In Fig. 2, we provide the pseudo code for rule-based Gujarati stemmer.

As shown in the Fig. 2, the rule-based stemmer consists of major three steps: token generator, stop word removal and suffix stripping. First tokens are generated from given document using lexical analyzer. Then, generated tokens are matched with the stop words. If match is found, then corresponding tokens is removed. Subsequently, the remaining tokens are passed on to the rule-based suffix stripping approach. Next, most appropriate rule is applied to the token to generate the actual stem word.

Table 2. List of notations

Notation	Description
D	Text document in Gujarati language
n	Total number of tokens
T = {T1, T2, ..., Tn}	Set of tokens generated from D
m	Total number of stop words
S = {S1, S2, ..., Sm}	Set of stop words
z	Total number of tokens left after removal of stop words
L = {L1, L2, ..., Lz}	Set of tokens left after stop word removal process
M = {M1, M2, ..., M8}	Set of suffix stripping rules in order of rule length, where M1 represents rule of maximum length and M8 represents rule of minimum length
r	Total number of root words
R = {R1, R2, ..., Rr}	Dictionary of root words
q	Total number of stem words
SW = {SW1, SW2, ..., SWq}	Set of stem words
s	Total number of non-stem words
NSW = {NSW1, NSW2, ..., NSWs}	Set of non-stem words
p	Total number of words left after dictionary lookup
N = {N1, N2, ..., Np}	Set of words left after dictionary lookup

4   Dictionary-Based Gujarati Stemmer

Numerous dictionary-based Gujarati stemmers are available. Dictionary approach deals with sound and efficient creation of dataset of Gujarati root words in the form of text, word document or excel document. Offline dictionary approach has been used in [15], whereas online dictionary approach has been used in [19]. In [17,18], dictionary approach has been used to implement inflectional and derivational hybrid stemmer.

Gujarati wordnet deals with lexical contents of Gujarati language. It assemblies Gujarati words into sets of synonyms called synsets, provides short explanation and their usage examples [35]. Further, contributors have claimed that resources like 'Bhagavad-Go-Mandal' [36] and 'Gujarati Lexicon' [37] were found to be very useful in synset development process [38]. Dictionary-based approach will be more helpful if existing Gujarati wordnets are accessible through API call. We have designed the dictionary in simple text form. It contains only root words in their base form and does not include POS approach. The procedure to deal with dictionary-based approach is discussed in Fig. 3.

```
Input: Text document D in Gujarati language
Output: Stem words as root words, non-stem words
BEGIN
    1.  Generate tokens from D using lexical analyser and store them in T
    2.  Create a list S of stop words
    3.  z = 0
    4.  FOR i = T₁ to Tₙ
    5.    FOR j = S₁ to Sₘ
    6.      IF Tᵢ = Sⱼ
    7.        Remove Tᵢ
    8.      ELSE
    9.        L[z] = Tᵢ
   10.        z = z+1
   11. q = 0, s = 0
   12. FOR i = L₁ to L_z
   13.    Apply M₁ to M₈ (in order of length – higher to lower) to Lᵢ to extract the stem word
   14.    IF rule is applicable to Lᵢ
   15.      SW[q] = stem word derived from Lᵢ
   16.      q = q+1
   17.    ELSE
   18.      NSW[s] = Lᵢ
   19.      s = s+1
   20. Display sets SW and NSW
END
```

Fig. 2. Pseudo code for rule-based Gujarati stemmer

As it can be observed from Fig. 3, the first 5 steps of Dictionary-based approach are similar to that of rule-based stemmer presented in Fig. 2. The only difference between these two stemmers is that in dictionary-based stemmers, tokens are directly compared with dictionary and declare the resulting token as a stem word or non-stem word.

5 Hybrid Gujarati Stemmer

Hybrid stemmer may be a combination of dictionary-based approach and rule-based approach or a combination of inflectional stemmer and derivational stemmer. A lightweight hybrid inflectional stemmer and heavyweight derivational stemmer have been discussed in [17]. Hybrid stemmer for Gujarati language including handcrafted rules and excluding handcrafted rules has been proposed in [18]. It uses Goldsmith's [39] take-all-splits method to identify suitable suffixes. It has been evaluated considering 8,525,649 words retrieved from EMILLE corpus [31]. Later, this method was extended by authors to create hybrid inflectional stemmer and rule-based derivational stemmer using POS approach [17]. Authors have also used take-all-splits methods that works for example, the suffix ની from રજની should not removed but ની from રજનિની must be removed. In Fig. 4, we present the pseudo code of hybrid stemmer.

Input: Text document D in Gujarati language
Output: Stem words as root words, non-stem words
BEGIN
    1.   Generate tokens from D using lexical analyser and store them in T
    2.   Create a list S of stop words
    3. $z = 0$
    4.   **FOR** $i = T_1$ to T_n
    5.   **FOR** $j = S_1$ to S_m
    6.   **IF** $T_i = S_j$
    7.   Remove T_i
    8.   **ELSE**
    9. $L[z] = T_i$
    10. $z = z+1$
    11. $q = 0, s = 0$
    12. **FOR** $i = L_1$ to L_2
    13.   **FOR** $j = R_1$ to R_r
    14.   **IF** $L_i = R_j$
    15. $SW[q]$ = stem word derived from L_i
    16. $q = q+1$
    17.   **ELSE**
    18. $NSW[s] = L_i$
    19. $s = s+1$
    20. Display sets SW and NSW
END

Fig. 3. Pseudo code for dictionary-based Gujarati stemmer

As shown in the Fig. 4, the hybrid stemmer consists of the major four steps: token generator, stop word removal, dictionary search and suffix stripping. The combination of dictionary search and suffix stripping help in producing adequate results.

6   Results and Discussion

In this section, we present the comparative experimental analysis of above discussed rule-based, dictionary-based and hybrid stemmers. The experiments are conducted considering the dataset from Ministry of Electronics and Information Technology (MeitY), Government of India's "Language Technology Proliferation and Development Centre"[1]. This dataset belongs to Hindi-Gujarati parallel chunked text corpus ILCI-II and is available for four different domains: entertainment, tourism, health and agriculture. We have tested performance of all three stemmers for each domain and the performance of is compared in terms of accuracy. Accuracy is defined as the correctly predicted stemmed words.

---

[1] https://www.tdil-dc.in/index.php?option=com_download&task=fsearch&lang=en.

Input: Text document D in Gujarati language
Output: Stem words as root words, non-stem words
BEGIN
    1.   Generate tokens from D using lexical analyser and store them in T
    2.   Create a list S of stop words
    3. $z = 0$
    4.   **FOR** $i = T_1$ to T_n
    5.   **FOR** $j = S_1$ to S_m
    6.   **IF** $T_i = S_j$
    7.   Remove T_i
    8.   **ELSE**
    9. $L[z] = T_i$
  10. $z = z+1$
  11. $q = 0, s = 0, p=0$
  12. **FOR** $i = L_1$ to L_z
  13.   **FOR** $j = R_1$ to R_r
  14.   **IF** $L_i = R_j$
  15. $SW[q]$ = stem word derived from L_i
  16. $q = q+1$
  17.   **ELSE**
  18. $N[p] = L_i$
  19. $p = p+1$
  20. **FOR** $i = N_1$ to N_p
  21.   Apply M_1 to M_8 (in order of length – higher to lower) to N_i to extract the stem word
  22.   **IF** rule is applicable to N_i
  23. $SW[q]$ = stem word derived from N_i
  24. $q = q+1$
  25.   **ELSE**
  26. $NSW[s] = N_i$
  27. $s = s+1$
  28. Display sets SW and NSW
END

Fig. 4. Pseudo code for hybrid Gujarati stemmer

6.1 Performance Evaluation Considering Entertainment Dataset

Entertainment dataset contains total 13013 words. Applying duplicate removal process on these 13013 words, we have identified 4916 unique words. We depict the performance of above discussed stemmers for 4916 words in Fig. 5. The accuracy provided by rule-based stemmer, dictionary-based stemmer and hybrid stemmer is 39.25%, 45.15% and 81.12% respectively. As it can be observed, the accuracy provided by rule-based stemmer is less and the accuracy provided by hybrid stemmer is higher amongst three stemmers.

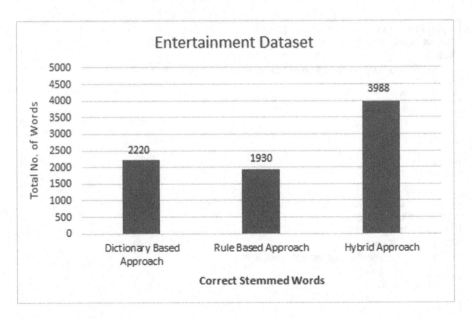

Fig. 5. Performance of rule-based, dictionary-based and hybrid Gujarati stemmers considering entertainment dataset

6.2   Performance Evaluation Considering Health Dataset

Health dataset consists of total 12815 words. We have identified 4201 unique words out of 12815 words. Stemming process is applied on unique words. The accuracy achieved for dictionary-based and rule-based stemmers is 46.13% and 38.99% respectively. However, the hybrid stemmer has emerged out as an efficient stemmer with 83.14% accuracy. Figure 6 depicts the results produced considering health dataset.

6.3   Performance Evaluation Considering Agriculture Dataset

Agriculture dataset comprises of total 12373 words, out of which total 3789 are unique words. As with previous two datasets, hybrid stemmer outperforms the rule-based and dictionary-based stemmers for agriculture dataset too. The results of three stemmers are depicted in Fig. 7. The accuracy provided by dictionary-based stemmer, rule-based stemmer and hybrid stemmer is 43.83%, 40.98% and 82.18% respectively.

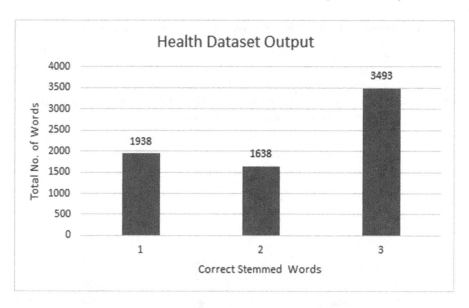

Fig. 6. Performance of rule-based, dictionary-based and hybrid Gujarati stemmers considering health dataset

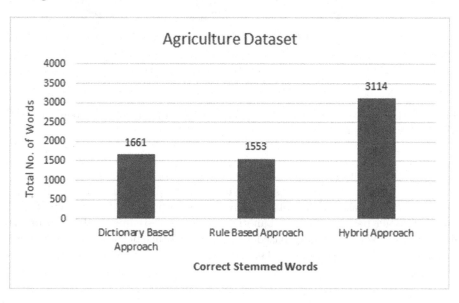

Fig. 7. Performance of rule-based, dictionary-based and hybrid Gujarati stemmers considering agriculture dataset

6.4   Performance Evaluation Considering Tourism Dataset

In tourism dataset, we have identified 4122 unique words out of total 10578 words. The results of three stemmers are depicted in Fig. 8. The accuracy provided by dictionary-based stemmer, rule-based stemmer and hybrid stemmer is 38.98%, 68.72% and 74.91% respectively. As with previous experiments, hybrid stemmer gives higher accuracy amongst all stemmers.

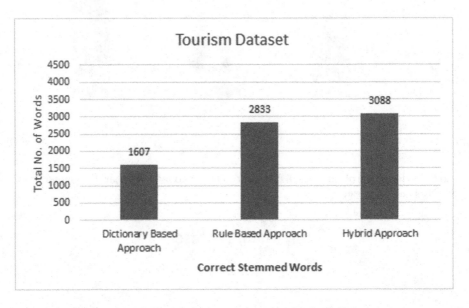

Fig. 8. Performance of rule-based, dictionary-based and hybrid Gujarati stemmers considering tourism dataset

Figure 9 presents results of all three stemmers for all four datasets. Results show that for all three datasets, hybrid stemmer outperforms the dictionary-based and rule-based stemmers and produces average 80% accuracy. Rule-based stemmer performs better than dictionary-based stemmer only for tourism dataset. However, dictionary-based stemmer performs better than rule-based stemmer for all three remaining datasets.

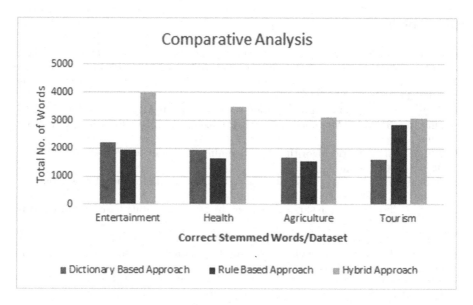

Fig. 9. Comparative Performance Analysis

7   Conclusion

In this paper, firstly we illustrated stemming methods available for regional as well as foreign languages. Subsequently, we discussed different types of stemmers available for Gujarati language. Secondly, we have designed and presented rule-based, dictionary-based and hybrid stemmers for Gujarati language. Thirdly, we depicted empirical analysis of these stemmers using Gujarati-Hindi parallel corpus from Language Technology Proliferation and Development Center, MeitY, Government of India. This corpus includes four different datasets: entertainment, health, agriculture and tourism. The performance of three Gujarati stemmers have been tested for all four datasets. It has been observed that hybrid stemmer outperforms the rule-based and dictionary-based stemmers for all four datasets providing 80% accuracy on an average. Moreover, we observed that the words in datasets are highly domain specific and hence, dictionary-based stemmer produces comparatively less accuracy. This work can further be extended by adding more domain specific root words in dictionary as well as by developing more number of stemming rules to achieve better accuracy for dictionary-based and rule-based stemmers respectively. A little amount of work has been done for derivational Gujarati stemmer that can be focused in future research work.

References

1. Bijal, D., Sanket, S.: Overview of stemming algorithms for indian and Non-Indian languages. Int. J. Comput. Sci. Inf. Technol. **5**(2), 1144–1146 (2014)
2. Lovins, J.B.: Development of a stemming algorithm. Mech. Transl. Comput. Linguist. **11**(1–2), 22–31 (1968)
3. Porter, M.F.: Readings in information retrieval. In: An Algorithm for Suffix Stripping, pp. 313–316. Morgan Kaufmann Publishers Inc., San Francisco, CA, USA (1997)
4. Sitaula, C.: A hybrid algorithm for stemming of Nepali text. Intell. Inf. Manag. **05**(04), 136–139 (2013)
5. Atwan, J., Wedyan, M., Al-Zoubi, H.: Arabic text light stemmer. Int. J. Comput. Acad. Res. **8**(2), 17–23 (2019)
6. Paul, S., Tandon, M., Joshi, N., Mathur, I.: Design of a rule based Hindi lemmatizer, pp. 67–74 (2013)
7. Ramanathan, A., Rao, D.D.: A lightweight stemmer for Hindi. In: Proceedings of the EACL 2003 Workshop on Computational Linguistics for South Asian Languages, pp. 43–48 (2003)
8. Mahmud, M.R., Afrin, M., Razzaque, M.A., Miller, E., Iwashige, J.: A rule based Bengali stemmer. In: Proceedings of the 2014 International Conference on Advances in Computing, Communications and Informatics, ICACCI 2014, pp. 2750–2756 (2014)
9. Sarkar, S., Bandyopadhyay, S.: Design of a rule-based stemmer for natural language text in Bengali. In: Proceedings of the IJCNLP-2008 Workshop on NLP for Less Privileged Languages, January, pp. 65–72 (2008)
10. Thangarasu, M., Manavalan, R.: Stemmers for Tamil language : performance analysis. Int. J. Comput. Sci. Eng. Technol. **4**(07), 902–908 (2013)
11. Kansal, R., Goyal, V., Lehal, G.S.: Rule based Urdu stemmer. In: Proceedings of COLING 2012: Demonstration Papers, December 2012, pp. 267–276 (2012)
12. Chauhan, K., Patel, R., Joshi, H.: Towards improvement in Gujarati text information retrieval by using effective Gujarati stemmer. J. Inf. Knowl. Res. Comput. Eng. **2**(2), 499–599 (2013)
13. Aswani, N., Gaizauskas, R.J. Developing morphological analysers for South Asian languages: experimenting with the Hindi and Gujarati languages. In: LREC, pp. 811–815 (2010)
14. Sheth, J., Patel, B.: Stemming techniques and naïve approach for Gujarati stemmer. Int. J. Comput. Appl. (IJCA) **2**, 9–11 (2012)
15. Desai, N., Dalwadi, B.: An affix removal stemmer for Gujarati text. In: 2016 3rd International Conference on Computing for Sustainable Global Development (INDIACom), pp. 2296–2299. IEEE (2016)
16. Patel, C.D., Patel, J.M.: Improving a lightweight stemmer for Gujarati. Int. J. Inf. **6**(1/2), 135–142 (2016)
17. Suba, K., Jiandani, D., Bhattacharyya, P.: Hybrid inflectional stemmer and rule-based derivational stemmer for Gujarati. In: Proceedings of the 2nd Workshop on South Southeast Asian Natural Language Processing (WSSANLP), pp. 1–8 (2011)
18. Patel, P., Popat, K., Bhattacharyya, P.: Hybrid stemmer for Gujarati. In: Proceedings of the 1st Workshop on South and Southeast Asian Natural Language Processing, pp. 51–55 (2010)
19. Patel, C.D., Patel, J.M.: Gujster: a rule based stemmer using dictionary approach. In: 2017 International Conference on Inventive Communication and Computational Technologies (ICICCT), pp. 496–499. IEEE (2017)

20. Baxi, J., Patel, P., Bhatt, B.: Morphological analyzer for Gujarati using paradigm based approach with knowledge based and statistical methods. In: Proceedings of the 12th International Conference on Natural Language Processing, pp. 178–182 (2015)
21. Boradia, A.: A study of different methods & techniques for stemming in Gujarati text mining. Int. J. Manag. Technol. Eng. **8**(2178), 2178–2187 (2018)
22. Sheth, J., Patel, B.: Dhiya: a stemmer for morphological level analysis of Gujarati language. In: 2014 International Conference on Issues and Challenges in Intelligent Computing Techniques (ICICT), pp. 151–154. IEEE (2014)
23. Ameta, J., Joshi, N., Mathur, I.: A lightweight stemmer for Gujarati. In: Proceedings of 46th Annual National Convention of Computer Society of India, pp. 1–4 (2011)
24. Sheth, J.: Saaraansh : Gujarati text summarization system. Int. J. Comput. Sci. Inf. Technol. Secur. (IJCSITS) **7**(3), 46–53 (2017)
25. Tikarya, A.B., Mayur, K., Patel, P.H.: Pre-processing phase of text summarization based on Gujarati Language. Int. J. Innovative Res. Comput. Sci. Technol. (IJIRCST) **2**(4), 1–5 (2014)
26. Hal, S.Y., Virani, S.H.: Improve accuracy of parts of speech tagger for Gujarati language. Int. J. Adv. Eng. Res. Dev. **2**(5), 187–192 (2015)
27. Patel, C., Gali, K.: Part-of-speech tagging for Gujarati using conditional random fields. In: Proceedings of the IJCNLP-2008 Workshop on NLP for Less Privileged Languages, January, pp. 117–122 (2008)
28. Ameta, J., Joshi, N., Mathur, I.: Improving the quality of Gujarati-Hindi machine translation through Part-Of-Speech tagging and stemmer assisted transliteration. Int. J. Nat. Lang. Comput. (IJNLC) **2**(3), 49–54 (2013)
29. Jivani, A.G., et al.: A comparative study of stemming algorithms. Int. J. Comput. Technol. Appl. **2**(6), 1930–1938 (2011)
30. Patil, H.B., Pawar, B.V., Patil, A.S.: A comprehensive analysis of stemmers available for Indic languages. Int. J. Nat. Lang. Comput. **5**(1), 45–55 (2016)
31. Distribution. http://www.emille.lancs.ac.uk/distribution.php
32. Patel, M., Balani, P.: Clustering algorithm for Gujarati language (2013)
33. Kapadia, U., Desai, A.: Morphological rule set and lexicon of Gujarati grammar : a linguistics approach. VNSGU J. Sci. Technol. **4**(1), 127–133 (2015)
34. Kothari, A.: Practical Gujarati Grammar, 2nd edn. Arunoday Publication, Ahmedabad (2010)
35. Panchal, P., Panchal, N., Samani, H.: Development of Gujarati wordnet for family of words. Int. Res. J. Comput. Sci. (IRJCS) **1**(4), 28–32 (2014)
36. Bhagvad-go-mandal. http://www.bhagavadgomandalonline.com/
37. Gujarati lexicon. http://www.gujaratilexicon.com/
38. Bhensdadia, C.K., Bhatt, B., Bhattacharyya, P.: Introduction to Gujarati wordnet. In: Proceedings Third National Workshop on IndoWordNet, pp. 1–5 (2010)
39. Goldsmith, J.: Unsupervised learning of the morphology of a natural language. Comput. Linguist. **27**(2), 153–198 (2001)

Predictive Analytics in Medical and Agricultural Domains

Suśruta: Artificial Intelligence and Bayesian Knowledge Network in Health Care – Smartphone Apps for Diagnosis and Differentiation of Anemias with Higher Accuracy at Resource Constrained Point-of-Care Settings

Shubham Yadav[1], Sakthi Ganesh[2], Debanjan Das[1], U. Venkanna[1],
Rajarshi Mahapatra[1], Ankur Kumar Shrivastava[3],
Prantar Chakrabarti[4], and Asoke K. Talukder[2(✉)]

[1] DSPM IIIT Naya Raipur, Atal Nagar, Chhattisgarh, India
{shubham16100,debanjan,
venkannau,rajarshi}@iiitnr.edu.in
[2] SRIT India Pvt. Ltd., Bangalore, India
sakthi.g@sritindia.com,
asoke.talukder@renaissance-it.com
[3] AIIMS Raipur, Raipur, Chhattisgarh, India
shrivastavadrankur@gmail.com
[4] NRS Medical College and Hospital, Kolkata, India
prantar@nrsmc.edu.in

Abstract. Anemia in India carries a major disease burden. This includes both nutritional anemias, of which iron deficiency anemia (IDA) is the commonest and inherited hemolytic anemias like β thalassemias (β-TT). In Eastern and North-Eastern India about 25–40% of the public blood bank blood is consumed by patients with β Thalassemia. Moreover, 51% of Indian women in the reproductive age suffer from IDA. Most vulnerable group of anemic patients is in the rural underserved regions. This underserved population can only be served through the use of artificial intelligence (AI), automation, supported by telemedicine. To combat the problems of both IDA and β-thalassemia by early diagnosis at the point-of-care– we have developed Suśruta - an Artificial Intelligence (AI) driven robust smartphone-based health care application. This App uses five major components of AI; namely: (1) Natural language processing (NLP) to analyze the unstructured clinical data and translate it into computer understandable 3rd Generation SNOMED and ICD10 ontologies; (2) Speech Synthesis; (3) Artificial Neural Network (ANN) with Machine Learning and Deep Learning (ML/DL) on 60,283 labelled common blood counts (CBC) and High Performance Liquid Chromatography (HPLC) data collected over 8 years by a teaching hospital in Kolkata for β-TT screening; (4) Computer Vision and Image Processing techniques to interpret hemoglobin content in blood through non-invasive analysis of conjunctiva and nailbed images; and (5) NoSQL and Big-data Graph database-driven Bayesian Knowledge Network for Evidence Based Medicine and Bayesian Outcome Tracing for Predictive Medicine. Unlike

© Springer Nature Switzerland AG 2019
S. Madria et al. (Eds.): BDA 2019, LNCS 11932, pp. 159–175, 2019.
https://doi.org/10.1007/978-3-030-37188-3_10

previous systems, the ML/DL technique of β-Thalassemia carrier screening with CBC improved the accuracy of screening by two folds compared similar approaches analyzing CBC with Mentzer Index. Moreover, the uniqueness of Suśruta is that the App is robust and works both in an offline and online mode at resource constrained point-of-care. This is the first time AI is used for comprehensive anemia care by early diagnosis, which empowers the ordinary health workers in rural underserved communities. Furthermore, it will introduce the concept of patient empowerment and person centered care by changing the definition of point-of-care in rural India.

Keywords: Big data · Artificial Intelligence · NoSQL · Healthcare · CDSS · Anemia · Thalassemia · AI · ML/DL · ANN · Bayesian Knowledge Network · Bayesian Outcome Tracing · Smartphone apps

1   Introduction

Red blood cells or erythrocytes carry oxygen and collect carbon dioxide through the help of hemoglobin. Hemoglobin is an iron-rich protein that gives the blood cells its red color and facilitates transportation of inhaled oxygen from the lungs to tissues and carbon dioxide from tissues to the lungs to be exhaled. Red blood cells are the most abundant cell in the blood, accounting for about 40–45% of its volume [1]. Oxygen is used by the cells to generate energy from the foods we eat. When the hemoglobin is unable to carry oxygen, cells are deprived of energy production. This may happen due to lack of hemoglobin or if the hemoglobin is defective due to genetic causes. The most common cause of lack of hemoglobin production in the body is due to insufficient iron in the body. Thalassemia also causes anemia – it is the most common genetic disorder (Hemoglobinopathy) in the world caused by defective or nonfunctional hemoglobin.

Anemia is one of the major public health challenges in the world. Globally, anemia affects 1.62 billion people, which corresponds to 24.8% of the population. The highest prevalence is in preschool-age children (47.4%), and the lowest prevalence is in men (12.7%). However, the population group with the greatest number of individuals affected is non-pregnant women (468.4 million) [2]. In India the burden is higher compared to the world average. According to WHO Global Nutrition Report 2017, 51% of Indian women in reproductive age are suffering from iron deficiency anemia. During pregnancy, anemia is likely to put both mother and the baby at risk of premature delivery and low birth weight of the baby – this carries a high risk of stunted or underdeveloped children.

On the other hand, thalassemia is the most common genetic disorder that is a life restricting chronic disease. According to the type of globin chain involved, two main types, i.e., the α- and β-thalassemia can be distinguished [3]. β-thalassemia requires specialized treatments (blood transfusion and iron chelation) causing severe distress and financial loss to the family. It is estimated that there is almost 3.6 to 3.9 crore (36 to 39 million) carriers of β-thalassemia patients in India [4]. The average prevalence of β-thalassemia carriers in India is about 3–4% of the general population. Several ethnic groups in India have a much higher prevalence ranging around 4–17% [5]. Prevention of β-thalassemia major is achieved through limiting carrier marriages. Several at-risk populations in the Mediterranean area of Europe [Cyprus, Sardinia, several regions of Continental Italy (Delta Po area, Sicily), and Greece] have reduced the prevalence of thalassemia by thalassemia screening and proper counseling. In Lebanon, Iran, Saudi Arabia, Tunisia, United Arab Emirates, Bahrain, Qatar, and Gaza Strip, the national premarital programs are mandatory and aimed at limiting carrier marriages [3]. In the traditional way, β-thalassemia screening requires high performance liquid chromatography (HPLC) tests that are expensive, proprietary and not widely available. Therefore, our goal is to replace the HPLC test with complete blood counts (CBC) that are cheaper with higher accessibility at least for public health purposes. Artificial Intelligence (AI) supported by automation is the only way for mass screening of β thalassemia carriers that will facilitate speed, accuracy, and reach.

The average distances to the nearest primary health centers (PHC), community health centers (CHC) and district hospitals (DH) in India are 9.06 km, 17.66 km and 32.57 km, respectively [6]. A report in July 2019 reveals that India has 82% shortfall of specialist physicians like surgeons, gynecologists, pediatricians, etc. in its public health system. Moreover, India has 40% shortfall in laboratory technicians, and 12–16% shortfall in nurses and pharmacist at the CHC and PHC level [7]. Therefore, nearly 86% of all the medical visits in India are made by ruralites with majority of the patients and their families traveling more than 100 km to avail health care facility of which 70–80% are out-of-pocket (OOP) [8]. In 2011–2012, 55 million people in India were pushed into poverty by OOP expenses; out of which, approximately 38 million people became poor only because of purchase of medicines through OOP payments [9]. Unless anemia diagnosis and care are taken to the resource restricted point-of-care, it will impact severely on the overall development of the rural population in India.

There is a concern in the accuracy of diagnosis in primary care. A study in USA found that only 12% of the cases were accurately diagnosed in primary care. In 21% of cases the diagnoses in the primary care were wrong. In remaining 66% of the cases the diagnosis was improved [10]. In countries in South Asia or Africa where the number of physicians is inadequate and the doctors spend only a few seconds per patient per encounter in the primary care, the effectiveness of future healthcare relies on the adoption of automation and artificial intelligence. In the UK the use of CDSS (Clinical Decision Support System) increased both efficiency and effectiveness of the primary health clinics. There was a 15% reduction in referrals (expert consultant's advice) [11].

We have constructed a mobile AI anemia care system named Suśruta. Our contributions in Suśruta are multifold. To the best of our knowledge, this is the first time 60,000 + persons' β-thalassemia screening data has been used in AI (Artificial Neural Network). This is the first time we have showed that AI and automation can be used in β-thalassemia screening bypassing the HPLC tests. Furthermore, the accuracy of our AI system with CBC is more than double compared to the clinical Mentzer Index – this is a significant achievement. Because thalassemia is a genetic disorder and inherited at birth, it is agnostic to time period; therefore, even an old blood test report can be used for our β-thalassemia screening.

Melanin is the pigment beneath the skin that gives human skin, hair, and eyes their color. Conjunctiva, nailbeds, lips, oral mucosa, certain other parts of the body do not have this pigment. Clinical signs, such as pallor of these parts of the body have traditionally been used by physicians in the diagnosis of anemia. Studies proved a statistically significant correlation between hemoglobin concentration and the color tint of the lower eyelid conjunctiva, nailbed rubor, nailbed blanching, and palmar crease rubor [12]. There are few studies where researchers have developed non-invasive anemia diagnosis from conjunctiva or nailbeds [13–15]. We have already developed a similar non-invasive anemia diagnostic system through the use of conjunctiva images. All these studies in different part of the world including our earlier system used expensive cameras and an ideal laboratory setup with unlimited resources. Many of them have even used expensive software and hardware environments like MATLAB, and expensive cloud infrastructure. In our current system we offer AI driven smartphone-based mobile anemia care, which works in resource constrained real-life environment. It uses all open-domain platforms that work both online and offline (with or without Internet) with a commodity cloud. It is designed to be error resistant and can be used by ordinary health workers like ASHA or Anganwadi workers at the rural point-of-care. The app does two main functions, namely:

- Diagnosis of β Thalassemia carriers and patients using the automated CBC (β TT app)
- Diagnosis of the iron deficiency anemia (IDA) through non-invasive techniques.

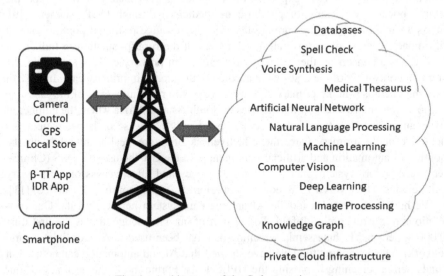

Fig. 1. Architecture of the Suśruta native app

2 Materials and Methods

2.1 Architecture of the Suśruta Native App

Figure 1 depicts the architecture of Suśruta. The user may be a doctor, nurse, or a health worker who will use the smartphone app for anemia care. The mobile applications in the smartphone are native Android apps[1]. The apps use the phone camera to take pictures of conjunctiva and nailbed along with data collection of Complete Blood Count. It also uses the phone GPS (Global Positioning System) with Google map to find out the location of the data collection point. The user does not need any software to be installed other than the Suśruta app on the mobile. The server side is deployed in a commodity cloud. The app works in offline mode and does not depend on availability of the Internet. When internet access is available, the data is synchronized and transferred to the main database on the server in the cloud directly, where all the AI components are running.

2.2 β-Thalassemia Trait (β-TT) Data Collection

In many European countries, and Middle Eastern countries thalassemia has been controlled through limiting carrier marriages. This requires mass screening and counseling. With the same goal in mind, the Government of West Bengal initiated the State Thalassemia Control Programme as a follow-up to the Jay Vigyan Mission Project on Community Control of Thalassemia syndromes in 2007. A leading teaching hospital Nil Ratan Sircar Medical College and Hospital (NRSMCH) in Kolkata, as a nodal center for this program had a dedicated team of doctors, medical laboratory technologists and data entry operators to conduct the screening. From 2008 to 2017, a total of 60,283 individuals were screened by the Department of Hematology, NRS Medical College, Kolkata. The target populations were antenatal mothers and the about to marry population like school and college students. The database also included data of thalassemia patients and their family members. CBC and Hb HPLC tests were conducted on each individual and the data was electronically captured. We used ANN (Artificial Neural Network) on this data to arrive at a classification algorithm. We used TensorFlow with Keras libraries for the deep learning and the classification of individuals screened.

In the program, 47 variables that included many clinical, pathological, and demographic information were recorded. Some of these screening data contained missing data points. Because our interest was to arrive at a classification algorithm using only complete blood count (CBC) data – we focused only on five types of blood variables viz., RBC (Red Blood Cell Count), HB (Hemoglobin), MCV (Mean Corpuscular Volume), MCH (Mean Corpuscular Hemoglobin), and RDW (Red cell Distribution Width). Along with these five blood parameters we took Age and Gender

---

[1] Future releases will support IOS and other operating environments.

(Sex). All the individuals had a definite diagnosis, validated by the Department of Hematology, NRS Medical College, Kolkata. Out of 60,283 individuals' data there were total 55,933 individuals who had all nine fields of interest, namely, Age, Sex, Diagnosis (label), RBC, HB, MCV, MCH, RDW and HPLC. This cohort of 55,933 individuals was our thalassemia data cohort. The average age of screened individuals was 23 with minimum 1 year and maximum 90 years. Number of Females in the cohort was 39847 (71.2%). Out of the entire cohort, 9761(17.5%) individuals have been diagnosed as Carriers, 1436(2.6%) individuals with confirmed thalassemia disease and 44,736 as normal (non-thalassemic) individuals.

2.3   Non-invasive Iron Deficiency Anemia Data Collection

A set of 12 eye images were collected using the high resolution iPhone camera of a smartphone in ideal light conditions. These 12 subjects subsequently, underwent Hemoglobin estimation in their blood within 48 h of data collection. The custom made app captured the anterior conjunctiva part of the eye after pulling the lower eyelid. The captured image is then analyzed using Image Processing and Computer Vision AI techniques for scoring of level of anemia. This preliminary data used to validate our idea, which will be further extended with a large number of data set collected in collaboration with the Ophthalmology department of another teaching hospital namely All India Institute of Medical Sciences (AIIMS) Raipur.

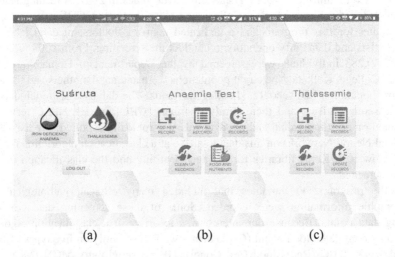

(a)   (b)   (c)

Fig. 2. (a) Main screen contains the two major modules of IDA and β-TT screening; (b) sub-screen of anaemia and (c) sub-screen of thalassemia test

Table 1. Functionalities of different facilities under anemia test

Facility	Screen of apps	Functionalities
Add Record		The health worker will collect demographical, clinical, and medical data of the person. The phone camera is used for taking 4 pictures including two pictures of eyes (left and right) and two pictures of fingernails (left and right). **Comorbidity:** Along with other data, comorbidity data also will be collected and for each disease, the respective ICD code (ICD 10) will be added.
		Food Habits: Food habits play a major role in iron deficient states. So food habits of the patient will also be recorded. **FACIT Score:** Fatigue is a burdensome symptom in iron deficiency anemia (IDA). FACIT Score is designed to capture the severity and impact of fatigue appropriately [16]. FACIT score will be calculated and collected using a questionnaire. An algorithm is used in the backend to calculate the score.
View Record		In the View Records function we examine the records that have been collected and saved in the local Android database.
Cleanup		This is to free up the space in phone's memory. All the server synchronized records can be removed from the server using this function
Nutrients		This is a static page which provides the diet to prevent anemia.

Table 2. Functionalities of different facilities under Thalassemia test

Facility	Screen of apps	Functionalities
Add Record		This screen is used to save all the required details. After the data is entered, the record will be stored in the device and also the results will be calculated and shown using a bullet graph.
View Record		In the View Records function we examine the records that have been collected and saved in the local Android database.
Cleanup		This is to free up the space in phone's memory. All the server synchronized records can be removed from the server using this function
Result in a Graph		Susruta always strives to focus on the best things. Here is an example of that. Here for displaying the results the bullet graph is used. Compared to normal bar graph/ line graph bullet graph give much more details at a single glance. All the ranges are present in the background so it will help the doctors to take a much better medical decision at a glance.

2.4   NoSQL and Big-Data Graph Database

Knowledge networks are generally represented through ontology-driven semantic networks. One such ontology-based medical database is SNOMED-CT (Systematized Nomenclature of Medicine – Clinical Terms). It is a multilingual vocabulary of clinical terminology that is developed by National Institute of Health (NIH) of USA to make healthcare data machine understandable. SNOMED CT is used in HL7 (Health Level Seven International) International Patient Summary (IPS). In September 2013 SNOMED-CT is standardized as the EHR data interoperability standards in India by Ministry of Health & Family Welfare (MoHFW). It is also a vehicle for CDSS (Clinical Decision Support System).

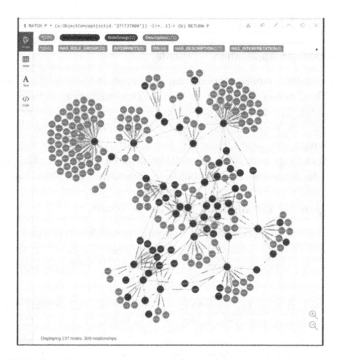

Fig. 3. The SNOMED-CT knowledge network of Anemia (Code: 271737000) in Neo4j graph database with a maximum parents path length of 2

SNOMED-CT is a medical terminology database. All medical terminologies from disease to findings are stored in this database with very complex relationships. SNOMED-CT is released in MySQL and Oracle relational databases formats. For efficiency and predictive medicine we have converted the SNOMED-CT database into a cloud-based Neo4j Graph database. Figure 3 shows the SNOMED-CT knowledge network for Anemia (SNOMED-CT code: 271737000) in SNOMED-CT Graph database with parents path length of 2. The SNOMED-CT Graph database has about 2.3 million nodes (terms) and 9.0 million relationships. The patient information in our application is stored in MongoDB NoSQL as unstructured patient record. As we examine more and more individuals, our knowledge about β-TT and IDA will only increase. We will know the environmental conditions and epidemiological factors that influence different phenotypes of anemia. This data will be converted as frequency and added in the knowledge network as weights. Bayesian probability is computed on this data for predictive diagnosis of anemia and its effect. This new data will be added into the AI engine as refined knowledge.

3 Results and Discussion

3.1 Anemia Mobile Application – Suśruta

The smartphone-based mobile application Suśruta is used to predict the cases of Anemia and Thalassemia as shown in Fig. 2. This prediction is based on scoring algorithms and Artificial Intelligence. For this app to get results for Anemia cases, no blood test is required. Following successful authentication, user is presented with the main service screen with Iron Deficiency Anemia (IDA) and Thalassemia. There are two primary modules present in this app- Anemia and Thalassemia. The functionality of the different facilities under each module is summarized in Tables 1 and 2.

3.2 AI Components and Big Data Used in the System

The basic goal of the smartphone app is to service the underserved rural population at the point-of-care. These underserved regions are resource limited from the health care perspective. Suśruta smartphone app uses many Artificial Intelligence (AI) algorithms like:

1. NLP (Natural Language Processing) to convert unstructured medical notes into 3rd Generation medical ontology terms of SNOMED and ICD10. Incidentally, SNOMED in India is maintained by NRCeS (National Resource Centre for EHR Standards) and has been standardized by the Government of India in 2014 to be the national standard. This interface is used to extract medical clinical terms from the patients' chief complain, which is enter into the system as free text.
2. Speech Synthesis tools to offer a voice prompt in Chief Complaints and medical notes. This interface is used to preset error messages in audio form.
3. Machine learning and deep learning (ML/DL) has been used on 55,933 individual's thalassemia screening data. This data has been collected by a teaching hospital in Kolkata over eight years. We used ANN (Artificial Neural Networks) that used Google's TensorFlow for β-thalassemia prediction. Public health personnel has recommended the use of Mentzer Index for differential diagnosis of β-TT from IDA. We therefore used Mentzer Index as benchmark to compare the outcome of our algorithms. Results from our ANN model were found to be more accurate than the Mentzer Index.
4. For non-invasive anemia detection from conjunctiva, we used Image Processing and Computer Vision AI techniques. We also used ANN for feature extraction and ML/DL. This module of the app is non-invasive; this means that to get the anemia diagnosis, no blood test is required. This app will be able to instantly predict the hemoglobin level.

5. Bayesian Knowledge Network (BKN) for Evidence Based Medicine (EBM) and Bayesian Outcome Tracing (BOT). BKN is a probabilistic directed graph that presents EBM through the aggregation of Chief complains, Comorbidity, EMR & Lab records, and Biomedical knowledge base. BOT models each patient outcome in a Hidden Markov Model as a latent variable, updated by observing the correctness of each patient's treatment plan and the outcome. This is used for reasoning of the anemia diagnosis.

3.3 β-Thalassemia Trait (β-TT) Screening

Figure 4 shows the characteristics of the cohort. It shows the correlation of different variables (or features) in the data of the entire cohort with 6 variables that include Age, RBC, HB, MCV, MCH, and RDW. There is no significant correlation between age and any other variable to indicate that anemia can occur at any age. There is a strong positive correlation between MCV and MCH which is logical. There is a negative correlation between RDW and MCH. There is a weak positive correlation between HB and MCH. There is a weak negative correlation between RDW and HB and RDW and MCV.

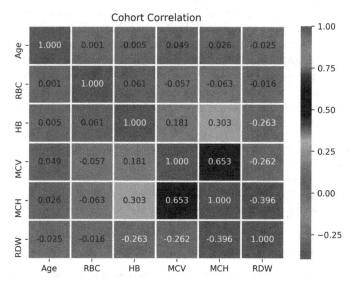

Fig. 4. The correlation matrix of the cohort with 6 variables that include Age, RBC (Red Blood Cells), HB (Hemoglobin), MCV (Mean Corpuscular Volume), MCH (Mean Cell Hemoglobin), and RDW (Red blood cell Distribution Width)

We have used Keras for our Artificial Neural Network (ANN) model. Keras offers a set of high-level user friendly neural networks APIs. It is written in Python and works as a wrapper over many backend ANN. The most widely used backends are Google's TensorFlow, Microsoft's CNTK (Cognitive Toolkit), Theano, etc. We used the sequential model Keras. The motivation for using Keras is that it makes use of GPU (Graphics Processing Unit) as well. We used 80% of the data as training and remaining 20% for test.

We have compared our CBC β-TT results with CBC plus Mentzer index results. Mentzer index was proposed by William C. Mentzer [17] in 1973, for the use of physicians for differential diagnosis of beta-thalassemia from iron deficiency anemia. Mentzer Index is used by health personnel globally. According to various studies Mentzer index was found to be the most reliable index with highest sensitivity and specificity [18]. Unlike our approach where we have taken five blood counts, Mentzer index used only two blood features namely, MCV and RBC. Mentzer hypothesized that when the quotient of the MCV (in fL) to the red blood cell count RBC (in Millions per microLiter) is less than 13, thalassemia is expected to be more likely. In mathematical terms, if MCV/RBC < 13, it is likely to be a thalassemia carrier. If the result is greater than 13 (MCV/RBC > 13), then iron-deficiency anemia is said to be more likely. We used Mentzer index for binary classification for identifying thalassemia carriers from non-thalassemia cases. In the current study most of the cases were healthy adults – this in other words means that only few of the 60,283 individuals showed any symptoms. Therefore, it is appropriate to consider the Mentzer Index using CBC as the comparator. Because our labeled data classified non-thalassemic cases as normal, we generalized the Mentzer Index such that: if the Mentzer score is less than 13, it is thalassemia carrier; however, if the score is greater than 13 it is not thalassemia carrier. The Mentzer index for the 20% test data is shown in the confusion matrix in Fig. 4a.

With Mentzer index for 20% test data (N = 11187) validated by Hb HPLC, the classification results are as follows:

1. Thalassemia patients/carriers classified as thalassemia patients/carriers 739 out of 2208 actual patients/carriers
2. Thalassemia patients classified as Healthy persons are: 1469
3. Healthy persons classified as thalassemia patients are: 155
4. Healthy persons classified as Healthy persons are: 8824

With our artificial neural network model (with CBC tests), for 20% test data (N = 11187), the classification results are as follows:

1. Thalassemia patients/carriers classified as thalassemia patients/carriers are: 1591 out of 2208 actual patients/carrier
2. Thalassemia patients classified as Healthy persons are: 617
3. Healthy persons classified as thalassemia patients are: 547
4. Healthy persons classified as Healthy persons are: 8432

To summarize: using Mentzer index on the test population of 11,187 samples subsequently validated by Hb HPLC, we find that 739 thalassemia carriers have been identified correctly. In quantitative terms this offers 33.47% accuracy. In contrast, when we use the AI model on the same test population of 11,187 β-TT individuals with CBC,

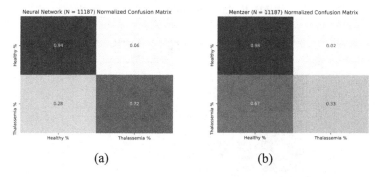

(a)   (b)

Fig. 5. (a) Normalized confusion matrix using CBC and Artificial Neural Network (ML/DL) for the test population of 11,187 carrier/patients. (b) Normalized confusion matrix for Mentzer index for the same population. The accuracy of the ANN model (4a) is 72.05% compared to the gold standard algorithm of Mentzer Index (4b), which stands at 33%. This indicates that the AI accuracy is more than double. Moreover, it reduces cost of test by use of CBC.

the correct classification is 1591 carriers/patients, which stand at 72.05% accuracy. From this study it can be observed that the selection of five features of RBC, HB, MCV, MCH, and RDW with AI increases the accuracy of the classification by more than double. Most interesting part of our innovation is that the accuracy increases with lower cost of the blood test. Moreover our algorithm works in a smartphone that can work offline. This, in other words, indicates that it can be used anywhere anytime. The mobile app is institutive and can be used by anybody with little training. Figure 5 shows the results through confusion matrix.

3.4 Scoring the Non-invasive Iron Deficiency Anemia

In this preliminary study eye images from 12 subjects have been collected and subsequently computer vision and image processing technique used for segmenting the conjunctival pallor. After cropping the conjunctiva pallor the intensity of red pixel and intensity of green pixel were measured. The difference between both logarithm values determines the Diffey's Erythema Index (EI) [19]

$$EI = \log(S_{red}) - \log(S_{green}) \qquad (1)$$

where S is the brightness of the conjunctiva in the relevant color channel.

Table 3. Range of EI with its respective anemia category

EI range	Anemia category
>0.10	Non-Anemic
0.08–0.10	Acute-Anemic
<0.08	Severe-Anemic

Table 4. Summary of all the results

S.No	Pictures	ROI	Haemoglobin Level	Erythema Index value	Category
1.			13.4	0.155261	Non-Anaemic
2.			13.6	0.1745	Non-Anaemic
3.			12.8	0.1073	Non-Anaemic
4.			11.4	0.1121	Non-Anaemic
5.			11.5	0.1142	Non-Anaemic
6.			7.2	0.0768	Severe-Anaemic
7.			8.8	0.0887	Acute-Anaemic
8.			13.2	0.1550	Non-Anaemic
9.			15.2	0.1807	Non-Anaemic
10.			11	0.1025	Non-Anaemic
11.			9.6	0.0905	Acute-Anaemic
12.			11.6	0.1281	Non-Anaemic

Here, we have classified the patient in one of the three categories- Non-Anemic, Severely Anemic and Acute Anemic using the boundary value obtained from the threshold value of EI. To decide the value of the threshold of EI, several eye images were examined whose Hemoglobin levels were already known. We selected a threshold difference of means which was based on the data of non-anemic, acute anemic and severe anemic. Table 3 depicts the EI range and respective Anemia category. The similar analysis has been carried out in all the collected images and Table 3 summarizes the results. The objective of the present work is to develop an AI and ML/DL enabled Android platform for offline and online data collection and classification of anemia in a resource constrained environment. In this work, as a proof of concept the initial data processing and analysis has been performed on a remote system under MATLAB platform. Now further implementation is under process to perform the analysis on Cloud and open-source platform. Although the existing non-invasive mode anemia detection technique uses a similar kind of image processing approach, the major limitation of those is that they presented under laboratory conditions. The existing system may not work under resource constrained environment like rural villages. On the other hand, our proposed platform totally aimed for limited resource cases. Further, the entire data processing and analysis will be performed on Cloud, which can be accessed from anywhere, anytime and any system (Table 4).

4 Conclusion

There has not been any study previously with such a large population of 60,283 individual's CBC data. We applied AI algorithms on this dataset to bring in speed, accuracy, and efficiency. Our AI algorithm uses CBC instead of Hb HPLC. In the process we not only reduced the cost of the screening but have also increased the accuracy of the screening more than double compared to existing methodologies of analyzing CBCs. This is a significant achievement.

In IDA, we had already developed a non-invasive technique of anemia diagnosis. This system used some ideal laboratory conditions. In the current system we have enhanced the IDA system to work in resource constrained environment. Moreover, we added additional features of FACIT and clinical information that increases its accuracy.

Furthermore, we used five different types of AI tools and techniques, namely

1. Natural Language Processing for disease ontology and SNOMED/ICD integration
2. Speech Synthesis
3. Machine Learning/Deep Learning/Artificial Neural Networks for β-TT screening. This module is in the production state.
4. Computer Vision and Image Processing for non-invasive iron deficiency anemia diagnosis. This module is at the calibration and trial phase.
5. Bayesian Knowledge Network for Evidence Based Medicine and Bayesian Outcome Tracing for Predictive Medicine.

This will empower a health worker in rural India to control the burden of anemia by early diagnosis at the point of care.

Acknowledgment. 1. The software was developed by SRIT engineers and students from IIITNR as part of internship program at SRIT facility at Bangalore. The project was funded by SRIT Healthcare division and supervised by Dr Asoke K Talukder from SRIT.

2. Thalassemia Control Programme which was part of Jay Vigyan Mission Project on Community Control of Thalassemia syndromes started with proper ethical approvals in 2007. The anonymized data used for deep learning was obtained in 2018.

3. The anemia data was collected following proper ethical committee approval for the present study. The data was shared for algorithm development by IIIT NR & SRIT.

References

1. Wiki: Blood cell (2019). https://en.wikipedia.org/wiki/Blood_cell. Accessed on June 2019
2. World Health Organization: Global anemia prevalence and number of individuals affected. Reference Source (2015). https://www.who.int/vmnis/anaemia/prevalence/summary/anaemia_data_status_t2/en/
3. Kim, S., Tridane, A.: Thalassemia in the United Arab Emirates: Why it can be prevented but not eradicated. PLoS ONE **12**(1), e0170485 (2017)
4. Ministry of Health and Family Welfare, Government of India. Policy For Prevention and Control of Hemoglobinopathies – Thalassemia, Sickle Cell Disease and variant Hemoglobins In India. New Delhi (2018)
5. Colah, R., Italia, K., Gorakshakar, A.: Burden of thalassemia in India: the road map for control. Pediatric Hematol. Oncol. J. **2**(4), 79–84 (2017)
6. Kumar, S., Dansereau, E.A., Murray, C.J.: Does distance matter for institutional delivery in rural India? Appl. Econ. **46**(33), 4091–4103 (2014)
7. Dey, S.: News article, 82% shortfall in specialists puts healthcare on sick bed. Time of India, TNN, July 1, 2019 (2019)
8. Kurukshetra, Ministry of rural development, vol. 65(9), p. 22, July 2017
9. Selvaraj, S., Farooqui, H.H., Karan, A.: Quantifying the financial burden of households' out-of-pocket payments on medicines in India: a repeated cross-sectional analysis of National Sample Survey data, 1994–2014. BMJ Open **8**(5), e018020 (2018)
10. Van Such, M., Lohr, R., Beckman, T., Naessens, J.M.: Extent of diagnostic agreement among medical referrals. J. Eval. Clin. Prac. **23**(4), 870–874 (2017)
11. Foot, C., Naylor, C., Imison, C.: The Quality of GP Diagnosis and Referral. The King's Fund, London (2010)
12. Strobach, R.S., Anderson, S.K., Doll, D.C., Ringenberg, Q.S.: The value of the physical examination in the diagnosis of anemia: correlation of the physical findings and the hemoglobin concentration. Archives of Internal Med. **148**(4), 831–832 (1988)
13. Collings, S., Thompson, O., Hirst, E., Goossens, L., George, A., Weinkove, R.: Non-invasive detection of anemia using digital photographs of the conjunctiva. PLoS ONE **11**(4), e0153286 (2016)
14. Mannino, R.G., et al.: Smartphone app for non-invasive detection of anemia using only patient-sourced photos. Nat. Commun. 9 (2018)
15. Tamir, A., et al.: Detection of anemia from image of the anterior conjunctiva of the eye by image processing and thresholding. In: 2017 IEEE Region 10 Humanitarian Technology Conference (R10-HTC), pp. 697–701. IEEE (2017)
16. Acaster, S., Dickerhoof, R., DeBusk, K., Bernard, K., Strauss, W., Allen, L.F.: Qualitative and quantitative validation of the FACIT-fatigue scale in iron deficiency anemia. Health and Qual. Life Outcomes **13**(1), 60 (2015)

17. Mentzer Jr., W.C.: Differentiation of iron deficiency from thalassaemia trait. Lancet **1**(7808), 882 (1973)

18. Vehapoglu, A., et al.: Hematological indices for differential diagnosis of beta thalassemia trait and iron deficiency anemia. Anemia **2014**, 576738 (2014)

19. Brian Riordan, M.E.B.M.E., Sprigle, S., Linden, M.: Testing the validity of erythema detection algorithms. Development, 38(1), 13–22 (2001)

Analyzing Domain Knowledge for Big Data Analysis: A Case Study with Urban Tree Type Classification

Samantha Detor[1,2], Abigail Roh[1,2], Yiqun Xie[2(✉)],
and Shashi Shekhar[2]

[1] Breck School, Minneapolis, MN 55422, USA
`{detosa,rohab}@student.breckschool.org`
[2] Department of Computer Science and Engineering, University of Minnesota,
Minneapolis, MN 55455, USA
`{xiexx347,shekhar}@umn.edu`

Abstract. The goals of this research were to create a labeled dataset of tree shadows and to test the feasibility of shadow-based tree type identification using aerial imagery. Urban tree big data that provides information about individual trees can help city planners optimize positive benefits of urban trees (e.g., increasing wellbeing of city residents) while managing potential negative impacts (e.g., risk to power lines). The continual rise of tree type specific threats, such as emerald ash borer, due to climate change has made this problem more pressing in recent years. However, urban tree big data are time consuming to create. This paper evaluates the potential of a new tree type identification method that utilizes shadows in aerial imagery to survey larger regions of land in a shorter amount of time. This work is challenging because there are structural variations across a given tree type and few verified tree type identification datasets exist. Related work has not explored how tree structure characteristics translate into a profile view of a tree's shadow or quantified the feasibility of shadow-only based tree type identification. We created a consistent and accurate dataset of 4,613 tree shadows using ground truthing procedures and novel methods for ensuring consistent collection of spatial shadow data that take binary and spatial agreement between raters into account. Our results show that identifying trees from shadows in aerial imagery is feasible and merits further exploration in the future.

Keywords: Aerial imagery · Urban trees · Labeled data creation · Ground truthing · Smart cities · Tree disease · Urban tree big data

1 Introduction

The goal of this study was to maximize the accuracy of classification of tree types based on corresponding shadows in aerial imagery. The study flow is depicted in Fig. 1. The image on the left displays a subsection of the high spatial resolution aerial

---

S. Detor and A. Roh—These authors contributed equally to this work.

S. Madria et al. (Eds.): BDA 2019, LNCS 11932, pp. 176–192, 2019.
https://doi.org/10.1007/978-3-030-37188-3_11

imagery used as an input in this study. The image in the center shows the same region with its corresponding final shadow annotations, which were part of the dataset created and utilized in this research. The third box shows the output of this research, which was synthesized through the analysis of the dataset created in the intermediate step of this research.

Rules and guidelines for identifying trees based on their shadows

Fig. 1. Study flow.

Often overlooked, trees are a key part of a sustainable, content, and safe city. Urban trees provide shade, clean air, and increase the wellbeing of city residents. A study in Chicago showed that trees help build community and lessen crime, and a view of trees from an office window increases job satisfaction and well-being [15, 22]. Information on the location of specific types of trees can assist city leaders in making sustainable decisions that benefit urban trees and protect urban infrastructure. For example, a map of tree types in urban areas could assist in the diversification of urban forests to prevent tree pests and diseases, such as emerald ash borer, from devastating urban tree canopies [3, 4, 14, 27]. Tree maps could also help electrical companies monitor trees located near power lines. Trees falling onto power lines have caused many power outages and forest fires in recent years [1, 8]. In 2018, trees close to a powerline were the cause of the Camp Fire in California, a wildfire that killed 88 people [7, 28]. Some types of trees have lower resistance to wind than others, making them a hazard to surrounding infrastructure [28]. Knowing where different types of trees are located in relation to power lines can help cities better manage risks to public safety through tree trimming and removal efforts.

This problem is challenging for several reasons. Manual field surveys are time consuming and labor intensive. Therefore, they are not feasible for tree type identification over a large geographic extent. There is also limited dataset availability. Although extremely high spatial resolution LiDAR (i.e., centimeter resolution) and hyperspectral (i.e., sub-meter resolution) data have been used in some studies, these data do not cover expansive geographic regions [28]. Additionally, there is a lack of labeled training data. Thus, it is hard for machine learning approaches to be applied to solve this problem. Most previous literature has not made an effort to address this issue. Because of the large variation among trees of the same type, a large labeled training dataset is an essential aspect of solving this problem. Additionally, the process machine learning and deep learning methods depend on to produce their output is hard to interpret, so it can be difficult to improve model performance [9, 27].

There have been some successes in geospatial image detection using deep learning [10, 20, 25, 26, 30]. However, this method has been applied to detect objects that are clearly distinguishable by the human eye. In satellite imagery, trees appear to be green circles, and not much distinction is visible on the tree type level [24]. Other studies have classified trees based on hyperspectral imagery [5, 13, 29, 31], but these data are not available over a large geographic extent at a sufficiently high spatial resolution to be useful in widespread individual tree type classification. Likewise, high spatial resolution LiDAR data (i.e., centimeter resolution) has been used to study tree architecture [2, 12], but these datasets are also limited to custom studies. The most widely available remote sensing dataset with sufficient spatial resolution for studying individual trees remains RGB aerial imagery. Aerial imagery is collected in many urban counties, such as Hennepin and Ramsey Counties in Minnesota, at a 4–8 cm resolution, which is considered high spatial resolution [6, 16].

Our previous work has suggested that shadows could be used to classify trees based on their type, with some promising preliminary results using deep learning to classify trees in three guesses [28]. However, the actual differences in signatures of the shadows have not been characterized and their effectivity has not been quantified.

To overcome limitations of related work, our approach decomposes this problem down into two steps. First, a high-quality dataset of labeled tree shadows was created. Then, visual signatures of nine different types of trees in the dataset—ash, elm, hackberry, linden, locust, maple, oak, pine, and spruce—were analyzed and summarized. These observations were used by two human raters to test the feasibility of tree type identification based on shadows.

This work makes three major contributions. First, it offers a method to ensure the consistency of spatial shadow data. To create the dataset, shadows were annotated in aerial imagery. The tree type identification field survey used to create the dataset in this work was also verified manually to ensure its accuracy. Second, this work creates a labeled dataset of 4,613 trees based on the aforementioned methods to improve consistency between human raters. Third, this work details unique and distinct characteristics of different tree types based on shadow-only identification.

Relevance to Big Data. The aerial imagery used in this study had a very high spatial resolution (i.e., 4–8 cm) and required a large amount of storage space. For example, Ramsey County 7.6 cm resolution imagery covering 440 square kilometers was 12.8 GB. By this estimation, the continental United States, which spans approximately 8 million square kilometers, could require around 0.23 petabytes of storage. Furthermore, due to the large volume of trees in urban environments, an inventory of individual tree types that covers a large geographic region would be very big. The spatial and temporal resolution of aerial imagery will continue to improve in the future, making this problem even more relevant to big data forums in the coming years.

Scope. This paper focuses on the creation of a consistent and accurate dataset of tree shadows that can be used to train and test machine learning and deep learning models in the future. The following topics are outside the scope of this paper: analysis of extremely high spatial resolution hyperspectral and LiDAR data for tree type identification and application of machine learning and deep learning methods. Classification here is limited to tree type classification by human raters, which is an important preliminary step to understanding the feasibility of machine learning-based approaches.

Outline. Section Two formally defines the problem. Section Three describes the approach we followed to create the labeled dataset and identify tree signatures. Section Four discusses the results of each phase of the approach. Section Five concludes the paper and suggests next steps of this study.

2 Problem Definition

Input

- High spatial resolution RGB aerial imagery (i.e., 4–8 cm resolution).
- Low spatial resolution LiDAR data (i.e., 1 m resolution).
- Field survey point datasets covering limited regions of Hennepin County and Ramsey County (Minnesota, USA).
- Pre-existing domain knowledge (i.e., information from published field guides).

Output

- Detailed rules and guidelines for classifying shadows by type.

Objective

- Maximize tree type identification accuracy.

Constraints

- Date and collection time of input aerial imagery was not available.
- Some datasets (i.e., high spatial resolution hyperspectral and LiDAR data) considered important in previous research aiming to identify trees was not available to the authors for the study region (as of September 23, 2019).
- Dataset of labeled tree shadows must be representative of the entire population.

3 Methods

The problem was broken down into two sub-problems. To determine important characteristics for distinguishing among nine different types of trees, the raters had to observe many samples of different types of trees. Thus, the first phase was creating a dataset with a high accuracy of labeled tree types and a high rate of consistency across shadow annotations. This phase consisted of ground truthing, dataset consistency testing, and data annotation (Fig. 2), described in further detail later in Sect. 3.1.

The goal of the second phase was to determine important characteristics for identifying tree types from shadows and to test the accuracy of identification of tree types based on their observed signatures. During phase two, certain tree types were chosen for closer observation based on their presence in the dataset: then signatures of these tree types' shadows were observed and validated (Fig. 2). This is described in further detail in Sect. 3.2.

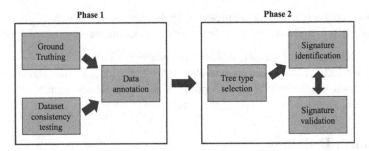

Fig. 2. Workflow of this study.

3.1 Labeled Dataset Creation

In the first phase of this study, pre-existing field survey datasets used to label the type of individual trees seen in aerial imagery were ground truthed (verified manually) to ensure accurate genus-level identifications. Dataset consistency testing was conducted to ensure consistent annotations between raters and guidelines for consistency testing were established. Using dataset consistency testing guidelines, a dataset of 4,613 shadows was annotated, and each shadow was labeled based on the verified field survey.

Ground Truthing. To assess the accuracy of the tree type identification dataset, two raters visited the locations of 100 tree points and identified them independently from the field survey. For safety reasons, the trees were selected from a 0.39 square kilometer area of the East Bank on the University of Minnesota campus. All unobstructed tree shadows within the test area were annotated. The 271 total annotated data points became the sample space for this experiment.

In ArcGIS, the genus and species data were removed from the feature class's attribute table and 100 of the 271 points were randomly selected. Each selected tree was visited, and its leaves, bark, and fruit, if present, were photographed. Figure 3 displays data collected for tree 0. Corresponding data were collected for every tree visited.

Fig. 3. Ground truthed data point.

After photographing the trees, raters identified their type using the following two field guides: *Minnesota Trees* and *The Sibley Guide to Trees* [19, 21]. The tree genus and species identifications from the original dataset were then matched with the raters' identifications. Percent agreement calculations were used to evaluate dataset accuracy based on ground truthed identifications.

Dataset Consistency Testing. To optimize consistency in data annotation between raters, a two-part dataset consistency test was designed to measure both binary agreement and spatial overlap. A fishnet, or grid covering the point dataset overlaid on the aerial imagery, was created and used as a coordinate system to randomly select data points for sampling. Grid extents were standardized. Two random numbers, representing the x and y coordinates of the grid, were generated. If a randomly generated cell in the fishnet contained points, all points within the cell were moved to the base of the corresponding tree trunk. Points were deleted if there were no local trees to which they could be assigned. Random cells were generated until the Hennepin 2015 dataset contained approximately 40 points, the Ramsey County 2015 dataset contained approximately 30 points, and the Ramsey County 2017 dataset contained approximately 30 points.

Each rater recorded a binary measure (1 or 0) for each data point indicating whether the tree was satisfactory for data annotation based on the clarity of the tree's branch structure. If determined suitable, the shadow was outlined according to guidelines established by the raters. Examples of annotations collected by each rater are shown in Fig. 4. Original annotations were traced for image clarity.

Fig. 4. Examples of shadow annotations collected by each rater during dataset consistency testing.

Once each rater had completed the dataset consistency test, binary agreement was assessed using the Cohen's Kappa Statistic [23, 32]. Cohen's Kappa is a statistic that accounts for chance when measuring agreement between raters on a scale of 0–1 where 0 is the probability of agreement based on random chance and 1 is perfect agreement. Equation (1) is the Cohen's Kappa statistic [23]. P_o is the agreement between the two raters and p_e is the chance agreement expected to happen for the given experiment. Additionally, spatial agreement between tree shadow annotations was measured through a calculation of percent overlap between the outlines of each rater (Eq. (2)).

$$\kappa = \frac{\text{po} - \text{pe}}{1 - \text{pe}} \qquad (1)$$

$$\frac{\sum areas\ of\ intersection}{\sum areas\ of\ intersection + \sum(area\ of\ annotations - intersection\ area)} \times 100\% \quad (2)$$

Multiple phases of analysis were conducted until the Cohen's Kappa value met or exceeded a value of 0.800 and the average percent overlap over all three datasets met or exceeded a value of 75%.

After each phase of analysis, the raters reviewed and adjusted the codebook to increase their accuracy and consistency. For example, raters decided to annotate shadows as close to the tree as possible, and trees without distinct branches or canopies were no longer included in the dataset.

A similar procedure was employed to confirm that each rater's annotations remained consistent over time. The raters annotated randomly selected points and ensured that the agreement between raters met preset thresholds. Then, no less than two weeks later, each rater re-annotated the randomly selected points. Each rater's self-agreement was evaluated using the same methods as inter-rater consistency testing. If binary or spatial agreement was not satisfactory (i.e., less than 0.800 and 75%, respectively), the process was repeated.

Annotation Methods. Shadow annotation consisted of two phases. First, annotations were auto-generated for each point in the field survey dataset. Then, each annotation was reviewed and revised to ensure its quality and adherence to guidelines established in the dataset consistency phase of this research.

To auto-generate annotations for points in the field survey dataset, normalized height models (NHMs) were created in ArcGIS using digital elevation models and digital spatial models, both of which are LiDAR derivatives. The aerial imagery dataset used in this study was integrated, so the dates and times of collection were not available. Thus, shadow size and direction information had to be approximated for each NHM used. To do this, five shadows were randomly selected from the extent of each NHM. The quality of the shadow and clarity of the tree's canopy in the NHM were reviewed. If the height and width of the shadow were distinct and the range of the tree's canopy in the NHM was clear and relatively even, then the shadow was annotated. Otherwise, a new point was randomly selected from the field survey. Based on the shadow annotations and the canopies in the NHM, the angle of the shadow (azimuth), the height-of-tree to length-of-shadow ratio, and the width-of-shadow to height-of-tree ratio were calculated. The angles and ratios were averaged across the five randomly selected trees, and the extent of the NHM was noted. This information and the field survey dataset were used as inputs into a MATLAB script, which generated an initial set of shadow annotations.

Each annotation was then reviewed by the raters. Based on the dataset consistency guidelines developed earlier in this work, annotations corresponding to shadows with significant obstruction by other objects (i.e., cars, houses, other shadows), low contrast with their background, or a significant number of branches too narrow to be visible in the aerial imagery (i.e., the tree is very young) were deleted. Annotations corresponding to trees removed before the aerial imagery was collected were also deleted. The size of each pre-generated annotation was checked and edited if necessary to ensure that it adequately suited the shape and extent of the tree's shadow. The completed dataset contained 4,613 shadows.

3.2   Identifying Tree Shadow Signatures in Aerial Imagery

To determine the differences among shadows of different types of trees, each tree type was observed in aerial imagery. Distinct characteristics were noted and, where possible, field guides with limited information about the characteristics of the silhouettes of different types of trees were reviewed to refine the observations and to guide further exploration.

Tree Type Selection. Nine types of trees were selected for identification in this study: ash, maple, linden, elm, oak, hackberry, locust, pine, and spruce. A tenth category included negative examples. The types of trees included in this analysis were the most common tree types in the initial 1,500 shadow dataset that the raters annotated, which included shadows from Hennepin County 2015 and Ramsey County 2015 data. Tree types were selected for identification based on their presence in the original dataset so that each category had an adequate number of samples.

Visual Signature Identification Methods. Raters reviewed samples from a small subsection of the dataset and created guidelines for distinguishing the nine chosen types of trees. To improve their ability to identify trees, raters predicted what type of tree they were annotating while growing the dataset and checked the accuracy of their guess with the field survey dataset. If they noticed other patterns when they were expanding the dataset, they added those patterns to the list of rules. Field guides were reviewed in order to gain additional insight into potential structural differences among the nine chosen types of trees.

Feasibility Testing. Feasibility testing was conducted to verify the observations made in the second phase of this study. In this phase of the study, two human raters identified tree types based on shadows. As they conducted feasibility testing, they refined tree type characteristics based on their performance (Fig. 5).

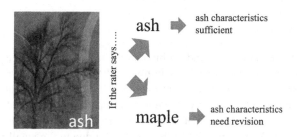

Fig. 5. Refinement based on feasibility testing.

Ten-Tree-Type Feasibility Testing. Each rater began by annotating a sample of shadows from the same aerial imagery dataset (Ramsey County 2018 7.6 cm resolution data). Information about the annotations was extracted to a separate file, where a portion of samples from each type of tree was randomly selected so that all ten categories of trees (maple, ash, pine, spruce, hackberry, locust, oak, linden, elm, and miscellaneous) were evenly represented. However, the distribution of tree types in the miscellaneous category was not even. From the evenly distributed dataset, 20 shadows were randomly selected, and a screenshot of each shadow was then taken at a consistent zoom (1:275). The screenshot and its identifying information were then put into a presentation, which became a test for the other rater. Each rater observed the 20 shadows that the other rater had annotated and noted distinct characteristics about the shadows, their top two identifications for the type of tree, and their final identification for the type of tree. This information was then compared to the identification included in the field survey point dataset, and the accuracy and Cohen's Kappa of the rater's classifications were calculated.

Ash-Tree Feasibility Testing. Methods followed for ash-tree feasibility testing were similar to methods followed in the ten-tree-type feasibility test except that there were only two categories, Ash and Miscellaneous.

4   Results

4.1   Experimental Goals

There were three goals of this research:

1. Create a substantial, accurate, and consistent dataset that can be used as a basis for machine learning and deep learning models in future research.
2. Observe and detail characteristics that are distinct among the shadows (profile geometries) of nine different types of trees and could be used to identify types of trees from their shadows in aerial imagery.
3. Test the feasibility of shadow-only-based tree type identification by human raters to validate the reliability of the characteristics observed in the second phase of this study and to better understand the potential for machine learning and deep learning classifiers to accurately classify trees by their shadows.

4.2   Labeled Dataset

To create an accurate dataset, ground truthing methods were employed to determine the accuracy level of the field surveys used in this study. Table 1 displays the agreement between the tree type identification point dataset and the tree genera and species data collected by the raters. The accuracy of the overall type level agreement was extremely high—97.9% (Table 1). One outlier is the accuracy of the elm type identification, which was very low. However, this can be discounted due to the small number of elm trees in the sample space. This shows that the dataset labels were sufficiently accurate to observe structural trends among different types of trees. These results also show that there are inaccuracies, although small, in manual identification.

Table 1. Tree type identification dataset accuracy assessment.

Genus	Genus (type) agreement (%)	Species agreement (%)	Total number of trees
Ash	100.	0.00	1
Elm	50.0	50.0	2
Hackberry	100.	100.	1
Linden	100.	75.0	8
Locust	100.	100.	18
Maple	100.	76.2	21
Oak	100.	80.0	15
Pine	100.	85.7	7
Spruce	100.	100.	7
Miscellaneous	92.9	64.0	14
Inaccessible trees	–	–	6
Total	**97.9**	**83.3**	**100**

Dataset consistency methods were conducted to ensure that shadow annotations had consistent qualities and dimensions in relation to the size of the tree. First, these methods were utilized to measure the agreement between annotations created by each rater. The two raters met the preset consistency thresholds (based on averages) in their third phase of testing. The results for the third phase of testing are shown in Table 2.

Table 2. Results from third phase of dataset consistency testing.

Metric	Hennepin County 2015	Ramsey County 2015	Ramsey County 2017	Total
Cohen's Kappa	0.867	0.878	0.742	0.873
Percent agreement	93.5	93.9	90.9	93.7
Percent overlap	78.9	73.7	77.8	76.3

Testing was also conducted to measure the consistency of a rater's annotations over time. To do this, each rater annotated a set of 100 points that had met the agreement standards between raters. Two weeks later, the rater re-annotated the same subsection of the dataset and compared their annotations to the annotations they had collected two weeks prior using the same statistics mentioned previously. Both raters met the required standards for self-agreement.

5   Visual Signature Identification and Validation

Signature Identification. Observations made about each of the nine types of trees analyzed in this research are shown in Table 3, which displays an example, icon, and description of the nine types of trees in this study. The grey area in the icon represents "fuzziness", while the black lines represent distinct branches. Icons are meant to highlight certain characteristics described in the narrative and are not to scale. The comments and examples included in the table were the most representative of the trees contained in the dataset.

Authors made several observations throughout this analysis. Previous research has noted that maple trees and ash trees are notable for their opposite branching. However, distinguishing whether a tree had opposite or alternate branching based on its shadow in aerial imagery was challenging and led to many incorrect tree type classifications. Additionally, young trees look different from their fully matured counterparts [11, 12]. As a result, the appearance of younger trees is not included in Table 3. One of the most distinct characteristics among tree types was "fuzziness" of the tree. Fuzziness is seen in aerial imagery because many small branches are not clearly visible due to the resolution. However, this characteristic is especially noteworthy in ash trees, which have a lot of fuzziness around their terminal branches.

Feasibility Testing. The purpose of feasibility testing was to verify whether the signatures observed earlier in this study were valid. Two types of tests were conducted: ten-tree-type feasibility tests and ash-tree feasibility tests. The results of the ten-tree-type feasibility tests are shown in Figs. 6 and 7.

Over the course of ten-tree-type feasibility testing, each rater improved their accuracy in one and two type identification predictions. During trials 1 and 2, raters strictly followed the guidelines they had created. However, after viewing many more shadow samples for each type of tree and noticing heavy variation among shadows of the same tree type, raters attempted a more "instinct-based" approach, which likely accounted for the dip in accuracy in the third trial. In trial 4, raters used a tree type narrative they had created. The narrative included descriptions and example shadows for each type of tree. Comparing the tree shadows they were observing to other tree shadows in the dataset with known labels helped each rater improve their identification accuracy, as shown in the figure. Accuracy improved when comparing unknown samples to known samples, which demonstrates the potential for machine learning techniques to be applied to this problem.

Additionally, the identification of the 100 trees during the ground truth phase of this study took 25 h to complete. On the other hand, the identification of 20 trees during the feasibility testing phase of this study took approximately one hour to complete. Thus, if this method was scaled up, trees could potentially be classified up to five times faster than traditional methods, even before the application of machine learning methods.

Results for ash-tree feasibility testing are shown in Figs. 8 and 9. In Fig. 8, the blue columns represent identification accuracies. The horizontal blue line represents the percent accuracy achieved based on a random guess. Results improved after the first trial and demonstrate potential for identification of ash trees from their shadows. In Fig. 9, blue columns represent Cohen's Kappa values. The Cohen's Kappa values for

Table 3. Tree structure observations.

Type	Example	Icon	Distinctive Characteristics in Shadows
Ash			• Trunk splits at start of canopy [17] • Fuzzy, feather-like ends of distinct terminal branches • Uneven canopy shape surrounding the main branches
Elm			• Trunk splits into many distinct branches [17] • Wavy branches • Light fuzziness • Branches grow at steep upward angle, but may angle down at ends (water fountain shape) [17]
Hackberry			• Distinct U-shaped fork in trunk occurs near the initial branches of the tree canopy [21] • Straighter branches than locust trees, which resemble hackberry trees • Many distinct branches create smooth canopy outline
Linden			• Very round, smooth, oval canopy, especially at top [21] • Relatively flat branch angle. • Trunk usually doesn't split, but if it does, the split is narrow • Few dark, distinct branches • High degree of uniform fuzziness
Locust			• Wide trunk split, sometimes U-shaped • Many distinct branches • Squiggly, spirally fuzziness • Distinct branches are squiggly and may have a few significantly curved gnarls [17, 21]

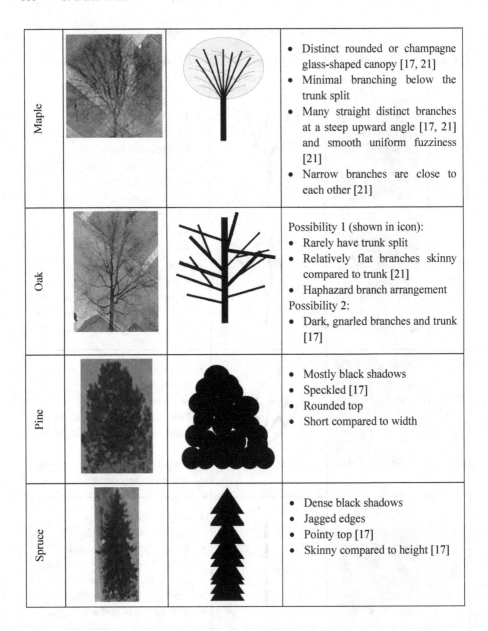

Maple			• Distinct rounded or champagne glass-shaped canopy [17, 21] • Minimal branching below the trunk split • Many straight distinct branches at a steep upward angle [17, 21] and smooth uniform fuzziness [21] • Narrow branches are close to each other [21]
Oak			Possibility 1 (shown in icon): • Rarely have trunk split • Relatively flat branches skinny compared to trunk [21] • Haphazard branch arrangement Possibility 2: • Dark, gnarled branches and trunk [17]
Pine			• Mostly black shadows • Speckled [17] • Rounded top • Short compared to width
Spruce			• Dense black shadows • Jagged edges • Pointy top [17] • Skinny compared to height [17]

ash-tree-feasibility testing are more consistent than the ten-tree-type feasibility test Cohen's Kappa value, suggesting that the differences are easier to observe. These clear differences may make ash classification more conducive to machine learning.

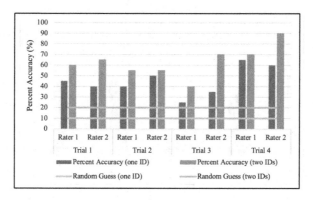

Fig. 6. Percent accuracy for each rater for ten-tree-type feasibility testing.

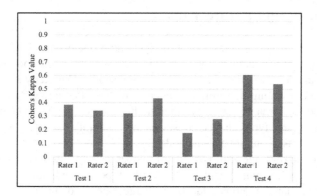

Fig. 7. Cohen's Kappa value achieved by each rater for ten-tree-type feasibility testing.

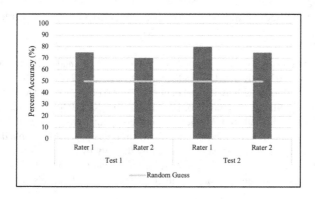

Fig. 8. Percent accuracy by each rater for ash-tree feasibility testing.

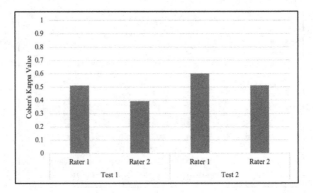

Fig. 9. Cohen's Kappa value achieved by each rater for ash-tree feasibility testing. (Color figure online)

6   Conclusions and Future Work

In this study, we created a dataset of 4,613 clear and distinct labeled tree shadows, established rules and guidelines for identifying trees based solely on shadows, and refined these rules to maximize feasibility accuracy scores. Our study shows there are distinct differences among the profile geometries of trees that allow them to be accurately classified by human raters faster than traditional field survey methods. Distinct differences among tree types also demonstrates the feasibility of classifying trees based on their shadows using automated methods.

The domain-knowledge outlined in this paper can be used to create machine learning and deep learning models that identify types of individual trees in aerial imagery covering large regions. Due to the large amount of high spatial resolution aerial imagery as well as the large volume of individual trees, supercomputers or cloud computers may be necessary to improve the scalability of this approach [18]. Additionally, as higher resolution hyperspectral and LiDAR data become more widely available, these methods could be explored in tandem to create more reliable classification results.

Acknowledgements. This study is supported by the US NSF under Grants No. 1901099, 1737633, 1541876, 1029711, IIS-1320580, 0940818 and IIS-1218168, the USDOD under Grants HM0210-13-1-0005, USDA under Grant No. 2017-51181-27222, ARPA-E under Grant No. DE-AR0000795, NIH under Grant No. UL1 TR002494, KL2 TR002492 and TL1 TR0024-93, and the OVPR U-Spatial and Minnesota Supercomputing Institute at the University of Minnesota. Aerial imagery used in this work and shown in this paper was supplied by Hennepin and Ramsey County. Field surveys were provided by the City of St. Paul Forestry Unit and the University of Minnesota. We thank Kim Koffolt for improving the readability of this paper. We also thank the Spatial Computing Group for their feedback throughout this work.

References

1. Acevedo, N.: Single fallen tree on power line leaves 900K without power. NBC News. Associated Press, Puerto Rico (2018). https://www.nbcnews.com/storyline/puerto-rico-crisis/puerto-rico-fallen-tree-power-line-leaves-900k-without-power-n865506. Accessed 1 Feb 2019
2. Åkerbloma, M., Raumonena, P., Mäkipääb, R., Kaasalainenaa, M.: Automatic tree species recognition with quantitative structure models. Remote Sens. Environ. **191**, 1–12 (2017). https://doi.org/10.1016/j.rse.2016.12.002
3. BenDor, T., Metcalf, S., Fontenot, L., Sangunett, B., Hannon, B.: Modeling the spread of the Emerald Ash Borer. Int. J. Ecol. Model. Syst. Ecol. **197**(1), 221–236 (2006). https://doi.org/10.1016/j.ecolmodel.2006.03.003
4. Citywide EAB management strategies. https://www.stpaul.gov/departments/parks-recreation/natural-resources/forestry/disease-pest-management/citywide-eab. Accessed 9 Nov 2018
5. Dalponte, M., Ørka, H.O., Gobakken, T., Gianelle, D., Næsset, E.: Tree species classification in boreal forests with hyperspectral data. IEEE Trans. Geosci. Remote Sens. **51**(5), 2632–2645 (2012). https://doi.org/10.1109/TGRS.2012.2216272
6. Elevation and Imagery. https://gis-hennepin.opendata.arcgis.com/pages/imagery. Accessed 25 Aug 2018
7. Fuller, T.: Three Weeks After Fire, Official Search for Dead is Completed. The New York Times. https://www.nytimes.com/2018/11/29/us/victims-california-fires-missing.html. Accessed 1 Feb 2019
8. Gold, R., Blunt, K., Smith, R.: PG&E sparked at least 1,500 California fires: now the utility faces collapse. Wall Street J. (2019). https://www.wsj.com/articles/pg-e-sparked-at-least-1-500-california-fires-now-the-utility-faces-collapse-11547410768. Accessed 1 Feb 2019
9. Hutson, M.: Has artificial intelligence become alchemy? Science **360**(6388), 478 (2018). https://doi.org/10.1126/science.360.6388.478
10. Lee, S.H., Chan, C.S., Wilkin, P., Remagnino, P.: Deep-plant: plant identification with convolutional neural networks. In: 2015 IEEE International Conference on Image Processing (ICIP), Quebec City, QC, Canada, 27–30 September 2015. https://doi.org/10.1109/ICIP.2015.7350839
11. Leonardi, C., Stagi, F.: The Architecture of Trees. Princeton Architectural Press, Hudson (2019). https://doi.org/10.1109/ICIP.2015.7350839
12. Malhi, Y., et al.: New perspectives on the ecology of tree structure and tree communities through terrestrial laser scanning. Interface Focus. **8**(2), 1–10 (2018). https://doi.org/10.1098/rsfs.2017.0052
13. Murfitt, J., He, Y., Yang, J., Mui, A., De Mille, K.: Ash decline assessment in emerald ash borer infested natural forests using high spatial resolution images. Remote Sens. **8**(3), 256 (2016). https://doi.org/10.3390/rs8030256
14. Nisley, R.G.: Emerald ash borer research: a decade of progress on an expanding pest problem. Northern Res. Stat. Res. Rev. **20**, 1–5 (2013)
15. Nowak, D.J., Dwyer, J.F.: Understanding the benefits and costs of urban forest ecosystems. In: Kuser, J.E. (ed.) Urban and Community Forestry in the Northeast, pp. 25–46. Springer, Dordrecht (2007). https://doi.org/10.1007/978-1-4020-4289-8_2
16. Orthophotos/Aerial 2015 (MapServer). https://maps.co.ramsey.mn.us/arcgis/rest/services/OrthoPhotos/Aerial2015/MapServer. Accessed 25 Aug 2018
17. Petrides, G.A., Wehr, J.: A Field Guide to Eastern Trees: Eastern United States and Canada, Including the Midwest. Houghton Mifflin Company, New York (1998)

18. Prasad, S.K., et al.: Parallel processing over spatial-temporal datasets from geo, bio, climate and social science communities: a research roadmap. In: 2017 IEEE International Congress on Big Data (BigData Congress), Honolulu, HI, USA, 25–30 June 2017. https://doi.org/10.1109/BigDataCongress.2017.39

19. Rathke, D.M.: Minnesota Trees. University of Minnesota Extension Service, Minnesota (2006)

20. Redmon, J., Farhadi, A.: YOLO9000: better, faster, stronger. In: Proceedings of the IEEE Conference on Computer Vision and Pattern Recognition, Honolulu, HI, 21–26 July 2017. https://doi.org/10.1109/CVPR.2017.690

21. Sibley, D.A.: The Sibley Guide to Trees. Alfred A. Knopf Inc., New York (2009)

22. Tyrväinen, L., Pauleit, S., Seeland, K., de Vries, S.: Benefits and uses of urban forests and trees. In: Konijnendijk, C.C., Nilsson, K., Randrup, T.B., Schipperijn, J. (eds.) Urban Forests and Trees, pp. 81–144. Springer, New York (2005). https://doi.org/10.1007/3-540-27684-X_5

23. Viera, A.J., Garrett, J.M.: Understanding interobserver agreement: the kappa statistic. Family Med. 37(5), 360–363 (2005)

24. Xie, Y., Bao, H., Shekhar, S., Knight, J.: A TIMBER framework for mining urban tree inventories using remote sensing datasets. In: Proceedings of 2018 IEEE International Conference on Data Mining (ICDM), Singapore, 17–20 November 2018. https://doi.org/10.1109/ICDM.2018.00183

25. Xie, Y., Bhojwani, R., Shekhar, S., Knight, J.: An unsupervised augmentation framework for deep learning based geospatial object detection: a summary of results. In: Proceedings of the ACM SIGSPATIAL International Conference on Advancements in Geographic Information Systems (SIGSPATIAL 2018), Seattle, WA, 6–9 November 2018. https://doi.org/10.1145/3274895.3274901

26. Xie, Y., Cai, J., Bhojwani, R., Shekhar, S., Knight, J.: A locally constrained YOLO framework for detecting small and densely distributed building footprints. Int. J. Geograph. Inf. Sci. 1–25 (2019). https://doi.org/10.1080/13658816.2019.1624761

27. Xie, Y., Gupta, J., Li, Y., Shekhar, S.: Transforming smart cities with spatial computing. In: 2018 IEEE International Smart Cities Conference (ISC2), Kansas City, MO, USA, 16–19 September 2018. https://doi.org/10.1109/ISC2.2018.8656800

28. Xie, Y., Shekhar, S., Feiock, R., Knight, J.: Revolutionizing tree management via intelligent spatial techniques. Paper presented at the 27th ACM SIGSPATIAL International Conference on Advances in Geographic Information Systems (SIGSPATIAL 2019), Chicago, IL, USA, 5–8 November 2019. https://doi.org/10.1145/3347146.3359066

29. Yu, C., Li, M., Zhang, M.: Classification of dominant tree species in an urban forest park using the remote sensing image of WorldView-2. In: 8th International Congress on Image and Signal Processing (CISP), pp. 742–747 (2015). https://doi.org/10.1109/CISP.2015.7407976

30. Yuan, J.: Learning building extraction in aerial scenes with convolutional networks. IEEE Trans. Pattern Anal. Mach. Intell. 40(11), 2793–2798 (2017). https://doi.org/10.1109/TPAMI.2017.2750680

31. Zhang, C., Qiu, F.: Mapping individual tree species in an urban forest using airborne lidar data and hyperspectral imagery. Photogram. Eng. Remote Sens. 78(10), 1079–1087 (2012). https://doi.org/10.14358/PERS.78.10.1079

32. Freelon, D.: ReCal2: Reliability for 2 Coders. https://dfreelon.org/utils/recalfront/recal2/. Accessed 21 Sept 2018

Market Intelligence for Agricultural Commodities Using Forecasting and Deep Learning Techniques

Swapnil Shrivastava[1]([⊠]), Supriya N. Pal[1], and Ranjeet Walia[2]

[1] Centre for Development of Advanced Computing (CDAC),
No. 1, Old Madras Road, Byappanahalli, Bangalore 560038, India
{swapnil,supriya}@cdac.in
[2] National Institute of Technology, Anu, Hamirpur 177005,
Himachal Pradesh, India
ranjeetwalia2000@gmail.com

Abstract. About two third of the Indian population continues to live in villages and depends on agriculture as main source of livelihood. The agrarian distress that took decades to build up, surfaced in the form of decline in the proportion of rural population and increase in the number of farmer's suicide cases. There is a strong hypothesis that in addition to production techniques, the widespread availability of Market Intelligence would bring substantial improvement in financial condition of farmers. Hence there is a need for comprehensive system, which would fetch, interlink, transform and analyze relevant data from various ministries/departments/organizations spread across the country to generate precise, appropriate and timely Market Intelligence. We took a step in this direction by design and implementation of Market Intelligence System Proof of Concept (PoC) using available datasets for few agricultural commodities. This PoC takes daily market price and weather data as input, transforms it into information and generates actionable intelligence by applying forecasting and deep learning techniques. The system provides trend analysis, short term as well as long term commodity price prediction and market selection as insights for farmers. The Auto Regressive Integrated Moving Average (ARIMA) forecasting technique and Recurrent Neural Network (RNN) deep learning techniques are applied for short term and long term agricultural commodity price prediction respectively. The study results demonstrate intended utility of forecasting and deep learning techniques for generating Market Intelligence System. The paper concludes with benefits of comprehensive Market Intelligence system, challenges and future work.

Keywords: Descriptive analytics · Predictive analytics · Forecasting · Market Intelligence · Deep learning

1 Introduction

India is widely known as the land of villages and agriculture is the largest source of livelihood for its rural population. As per 2010–11 census report, 70% of its population lives in rural area and depends primarily on agriculture for their livelihood [1]. Time to

© Springer Nature Switzerland AG 2019
S. Madria et al. (Eds.): BDA 2019, LNCS 11932, pp. 193–205, 2019.
https://doi.org/10.1007/978-3-030-37188-3_12

time actions taken by the Indian government has increased agricultural production in the country, made India self-sufficient in food and a major food exporting nation. According to Agriculture census 2010–11 report released by Department of Agriculture, Cooperation and Farmers Welfare, Government of India, approximately 85% of the farmers in the country are small and marginal land holdings farmers i.e. with land holding less than two hectares [2]. This number of small and marginal agricultural land holdings in the country has further increased in 2015–16 as compared to 2010–11 [3]. Presently due to unaffordability and unavailability of proper facilities, these farmers sell their agricultural produce to intermediaries in the village. Intermediaries in turn sell produce to traders at market located in nearby town or city. Due to lack of market information, farmers are not aware of the actual price for their produce prevailing in the market. They cannot take informed decisions such as where to sell, when to sell and what price would fetch high reward for their farm produce. Intermediaries take advantage of their ignorance and offer them un-remunerative bargain for agricultural commodities. Through the decades, level of disparity between farmer's income and income of those employed in non-agriculture sectors kept deteriorating [4]. In the recent times, authorities have notified rise in farmer's suicide cases [5] and decline in the proportion of rural population [1]. Due to unregulated agricultural marketing practices, disinterest in farming and farm investments is growing amongst farmers. Hence, in addition to production techniques, widespread availability of Market Intelligence would bring substantial improvement in financial condition of farmers [6].

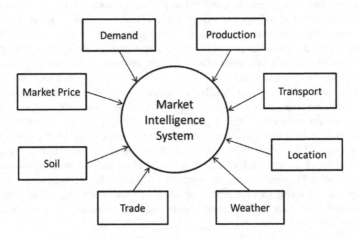

Fig. 1. Market Intelligence System

Market Intelligence is not a technique but a process that consist of collecting agricultural data, transforming it into information, extracting insights and distributing them to farmers, agriculture departments and market functionaries (e.g. traders) for decision making. The various states in India bank on interstate trade for agricultural commodities supply. In such a scenario, diffusion of Market Intelligence in terms of demand, production and market price could play a vital role. It would help the farmers to take decisions like farm produce should be sent to which market and whether to sell now or later. The data about these driving factors of market price exists as silos with various central/state government departments and organizations. For instance, daily market price and arrival information of agricultural commodities from regulated markets spread across the country are available on Agmarknet [7] and eNAM [8] portals. However to the best of our understanding, there is no comprehensive system in the country, which interlinks data from these wide varieties of data silos for generating Market Intelligence. The emergence of Big Data Analytics and Deep Learning platforms has made creation of such a system technically feasible. As shown in Fig. 1, Market Intelligence System requires real time processing and storage of heterogeneous aspects like trade reports, market price, demand, production and location.

Market Intelligence consist of insights for crop selection, market demand, commodity price prediction and so on. The most important aspect of Market Intelligence for farmer is market price [9]. The market price for a commodity is found to be highly uncertain and depends upon several environmental (soil, rainfall, temperature) as well as economic (production, demand) factors. Hence we designed and implemented a Market Intelligence System Proof of Concept for market price using forecasting and deep learning techniques on available daily market price of few agricultural commodities and weather data (temperature, rainfall) as input. The data is fetched from these input data sources, integrated and transformed into appropriate form (information) for analysis and prediction. The system provides trend analysis, short term as well as long term commodity price prediction and market selection as actionable insights from this information. These outcomes or insights are believed to be effective and applicable for farmer's decision-making.

The remaining paper reviews existing systems that collects agricultural data, prevalent market intelligence dissemination practices, and various forecasting as well as deep learning studies for commodity price prediction in the Literature Survey section. The Sect. 3 describes the design, implementation and outcome of a Market Intelligence System Proof of Concept. The Sect. 4 mentions benefit of the system, Sect. 5 discusses challenges to be addressed and Sect. 6 concludes the paper.

2 Literature Survey

In comparison to Healthcare, Banking and eGovernance, the propagation of technology in Agriculture sector until recently was in elementary stage. Soil Health Card, Agmarknet and Kisaan Call Centre are ICT initiatives by government for the benefit of farmers. The data about land utilization consisting of area, yield and production as well as demand and supply projection of various agricultural commodities are available on Open Government Data (OGD) platform [10]. The weather information and alert generation mechanism in case of natural calamity are provided by Agricultural Meteorology Division portal [11]. The central and state government departments collect agricultural data on wholesale and retail prices from various markets. This information is circulated to farmers through periodical bulletins and notice boards in villages. However this is not an efficient information dissemination method because of information lag. In other words, by the time market price reaches farmers, actual price in the market would have changed. With emergence of new communication medium, market information was also provided via television, radio and print media. However, there is a strong need for comprehensive solution to fetch data from various agricultural departments/agencies, integrate and convert this vast volume of data into information for actionable insight generation. This market intelligence and information should be personalized and easily accessible to the farmer. In the recent times, there has been a great enthusiasm for the notion of Big Data. The potential of data-driven application and decision-making is now being recognized by academia, industry and government. The challenges and opportunities introduced by Big Data are modeled across three dimensions i.e., variety, velocity and volume [12]. This definition was revised by Gartner to bring out the need for new ways of storage and processing that supports these three characteristics of Big Data to enable enhanced decision making and novel insight discovery [13]. Big Data Analytics performs processing and analysis on huge volume of data coming at different speed from wide variety of data sources to derive novel insights from them for decision making.

The agricultural and related data such as daily market prices, weather information and soil quality are time series in nature. Time series data is a sequence of well-defined data points measured at certain interval of time. Agmarknet and eNAM provide functionalities for trend analysis on daily market price of commodities. For example, they provide information on variation in daily market price with respect to time, market and commodity. However Market Intelligence is not just about monitoring the past but also making predictions about the future with certain level of confidence. The forecasting and deep learning techniques are appropriate for prediction based on time series data. These techniques discover a pattern in the historical time series data and then extrapolate this pattern into the future values of the time series. There are several studies that discuss application of forecasting [14] and deep learning [15] techniques for agricultural commodity price prediction. The Auto Regressive Integrated Moving Average (ARIMA) model was applied on monthly Onion price at Belgaum district of Karnataka, India for forecasting Onion price [16]. The monthly Onion price for 15 years duration starting from 1997 to 2012 was considered in this study. The forecast values for price showed an increasing trend, which meant that farmers could expect higher return from onion production. In a separate study [17], market price data was collected for potatoes from an organized market in the Nagaon district of Assam state in India. Markov Chain Model was built on this data of three year duration to predict the future price for short period of next fifteen days. The forecasted values obtained from this study were highly accurate and almost same as the actual values. A comparative study of various forecasting and deep learning techniques is performed for price prediction over fruit and vegetable market data. They were ARIMA, Partial Least Square (PLS), Artificial Neural Network (ANN) and RSMPLS obtained by integration of Response Surface Methodology (RSM) to PLS algorithms. Of these PLS and ANN algorithms had high accuracy and were recommended for short term as well as long term forecasting, respectively [18]. In a separate paper, ARIMA model and PLS regression method are combined in the form of mixed model. It also provides price warning functionality using neural network [19]. The historical market price data from different markets was input to mixed model to forecast weekly prices. The accuracy of mixed model is more as compared to each individual model. Usage of Neural Network increased the accuracy of warning values. An optimization model is devised to identify the optimal place and optimal time for farmers to sell their farm produce [20]. The factors like distance of the market from the farmers location, type of produce, transportation cost and time are instrumental in influencing farmer's choice of market to sell agricultural commodities. As mentioned in previous section, price fluctuations of agricultural commodities are influenced by several economic (e.g. market price, demand) and environmental (e.g. climate, irrigation type) factors. Hence forecasting or prediction techniques that involves application of single model or considers only one factor i.e. market price are inadequate for agricultural commodity price prediction. The prediction or forecasting of market price or demand of agricultural commodity is a complex problem that requires various parameters and novel solutions. As a step in this direction, design and implementation of Market Intelligence System PoC for few agricultural commodities that considers daily market price along with weather data for market price prediction is discussed in the next section.

3   Market Intelligence System PoC: Design and Implementation

Market Intelligence is not just a technique but a process that transforms agricultural data to agricultural information, to actionable intelligence and disseminates them to farmers, government departments and market functionaries (e.g. traders) for decision making. The high level design of Market Intelligence System Proof of Concept is as shown in Fig. 2. It is composed of the following main components viz. Data Source, Data Pre-processing, Data Repository, Market Insights and User Interface.

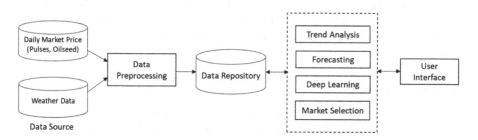

Fig. 2. High level design of Market Intelligence System PoC

3.1   Data Source

The daily market price and weather datasets are two data sources of agricultural data that are taken as input to this system. The first data source is daily market price dataset with data from 2012 to 2017 provided by Agmarknet [8]. This dataset contains daily market price data for two agricultural commodity types i.e. Oilseed and Pulses. This dataset is comprised of daily market price records collected from 1844 agricultural markets spread across 435 cities in 28 states/union territories in the country. The dataset contains agricultural commodity name, state, district, market, date of arrival, minimum price of the day, maximum price of the day, average price of the day and total quantity that arrived on the day, as attributes. As mentioned in the previous section, market price for a commodity is highly uncertain and depends upon several factors like rainfall, temperature, production, demand etc.: In this PoC, we considered freely available weather data for same duration and locations (districts or cities) as that of Agmarknet dataset. The weather data has local time, cloudiness, precipitation, pressure, temperature, humidity, wind (speed, gusts, direction), sunrise/sunset, moon rise/moonset and moon phase as attributes [21].

3.2 Data Pre-processing

This component receives data from data sources discussed in previous sub section, performs various data preprocessing techniques and sends transformed data i.e. information to the Data Repository for storage. The daily market price was provided year wise in separate csv files by Agmarknet. We observed that number of records increased whereas number of missing entries reduced from 2012 to 2017. The NA value was also observed in market price attribute of many records. To get accurate and quality results, analytics is required to be performed on continuous and complete data. Hence we performed data cleaning techniques for filling of missing values, smoothing noisy data and filling missing records. The missing record for a day of month was filled by generating random market price values based on records available for that month. The data integration scripts were written to append market information from different csv files. In this appended market price dataset, temperature and rainfall attributes of weather data were merged for each day. The data transformation technique was used to normalize or adjust values measured on different scales to a common scale. For example market price values were normalized for continuous compounding of returns over time required in RNN model to eliminate false result. The data reduction techniques like correlation were applied to select attributes that were relevant for further analysis.

3.3 Data Repository

The extracted, cleaned and integrated data from data preprocessing component is loaded into the Data Repository. As shown in Fig. 3, data is stored in a data model appropriate for the analytics and visualization requirement of the system. The date and daily market price of agricultural commodity are required for forecasting using ARIMA model. Whereas model for market price prediction using deep learning techniques is comprised of multiple features including market price, rainfall and temperature. The Daily Market Price table in this component contains unique row id, commodity id, market id, date of arrival, average price of the day, arrival quantity, temperature and rainfall attributes. The Market, District and State tables capture the parent child hierarchy amongst them i.e. market belongs to a district and district belongs to a state. The Commodity table and Market table are referenced by the Daily Market Price table using commodity id and market id respectively. The commodity name from Commodity table has a commodity type. For example, Arhar/Tuvar Dal commodity belongs to commodity type Pulses.

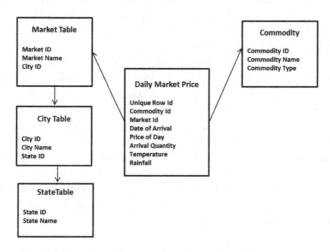

Fig. 3. Curated and integrated market price and weather data

3.4   Market Insights

This component consist of descriptive analytics, forecasting and deep learning techniques for analyzing the data stored in the data repository to generate insights for informed decision making. These insights are broadly categorized as short term forecasting, long term forecasting, trend analysis and market selection.

The Auto Regressive Integrated Moving Average (ARIMA) model is a statistical method used for analyzing and forecasting time series data. It has been used in this system for short term (daily, weekly) market price forecasting for agricultural commodities. The ARIMA model is built for date of arrival and market price attribute values for a commodity sold at a market located in a district/city of a state. ARIMA model is specified by three parameters viz p, d and q. Where p is the number of auto references needed to achieve stationarity and q is the number of lagged forecast errors in the prediction equation. Autoregressive refer to representing current value of an attribute in terms of prior value of the same attribute. Autocorrelation functions (ACF and PACF) are used to study characteristics of time series (stationary/not stationary) and also to obtain the values of p and q. ARIMA model works on the assumption that the data is stationary i.e. there is no trend and seasonality component present in the data. If a time series is not stationary, then it can be made nearly stationary by taking the first level of differencing as one. This removes trend and seasonality components from the time series data to make the data stationary. In this PoC, p, d and q values were fetched by applying Auto ARIMA function on the required dataset. The ARIMA model is specified with lag value of 2 i.e. previous two values is used to forecast next price. The degree of differencing is given as one and two lagged value of the error in the model. The ACF and PACF function graphs for this model were plotted. The PACF graph showed a definite pattern, which did not repeat. This means that the time series does not show any seasonality. The ARIMA model is said to work well for short term prediction. This model was tested to forecast market price for next day to next week of certain commodity (e.g. Arhar Dal) for certain market (e.g. Garhwal) in some city (e.g. Garhwal) of a state (e.g. Jharkhand). The forecast result would help the farmer to take informed decision on when to sell the commodity.

Recurrent Neural Network (RNN) is a deep learning technique that is used for processing sequential data. It is a standard feed forward network and due to this underlying characteristic, it exhibits temporal dynamic behavior. RNN can use its internal state (memory) to process sequences of inputs. RNN is a network of neurons that are interconnected and organized in the form of layers. There are three layers viz input layer that receives data from external data source, one or more hidden layers that modifies or processes data while enroute from input layer to output layer, and output layer that yields the output. Long Short Term Memory (LSTM) is a type of RNN that replaces traditional artificial neurons in hidden layer with memory cell that is a unit of computation. This cell helps the network to trace the past time series records and hence can comprehend the structure of data dynamically over time with high prediction capacity.

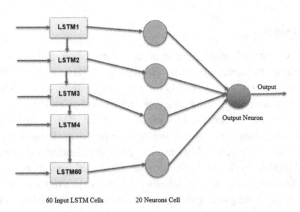

Fig. 4. Commodity price prediction using RNN

The date of arrival along with multiple features including market price, rainfall and temperature are fetched from data repository as input for RNN model for this system. RNN model built in this system for commodity price prediction like any other RNN model is composed of three layers as shown in Fig. 4. LSTM is used in input layer to process the data fetched from data repository. The data fetched from data repository is normalized (brought in range of 0 and 1) before sending it as input to LSTM. To predict market price for nth day, a batch of size 40 with records from n-1 to n-40th day is formed. There are 60 cells in LSTM (input layer) that performs three key tasks i.e. forget, update and output of memory cell. The output of LSTM is sent to hidden layer comprising of 20 neurons. The hidden layer is fully-connected dense layer. These two layers provide back-propagation that sustains efficient learning rate and minimization of network loss. The output layer has one neuron which provides predicted price value for the next day. The model has been trained over 200 epochs to calculate the loss and accuracy. The calculated loss is Mean Absolute Percentage Error (MAPE) between predicted and actual price. To minimize loss, this RNN-LSTM model has been trained with 80% of data from dataset and tested with the remaining 20% of data. This model was tested to predict market price value for 30 day duration for certain commodity (e.g. Arhar Dal) for certain market (e.g. Garhwal) in some city (e.g. Garhwal) of a state (e.g. Jharkhand).

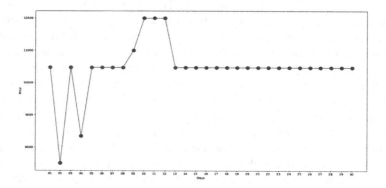

Fig. 5. Trend analysis of daily market price

Descriptive Analytics includes statistical and visualization methods to derive summary from the historical data. The line graph plotting daily market price w.r.t. time provides trend analysis of agricultural commodity price in a market as shown in Fig. 5. This trend analysis would help the farmer and authorities to monitor the variation in daily market price.

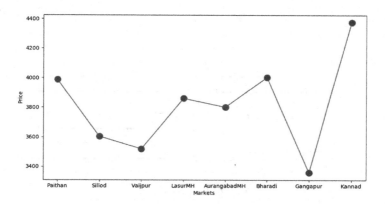

Fig. 6. Market selection for highest remuneration

The Market Selection feature would assist the farmer to identify the market that offers highest price for their agriculture produce. The next day price predictions for Groundnut oil in various markets in Aurangabad city of Maharashtra state is shown in Fig. 6. The market named Kannad is showing the largest predicted return value for Groundnut Oil on the next day. Hence the farmer would fetch highest remuneration by selling Groundnut Oil in that market.

3.5   User Interface

The user interface of this Proof of Concept is a standalone client as shown in Fig. 7. However the convenient way of accessing functionalities of Market Intelligence System by farmers would be through mobile app. This component provides end users of the system i.e. farmers with functionalities to perform short term and long term price predictions. Initially list of states and union territories are displayed. When user selects a state/union territory, the list of all the districts or cities in that state are shown. On selection of a city or district, the list of markets in the selected district or city would be displayed. The user can also choose the commodity type followed by commodity name. The user can then click long term button to perform long term prediction or short term button for short term prediction of market price for chosen market and commodity name. The user can also select all-markets option for market and press market selection button to view market price prediction for next day at all the markets in a city or district. The interface also provides functionality for trend analysis. The user can select a commodity name, market name, start date and end date followed by trend analysis button to view the market price trend in that particular duration.

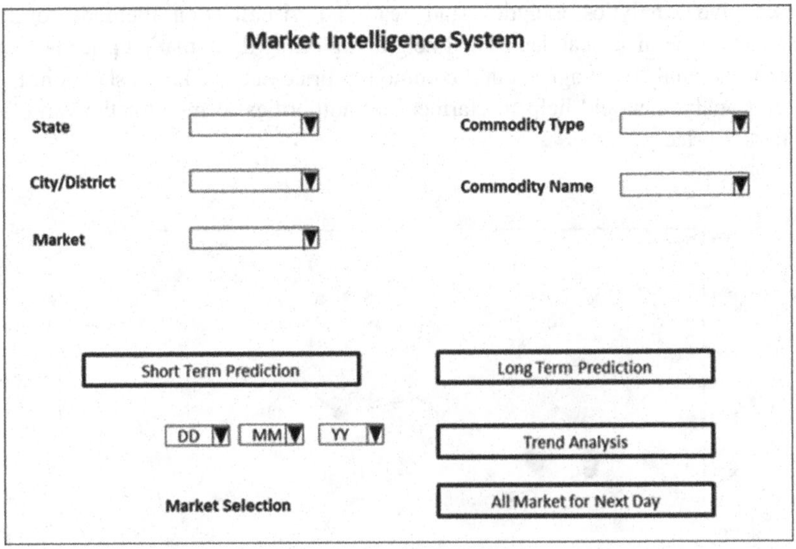

Fig. 7. User interface

The design, implementation and outcome of a Market Intelligence System PoC for few agricultural commodities using available dataset were discussed in this section. The study results demonstrate intended utility of forecasting and deep learning techniques for generating Market Intelligence System.

4 Benefits

We strongly believe that the development of comprehensive Market Intelligence System would help farmers, government authorities and market functionaries in decision making. Market Intelligence system would make farmers aware of the prevalent market prices for their agricultural commodities. This would enable farmers to strongly bargain with traders and fetch remunerative prices for their farm produce. The forecast of market price could help farmers to take decisions like whether to sell the commodity now or wait for some more time. The market selection functionality provides farmers with timely and reliable information to decide which market they should send their produce to maximize returns. The long term predictions can help farmers to make decisions related to crop selection. For example, analysis of market price for various agricultural commodities would help farmer to choose cropping patterns that results in higher value crop produce. The sharp changes in commodity price detected using trend analysis would alert authorities to investigate reason(s) for below Minimum Support Price sale. The system has got the potential to enable market functionaries to take decisions for sending the produce to markets having high demand for it. They can also take decision with objective to cut down marketing channels, minimize transport costs and so on.

5 Challenges

There are challenges from technical, societal and policy aspects foreseen in implementation of comprehensive Market Intelligence System. To the best of our knowledge, building highly accurate price and demand prediction model that depends upon several environmental and economic factors is still an unattended task. The prediction or forecasting of market price or demand of agricultural commodity is a complex problem that requires various parameters and novel solutions. The selection of market in this system is presently done based on single criteria i.e. market price. However there are several environmental factors like transportation cost, time taken to transport and the distance of the market from the farmer's location that influences the choice of market. The user interface discussed in this PoC is a standalone client. Efforts are required to take advantage of deep rooted mobile phone network in the country and make market intelligence available in the form of a mobile app. There is a need to spread awareness in various agricultural departments regarding potential of agricultural data lying in their custody. The departments should take steps for provisioning of up-to-date, curated and trustworthy data from these data silos for generation of accurate

Market Intelligence. The inclusion of technology in farming and marketing practices would also require a change in mind-set of the farmers to accept them. The creation of awareness about the available resources requires efforts at the grass root level.

6 Summary

In addition to production techniques, widespread availability of Market Intelligence would bring substantial improvement in financial condition of farmers. To the best of our understanding, there is no comprehensive system in the country, which interlinks agricultural data from various agricultural units spread across the country for generating market intelligence. We designed and implemented a preliminary Market Intelligence System with available daily market price and weather dataset for few agricultural commodities. The system provides trend analysis, short term as well as long term commodity price prediction and market selection as insights for the farmers. The Auto Regressive Integrated Moving Average (ARIMA) forecasting technique and Recurrent Neural Network (RNN) deep learning technique (RNN) are applied for short term (daily, weekly) and long term (monthly) agricultural commodity price prediction respectively. The study results demonstrate intended utility of this Market Intelligence System. We conjecture that the system could be made efficient, robust and comprehensive by onboarding various agricultural units for real time integration of factors like market price, demand, production and land utilization. The system discussed in this paper was initiated for farmers as end users. However it can be extended to enable government authorities, traders and market functionaries to take decisions.

References

1. Chandramouli, C.: Rural Urban Distribution of Population, Census of India (2011). http://cen-susindia.gov.in/
2. Agriculture Census 2010-11, All India Report on Number and Area of Operational Holdings, Ministry of Agriculture, Government of India (2014). https://agcensus.nic.in/
3. Agriculture Census 2015-16 (Phase-I) Provisional Results, All India Report on Number and Area of Operational Holdings, Ministry of Agriculture & Farmer Welfare, Government of India (2018). https://agcensus.nic.in/
4. Chand, R.: Doubling Farmer's Income: Rationale, Strategy, Prospects and Action Plan. Niti Aayog (2017)
5. Chapter 2a: Suicides in Farming Sector, Accidental Deaths and Suicides in India Report, National Crime Records Bureau (2015). http://ncrb.gov.in/
6. Kumar, M.: Problems of agriculture marketing in India. Int. J. Res. Anal. Rev. (IJRAR) 5(4), 22–27 (2018)
7. Agriculture Marketing Portal. http://agmarknet.gov.in/
8. National Agriculture Market. http://www.enam.gov.in/NAM/home/index.html
9. Swaminathana, B., Sivabalanb, K.C.: Role of Agricultural Market Intelligence in uplifting small and marginal farmers: aspects, prospects and suspects, Agricultural and Rural development for sustainable agriculture and all round welfare of rural community (2016)
10. Open Government Data (OGD) Platform, India. https://data.gov.in/

11. Agricultural Meteorology Division. http://imdagrimet.gov.in/
12. Laney, D.: 3-D Data Management: Controlling Data Volume, Velocity and Variety, META Group Original Research Note (2001)
13. Beyer, M.A., Laney, D.: The Importance of Big Data: a Sefinition. Gartner, Stamford (2012)
14. Kumar, U.D.: Business Analytics: The Science of Data-Driven Decision Making. Wiley, New York (2018)
15. Larose, D.T., Larose, C.D.: Data Mining and Predictive Analytics. Wiley, New York (2017)
16. Jalikatti, V.N., Chourad, R., Ahmad, D.G.N., Amarapurkar, S., Sheikh, S.: Forecasting the prices of onion in Belgaum market of Northern Karnataka using ARIMA technique. Int. Res. J. Agric. Econ. Stat. 5(2), 153–159 (2014)
17. Bairagi, A., Kakaty, S.C.: Markov chain modelling for prediction on future market price of potatoes with special reference to Nagaon District. IOSR J. Bus. Manage. 19(12), 25–31 (2017). Version I
18. Peng, Y.-H., Hsu, C.-S., Huang, P.-C.: Developing Crop Price Forecasting Service Using Open Data from Taiwan Markets. TAAI, Taiwan (2015)
19. Wu, H., Wu, H., Zhu, M., Chen, W., Chen, W.: A new method of large scale short term forecasting of agricultural commodity prices: illustrated by the case of agricultural markets in Beijing. J. Big Data 4, 1 (2017)
20. Kopparapu, S.K., Saxena, V.: Identifying the best market to sell: A cost function formulation. Sadhana 39(6), 409–1423 (2014)
21. Open Weather Data Portal, Russia. https://rp5.ru/Weather_in_India

Graph Analytics

TKG: Efficient Mining of Top-K Frequent Subgraphs

Philippe Fournier-Viger[1]($\boxtimes$), Chao Cheng[2], Jerry Chun-Wei Lin[3], Unil Yun[4], and R. Uday Kiran[5,6]

[1] School of Natural Sciences and Humanities,
Harbin Institute of Technology (Shenzhen), Shenzhen, China
philfv@hit.edu.cn
[2] School of Computer Sciences and Technology,
Harbin Institute of Technology (Shenzhen), Shenzhen, China
tidescheng@gmail.com
[3] Department of Computing, Mathematics and Physics, Western Norway
University of Applied Sciences (HVL), Bergen, Norway
jerrylin@ieee.org
[4] Department of Computer Engineering, Sejong University, Seoul, Republic of Korea
yunei@sejong.ac.kr
[5] The University of Tokyo, Tokyo, Japan
uday_rage@tkl.iis.u-tokyo.ac.jp
[6] National Institute of Information and Communications Technology, Tokyo, Japan

Abstract. Frequent subgraph mining is a popular data mining task, which consists of finding all subgraphs that appear in at least *minsup* graphs of a graph database. An important limitation of traditional frequent subgraph mining algorithms is that the *minsup* parameter is hard to set. If set too high, few patterns are found and useful information may be missed. But if set too low, runtimes can become very long and a huge number of patterns may be found. Finding an appropriate *minsup* value to find just enough patterns can thus be very time-consuming. This paper addresses this limitation by proposing an efficient algorithm named TKG to find the top-k frequent subgraphs, where the only parameter is k, the number of patterns to be found. The algorithm utilizes a dynamic search procedure to always explore the most promising patterns first. An extensive experimental evaluation shows that TKG has excellent performance and that it provides a valuable alternative to traditional frequent subgraph mining algorithms.

Keywords: Graph mining · Frequent subgraphs · Top-k subgraphs

1 Introduction

In the last decades, many studies have been carried out on designing efficient algorithms to discover interesting patterns in different types of data such as customer transactions [8] and sequences [6]. One of the most popular pattern

© Springer Nature Switzerland AG 2019
S. Madria et al. (Eds.): BDA 2019, LNCS 11932, pp. 209–226, 2019.
https://doi.org/10.1007/978-3-030-37188-3_13

mining task is Frequent Subgraph Mining (FSM) [10,14–16,18,20]. It consists of finding all subgraphs that appear in at least *minsup* graphs of a graph database, where *minsup* is a parameter set by the user. The number of graphs containing a pattern is called its *support*. FSM has several applications such as to analyze collections of chemical molecules to find common sub-molecules [10], and to perform graph indexing [22].

But discovering all frequent subgraphs in a set of graphs is a difficult task. To perform this task efficiently, various algorithms have been proposed using various data structures and search strategies [10]. However, traditional FSM algorithms have an important limitation, which is that it is often difficult for users to select an appropriate value for the *minsup* threshold. On one hand, if the threshold is set too low, few patterns are found, and the user may miss valuable information. On the other hand, if the threshold is set too high, millions of patterns may be found, and algorithms may have very long execution times, or even run out of memory or storage space. Since users typically have limited time and storage space to analyze patterns, they are generally interested in finding enough but not too many patterns. Finding a suitable *minsup* value that will yield just enough patterns is difficult because it depends on dataset characteristics that are generally unknown to the user. Thus, many users will run an FSM algorithm several times with different *minsup* values using a trial-and-error approach until enough patterns are found, which is time-consuming.

To address this issue, Li et al. [13] proposed the TGP algorithm to directly find the k most frequent closed subgraphs in a graph database, where k is set by the user instead of the *minsup* threshold. This approach has the advantage of being intuitive for the user as one can directly specify the number of patterns to be found. However, a major issue is that TGP explicitly generates all patterns to then find the top-k closed patterns. Since the number of patterns can increase exponentially with the size of a graph, this approach is inefficient even for moderately large graph databases. In fact, Li et al. [13] reported that the TGP algorithm could not be applied on the Chemical340 dataset, although it had been commonly used to evaluate prior FSM algorithms [20]. To cope with the fact that top-k subgraph mining is more difficult than traditional FSM, researchers have then developed approximate algorithms. The FS^3 algorithm can find an approximate solution to the top-k frequent subgraph mining problem using sampling. Moreover, two approximate top-k frequent subgraph mining algorithms based on sampling were proposed for mining a restricted type of graphs called induced subgraphs [3,4]. However, a major problem is that approximate algorithms cannot guarantee finding all patterns, and may thus miss important information.

To provide an efficient algorithm for top-k frequent subgraph mining that can guarantee finding all frequent subgraphs, this paper proposes an algorithm named *TKG* (Top-K Graph miner). It starts searching for patterns using an internal *minsup* threshold set to 0 and gradually raises the threshold as patterns are found. To ensure that the threshold can be raised as quickly as possible and efficiently reduce the search space, TKG relies on a search procedure that dynamically selects the next promising patterns to be explored. As it will be shown in the experimental evaluation of this paper, TKG has excellent performance on standard benchmark datasets, including the Chemical340 dataset for which the TGP algorithm could not run. Moreover, it was observed that the performance of TKG is close to that of the state-of-the-art gSpan algorithm for FSM, even though top-k subgraph mining is a more difficult problem than FSM. Hence, TKG provides a valuable and efficient alternative to traditional FSM algorithms.

The rest of this paper is organized as follows. Section 2 reviews related work. Section 3 describes the problems of (top-k) frequent subgraph mining. Then, Sect. 4 presents the proposed algorithm, Sect. 5 describes the experimental evaluation, and Sect. 6 draws a conclusion.

2 Related Work

The problem of frequent subgraph mining was introduced by Inokuchi et al. [9]. They proposed an algorithm named AGM that can discover all frequent connected and disconnected sub-graphs. It utilizes a breadth-first search where pairs of subgraphs of a size u are combined to generate candidate subgraphs of size $(u + 1)$. A similar breadth-first search is used by the FSG algorithm [11]. A drawback of this approach is that it can generate numerous candidates that are infrequent or do not exist in the database, and thus these algorithms may waste a considerable time evaluating infrequent subgraphs. To address these issues, the gSpan algorithm [20] was proposed. To avoid generating candidates, gSpan adopts a pattern-growth approach, which recursively grows patterns by scanning the graph database. Furthermore, to efficiently detect if a newly found subgraph is isomorphic to an already found subgraph, a novel representation of graphs called *Depth-First-Search code* (DFS code) was introduced. Though several other FSM algorithms have then been proposed [10,16], gSpan remains by far the most popular due to its efficiency and because it can be easily extended to handle other subgraph mining problems and constraints [10]. For example, CloseGraph is a popular extension of gSpan [21] to mine a subset of frequent subgraphs called closed patterns (subgraphs that have no supergraph having the same support).

Although traditional FSM algorithms have many applications, how to set the *minsup* threshold is not intuitive. To address, this issue, the TGP [13] algorithm was designed to find the top-k closed subgraphs. For this problem, two key challenges are how to find top-k patterns and how to determine if a pattern is closed. The solution proposed in TGP is to initially scan the database to calculate the DFS codes of all subgraphs of each input graph, and combine all these DFS code in a huge structure called the *Lexicographical pattern net*. In this structure each subgraph is linked to its immediate super-graphs, which allows to quickly check if a subgraph is closed. Then, TGP starts to search for the top-k closed subgraphs using that structure, while gradually raising an internal *minsup* threshold initially set to 0. Though, this approach guarantees finding the top-k closed subgraphs, it is inefficient in time and memory because the DFS codes of all patterns must be calculated and stored in memory. This structure is huge because the number of subgraphs can increase exponentially with graph size. As a result, TKG is unable to run on moderately large datasets such as the Chemical Compound benchmark dataset (also known as Chemical340), where the largest graph has 214 edges and 214 vertices. But traditional FSM algorithms relying on DFS codes such as gSpan and CloseGraph can run very efficiently on this dataset.

Then, the FS^3 (Fixed Sized Subgraph Sampler) algorithm was proposed to find an approximate set of top-k frequent subgraphs [17]. This algorithm was designed with the idea of trading result completeness and accuracy for efficiency. To apply FS^3, the user must specify a number of iterations, a fixed size p for subgraphs to be found, and the number of patterns k to discover. To find frequent patterns, FS^3 performs two phase sampling: (1) it first samples a graph from the database, and then (2) samples a p-size subgraph biased toward frequent subgraphs in the whole database using the Markov Chain Monte Carlo method. This process is repeated until the maximum number of iterations is reached, and a priority queue structure is used to maintain a list of the k best (most frequent) sampled subgraphs. An advantage of this approach is that it is very fast as it avoids calculating subgraph isomorphism, one of the costliest operations in FSM that is NP-complete. But an important drawback is that the support of patterns is approximately calculated. As a result, the FS^3 algorithm can not only miss frequent or top-k patterns due to sampling, but it may also return infrequent patterns. Moreover, another serious limitation is that a fixed subgraph size must be set by the user. Setting this parameter is not intuitive, and restricting the search to a fixed size can result in missing several interesting patterns.

Then, another approximate algorithm for mining top-k fixed size frequent subgraphs was proposed, named kFSIM [3]. It adopts a similar sampling approach as FS^3, which also avoids subgraph isomorphism checks but may incorrectly calculate the support of patterns. kFSIM relies on a novel measure called $indFreq$ to accelerate support calculation and improves its accuracy. kFSIM is designed for handling a restricted type of graphs called induced subgraphs [3,4], and was shown to outperform FS^3 in terms of accuracy and runtimes on real datasets. Then, the authors of kFSIM proposed a similar algorithm named kFSIM [3] for finding top-k frequent fixed size induced subgraphs in a stream using sampling and a window [3]. However, it also an approximate algorithm that cannot guarantee result completeness and accuracy.

To address the aforementioned drawbacks of previous algorithms, this paper present an efficient algorithm named TKG that is exact (find all top-k frequent subgraphs) and has runtimes that are close to those of the gSpan algorithm for traditional frequent subgraph mining. As it will be shown in the experimental evaluation section, TKG runs efficiently on the Chemical340 dataset, where TGP could not run. The next section introduces preliminaries and formally defines the problems of FSM and top-k FSM. Then, the next section presents the proposed algorithm.

3   Preliminaries and Problem Definition

Frequent subgraph mining is applied on a database of labeled graphs [9,10,20].

Definition 1 (Labeled graph). *Formally, a labeled graph is defined as a tuple $G = (V, E, L_V, L_E, \phi_V, \phi_E)$ in which V, E, L_V and L_E are the sets of edges, vertices, vertex labels and edge labels, respectively. Furthermore, ϕ_V and ϕ_E are functions that map vertices and edges to their labels, respectively ($\phi_V : V \to L_V$ and $\phi_E : E \to L_E$).*

Furthermore, it is assumed that graphs are connected (one can follows graph edges to reach a vertex from any other vertex), do not contain self-loops (an edge from a vertex to itself) and multiple edges between pairs of vertices.

Definition 2 (Graph database). *A graph database $GD = \{G_1, G_2 \ldots G_n\}$ is defined as a set of n labeled graphs.*

For example, Fig. 1 shows a graph database containing three graphs denoted as G_1, G_2 and G_3. The graph G_3 contains four vertices and four edges. The edge labels are $L_E = \{x, y, z, w\}$ and the vertex labels are $L_V = \{A, B\}$. This database will be used as running example.

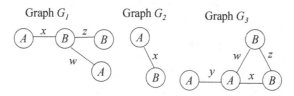

Fig. 1. A graph database containing three graphs

The goal of frequent subgraph mining is to find patterns having a high support (occurring in many graphs). The support is defined based on the concept of graph isomorphism and subgraph isomorphism.

Definition 3 (Graph isomorphism). *Let there be a labeled graph $G_x = (V_x, E_x, L_{xV}, L_{xE}, \phi_{xV}, \phi_{xE})$ and another labeled graph $G_y = (V_y, E_y, L_{yV}, L_{yE}, \phi_{yV}, \phi_{yE})$. It is said that the graph G_x is isomorphic to G_y if there is a bijective mapping $f : V_x \to V_y$ meeting two conditions. First, for any vertex $v \in V_x$, it follows that $L_{xV}(v) = L_{yV}(f(v))$. Second, for any pair $(u, v) \in E_x$, it follows that $(f(u), f(v)) \in E_y$ and $L_{xE}(u, v) = L_{yE}(f(u), f(v))$.*

Intuitively, if G_x is isormorphic to G_y, it means that the two graphs are equivalent because labels from nodes and edges from one graph can be mapped to the other while preserving the same graph structure. To check if a subgraph appears in a graph, the relationship of subgraph isomorphism, and the concept of support are defined as follows.

Definition 4 (Subgraph isomorphism). *Let there be two graphs $G_x = (V_x, E_x, L_{xV}, L_{xE}, \phi_{xV}, \phi_{xE})$ and $G_z = (V_z, E_z, L_{zV}, L_{zE}, \phi_{zV}, \phi_{zE})$. It is said that G_x appears in the graph G_z, or equivalently that G_x is a subgraph isomorphism of G_z, if G_x is isomorphic to a subgraph $G_y \subseteq G_z$.*

Definition 5. *Let there be a graph database GD. The support (occurrence fre-quency) of a subgraph G_x in GD is defined as $sup(G_x) = |\{g|g \in GD \land G_x \sqsubseteq g\}|$.*

In other words, the support of a subgraph g in a database is the number of graphs that contains g. For example, the graph (A)—x—(B) appears in the graph G_1, G_2 and G_3, and thus has a support of 3. The graph (A)—w—(B) only appears in G_1 and G_3, and hence has a support of 2.

It is to be noted that the support of a subgraph remains the same if a subgraph appears once or multiple times in the same graph. The problem of frequent subgraph mining is defined as follows.

Definition 6 (Frequent Subgraph mining). *Let there be a user-defined threshold minsup > 0 and a graph database GD. The problem of frequent sub-graph mining consists of finding all subgraphs that have a support no less than minsup.*

For example, Fig. 2 shows the seven frequent subgraphs found in the graph database of Fig. 1 for $minsup = 2$, denoted as $g_1, g_2, \ldots, g_7$.

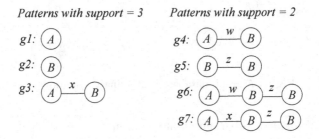

Fig. 2. Frequent subgraphs for $minsup = 2$, i.e. top-k frequent subgraphs for $k = 7$

The problem of mining the top-k frequent subgraphs addressed in this paper is a variation of the problem of frequent subgraph mining where $minsup$ is replaced by the parameter k.

Definition 7 (Top-k Frequent Subgraph mining). *Let there be a user-defined parameter $k \geq 1$ and a graph database GD. The problem of top-k frequent subgraph mining consists of finding a set T of k subgraphs such that their support is greater or equal to that of any other subgraphs not in T.*

For example, Fig. 2 shows the top-k frequent subgraphs found in the graph database of Fig. 1 for $k = 7$. If $k = 3$, only the subgraphs g_1, g_2 and g_3 are found. These subgraphs are the top-3 frequent subgraphs because no other subgraphs have a higher support.

It is important to note that in some cases, more than k patterns could be included in the set T, and thus that there can be several good solutions to top-k frequent subgraph mining problem. This is for example the case if $m > k$

patterns have exactly the same support. Moreover, it is possible that T contains less than k patterns for very small graph databases where the number of possible patterns is less than k.

The problem of top-k frequent subgraph mining is more difficult than the problem of frequent subgraph mining because the optimal minimum support value to obtain the k most frequent patterns is not known beforehand. As a consequence, all patterns having a support greater than zero may have to be considered to find the top-k patterns. Thus, the search space of top-k frequent subgraph mining is always greater or equal to that of frequent subgraph mining when the minimum support threshold is set to the optimal value.

To find the top-k frequent subgraphs efficiently, the next section presents the proposed TKG algorithm.

4   The TKG Algorithm

The designed TKG algorithm performs a search for frequent subgraphs while keeping a list of the current best subgraphs found until now. TKG relies on an internal *minsup* threshold initially set to 1, which is then gradually increased as more patterns are found. Increasing the internal threshold allows to reduce the search space.

To explore the search space of subgraphs, TKG reuses the concept of rightmost path extension and DFS code, introduced in the gSpan algorithm [20]. A reason for using these concepts is that it allows to explore the search space while avoiding generating candidates[1]. Moreover, it allows to avoid using a breadth-first search, which is key to design an efficient top-k algorithm. The reason is that if a top-k pattern has a very large size, a top-k algorithm based on a breadth-first search could be forced to generate all patterns up to that size, which would be inefficient. Using rightmost path extension and DFS codes allows to search in different orders such as using a depth-first search. Moreover, these concepts were shown to be one of the best way for tackling subgraph mining problems [10,21,24]. It is to be noted that while the gSpan algorithm utilizes a depth-first search for frequent subgraph mining, the proposed algorithm utilizes a novel search space traversal approach called *dynamic search* to always explore the most promising patterns first. This allows to guide the search towards frequent subgraphs, raise the internal *minsup* threshold more quickly, and thus reduce a larger part of the search space. As it will be shown in the experimental evaluation, the proposed dynamic search traversal greatly improves the efficiency of top-k frequent subgraph mining compared to using a depth-first search.

Before presenting the details of the proposed algorithm, the next subsection introduces key concepts related to rightmost path extension and DFS codes.

---

[1] *Here, generating a candidate* means to combine two subgraphs to obtain another subgraph that may or may not exist in the database [20]. This is done by algorithms such as AGM [9] and FSG [11] to explore the search space.

4.1   Rightmost Path Extensions and DFS Codes

A key challenge in frequent subgraph mining is to have a method that allows to systematically enumerate all subgraphs appearing in a graph. A popular solution to this problem is to use the concept of rightmost path extension [20], which allows to consider all edges of a graph using a depth-first search, without considering the same edge twice. In that context, an *extension* means an edge of a graph that can extend a subgraph.

Definition 8 (Rightmost path extension). *A depth-first search can be performed over a graph using a recursive stack. Vertices in the recursive stack are said to form the rightmost path in the graph, and the currently processed vertex is called the rightmost vertex. Rightmost path extension consists of performing two types of extensions: forward extensions and backward extensions. Backward extensions are performed before forward extensions and are used for visiting edges that will form cycles (larger cycles are preferred). Forward extensions are used for visiting edges that lead to new vertices. For forward extensions, extension of the rightmost vertex is considered first, and then extensions of vertices on the rightmost path (which makes it a depth-first search).*

For instance, consider a depth-first search over the graph G_3 of the running example, where nodes are visited according to the order depicted on Fig. 3 (left), where numbers '1', '2', '3' denote the visiting order of vertices. When the depth-first search reaches node '3', vertex 3 is the rightmost vertex, (A)—x—(B) is a backward edge and (A)—y—(A) is a forward edge. Thus, the edge (A)—x—(B) will be considered next to pursue the search, and then (A)—y—(A).

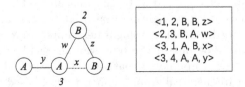

```
<1, 2, B, B, z>
<2, 3, B, A, w>
<3, 1, A, B, x>
<3, 4, A, A, y>
```

Fig. 3. (left) A rightmost path over the graph G_3 of Fig. 1, where '1', '2' and '3' is the vertex visiting order, $\langle 1, 2, 3 \rangle$ is a rightmost path and vertex 3 is the rightmost vertex. (right) The DFS code obtained after extending that path with the backward edge A—x—B and then the forward edge A—y—A.

For a subgraph g and a graph G_i of a graph database, the gSpan algorithm [20] first finds an isomorphic mapping from g to G_i. Then, it applies the concept of rightmost path extension to finds subgraphs that can extend g with an additional edge, and at the same time calculates its support. Recursively applying the concept of rightmost extension ensures that all subgraphs can be eventually considered. However, two extensions may still yield two subgraphs that are isomorphic (that are equivalent). It is thus important to identify all such duplicates during the mining process to avoid considering a same subgraph multiple times. To solve this problem, a code called DFS code was proposed to represent each subgraph [20], which allows to identify duplicates.

Definition 9 (Extended edges). *Let there be an edge between two vertices v_i and v_j and ϕ be the labeling function (ϕ_V or ϕ_E). A tuple $\langle v_i, v_j, \phi(v_i), \phi(v_j), \phi(v_i, v_j) \rangle$ representing the edge, its label and the vertex labels is called an extended edge.*

For example, in Fig. 3, $\langle 1, 2, B, B, z \rangle$ and $\langle 2, 3, B, A, w \rangle$ are extended edges. Moreover, the edges (A)—x—(B) and (A)—y—(A) are represented by the extended edges $\langle 3, 1, A, B, x \rangle$ and $\langle 3, 4, A, A, y \rangle$, respectively.

Definition 10 (DFS code). *The DFS code of a graph is a sequence of extended edges, sorted in depth-first search order.*

Continuing the previous example of Fig. 3 (left), if the backward edge (A)—x—(B) and forward edge (A)—y—(A) are used as extension, the DFS code of the resulting subgraph is the sequence of four extended edges, shown in Fig. 3 (right).

From a DFS code, one can recover the corresponding graph in the original visiting order. A graph can have many different DFS codes. To consider a single DFS code for each graph, a total order on extended edges is defined [20].

Definition 11 (Total order of extended edges). *Let t_1 and t_2 be two extended edges:*

$$t_1 = \langle v_i, v_j, L(v_i), L(v_j), L(v_i, v_j) \rangle$$

$$t_2 = \langle v_x, v_y, L(v_x), L(v_y), L(v_x, v_y) \rangle$$

The edge t_1 is said to be smaller than t_2 if and only if (i) $(v_i, v_j) <_e (v_x, v_y)$ (ii) $(v_i, v_j) =_e (v_x, v_y)$ and $\langle L(v_i), L(v_j), L(v_i, v_j) \rangle <_l \langle L(v_x), L(v_y), L(v_x, v_y) \rangle$. Relationship $<_e$ is consistent with the rule for rightmost path extension, that is, for $e_{ij} = (v_i, v_j)$ and $e_{xy} = (v_x, v_y)$, $e_{ij} <_e e_{xy}$ if and only if (a) e_{ij} and e_{xy} are both forward edges, then $j < y$ or $j = y$ and $i > x$; (b) e_{ij} and e_{xy} are both backward edges, then $i < x$ or $i = x$ and $i > x$; (c) e_{ij} is a forward edge and e_{xy} is a backward edge, then $j \leq x$; (d) e_{ij} is a backward edge and e_{xy} is a forward edge, then $i < y$. $<_l$ is consistent with the lexicographic order.

This total order allows to order DFS codes. For example, in the simple graph (A)—x—(B)—z—(B), a DFS code beginning with $\langle 0, 1, A, B, x \rangle$ is smaller than those beginning with $\langle 0, 1, B, A, x \rangle$ or $\langle 0, 1, B, B, z \rangle$.

Definition 12 (Canonical DFS code). *A DFS code is called canonical if and only if it has the least order among all DFS code corresponding to the same graph.*

The property that each graph has only one canonical DFS code allows to efficiently detect duplicate subgraphs. During the search for subgraphs, a graph can be checked for canonicity and if it is non canonical, it can be ignored. This eliminates the need of comparing a subgraph with previously considered subgraphs to determine if it is a duplicate [20].

4.2   The Algorithm

The proposed TKG algorithm takes as input a graph database and a parameter k. It outputs the set T of the top-k frequent subgraphs. The pseudocode is shown in Algorithm 1.

Algorithm 1. The TKG algorithm

input : GD: a graph database, k: a user-specified number of patterns
output: the top-k frequent subgraphs

1 Initialize a priority queue Q_K for storing the current top-k frequent subgraphs, where subgraphs with smaller support have higher priority.
2 Initialize a priority queue Q_c for storing candidate subgraphs for next extension, where subgraphs with higher support have higher priority. Initially, contains an empty graph.
3 $minsup = 1$
4 **while** Q_c *is not empty* **do**
5     $g \leftarrow$ pop highest priority subgraph from Q_c
6     $\varepsilon \leftarrow rightMostPathExtensions(g, GD)$ // Finds edges that can extend g and compute their support values.
7   **foreach** $(t, sup(t)) \in \varepsilon$ **do**
8        $g' \leftarrow g \cup \{t\}$ // Add the edge t to the DFS code of graph g
9        $sup(g') \leftarrow sup(t)$
10   **if** $sup(g') \geq minsup$ *and* $isCanonical(g')$ **then**
11   // Save pattern g' in list of current top-k patterns
12   Insert g' into Q_K
13   **if** $Q_K.size() \geq k$ **then**
14   // Raise the internal threshold
15   **if** $Q_K.size() > k$ **then** pop the highest priority (least support) subgraph from Q_K;
16             $minsup = sup(Q_K.peek())$
17   **end**
18   // Save g' as a candidate for future extension instead of doing a depth-first search
19   Insert g' into Q_c
20   **end**
21   **end**
22 **end**
23 Return Q_K

To dynamically search for top-k frequent subgraphs, the algorithm relies on two priority queues. The first one, Q_K, is used for storing at any time the k most frequent subgraphs found until now (Line 1). In that queue, subgraphs with lower support have higher priority. The second queue, Q_c stores subgraphs that may be extended to find larger subgraphs (Line 2). In that queue, graphs with higher support have higher priority. Initially, this queue contains only one element that is an empty graph (without edges and vertices). The algorithm utilizes an internal *minsup* threshold, initially set to the lowest value (e.g. 1) (Line 3). While Q_c is not empty (Line 4), the algorithm considers extending the most promising subgraph g (the one that has the largest support) in the queue Q_c of subgraphs to be extended (Line 5). The assumption is that subgraphs having a high support should be extended first because they are more likely to yield subgraphs having a high support, and thus to help increase the internal *minsup* threshold more quickly to reduce the search space. This graph g is popped from Q_c (Line 5). Then, the procedure $rightMostPathExtensions()$ is called with g to find all of its extensions (extended edges) and their supports (Line 6). For each extension t, the algorithm combine the extension with the original subgraph g to form a one edge larger subgraph g' (Line 7–8). If the support of g' is larger than the current *minsup* threshold, and if the newly formed g' is canonical (tested by calling the $isCanonical()$ procedure), g' is inserted into Q_K as one of the current k best frequent subgraphs (Line 9–12). Then, if the size of Q_K is larger than k, the subgraph having the highest priority (lowest support) in Q_K is popped from Q_k (Line 13–15). Moreover, if the size of Q_k is greater or equal to k, the *minsup* threshold is set to the support of the subgraph having the smallest support in Q_K (Line 16). Then, rather than immediately considering extending g', the algorithm stores g' in Q_c as a graph that may be eventually extended (Line 19). Then, the algorithm processes other extensions of g (Line 7–20). Then, the algorithm continues the while loop (Line 4–22) such that the graph having the highest priority in Q_c will next be considered for extensions. We name *dynamic search* this approach of extending subgraphs having the highest support first. Note that using this approach, subgraphs are generated using a different order than the depth-first search used by gSpan. In the experimental evaluation the performance of the two search order will be compared. When the algorithm terminates, the set Q_k contains top-k frequent subgraphs.

In the proposed algorithm, the procedure $rightMostPathExtension(g, GD)$ finds all extensions of a graph g (represented by its DFS code). This is done by finding all isomorphic mappings of that code to each graph G_i in the input database, and then by finding the forward and backward extensions of each mapping. The procedure $isCanonical(g')$ performs canonicality checking by recovering the graph corresponding to a DFS code g', generating the canonical DFS code g'' and comparing g' with g''. If the two codes are same, then g' is canonical. These two procedures are implemented as in the gSpan algorithm [20], and thus details about these procedure are omitted from this paper.

In terms of implementation, the TKG algorithm represents all graphs as DFS codes. In other words, g, g', and subgraphs in Q_K and Q_C are internally stored as DFS codes. And when the algorithm terminates, the DFS codes of the top-k subgraphs can be saved as graphs in an output file. In terms of data structures, priority queues can be implemented using heaps or other structures such as red-black trees. Such structures provide low complexity for inserting, deleting elements, and obtaining the element having the highest priority.

The proposed TKG algorithm is correct and complete since it relies on the concepts of DFS code and rightmost path extension introduced in gSpan to ensure that all patterns can be visited, to detect duplicates, and to calculate their support. Then, to ensure that top-k patterns are found, the algorithm starts from $minsup = 1$ and raises the minimum support threshold when at all least k patterns have been found, using the least support among the current top-k patterns. By doing so, a set of top-k most frequent subgraphs is found.

It is to be noted that a top-k problem may have more than one good solution (as explained in previous section). For example, if there is more than k patterns that have exactly the same support, more than k patterns may be considered as top-k patterns. In that case, TKG will return k of those subgraphs because the user wants k patterns. And this satisfies the problem definition. However, if one wants to keep more than k patterns, it is easy to modify Line 13 to 17 of TKG to keep more than k patterns.

Lastly, note that using the dynamic search instead of a depth-first search does not influence TKG's correctness and completeness since subgraphs are just visited in a different order. However, the dynamic search is useful to improve performance as it can help raising the internal $minsup$ threshold more quickly.

4.3   Additional Optimizations

Besides using a dynamic search to explore the search space of frequent subgraphs, two other optimizations are also proposed in the designed TKG algorithm for speeding up the pattern mining process.

The first optimization is called the *skip strategy*. Recall that for a candidate graph g for extension, the procedure $rightMostPathExtensions(g, GD)$ finds all extensions from each graph G_i in the input database. After processing a graph G_j, let $hsup$ be the highest support among the found extensions and rn be the number of remaining graphs. If $hsup + rn < minsup$, it indicates that the subgraph g cannot be used to find any frequent extensions. Therefore, the procedure $rightMostPathExtensions(g, GD)$ can stop processing the remaining graphs of the database and empty the current extension list to avoid further checking. This strategy can decrease runtimes.

The second optimization is to initially scan the database to calculate the support of all single edge graphs and then to use this information to update Q_k, $minsup$ and Q_c, before performing the dynamic search. Doing so decreases the processing time for single edge graphs.

5   Experimental Evaluation

To evaluate the performance of the proposed algorithm, extensive experiments have been done. The testing environment is a workstation running Ubuntu 16.04, equipped with an Intel(R) Xeon(R) CPU E3-1270 3.60 GHz, and 64 GB of RAM. The TKG and gSpan algorithms were implemented in Java. Both algorithm implementations use the same code for loading datasets, outputting patterns, performing canonical testing and generating DFS codes. In the experiments, runtime and memory usage were measured using the standard Java API. Four standard benchmark datasets have been used, which have varied characteristics. They are described in Table 1 in terms of number of input graphs, average number of nodes per graph, average number of edges per graph, total number of vertices and total number of edges.

The four datasets are all bio- or chemo-informatics datasets. The protein dataset [1] contains 1113 graphs, each representing information about the structures of proteins. The ncil dataset [19] contains 4110 graphs representing chemical compounds related to cancer research. The enzymes datasets [1] contains 600 graphs representing enzymes from a database called BRENDA. Lastly, the Chemical340 dataset contains 340 graphs [20], where each vertex represents an atom, and its label provides information about the atom element and type. Furthermore, each edge represents the bond between two atoms and the edge label indicates bond types. In that dataset, the largest graph contains 214 vertices and 214 edges.

Table 1. Dataset characteristics

Dataset	$|GD|$	Avg. nodes	Avg. edges	$|L_V|$	$|L_E|$
Protein	1113	39.05	72.82	3	1
ncil	4110	29.87	32.3	37	3
Enzymes	600	32.63	62.13	3	1
Chemical340	340	27.02	27.40	66	4

5.1 Influence of k on the Performance of TKG

In a first experiment, the parameter k was varied to evaluate its influence on the performance of TKG in terms of runtime and memory usage. Three versions of TKG are compared: (1) TKG (with all optimizations), (2) TKG without dynamic search (using the depth-first search of gSpan), and (3) TKG without the skip strategy. Results for runtime are shown in Fig. 4 for the four datasets. Results for memory usage are shown in Fig. 5.

It is first observed that as k is increased runtime and memory usage increase. This is reasonable since as k is increased, more patterns must be found. As a result, TKG may need to consider more patterns to fill Q_k and be able to raise the internal *minsup* threshold to reduce the search space.

It is also observed that the dynamic search strategy generally greatly decreases runtime, and considerably decrease memory usage on the protein and enzymes datasets. For example, on the protein dataset, when $k = 200$, TKG with dynamic search is up to 100 times faster than TKG using a depth-first search and consumes up to 8 times less memory. On the enzymes dataset, when

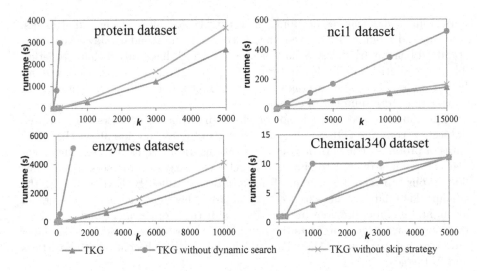

Fig. 4. Influence of k on runtime for different datasets.

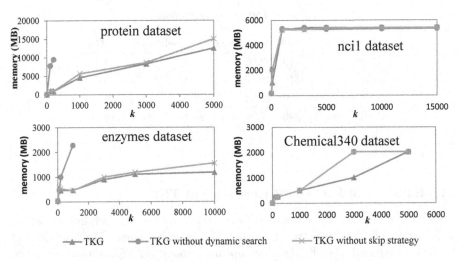

Fig. 5. Influence of k on peak memory usage for different datasets.

$k = 1000$, TKG with dynamic search is up to 40 times faster than TKG using a depth-first search and consumes up to 4 times less memory. On the ncil and Chemical340 datasets, not much memory reduction is achieved because although the dynamic search can help raise the internal *minsup* threshold more quickly and thus reduce the number of subgraphs considered, the queue Qc must be maintained in memory, which offsets those benefits.

It is also found that using the skip strategy is useful to reduce runtime, and can slightly reduce memory usage on some datasets. For example, on the protein dataset, when $k = 1000$, it can reduce runtime by up to 50% and memory by up to 25%. The skip counting strategy reduces runtime because when the strategy is applicable, less comparisons are made and less extensions are stored in memory.

It is also interesting to observe that the proposed TKG algorithm has a small runtime (less than 15 s) for k values of up to 5000 on the Chemical340 dataset, while the TGP algorithm for top-k closed subgraph mining was reported to be unable to terminate on that dataset [13]. The main reason, discussed in the related work section, is that TGP must calculate and store the DFS codes of all patterns in memory to then find the top-k patterns, which is inefficient for moderately large graphs. The TKG algorithm does not use this approach. It instead explores the search space using a dynamic search to guide the search toward the most promising patterns, and TKG reduces the search space using the internal *minsup* threshold to avoid generating the DFS codes of all patterns. Moreover, TKG does not need to keep all patterns in memory.

5.2   Performance Comparison with gSpan Set with an Optimal *minsup* Threshold

In another experiment, the performance of TKG was compared with that of gSpan. The goal of this experiment is to assess if top-k frequent subgraph mining using TKG can have similar performance to that of the traditional task of frequent subgraph mining. This question is interesting because top-k frequent subgraph mining is a more difficult problem than frequent subgraph mining. The reason is that in top-k frequent subgraph mining, the search for patterns must start from *minsup* = 1, while in frequent subgraph mining the *minsup* threshold is fixed beforehand by the user. The comparison of TKG with gSpan is also interesting because TKG reuses some techniques from gSpan.

TKG was run with k values from 1 to 5000 on each dataset. Then, the gSpan algorithm was run with the optimal *minsup* value to obtain the same number of patterns. The runtime and peak memory usage was measured. Tables 2 and 3 show results for the protein and enzymes datasets, respectively. Results for the other two datasets are not shown but similar trends were observed.

From these results, it is found that TKG and gSpan have very similar runtimes. This is a good result since the problem of top-k subgraph mining is more difficult than the traditional problem of frequent subgraph mining.

In terms of memory, TKG generally consumes more memory than gSpan (up to twice more). This is reasonable since TKG needs to keep a priority queue Q_k to store the current top-k patterns, and another priority queue Q_c to store patterns to be extended by the dynamic search.

It is important to note that this experiment was done by setting an optimal *minsup* value for gSpan to obtain the same number of patterns as TKG. But in real-life, the user typically don't know how to set the *minsup* threshold. Setting k is more intuitive than setting *minsup* because the former represents the number of patterns that the user wants to analyze. For example, consider that a user wants to find 200 to 1000 patterns in the protein dataset. According to Table 2, the range of *minsup* values that would satisfy the user is [0.5247, 0.6289], which has a length of $0.6289 - 0.5247 = 0.1042$. Thus, the user has about 10.4% chance of setting correctly the *minsup* threshold of gSpan. Now consider the same scenario for the enzymes datasets. In that case, the range of suitable *minsup* values is $[0.6650, 0.7817]$ according to Table 3. Hence, the user has about 11.7% chance of correctly setting the *minsup* threshold of gSpan. If the user sets the *minsup* threshold too low, gSpan may find too many patterns and may become very slow, while if the threshold is set too high, the user may need to run the algorithm again until a suitable value is found, which is time-consuming. To avoid such trial-and-error approach to find a suitable *minsup* value, this paper has proposed the TKG algorithm, which let the user directly specify the number of patterns to be found. Because the runtime of TKG is close to that of gSpan, TKG can be considered a valuable alternative to gSpan.

Table 2. Comparison of TKG and gSpan with optimal *minsup* threshold on the protein dataset

k	*minsup*	TKG runtime (s)	gSpan runtime (s)	TKG memory (MB)	gSpan memory (MB)
1	0.9227	1	1	85	85
100	0.6720	14	13	1020	507
200	0.6289	31	31	1019	976
1000	0.5247	321	275	4583	3503
3000	0.4618	1205	1198	4583	3503
5000	0.4367	2673	2650	8310	6182

Table 3. Comparison of TKG and gSpan with optimal *minsup* threshold on the enzymes dataset

k	*minsup*	TKG runtime (s)	gSpan runtime (s)	TKG memory (MB)	gSpan memory (MB)
1	0.9767	1	1	46	46
100	0.8067	12	8	527	276
200	0.7817	19	15	462	252
1000	0.6650	151	134	469	296
3000	0.600	625	612	1016	902
5000	0.5700	1280	1249	1113	1060

6   Conclusion

This paper has presented a novel algorithm named TKG to find the top-k frequent subgraphs in a graph database. The user only needs to set a parameter k, which controls the number of patterns to be found. To quickly raise the internal *minsup* threshold, the algorithm utilizes a dynamic search procedure that explores the most promising patterns first. Moreover, a skip strategy has been integrated in the algorithm to improve its performance. An extensive experimental evaluation has shown that TKG has excellent performance. In particular, the dynamic search and optimizations can decrease the runtime of TKG by up to about 100 times and its memory by up to 8 times. It was also found that TKG has similar runtimes to gSpan and thus that it provides a valuable alternative to traditional frequent subgraph mining algorithms.

The source code of the TKG and gSpan algorithms, as well as the datasets can be downloaded from http://www.philippe-fournier-viger.com/ spmf/tkgtkg/, and will also be integrated into the next release of the open-source SPMF data mining software [5].

For future work, designing efficient algorithms for other graph pattern mining tasks will be considered such as for discovering significant trend sequences in dynamic attributed graphs [2], subgraphs in graphs [12], high utility patterns [7] and rare subgraphs [23].

Acknowledgements. The work presented in this paper has been partly funded by the National Science Foundation of China.

References

1. Borgwardt, K.M., Ong, C.S., Schönauer, S., Vishwanathan, S.V.N., Smola, A.J., Kriegel, H.P.: Protein function prediction via graph kernels. Bioinformatics **21**(Suppl 1), 47–56 (2005)
2. Cheng, Z., Flouvat, F., Selmaoui-Folcher, N.: Mining recurrent patterns in a dynamic attributed graph. In: Kim, J., Shim, K., Cao, L., Lee, J.-G., Lin, X., Moon, Y.-S. (eds.) PAKDD 2017. LNCS (LNAI), vol. 10235, pp. 631–643. Springer, Cham (2017). https://doi.org/10.1007/978-3-319-57529-2_49
3. Duong, V.T.T., Khan, K.U., Jeong, B.S., Lee, Y.K.: Top-k frequent induced subgraph mining using sampling. In: Proceedings 6th International Conference on Emerging Databases: Technologies, Applications, and Theory (2016)
4. Duong, V.T.T., Khan, K.U., Lee, Y.K.: Top-k frequent induced subgraph mining on a sliding window using sampling. In: Proceedings 11th International Conference on Ubiquitous Information Management and Communication (2017)
5. Fournier-Viger, P., et al.: The SPMF open-source data mining library version 2. In: Berendt, B., et al. (eds.) ECML PKDD 2016. LNCS (LNAI), vol. 9853, pp. 36–40. Springer, Cham (2016). https://doi.org/10.1007/978-3-319-46131-1_8
6. Fournier-Viger, P., Lin, J.C.W., Kiran, U.R., Koh, Y.S.: A survey of sequential pattern mining. Data Sci. Pattern Recogn. **1**(1), 54–77 (2017)
7. Fournier-Viger, P., Chun-Wei Lin, J., Truong-Chi, T., Nkambou, R.: A survey of high utility itemset mining. In: Fournier-Viger, P., Lin, J.C.-W., Nkambou, R., Vo, B., Tseng, V.S. (eds.) High-Utility Pattern Mining. SBD, vol. 51, pp. 1–45. Springer, Cham (2019). https://doi.org/10.1007/978-3-030-04921-8_1
8. Fournier-Viger, P., Lin, J.C.W., Vo, B., Chi, T.T., Zhang, J., Le, B.: A survey of itemset mining. WIREs Data Min. Knowl. Discov. (2017)

9. Inokuchi, A., Washio, T., Motoda, H.: An apriori-based algorithm for mining frequent substructures from graph data. In: Zighed, D.A., Komorowski, J., Żytkow, J. (eds.) PKDD 2000. LNCS (LNAI), vol. 1910, pp. 13–23. Springer, Heidelberg (2000). https://doi.org/10.1007/3-540-45372-5_2

10. Jiang, C., Coenen, F., Zito, M.: A survey of frequent subgraph mining algorithms. Knowl. Eng. Rev. **28**, 75–105 (2013)

11. Kuramochi, M., Karypis, G.: Frequent subgraph discovery. In: Proceedings 1st IEEE International Conference on Data Mining (2001)

12. Lee, G., Yun, U., Kim, D.: A weight-based approach: frequent graph pattern mining with length-decreasing support constraints using weighted smallest valid extension. Adv. Sci. Lett. **22**(9), 2480–2484 (2016)

13. Li, Y., Lin, Q., Li, R., Duan, D.: TGP: mining top-k frequent closed graph pattern without minimum support. In: Proceedings 6th International Conference on Advanced Data Mining and Applications (2010)

14. Mrzic, A., et al.: Grasping frequent subgraph mining for bioinformatics applications. In: BioData Mining (2018)

15. Nguyen, D., Luo, W., Nguyen, T.D., Venkatesh, S., Phung, D.Q.: Learning graph representation via frequent subgraphs. In: Proceedings 2018 SIAM International Conference on Data Mining, pp. 306–314 (2018)

16. Nijssen, S., Kok, J.N.: The gaston tool for frequent subgraph mining. Electron. Notes Theor. Comput. Sci. **127**, 77–87 (2005)

17. Saha, T.K., Hasan, M.A.: FS3: a sampling based method for top-k frequent subgraph mining. In: Proceedings 2014 IEEE International Conference on Big Data, pp. 72–79 (2014)

18. Sankar, A., Ranu, S., Raman, K.: Predicting novel metabolic pathways through subgraph mining. Bioinformatics **33**(24), 3955–3963 (2017)

19. Wale, N., Watson, I.A., Karypis, G.: Comparison of descriptor spaces for chemical compound retrieval and classification. In: Proceedings 6th International Conference on Data Mining, pp. 678–689 (2006)

20. Yan, X., Han, J.: gSpan: graph-based substructure pattern mining. In: Proceedings 2nd IEEE International Conference on Data Mining (2002)

21. Yan, X., Han, J.: CloseGraph: mining closed frequent graph patterns. In: Proceedings of the 9th ACM SIGKDD International Conference on Knowledge Discovery and Data Mining (2003)

22. Yan, X., Yu, P.S., Han, J.: Graph indexing: a frequent structure-based approach. In: Proceedings of the 2004 SIGMOD Conference (2004)

23. Yun, U., Lee, G., Kim, C.H.: The smallest valid extension-based efficient, rare graph pattern mining, considering length-decreasing support constraints and symmetry characteristics of graphs. Symmetry **8**(5), 32 (2016)

24. Zhu, F., Yan, X., Han, J., Yu, P.S.: gPrune: a constraint pushing framework for graph pattern mining. In: Proceedings of the 11th Pacific-Asia Conference on Knowledge Discovery and Data Mining (2007)

Why Multilayer Networks Instead of Simple Graphs? Modeling Effectiveness and Analysis Flexibility and Efficiency!

Sharma Chakravarthy[✉], Abhishek Santra, and Kanthi Sannappa Komar

Information Technology Laboratory (IT Lab), Computer Science and Engineering Department, University of Texas at Arlington, Arlington, TX 76019, USA
sharmac@cse.uta.edu

Abstract. We are on the cusp of analyzing a variety of data being collected in every walk of life - social, biological, health-care, corporate, climate, to name a few. We are also in search for models and analytical techniques that can accommodate more complex and increasingly large size data (scalability). Our ability to analyze large complex, disparate data for a broad set of analysis objectives differentiates big data analytics from mining which is narrow in scope. Hence, flexibility of analysis (different from scalability) is important. Concomitantly, efficiency is important due to large number of analysis needs. Our ultimate goal is to go from vertical analysis of data individually (corresponding to one of the 4 V's) to holistically (also termed fusion-based) analyze that corresponds to all or a subset of V's!

In order to accomplish the above, we are always in search for more effective models to represent data and different analysis techniques that support flexibility of analysis, efficiency, and scalability. We want to use techniques that have worked well – whether it is for modeling, efficiency or scalability. We also want to extend these techniques and/or develop new and improved ones to accommodate more complex, diverse, and larger size data.

The goal of this paper is to provide the reader an understanding of data analysis approaches using graphs. Our thesis is that there are several ways in which a graph representation can be used – both for modeling and analysis. We will take the reader through the evolution of graph usage and relevance leading to the current state of the use of multilayer Networks (MLNs) or multiplexes for modeling and analysis. Graphs are not new, but how they are used for big data analytics is going through a transformation which is important to understand. The hope is that the reader understands the path that has led us to this juncture and how graph usage is extended!

Keywords: Graph-based modeling and analysis · Multilayer networks · Decoupling approach · Efficiency and scalability

S. Madria et al. (Eds.): BDA 2019, LNCS 11932, pp. 227–244, 2019.
https://doi.org/10.1007/978-3-030-37188-3_14

1 Introduction

This is a paper on big data analytics and science that is intended to capture the progress of graph mining (one form of analysis) to a broader analysis of complex data sets using graphs. We start with traditional graph mining and trace the path towards multilayer networks for effective modeling and efficient & flexible analysis.

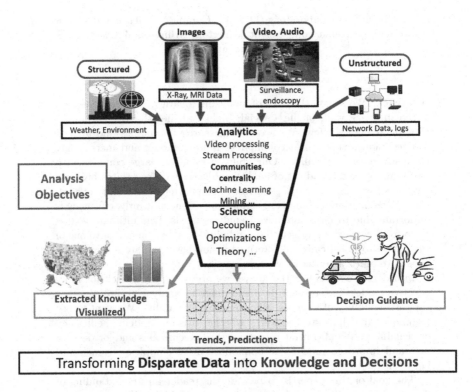

Fig. 1. High level architecture of big data analytics/science

Data mining is the process of automatically discovering useful information from large data repositories. Data collected and validated (termed labeled) has been used for generating models (Decision trees, SVM, termed supervised approach) that can be used for predicting outcomes for new data. Data has also has been processed in a number of ways to glean patterns without using labeled data (e.g., clustering, association rules, subgraph mining, termed unsupervised). Data mining is different from querying or generating different types of reports from managed data sets using a DBMS or a data warehouse. Big data analytics use or incorporate appropriate mining and other techniques for a broader holistic analysis. Starting with text mining, a number of traditional mining techniques have been developed (decision trees, clustering, neural nets, etc.) Both supervised and unsupervised approaches have shown to be useful for different applications. Graph mining although developed in the 80's have become popular recently due to their usage in modeling and processing internet and social network. More recently, association rule mining (and its flavors) became possible with the advances in data representation, availability of large real-world data, large and cheap storage availability, and relevant technical advances. Data Mining, as we have it today, became even more important from a business perspective (similar to Data warehouses, but with different requirements) when we progressed in our ability (storage, networking, processing, and algorithms) to handle vast amounts of *real business data* (in contrast to samples of representative data) for identifying non-intuitive nuggets with certain confidence for driving business goals.

In our view, the fundamental difference between mining and big data analytics is the scope and diversity of data. The holistic aspect of analysis and the breadth (or diversity) of data along with their characteristics are the challenges that need to be taken into account. Here, the goal is more ambitious than traditional mining in that this analytics is likely to need multiple approaches working in concert. Hence, not only the need for large number of formalism, techniques, and algorithms, but also a mechanism to combine or compose them in novel ways based on user-defined or user-specified analysis objectives (see Fig. 1). The long-term goal is even more ambitious in terms of requirements for holistically analyzing (and developing formalism for) disparate data that corresponds to 4V's (Volume, Velocity, Variety, and Veracity) or even 5V's (plus Value). This is the challenge currently faced by the community addressing big data analysis/science. The basic premise here is that dealing with each of the V's individually or in small combinations (as has been done up to this point to a large extent) is not sufficient, but need to include *all or combinations* of them as warranted for a holistic analysis leading to inferring better and concise **(actionable) knowledge** for decision making. Towards this end, we will present *our* previous and ongoing contributions towards big data analysis.

Figure 1 shows, at a high level, the problem of big data analysis and science. As shown in the figure, the ultimate goal is to synthesize meaningful and beneficial actionable knowledge with good confidence that can be used for decision making (what humans call wisdom which is culled from data and events based on a combination of nature and nurture (together as experience) as biologists put it) by using all available relevant disparate data coming from a variety of sources. The inverted triangle shows the reduction of large raw data into small nuggets of knowledge. Figure also shows some of the technologies that are available today (used for analytics), and a partial list of underpinnings (i.e., science) using which we develop to support these technologies. The way we see big data analysis is that instead of addressing each V (or small combinations of V), a holistic approach is the desired goal driven by analysis objectives/expectations. However, the problem is still the same as that of culling, filtering, aggregating, and inferring nuggets of (actionable) knowledge that can be used for decision making including real-time decision making. Most of the current approaches have addressed a small subset of this problem.

Although not explicitly shown in the figure, techniques for the visualization of data as well as the derived knowledge is quite important. Pictorial representation and multi-dimensional subject-oriented analysis of the results are also very useful for understanding the results of analysis.

Personalized health care is a good example of how it is critical to avail and process all types of data related to a single person (or even a community) over a period of time (lab results, X-rays, EKG, endoscopy video, MRI, etc.) to make a meaningful decision for that individual (or community) rather than using average cases which is how it is done today. Similar applications include climate change studies, monitoring earthquakes, pollution, and others.

The remainder of the paper is organized as follow. We briefly indicate our contributions to data mining over the years in Sect. 2. Then we focus on graph-based modeling and analysis that includes our current work in Sect. 3. Finally, Sect. 4 has conclusions.

2   Data Mining or Knowledge Discovery in Databases

Data mining aims at discovering important and previously unknown patterns from the data sets. Although not explicitly termed data mining and might not have used real-world business data, the concept of understanding data, in ways that are different from querying and analysis that was available through RDBMSs and data warehouses, predates them. Classification, clustering, prediction, deviation analysis, and neural networks were used by many businesses for selective marketing, credit card transaction approval, and mortgage and other types of lending. Supervised and unsupervised approaches were developed and multi-fold cross valida-

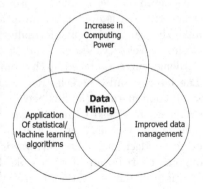

Fig. 2. Convergence of technologies that facilitated data mining

tion was widely used for establishing the accuracy of models. A number of algorithms were developed, some couched in expert systems used by businesses. Due to the limitations of storage and processing, the sizes of the data sets used were small and often statistically representative samples were used for processing (instead of all available data) and results extrapolated (or generalized) for larger data sets.

The rapid improvement in the size of the storage devices along with the associated drop in the cost in the 1990s, and increase in the computing power as well as the wide use of statistical approaches for processing data gave rise to the field of data mining as we know it today (see Fig. 2. Suddenly, it became feasible for organizations to store unprecedented amounts of organizational data and process it. These organizations, though having a gold mine of data, were not able to fully capitalize on its value mainly because the algorithms and approaches had to be scaled to very large data sizes. Typically, the data captures the business trends over a period of time and hence using real-world business data (rather than samples) became the goal. However, the nuggets of useful knowledge hidden were not so easy to discern. To compete effectively, decision makers felt they needed to identify and utilize "nuggets of knowledge" buried in the collected data and take advantage of the high return opportunities in a timely manner. Association rule mining is a good example of this trend.

While these developments were ongoing, although graph theory has been around for a very long time, graphs for data representation were not that popular. Graph theory belonged more to the realm of mathematicians rather than its application in computer science. A few researchers were using graphs for representing data in specific domains and were trying to identify patterns in graphs around the same time businesses were trying to do the same using their transactional data. Although association rule mining took off and became widely popular with many commercial implementations, graph usage and mining caught the attention of businesses much later. One of the early work on graph mining was Subdue [26] which developed main memory algorithms for identifying substructures in graphs (or forests) that were "interesting" based on some metric. They used a information theoretic metric termed minimum description length (or MDL) for this purpose. The data sets were drawn from chemical representations, CAD circuits, etc.

Data mining became a hot research area with the advent of association rule mining and graph mining. Association rule mining [6] started with market basket analysis for identifying items bought together with given support and confidence from actual point-of-sales data. These data sets were huge (for example, Walmart's point-of-sales data around that time was estimated to be around 1Gb per day) and multiple years of data could not be held in main memory for analysis. For association rule mining of this data, novel data structures (e.g., hash tree) and approaches to reduce the number of passes on the data (to minimize I/O's) were developed. Teradata developed a parallel processing system to facilitate analysis of very large amounts of transactional data. As the search space was prohibitively large, *a priori* and other properties were identified to reduce the number of item sets carried over from iteration to the next one. Partitioning approaches were developed to reduce the number of passes on data (stored in disks) and the response time. A large amount of work followed resulting in a number of mining systems marketed by almost every major vendor.

Along the same lines, the expressiveness of graphs became important with the advent of Internet and social networks. Graphs both for modeling and analysis (e.g., mining, PageRank, etc.) and their importance for identifying important and useful graph patterns became apparent. Frequent subgraphs, identification and counting of triangles and other substructures in very large graphs became important in addition to interesting substructures. Today, there is renewed interest in using graphs for modeling and analysis of complex data as discussed in Sect. 3.2.

2.1   IT Lab Contributions

Information Technology Laboratory (or IT Lab) at UT Arlington or UTA (also in its previous incarnation at the University of Florida or UFL, Gainesville) has been engaged in managing, processing, and analyzing large amounts of data using diverse techniques, such as semantic and multiple query optimization [12,21–24,57], real-time transaction scheduling [40–42], incorporating active capability [1,18,25,36,37,43,56] into DBMSs, etc. An active object-oriented DBMS termed **Sentinel** was developed at UFL incorporating **Snoop** as the event specification languages. This was later integrated with the stream processing system **MavStream** to provide both event and stream processing capabilities in a seamless manner in a single system termed, MavEStream. Currently it is being extended [11] to support continuous queries on videos by extending it to support object comparison, spatial, and temporal aggregate computations MavEStream.

Specifically, from the view point of mining, we have contributed to both association rule mining [34,50–54,71–73] and substructure discovery [7,15,20]. We have also developed a number of techniques for aggregating streaming, real-time relational data termed stream data processing/analysis [2–5,13–17,19,27,35,38,44–48,65,66,68,70]. A prototype data stream management system (DSMS) **MavEStream** was implemented at UTA and the work was consolidated into an authored book.

For association rules, we have extensively evaluated performance of database algorithms on different RDBMSs and compared them. This helped us to identify some of the quirks in the optimization of SQL queries by different vendors and understand the difficulties of optimizing queries with 10 to 20 joins, a large number of them being self joins. RDBMSs query optimization were not designed with those number and types of joins in mind.

On the graph mining side, we have tried to scale the main memory substructure mining algorithm of Subdue using a number of alternative approaches. The first one [7,15,55] mapped graphs into relations and the substructure discovery algorithm to SQL in order to leverage the built-in capabilities of a DBMS (buffer manager, query optimizer) instead of re-inventing them for mining. This allowed us to scale the size of the graphs to millions of nodes and edges. This approach has certain limitations due to large number of joins as well DBMS's inability to order a relation using columns. Recently, we have been able to successfully scale this algorithm even further to arbitrary sizes using the map/reduce paradigm.

We have achieved data scalability using divide and conquer with and without using map/reduce. We have developed generic Map/Reduce based algorithms for horizontal scalability of substructure discovery that can work with any partitioning strategy. The basic components of graph mining - subgraph expansion, duplicate removal and counting of isomorphic substructures were incorporated into the algorithms for the Map/Reduce paradigm by carefully orchestrating new representations. Vertical scalability was achieved by showing that these algorithms produce the same results (loss-less property) irrespective of the number of partitions. Experiments validated the advantage of using Map/Reduce based substructure discovery to scale to arbitrarily large graphs [28–30]. In an effort to analyze the partitioning strategies and associated algorithms from a performance standpoint, we have done a component cost analysis of substructure discovery in a distributed framework. The cost analysis identified places for improvements in using the range-based partitioning strategy over its counterpart. Theoretical justification along with experimental evaluation of the improvements were verified by varying a number of user parameters. The cost analysis also pointed out the portability of our algorithms to a different paradigm such as Spark to reap similar benefits.

Our approach to query processing on graphs developed a cost-based plan generator by defining and using a catalog that is relevant to graphs [31,39]. To process a plan on large graphs, partitioning of graphs was used for processing the plan on each partition separately and combining the results in a loss-less manner. Several heuristics were developed for optimizing the number of partitions loaded for this purpose [9,10].

3   Graph-Based Modeling and Analysis

To achieve the goals of big data analytics/science as explained in Sect. 1, a number of perspectives on how to analyze a data set as well as a number of approaches and their combinations need to be taken into consideration. Research is ongoing by a large number of scientists with a broad brush covering data sets from a variety of domains. In this section, we present some of the approaches that we are working on to address the big data analysis problem in a small way. This tutorial mainly focuses on this. Although they do not right now solve the big data analysis problem completely as is posited in Fig. 1, they address components whose solutions are likely to contribute to the overall solution.

3.1   Modeling Data Using Single Graphs

The basic idea of graphs were first introduced in the 18th century by Swiss mathematician Leonhard Euler. Interestingly, it seems to have been applied for a real-world problem (famous Konigsberg bridge problem) – to formulate and solve it. We have come a long way from there as a topic of mathematics and recently with the explosion of social networks and internet as a computer science topic. Graph theory is ultimately the study of entities as nodes (or vertices) and relationships as edges. Weights and labels can be associated with nodes, and can include directions as well for edges. Other properties such as cycles, spanning trees can be associated with graphs. This rich graph representation can be used to abstract a large number of real-world problems from city layouts to printed circuits to social and biological networks. Studying graphs through a framework provides answers to logistics, networking, optimization, matching, and operational problems.

For the purpose of modeling and analyzing complex data we have in mind, let us consider the following data sets and associated analysis objectives.

US-Based Airlines: This is a data set of six US-based airlines and their flight information among US cities. This information has been collected by us from multiple sources. The number of US cities is the same for all airlines. But their connectivity (direct flights between cities) varies.

Potential Analysis Questions: Airlines are always looking for expansion into new/existing markets (cities) to increase business. Using the above publicly available data, an interesting analysis objective would be: **Can we identify and rank cities for potential expansion as a hub for each airline taking all competitors into consideration?**

If we were to represent each airline and its direct flight connectivity, we can either do it as 3 separate simple graphs as shown in Fig. 3(a) or as one graph in which we show them combined into a single graph as shown in Fig. 4(a). It is easy to see that there are multiple edges in the combined graph and colors correspond to edge labels. If we do not keep relationships in each layer separate, there is information loss. We will not be able to know how many airlines have direct flights between the cities. If we keep them separate, there

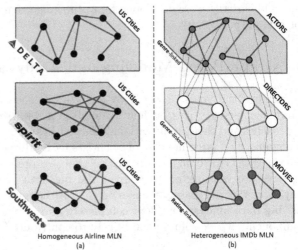

Fig. 3. MLN modeling alternatives

are no algorithms to compute communities (or hubs) on such a graph (termed an attribute graph)! This data set has the same entities in all layers and different relationships and is termed **Homogeneous Multilayer Network or HoMLN.**

However, if we represent this data set as 3 layers as shown in Fig. 3(a), the challenge is to compute network properties such as communities and hubs. Currently, it is done in a way where there is loss of information due to combining multiple edges into a single edge using Boolean operations. Or it is done on the entire MLN resulting in in-efficient algorithms. Our decoupling-based multilayer network (MLN) analysis provides a solution for this.

IMDb Data Set: The IMDb data set captures movies, TV episodes, actor, directors and other related information, such as rating. This is a large data set consisting of movie and TV episode data from their beginnings.

Potential Analysis Question: For a large data set such as above, is it possible to **infer, through data-driven analysis, well-known actors who have worked in several common genres,** *but have never worked together!* This would be of interest to directors and producers who are scouting for bankable and novel casting for their next project!

This data set is more complicated in that there are different entity types – actors, directors, and movies. Each entity is connected among themselves – actors acting in a specific genre, directors directing a specific genre, and movies whose ratings are in a range. In addition, there are also links (or inter-layer edges) between the actors and director entity nodes indicating which director has directed which actor at least once. Figure 3(b) shows each entity type separately and also the connections between different entity types as

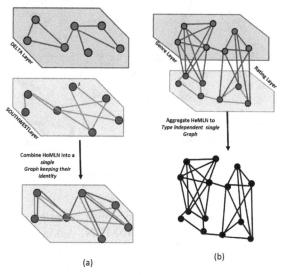

(a)

(b)

Fig. 4. Reducing MLN to single graphs

inter-layer edges. Note that HoMLNs have a single entity type and their connections across layers are implicit and hence not shown in Fig. 3(a). Figure 4(b) shows what happens when the HeMLN in Fig. 3(b) is converted into a single graph shown in Fig. 4(b) using type-independence approach. That is, all the type information (essentially labels of both nodes and edges) is lost. Although network properties can be computed on the resulting graph shown in Fig. 4(b), both structure and semantics of the original representation is lost.

On the other hand, if we keep edge and node labels in the aggregated graph, we will have an attribute graph that has node and edge labels. There are no algorithms for computing networking properties that we are dealing with such as community and centrality on such a graph. However, attributed graphs are

widely used for substructure discovery, querying etc. An attributed graph is
shown in Fig. 5 (source) for a better understanding.

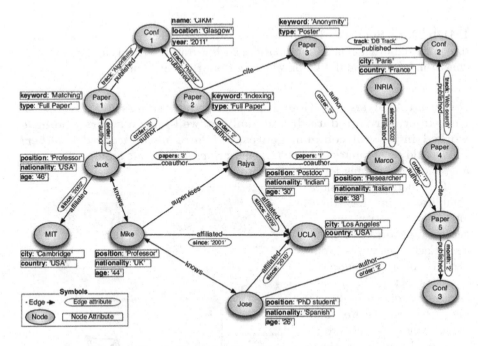

Fig. 5. Example of an attributed graph

This representation with multiple entity types is termed **Heterogeneous
multilayer network or HeMLN**. Again the issue is which representation is
more effective from a modeling perspective and which one from a computation
perspective.

The above are just a couple of examples to indicate the analysis complexity
and potential of the approach proposed in this paper. Data sets being analyzed
may also contain features that are derived from contents (e.g., posts in Facebook)
in addition to explicit ones. Such derived information can also be incorporated
into this approach as shown in [74].

3.2   Modeling and Analyzing Complex Data Using MLNs

In the previous section we illustrated the difficulties of modeling complex data
sets as a single graph without loss of information. In addition, computation of
network properties, such as community and centrality (e.g., hub), which are
available for single graphs, are not available for MLNs or multiplexes. Hence,
if one wants to use multiplexes for modeling, it is imperative that we develop

approaches and algorithms for computing network properties and efficiently. Network Decoupling approach is one such techniques which we will discuss in detail in this tutorial with its application to real-world data sets.

Before we go into modeling using MLNs (HoMLN or HeMLN) and computations on that, let us clearly understand the alternatives. Figure 7 shows three alternatives ways of modeling and computing network properties (currently, community and centrality). Figure 7(a) shows a HeMLN aggregated into a single graph removing node/edge identities resulting in losing entity and/or relationship identities. Computing network properties on this single graph is not preferred due to loss of information. On the other end of the spectrum, Fig. 7(c) shows the same HeMLN layers and computing the result using the entire MLN as a whole. Although this has been proposed in the literature (e.g., Infomap recently), this is likely to be computationally expensive as the number of layers and data sizes become large. The advantage of modeling is lost to some extent due to computational complexity.

Figure 7(b) on the other hand proposes an approach developed by us (termed networking decoupling) where network property for each layer is computed independently (possibly in parallel) and compose them using a binary operator Θ as shown. We have shown this to be effective, can be done for Boolean operations for HoMLNs and for HeMLNs as well without aggregating and losing type information. Furthermore, we have shown it be more efficient than the approaches shown in Fig. 6(a) or (c). Furthermore, the advantages of modeling is retained as well.

More clearly, current approaches, such as using the MLN as a whole [75], type-independent [33], and projection-based [8,69], do not accomplish this as they aggregate (or collapse) layers into a simple graph in different ways. More importantly, aggregation approaches are likely to result in some information loss [49], distortion of properties [49], or hide the effect of different entity types and/or different intra- or inter-layer relationships as elaborated in [32]. Structure-preservation is critical for understanding a HeMLN community and for drill-down analysis of communities.

From an analysis perspective, lack of structure- and semantics makes the drill down extremely difficult (or even impossible) and hence the understanding of results. Our computation results clearly show the community structure and how easy it is to drill down to see patterns in terms of original labels.

Figure 6(a) and (b) illustrate the difference between the current approaches and our proposed approach. Figure 6(a) shows type-independent aggregation[1] of two layers into a single graph on which extant community detection is applied. As can be seen, **both structure as well as entity and relationship labels – shown as colored nodes and edges – are lost in the resulting communities.** In contrast, the Fig. 6(b) shows the same layers and community detection using the proposed definition and the decoupling approach. As there is no aggregation, both structure and semantics are preserved.

---

[1] Other aggregation approaches have the same problem.

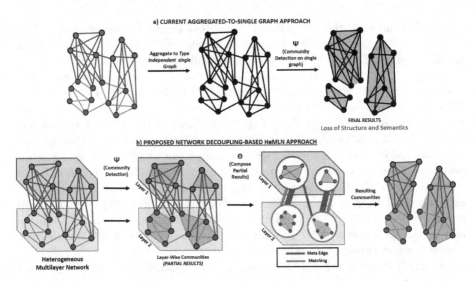

Fig. 6. Lossy traditional approach vs. structure- and semantics-preserving decoupling approach

3.3   Our Contributions Towards MLN Modeling and Analysis

Instead of modeling complex data as a monoplex, we have proposed multiplexes [49,60], In a multiplex, instead of creating a single graph with colored nodes and/or edges, a number of graphs are created, each representing one aspect or feature or perspective. For example, Fig. 3 shows both a homogeneous and a heterogeneous multiplex. In each layer, actors are nodes and one genre is used to create edges. We believe this model is better than a monoplex as the graph in each layer is smaller, will not have multiple edges, and easier to understand semantically.

Efficient Analysis of Multiplexes: Our ongoing research towards big data analytics includes modeling and efficient analysis of multiplexes. Applying community and hub detection algorithms for a multiplex-as-a-whole is being explored by the research community [67,76]. This, in our view, is not the best way as the complexity of community and hub detection increases and the decomposition of the problem into its components (layers) is not leveraged.

For analyzing multiplexes in a holistic and flexible way, we propose **decoupling** as the approach of choice. The basic idea of decoupling is to analyze **individual layers** using extant algorithms and *compose* the results of individual layers to obtain results for any combinations of layers in a loss-less manner [58,59,61–64].

Figure 7(b) shows our decoupled approach to HeMLN community detection. The steps of our decoupling approach is given below:

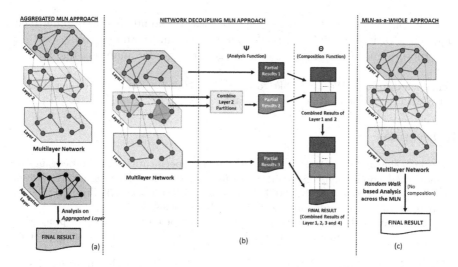

Fig. 7. Lossy traditional approach vs. structure- and semantics-preserving decoupling approach

(i) First use the function Ψ (e.g., community detection, but any network property in general) to find the property in each of the layers individually (can also be done in parallel),

(ii) for any two chosen layers, compose the partial results from each layer using the identified composition function Θ. For example, we have proposed vertex- and edge-based composition for Boolean composition of Homogeneous MLNs and a bipartite-graph and maximal matching for Hetereogeneous MLNs.

(iii) repeat this process for computing the network property for given number of layers in the order specified.

In this tutorial, we will discuss the MLN approach to modeling and analysis and contrast it with traditional approaches for efficiency and scalability. We will discuss how the objectives proposed in Sect. 1 can be efficiently computed using the proposed approach. We will also co consider other analysis objectives on different data sets.

4 Conclusions

In this paper, we have demonstrated why a multilayer network is better both from a modeling as well as analysis perspective. Multilayer networks provide an effective way of compartmentalizing the information in the data set using analysis objectives. Further, when appropriate composition mechanisms are available or developed (e.g., decoupling approach demonstrated here), computational efficiency as well as analysis flexibility can be accomplished. We have also been able to fold content extraction [74] into this framework in our analysis of Facebook data. We are currently working on approaches to extend this decoupling technique for a variety of analysis, in addition to community and centrality (such as substructure, motifs, querying), with the goal of broader analysis capability.

Acknowledgment. We would like to thank Dr. Sanjukta Bhowmick on her collaboration with us on the multilayer network analysis.

References

1. Adaikkalavan, R., Chakravarthy, S.: Event specification and processing for advanced applications: generalization and formalization. In: Wagner, R., Revell, N., Pernul, G. (eds.) DEXA 2007. LNCS, vol. 4653, pp. 369–379. Springer, Heidelberg (2007). https://doi.org/10.1007/978-3-540-74469-6_37
2. Adaikkalavan, R., Chakravarthy, S.: Events must be complete in event processing! In: Proceedings, Annual ACM SIG Symposium On Applied Computing, pp. 1038–1039 (2008)
3. Aery, M., Chakravarthy, S.: eMailSift: mining-based approaches to email classification. In: SIGIR, pp. 580–581 (2004)
4. Aery, M., Chakravarthy, S.: eMailSift: email classification based on structure and content. In: ICDM, pp. 18–25 (2005)
5. Aery, M., Chakravarthy, S.: InfoSift: adapting graph mining techniques for text classification. In: FLAIRS Conference, pp. 277–282 (2005)
6. Agrawal, R., Imielinski, T., Swami, A.: Database mining: a performance perspective. IEEE Trans. Knowl. Data Eng. **5**(6), 914–925 (1993)
7. Balachandran, R., Padmanabhan, S., Chakravarthy, S.: Enhanced DB-Subdue: supporting subtle aspects of graph mining using a relational approach. In: Ng, W.-K., Kitsuregawa, M., Li, J., Chang, K. (eds.) PAKDD 2006. LNCS (LNAI), vol. 3918, pp. 673–678. Springer, Heidelberg (2006). https://doi.org/10.1007/11731139_77
8. Berenstein, A., Magarinos, M.P., Chernomoretz, A., Aguero, F.: A multilayer network approach for guiding drug repositioning in neglected diseases. PLOS **10**, e0004300 (2016)
9. Bodra, J.: Processing Queries Over Partitioned Graph Databases: An Approach and it's Evaluation. Master's thesis, The University of Texas at Arlington, May 2016. http://itlab.uta.edu/students/alumni/MS/Jay_D_Bodra/JBod_MS2016.pdf
10. Bodra, J., Das, S., Santra, A., Chakravarthy, S.: Query processing on large graphs: scalability through partitioning. In: Ordonez, C., Bellatreche, L. (eds.) DaWaK 2018. LNCS, vol. 11031, pp. 271–288. Springer, Cham (2018). https://doi.org/10.1007/978-3-319-98539-8_21
11. Chakravarthy, S., Aved, A., Shirvani, S., Annappa, M., Blasch, E.: Adapting stream processing framework for video analysis. Procedia Comput. Sci. **51**, 2648–2657 (2015)
12. Chakravarthy, S.: Divide and conquer: a basis for augmenting a conventional query optimizer with multiple query processing capabilities. In: ICDE, pp. 482–490 (1991)
13. Chakravarthy, S., Adaikkalavan, R.: Ubiquitous nature of event-driven approaches: a retrospective view. In: Chandy, M., Etzion, O., von Ammon, R. (eds.) Event Processing. No. 07191 in Dagstuhl Seminar Proceedings, Internationales Begegnungs- und Forschungszentrum für Informatik (IBFI), Schloss Dagstuhl, Germany, Dagstuhl, Germany (2007). http://drops.dagstuhl.de/opus/volltexte/2007/1150

14. Chakravarthy, S., Adaikkalavan, R.: Event and streams: harnessing and unleashing their synergy. In: International Conference on Distributed Event-Based Systems, pp. 1–12, July 2008

15. Chakravarthy, S., Beera, R., Balachandran, R.: DB-Subdue: database approach to graph mining. In: Dai, H., Srikant, R., Zhang, C. (eds.) PAKDD 2004. LNCS (LNAI), vol. 3056, pp. 341–350. Springer, Heidelberg (2004). https://doi.org/10.1007/978-3-540-24775-3_42

16. Chakravarthy, S., Jiang, Q.: Stream Data Processing: A Quality of Service Perspective. Advances in Database Systems, vol. 36. Springer, Heidelberg (2009). https://doi.org/10.1007/978-0-387-71003-7

17. Chakravarthy, S., Pajjuri, V.: Scheduling strategies and their evaluation in a data stream management system. In: Bell, D.A., Hong, J. (eds.) BNCOD 2006. LNCS, vol. 4042, pp. 220–231. Springer, Heidelberg (2006). https://doi.org/10.1007/11788911_19

18. Chakravarthy, S., Sanka, A., Jacob, J., Pandrangi, N.: A learning-based approach for fetching pages in WebVigiL. In: Proceedings, Annual ACM SIG Symposium On Applied Computing, pp. 1725–1731 (2004)

19. Chakravarthy, S., Venkatachalam, A., Telang, A.: A graph-based approach for multi-folder email classification. In: ICDM, pp. 78–87 (2010)

20. Chakravarthy, S., Zhang, H.: Visualization of association rules over relational DBMSs. In: Proceedings, Annual ACM SIG Symposium On Applied Computing, pp. 922–926 (2003)

21. Chakravarthy, U.S., Grant, J., Minker, J.: Foundations of semantic query optimization for deductive databases. In: Foundations of Deductive Databases and Logic Programming, pp. 243–273. Morgan Kaufmann (1988)

22. Chakravarthy, U.S., Grant, J., Minker, J.: Logic-based approach to semantic query optimization. ACM Trans. Database Syst. **15**(2), 162–207 (1990)

23. Chakravarthy, U.S., Minker, J.: Multiple query processing in deductive databases using query graphs. In: VLDB, pp. 384–391 (1986)

24. Chakravarthy, U.S., Minker, J., Grant, J.: Semantic query optimization: additional constraints and control strategies. In: Expert Database Conference, pp. 345–379 (1986)

25. Chamakura, S., Sachde, A., Chakravarthy, S., Arora, A.: WEBVIGIL: monitoring multiple web pages and presentation of XML pages. In: ICDE Workshops, p. 1276 (2006)

26. Cook, D.J., Holder, L.B.: Substructure discovery using minimum description length and background knowledge. J. Artif. Intell. Res. **1**, 231–255 (1994)

27. Cuzzocrea, A., Chakravarthy, S.: Event-based compression and mining of data streams. In: Lovrek, I., Howlett, R.J., Jain, L.C. (eds.) KES 2008. LNCS (LNAI), vol. 5178, pp. 670–681. Springer, Heidelberg (2008). https://doi.org/10.1007/978-3-540-85565-1_83

28. Das, S.: Divide and Conquer Approach to Scalable Substructure Discovery: Partitioning Schemes, Algorithms, Optimization and Performance Analysis Using Map/Reduce Paradigm. Ph.D. thesis, The University of Texas at Arlington, May 2017. http://itlab.uta.edu/students/alumni/PhD/Soumyava_Das/SDas_PhD2017.pdf

29. Das, S., Chakravarthy, S.: Partition and conquer: map/reduce way of substructure discovery. In: Madria, S., Hara, T. (eds.) DaWaK 2015. LNCS, vol. 9263, pp. 365–378. Springer, Cham (2015). https://doi.org/10.1007/978-3-319-22729-0_28

30. Das, S., Chakravarthy, S.: Duplicate reduction in graph mining: approaches, analysis, and evaluation. IEEE Trans. Knowl. Data Eng. **30**(8), 1454–1466 (2018). https://doi.org/10.1109/TKDE.2018.2795003

31. Das, S., Goyal, A., Chakravarthy, S.: Plan before you execute: a cost-based query optimizer for attributed graph databases. In: Madria, S., Hara, T. (eds.) DaWaK 2016. LNCS, vol. 9829, pp. 314–328. Springer, Cham (2016). https://doi.org/10.1007/978-3-319-43946-4_21

32. De Domenico, M., Solé-Ribalta, A., Gómez, S., Arenas, A.: Navigability of interconnected networks under random failures. Proc. Natl. Acad. Sci. **111**, 8351–8356 (2014). https://www.pnas.org/content/early/2014/05/21/1318469111

33. Domenico, M.D., Nicosia, V., Arenas, A., Latora, V.: Layer aggregation and reducibility of multilayer interconnected networks. CoRR abs/1405.0425 (2014). http://arxiv.org/abs/1405.0425

34. Dudgikar, M., Chakravarthy, S., Liuzzi, R.A., Wong, L.: A layered optimizer for mining association rules over relational database management systems. In: IKE, pp. 422–430 (2003)

35. Elkhalifa, L., Adaikkalavan, R., Chakravarthy, S.: InfoFilter: a system for expressive pattern specification and detection over text streams. In: Proceedings, Annual ACM SIG Symposium On Applied Computing, pp. 1084–1088 (2005)

36. Eppili, A., Jacob, J., Sachde, A., Chakravarthy, S.: Expressive profile specification and its semantics for a web monitoring system. In: Atzeni, P., Chu, W., Lu, H., Zhou, S., Ling, T.-W. (eds.) ER 2004. LNCS, vol. 3288, pp. 420–433. Springer, Heidelberg (2004). https://doi.org/10.1007/978-3-540-30464-7_33

37. Garg, V., Adaikkalavan, R., Chakravarthy, S.: Extensions to stream processing architecture for supporting event processing. In: Bressan, S., Küng, J., Wagner, R. (eds.) DEXA 2006. LNCS, vol. 4080, pp. 945–955. Springer, Heidelberg (2006). https://doi.org/10.1007/11827405_92

38. Gilani, A., Sonune, S., Kendai, B., Chakravarthy, S.: The anatomy of a stream processing system. In: Bell, D.A., Hong, J. (eds.) BNCOD 2006. LNCS, vol. 4042, pp. 232–239. Springer, Heidelberg (2006). https://doi.org/10.1007/11788911_20

39. Goyal, A.: QP-SUBDUE: Processing Queries Over Graph Databases. Master's thesis, The University of Texas at Arlington, December 2015. http://itlab.uta.edu/students/alumni/MS/Ankur_Goyal/AGoy_MS2015.pdf

40. Hong, D., Chakravarthy, S., Johnson, T.: Locking based concurrency control for integrated real-time database systems. In: Proceedings, International Workshop on Real-Time Databases (RTDB), pp. 138–143 (1996)

41. Hong, D., Johnson, T., Chakravarthy, S.: Real-time transaction scheduling: a cost conscious approach. In: Proceedings, International Conference on Management of Data (SIGMOD), pp. 197–206 (1993)

42. Hong, D.K., Kim, M.J., Chakravarthy, S.: Incorporating load factor into the scheduling of soft real-time transactions for main memory databases. In: Proceedings, IEEE International Conference on Embedded and Real-Time Computing Systems and Applications (RTCSA), pp. 60–66 (1996)

43. Jacob, J., Sachde, A., Chakravarthy, S.: CX-DIFF: a change detection algorithm for XML content and change presentation issues for WebVigiL. In: Jeusfeld, M.A., Pastor, Ó. (eds.) ER 2003. LNCS, vol. 2814, pp. 273–284. Springer, Heidelberg (2003). https://doi.org/10.1007/978-3-540-39597-3_28

44. Jiang, Q., Adaikkalavan, R., Chakravarthy, S.: MavEStream: synergistic integration of stream and event processing. In: International Conference on Digital Communications, p. 29 (2007)

45. Jiang, Q., Chakravarthy, S.: Scheduling strategies for processing continuous queries over streams. In: Williams, H., MacKinnon, L. (eds.) BNCOD 2004. LNCS, vol. 3112, pp. 16–30. Springer, Heidelberg (2004). https://doi.org/10.1007/978-3-540-27811-5_3

46. Jiang, Q., Chakravarthy, S.: Anatomy of a data stream management system. ADBIS Res. Commun. **215**, 654–655 (2006)

47. Kendai, B., Chakravarthy, S.: Load shedding in MavStream: analysis, implementation, and evaluation. In: Gray, A., Jeffery, K., Shao, J. (eds.) BNCOD 2008. LNCS, vol. 5071, pp. 100–112. Springer, Heidelberg (2008). https://doi.org/10.1007/978-3-540-70504-8_10

48. Kendai, B., Chakravarthy, S.: Runtime optimization of continuous queries. In: COMAD, pp. 104–115 (2008)

49. Kivelä, M., Arenas, A., Barthelemy, M., Gleeson, J.P., Moreno, Y., Porter, M.A.: Multilayer networks. CoRR abs/1309.7233 (2013). http://arxiv.org/abs/1309.7233

50. Kona, H., Chakravarthy, S.: An SQL-based approach to incremental association rule mining. Special issue of the Foundations of Computing and Decision Sciences Journal (2006)

51. Kona, H., Chakravarthy, S.: Partitioned approach to association rule mining over multiple databases. In: Kambayashi, Y., Mohania, M., Wöß, W. (eds.) DaWaK 2004. LNCS, vol. 3181, pp. 320–330. Springer, Heidelberg (2004). https://doi.org/10.1007/978-3-540-30076-2_32

52. Mishra, P., Chakravarthy, S.: Performance evaluation and analysis of K-way join variants for association rule mining. In: James, A., Younas, M., Lings, B. (eds.) BNCOD 2003. LNCS, vol. 2712, pp. 95–114. Springer, Heidelberg (2003). https://doi.org/10.1007/3-540-45073-4_9

53. Mishra, P., Chakravarthy, S.: Performance evaluation of SQL-OR variants for association rule mining. In: Kambayashi, Y., Mohania, M., Wöß, W. (eds.) DaWaK 2003. LNCS, vol. 2737, pp. 288–298. Springer, Heidelberg (2003). https://doi.org/10.1007/978-3-540-45228-7_29

54. Mishra, P.: Performance Evaluation and Analysis of SQL-based Approaches for Association Rule Mining. Master's thesis, The University of Texas at Arlington (December 2002)

55. Padmanabhan, S.: HDB-Subdue: A Relational Database Approach to Graph Mining and Hierarchical Reduction. Master's thesis, The University of Texas at Arlington (December 2005)

56. Pandrangi, N., Jacob, J., Sanka, A., Chakravarthy, S.: WebVigiL: user profile-based change detection for HTML/XML documents. In: James, A., Younas, M., Lings, B. (eds.) BNCOD 2003. LNCS, vol. 2712, pp. 38–57. Springer, Heidelberg (2003). https://doi.org/10.1007/3-540-45073-4_5

57. Rosenthal, A., Chakravarthy, U.S.: Anatomy of a mudular multiple query optimizer. In: VLDB, pp. 230–239 (1988)

58. Santra, A., Bhowmick, S., Chakravarthy, S.: Efficient community re-creation in multilayer networks using Boolean operations. In: International Conference on Computational Science, Zurich, Switzerland, pp. 58–67 (2017). https://doi.org/10.1016/j.procs.2017.05.246

59. Santra, A., Bhowmick, S., Chakravarthy, S.: Hubify: efficient estimation of central entities across multiplex layer compositions. In: IEEE International Conference on Data Mining Workshops (2017)

60. Santra, A., Bhowmick, S.: Holistic analysis of multi-source, multi-feature data: modeling and computation challenges. In: Reddy, P.K., Sureka, A., Chakravarthy, S., Bhalla, S. (eds.) BDA 2017. LNCS, vol. 10721, pp. 59–68. Springer, Cham (2017). https://doi.org/10.1007/978-3-319-72413-3_4

61. Santra, A., Bhowmick, S., Chakravarthy, S.: Efficient community detection in Boolean composed multiplex networks. University of Texas at Arlington, June 2019. http://itlab.uta.edu/research/current/Multi%20Source%20Data%20Analysis/ArXiv2019-HoMLN-Final.pdf

62. Santra, A., Komar, K.S., Bhowmick, S., Chakravarthy, S.: An efficient framework for computing structure and semantics-preserving community in a heterogeneous multilayer network. University of Texas at Arlington, June 2019. http://itlab.uta.edu/research/current/Multi%20Source%20Data%20Analysis/ArXiv2019-HeMLN-Final.pdf

63. Santra, A., Komar, K.S., Bhowmick, S., Chakravarthy, S.: Making a case for mlns for data-driven analysis: Modeling, efficiency, and versatility. University of Texas at Arlington, August 2019. http://itlab.uta.edu/research/current/Multi%20Source%20Data%20Analysis/BigData_2019-final.pdf

64. Santra, A., Komar, K.S., Bhowmick, S., Chakravarthy, S.: Structure-preserving community in a multilayer network: definition, detection, and analysis. arXiv preprint arXiv:1903.02641 (2019)

65. Savla, S., Chakravarthy, S.: A single pass algorithm for detecting significant intervals in time-series data. In: ADMKD, pp. 49–60 (2006)

66. Savla, S., Chakravarthy, S.: An efficient single pass approach to frequent episode discovery in sequence data. In: International Conference on Intelligent Environments (IE08) (2008)

67. Solé-Ribalta, A., De Domenico, M., Gómez, S., Arenas, A.: Centrality rankings in multiplex networks. In: Proceedings of the 2014 ACM Conference on Web Science, pp. 149–155. ACM (2014)

68. Srinivasan, A., Bhatia, D., Chakravarthy, S.: Discovery of interesting episodes in sequence data. In: Proceedings, Annual ACM SIG Symposium On Applied Computing, pp. 598–602 (2006)

69. Sun, Y., Han, J.: Mining heterogeneous information networks: a structural analysis approach. ACM SIGKDD Explor. Newslett. **14**(2), 20–28 (2013)

70. Telang, A., Mishra, R., Chakravarthy, S.: Ranking issues for information integration. In: ICDE Workshops, pp. 257–260 (2007)

71. Thomas, S.: Architectures and Optimizations for Integrating Data Mining Algorithms with Database Systems. Ph.D. thesis, The University of Florida at Gainesville, December 1998

72. Thomas, S., Chakravarthy, S.: Performance evaluation and optimization of join queries for association rule mining. In: Mohania, M., Tjoa, A.M. (eds.) DaWaK 1999. LNCS, vol. 1676, pp. 241–250. Springer, Heidelberg (1999). https://doi.org/10.1007/3-540-48298-9_26

73. Thomas, S., Chakravarthy, S.: Incremental mining of constrained associations. In: Valero, M., Prasanna, V.K., Vajapeyam, S. (eds.) HiPC 2000. LNCS, vol. 1970, pp. 547–558. Springer, Heidelberg (2000). https://doi.org/10.1007/3-540-44467-X_50

74. Vu, X.S., Santra, A., Chakravarthy, S., Jiang, L.: Generic multilayer network data analysis with the fusion of content and structure. In: CICLing 2019, La Rochelle, France (2019)

75. Wilson, J.D., Palowitch, J., Bhamidi, S., Nobel, A.B.: Community extraction in multilayer networks with heterogeneous community structure. J. Mach. Learn. Res. **18**(1), 5458–5506 (2017). http://dl.acm.org/citation.cfm?id=3122009.3208030

76. Zhang, H., Wang, C.D., Lai, J.H., Philip, S.Y.: Modularity in complex multilayer networks with multiple aspects: a static perspective. Appl. Inform. **4**, 7 (2017)

Gossip Based Distributed Real Time Task Scheduling with Guaranteed Performance on Heterogeneous Networks

Moumita Chatterjee[1,2(✉)], Anirban Mitra[1,3], Sudipta Roy[5],
Somasis Roy[2], Hirav Shah[4], and Sanjit Kr. Setua[1]

[1] Department of Computer Science and Engineering, Calcutta University,
JD-2, Sector-III, Salt Lake, Kolkata 700098, India
moumitachatterji@gmail.com,
anirban.mitra.cse@gmail.com, sksetua@gmail.com
[2] Department of Computer Science and Engineering,
Aliah University, New Town, West Bengal, India
somasis.roy@gmail.com
[3] Department of Computer Science and Engineering,
Academy of Technology, Adisaptagram 712121, West Bengal, India
[4] Department of Computer Science and Engineering, Ganpat University,
Mehsana 384012, Gujarat, India
hiravjshah@gmail.com
[5] PRT2L, Washington University in St. Louis, St. Louis, MO 63110, USA
sudiptaroy01@yahoo.com

Abstract. This paper considers the scheduling of distributable real time tasks in dynamic networks which are prone to failures and do not have a fixed network infrastructure. We propose a distributable scheduling algorithm using gossip called GBTS-F for reliable and dynamic discovery of appropriate nodes which can execute the tasks. GBTS-F uses the slack time of the tasks for optimizing the gossiping duration and thus satisfies the each task timing constraints with probabilistic guarantee. Even though gossip protocols are usually fault tolerant but to handle byzantine faults and to control the high message complexity incurred during gossiping we propose to use an expander graph. Performance analysis and simulation results show that GBTS-F performs better than other state of art algorithms in terms of message complexity and task success probability.

Keywords: Time/utility functions · Real-time task · Gossip · Transition probability · Expander graphs

1 Introduction

Emerging real time systems on large scale distributed networks are often compelled to run on dynamic networks infrastructures – i.e. those networks that are prone to failures or do not have a fixed network structure including wireless and mobile ad-hoc networks. These networks are often subjected to arbitrary node and link failures and

© Springer Nature Switzerland AG 2019
S. Madria et al. (Eds.): BDA 2019, LNCS 11932, pp. 245–264, 2019.
https://doi.org/10.1007/978-3-030-37188-3_15

message losses which make direct communication among the nodes inefficient. In addition, as the number of nodes increases, maintaining link and route information about the entire network becomes infeasible. In spite of these uncertainties in communication, distributed real time systems demand deterministic (or probabilistic) timeliness assurances on end-to-end distributed task execution.

A number of research papers [1–7], have dealt with real time task scheduling and timing assurances in large scale unreliable networks. These algorithms mainly use gossip as a medium of communicating task parameters and for discovering nodes for hosting of tasks and for detecting node/link failures. Traditionally, gossip algorithms incur high message overheads. So a number of modifications were applied by these algorithms to the basic gossip approach to reduce the time and the message complexity of the protocol. RTGL [1] achieves good communication complexity but incur high message overhead. RTQG [2] extends RTGL using a quorum based approach to reduce the message complexity. RTSRD [4] which also builds on RTGL, adopts the reliable data delivery (RDD) for achieving real time point to point communication. RTG-DS [3] adds deadlock detection and notification in addition to scheduling to the already existing RTG protocol. Our previous work GBTS [13] which further extends the RTG-L protocol used a different message broadcasting approach on top of gossip protocol to achieve improved communication complexity compared to previous protocols.

Our proposed work builds on GBTS [13] which describes our previous version of distributed real time task scheduling. At its core, GBTS is a gossip based protocol. The most important difference is that our proposed work GBTS-F introduces an integrated solution for real time task scheduling and communication algorithm that provides end-to-end probabilistic timing guarantees on distributed task execution in presence of unreliable networks. GBTS-F consists of 2 parts: a task scheduling algorithm and a model for tolerating byzantine failures. The first part divides a distributed task's end-to-end timing constraints into local timing constraints on each executing node and executes its local scheduling algorithm on each local section of the task. The second part ensures connectivity of the network and reliable communication in presence of different failures of the network like byzantine failures. Byzantine failures can seriously affect real time resource management by malicious behaviors, disruption of communications and transmission of false messages. Nodes exhibiting byzantine behaviors are difficult to deal with as they exhibit normal behavior during the authentication process. We propose to utilize a highly connected sparse graph called the expander graph to guarantee reliable communication and ensure network connectivity in presence of random (Byzantine) failures of the network. Our algorithm can tolerate a linear number of byzantine faults while ensuring logarithmic number of communication rounds and polynomial number of messages.

The rest of the paper is organized as follows: In Sect. 2 we describe our network model and problem statement and objectives of our work. Our proposed work is explained in Sect. 3. Experimental results are described in Sects. 4 and 5 concludes the paper.

2 Models and Objectives of Work

2.1 Distributable Real Time Tasks

Distributable Real time tasks executes by invoking local and remote object operations using location-independent calls and returns. Nodes create tasks at random times in response to some events. A task T_i consists of a number of sections where each section requires a different set of objects for its execution. We define a distributable real time task T_i as a multi-tuple as follows: $<Taskid_i, <T_{ij}, O_k, Exe_{kj}, size_{ij}>, TotalExe_i^{EX}$, $NumofSec_i, Deadline_i>$ where

1. $Taskid_i$ is the identifier of the task T_i.
2. $<T_{ij}, O_k, Exe_{kj}, size_{ij}>$ signifies that the section of the task T_{ij} requires object O_k for its execution, the execution time for this section of the task at node k is Exe_{kj} and size of the task is $size_{ik}$.
3. $TotalExe_i^{EX}$ – Total Execution Time of the task T_i.
4. $NumofSec_i$ – total number of task sections of task T_i.
5. $Deadline_i$ – deadline of the task T_i.

The section of the task T_{ij} executing on object O_k is called the jth section of the task. A task's beginning section is termed its root and the current active section of the task is termed its head. The execution time of the task section on the objects and the operations to be performed by the task sections are assumed to be known in advance However the location of the objects are not known beforehand as the nodes may dynamically leave or join the network or may fail due to some reasons.

2.2 Network Model

We define a dynamic network of as $G(t) = (V(t), E(t))$ where $V(t) = \{1..n\}$ and E(t) denotes the bi-directional links among the nodes where V(t) and E(t) can change at any instant of time due to the arrival or departure of nodes in the network. We assume communication links failure during delivery of messages. Network partitioning can also occur causing nodes and links to fail concurrently [9].

The immediate neighbors of node v_i are the set of nodes that v_i is directly connected to at time t

$$\Gamma_i^1(t) = \{v_j \in V(t) : \{v_i, v_j\} \in E(t)\}. \tag{1}$$

The degree of a node v_i are the number of nodes in the immediate neighborhood of v_i at time t.

$$\beta_i(t) = |\Gamma_i^1(t)| \tag{2}$$

Let $\delta = \min_{i \in V(t)} \beta_i(t)$ be the minimum degree of the network. The average degree of the network is defined as $d = 2|E(t)|/n$ and the maximum degree is defined as $\Delta = \max_{i \in V(t)} \beta_i(t)$.

A path between two nodes v_i and v_j is a series of nodes $<v_0, v_1, \ldots v_{k-1}, v_k>$ where $\{v_{i-1}, v_i\} \in E(t)$ for all $i = 1 \ldots k$. The number of edges along the path is defined as the length of the path. We let $d_{ij}(t)$ denote the shortest path between any two nodes v_i and v_j at time t.

Let $\Gamma_i^k(t)$ denote the set of nodes at a distance of k from node v_i i.e.

$$\Gamma_i^k(t) = \left\{ v_j \in V(t) | d_{ij}(t) = k \right\} \tag{3}$$

Let D(t) denote the diameter of the graph which is the length of the shortest path denoted as:

$$D(t) = max_{v_i, v_j \in V(t)} \left\{ d_{ij}(t) \right\} \tag{4}$$

We define $N_i(t)$ as the number of actual edges in the first neighborhood of node v_i. i.e. $N_i(t)$ is given by

$$N_i(t) = \left| \{v_l, v_k\} : v_l, \ v_k \in \Gamma_i^1(t) \wedge \{v_l, v_k\} \in E(t) \right| \tag{5}$$

2.3 Timeliness Model and Utility Enhancement Scheduling

Time Utility function of a task is used for specifying time constraints [5]. TUF is used for representing the utility gained by the system from the task completion as a function of the deadline of the task. A TUF of a task indicates the urgency for task completion measured as deadline along the X axis and its significance measured as utility function along the Y-axis. TUF of a task T_i's is denoted by $U_i(t)$. A classical deadline is a downward step TUF and has unit value i.e. $U_i(t) = \{0, 1\}$. This downward step TUF is used to generalize the classical deadline where $U_i(t) = \{0, n\}$. A task having more than one section has the same TUFS for all the sections. The jth section's TUF is denoted by $U_{ij}(t)$.

In this work we focus on downward step TUFs. A TUF is defined by its arrival time and deadline. Arrival time denotes the value for which the function is defined and deadline is the time after which the task has no utility value.

Thus we can write

$$U_i(t) > 0 \text{ if } t \in \{Arrivaltime_i, Deadline_i\}$$
$$U_i(t) = 0 \text{ otherwise}$$

When a task cannot complete within its deadline then an exception is raised which causes the task to be aborted and release system resources.

2.4 Problem Definition

The problem addressed in this paper is the scheduling of distributable real time tasks in a dynamic network. A real time task requires start to end timeliness guarantee on its execution in addition to integrity for the entire duration for its execution. A task invoked by a node may request multiple objects for its execution. When a node invokes a task for execution, the task starts executing at that node until it requires a different set of objects for its execution. The node at which the task is currently executing is called the head node of the task. The current head node of the task determines the subsequent node on which the next segment of the task will execute. A task does not release the objects until it has completed its execution within its deadline. If several tasks compete for an object then the contention is handled by the node possessing the object as described in [14]. If a task fails to meet its deadline then the node at which the task's deadline expires must inform all the upstream nodes where the task was partially executed and all the objects locked by the task has to be released. As nodes in the network may fail or leave the network at any time during the computation, so the present head node of the task must dynamically determine the next head node on which the next section of the task can execute.

Definition 1: (Distributable Real Time Task Scheduling in Dynamic Distributed Systems). Given a set of nodes V belonging to a dynamic network $N = <V, E>$ where each node holds a different set of resources and $\forall <v_i, v_j> \in E$ signifies the presence of a network link among the nodes v_i and v_j at time t, the set of resources present in node v_i is represented as R_{vi} and the set of resources required by a real time task T is represented as R_T. If T is invoked by a node, the distributable real time task scheduling at N in presence of any number of failures can be defined as a mapping of task T to a set of functional nodes $V_T \in V$, such that the following conditions are satisfied:

(1) The resource requirements of each section of task T is satisfied i.e.

$$R_T \subseteq \forall_{v_i \in V_T} R_{vi}$$

(2) T's execution is completed within its deadline.
(3) The present head node of task segment T_i can dynamically determine the subsequent node in the set V_T on which the next segment T_{i+1} of the task will execute under the constraint of the network structure within a specified time.

2.5   Objectives and Contributions of the Proposed Work

The main objective of our present work is to develop an algorithm for scheduling real time tasks in a network with link/node failures and to demonstrate the efficiency of the new approach. The main contribution of the paper is as follows:

1. We propose to use *Time Utility Function* for locally scheduling sections of task at a node for increasing the total utility and minimizing the number of abortions.
2. We propose to use *Expander graphs* for communicating among the nodes to optimize the message complexity and the number of rounds.
3. We propose an algorithm for *distributed construction of expander graph* that can scale to any number of nodes and can tolerate node failures.
4. We consider by simulations that by *gossiping using expander graph* we can contact almost all the nodes in the network in spite of node/link failures and network partitioning as expander graph have high connectivity properties.

3 Proposed Method

In this section we introduce a new algorithm to schedule real time tasks in a failure prone network while maximizing the utility gained from executing the tasks and minimizing the number of task abortions. The algorithm has two phases:

1. Local phase: for locally scheduling task section at a node.
2. Global phase: for locating the object required by the next task section and handling node failures.

3.1 Local Phase

The local phase is concerned with scheduling tasks locally such that tasks satisfy their time constraint. For real time tasks the timing constraint is specified using TUF (Time/Utility Function) [6]. A TUF represents the utility which is achieved by the system on completion of the task. TUF indicates the urgency of the task computed as the deadline along X axis and the importance computed as a utility function along Y axis. A TUF of task T_i is represented by $U_i(t)$. Each TUF $U_i(t)$ has an arrival time Ari and deadline Dl$_i$. The steps for scheduling task section at a node for maximizing TUF for each task are given below:

Step 1: TUF Calculation for each task section t_{ik}.

At time t when a node j receives task section t_{ik} of task T_i, the following values are computed:

1. Number of task sections NS is decreased by 1 i.e. $NS = NS - 1$
2. Total Remaining Slack Time $SLACK_i^{Total}$ of task $T_i = DL - t - (ER_{ik} - t)$
3. Local Slack time LS_{ij} for task T_i is calculated as

   If $NS > 1$ then

   $$LS_{ij} = \frac{SLACK_i^{Total}}{NS}$$

   Else if $0 < NS \leq 1$ then

   $$LS_{ij} = SLACK_i^{Total} \tag{6}$$

4. Local termination time of task section t_{ik} is calculated as

   $$TER_k^i = t + ER_{ik} + LS_{ij} \tag{7}$$

5. The remaining slack time is equally divided among the remaining tasks sections so that each of them can complete its execution.

6. The algorithm uses the local termination time to check for schedule feasibility during local schedule construction.

Step 2: Resource Utilization of each task section t_{ik}.

ER_{ik} denotes the amount of CPU required by t_{ik}. $size_{ik}$ denotes the amount of memory required by t_{ik}. The product of CPU and memory utilization gives the utilization of a node's resource by t_{ik}. Higher value of this metric indicates a high consumption of CPU and memory resources. The metric is calculated using the formula:

$$UTIL^{CPU}_{memory} = ER_{ik} \times size_{ik} \qquad (8)$$

Step 3: U2R Calculation.

U^2R (Utility and Utilization Ratio) is defined to measure the relevance and the utilization of t_{ik}. U^2R determines the utility of executing t_{ik} per unit CPU and memory utilization. The U^2R of t_{ik} at time t is given by

$$U^2R_{ik}(t) = \frac{\left(t + TER^i_k\right)}{UTIL^{CPU}_{memory}}. \qquad (9)$$

Step 4: Building local schedules.

This step is concerned in constructing of schedules for the local task sections that minimizes the number of sections missing their local termination time and maximizes the total U^2R gained while executing the tasks. Algorithm 1 describes the local scheduling algorithm.

3.2   Global Phase

We first describe our communication structure using gossip before describing the global phase of the GBTS-F algorithm.

3.2.1   Expander Graph Construction

This section deals with describing the procedure used for controlling the messaging complexity of the proposed method. In the communication graph each node represents a processor and the edges represent the interconnection between the processors. A processor can send message to other processor using the edges of the communication graph. Nodes can fail or leave the network at any time during the communication. During crashes the nodes which represent the faulty processors are eliminated from the graph. Thus to maintain communication the neighborhood of the non faulty nodes alters dynamically to guarantee that there exists at least one connected component so that messages can reach maximum number of possible.

Expander graphs belong to the families of sparse graphs having good connectivity properties that are measured using vertex and edge expansion. We construct an expander graph with the vertex set V using our proposed algorithm for distributed expander construction that can scale to any number of nodes. New nodes can join the expander using low message complexity. The graph constructed should be fault tolerant such that it should have at least one "good" component and should be able to spread the messages all over the network in case of node and link failures.

Algorithm 1: Local Scheduling Algorithm

---

1. Initialize t: $=t_{curr}$
2. **For** $\forall t_k \in TL_j$ **Do**
3. $SLACK_i^{Total} = DL - t - (ER_{ik} - t)$
4. **If** NS> 1 **Then**
5. $LS_{ij} = \frac{SLACK_i^{Total}}{NS}$
6. **Else if** 0<NS≤ 1   **Then**
7. $LS_{ij} = SLACK_i^{Total}$
8. $TER_k^i = t + ER_{ik} + LS_{ij}$
9. **If** feasible (t_k) = false **Then**
10. Abort t_k
11. **Else**
12. Calculate $UTIL_{memory}^{CPU} = ER_{ik} \times size_{ik}$
13. $U^2R = \frac{(t+TER_k^i)}{UTIL_{memory}^{CPU}}$.
14. **End If**
15. **End For**
16. $Sch_{tmp} := sortbyUCR(TL_j)$
17. **For** $\forall t_k \in Sch_{tmp}$ in descending order of U^2R **Do**
18. **If** $t_k.U^2R > 0$ **Then**
19. Copy Sch into $Sch_{tent} : Sch_{tent} = Sch$
20. Insert t_k into Sch_{tent} at its termination time position $t_k.TER$
21. **If** feasible(Sch_{tent}) then $Sch := Sch_{tent}$ **Then**
22. **Else**
23. Remove t_k from Sch_{tmp}
24. **Break**
25. $t_{exe} :=$ headof(Sch)
26. Return t_{exe}

Definition 2: (Expander graphs): Let H denote the sub graph of $G(t)$ induced by the subset of V(t) (H $\subset$ V(t)). Let $N_G(H)$ denote the set of all nodes in H and all their neighbors. A graph G(t) is called an expander graph if there exists a positive constant

$b > 1$ such that $E(H, H^c) \geq b|H|$ whenever $|H| \leq |V(t)|/2$ where $E(H, H^c) = |\{(i, j) \in H, j \in H^c\}|$.

Expander graph maintains the following properties: [10]

P1 (Vertex Expansion Property): Any connected subset $S \subseteq V(t)$ $(3 \leq |S| \leq C_a n/d$ $(0 \leq C_a \leq \frac{d}{2}))$ have $|\Gamma(S) \backslash S| \geq C_b d|S|$ $(C_b \in (0, 1))$.

P2 (Edge Expansion Property): The number of vertices in S^c having at least $C_\delta d(|S|/n)$ $(C_\delta \in (0, 1)$ neighbors in S is less than or equal to $|S|^c - \frac{C_w n^2}{d|S|}$ $(C_w > 0)$.

P3 (Regularity Property): The degree of the vertices $d = \Omega(\Delta)$. If degree d is equal to $\omega(log\ n)$ then $d = O(\delta)$.

Property P1 ensures that the number of neighbors of a connected set is larger than the set itself. This means that information dissemination does not terminate in a small set but reaches to a large number of nodes. Property P2 guarantees that in information propagation process the vertices that are uninformed have a sufficient number of informed neighbors. Property P3 assures regularity of the graph. We require that the average degree of our graph be logarithmic or has $\delta = d = \Delta$. Expander graphs are those graphs which meets all the above properties of Definition 2.

Our objectives in constructing expanders are summarized as follows: We construct an expander graph with the vertex set V using our proposed algorithm for distributed expander construction that can scale to any number of nodes. New nodes can join the expander using low message complexity. The graph constructed should be fault tolerant such that it should have at least one "good" component and should be able to spread the messages all over the network in case of node and link failures.

Reiter et al. [8] constructed a (d, ε) - regular random graph in consistent with the refinement [10] of the configuration model. In [8], the nodes are sampled from a tree and added to $\Gamma_G(x)$ while maintaining a degree of d. We propose an algorithm for building an expander graph which is a modification of the algorithm presented in [8].

Our construction of expander graph focuses on the fact that the network may vary from time to time. Our algorithm permits node failure and leaving and can tolerate some period of time when a node may have less than d neighbors. A node runs the expander construction algorithm at any time when it detects any change in its neighborhood. The fault tolerant property of the expander graphs ensures that the network remains connected even when some of the nodes may have degree less than d. This allows the nodes with low degree to execute the expander construction algorithm and recover rapidly. Algorithm 2 describes the Expander construction algorithm.

The energy of a vertex is defined as the set of actual edges among the first neighbors of a vertex x divided by the set of edges that can possibly exist among the first-neighborhood of a vertex x. Formally, the energy of a vertex is given by

$$E_x^t = \begin{cases} \frac{2.N_x(t)}{\beta_x(t)(\beta_x(t)-1)} & if\ \left|\Gamma_x^1(t)\right| > 1 \\ 0 & otherwise \end{cases} \quad (10)$$

Our expander construction algorithm samples nodes from the graph based on the energy of a node. Let $\Gamma_G(x)$ be the set of neighbors of x in G. We define a random walk from x to its neighbor y as the one that goes from a low energy region to high energy region. The probability transition matrix for such a walk is defined by

$$P = \begin{cases} \frac{E_x^t}{\sum_{y \in \Gamma_G(x)} E_y^t} & if \ y \in \ \Gamma_G(x) \\ 0 & otherwise \end{cases} \tag{11}$$

P allows nodes to move towards other nodes having high energy. This mean the nodes will possess more neighbors to chose from in the next step.

Algorithm 2: Algorithm EXCONST

---

1. Upon receiving (ADD:y)
        If y=x or y∈ $\Gamma_G(x)$ Then
2.1.   Do nothing
Else
3.1 **If |$\Gamma_G(x)$|= d Then**
   3.1.1. Pick a node z from |$\Gamma_G(x)$ |with minimum energy
   3.1.2. **If $E_z^t < E_y^t$ Then**
   3.1.3. Remove z from $\Gamma_G(x)$
   3.1.4. Send (Remove: x) to z
   3.1.5. Add y to $\Gamma_G(x)$
   3.1.6. Send (Add: x) to y
Upon receiving (Remove : y)
   4.1 Remove y from $\Gamma_G(x)$
Upon receiving (failed:y)
   5.1 Remove y from $\Gamma_G(x)$

3.2.2   Byzantine Attacks Detection

The second part ensures reliable communication and connectivity of the network in presence of different failures of the network like byzantine failures. Byzantine failures can seriously affect real time resource management by malicious behaviors, disruption of communications and transmission of false messages. A node which exhibits byzantine behavior is known as byzantine node. Such nodes are difficult to deal with as they exhibit normal behavior during the authentication process.

Nodes in an unreliable network may display byzantine behavior in several ways: A node receiving a gossip message may stop transmit the message. A node may also transmit false messages trying to mislead other nodes in making a wrong decision. For example, a node may act as a head node and send Request messages to other nodes requesting object. The receiver nodes reply back thus increasing the message overhead in the network. Gossip messages are usually robust against message losses and node failures. In addition the fault tolerant property of the expander graphs ensures that the network remains connected even when some of the nodes may have less number of neighbors. This allows the nodes with low degree to execute the expander construction algorithm and recover rapidly.

3.2.3   GBTS-F Global Phase Description

At the beginning of each gossip round, the current head node selects a neighbor according to probability P and sends the gossip message. After forwarding the message, the current head node waits for a specified time interval that is bounded by the deadline of the task for receiving a reply. When a node receives a message, it forwards it along its neighbor with highest probability P. The gossiping process continues until the destination node is found. When the node that has the object and can accept the task section receives the message, it sends a reply to the head node. If the head node does not obtain a reply within the predefined deadline then the corresponding task section is aborted and all the preceding upstream nodes are informed of the decision. Otherwise the node containing the object is the next head node of the task. The GBTS-F protocol is illustrated in Algorithm 3.

Algorithm 3: GBTS-F Global Scheduling Algorithm

---

1. **Local Phase at node N_i**
2. $t := t_{curr}$
3. Execute task section t_k
4. Decide the next task section t_{k+1} and the required object
5. Calculate deadline for the next task section D_{next}
6. Calculate probability P for each neighbor and forwards gossip message along the edge with highest probability P and Wait till D_{next}
7. **If** ack = true from N_k and D_{next} not reached **Then**
8. Send task to N_k
9. **Else If** D_{next} reached
10. Abort (t_k)
11. Select neighbors with probability P and sends Abort(t_k) message.
12. **Remote Phase at each intermediate node N_k on receiving the gossip message**
13. **If** noofgossipmessages<=1 **Then**
14.   Forward gossip message.
15. **Remote Phase at node N_j holding object required for next task section**
16. Upon receiving gossip message
17. **If** accept=T_i .schedule **Then**
18. Send ack to N_i and Execute task section.

---

4   Analysis of the GBTS-F Protocol

Theorem 1: The graph constructed by Algorithm EXCONST is an expander graph.

Proof: We prove our theorem by using the probability transition matrix P as defined in (12). Such a P is symmetric and has a uniform stationary distribution. The conductance of P is defined as

$$\Phi(P) = \min_{H \subset V:|H| \le n/2} \frac{\sum_{i \in H; j \in H^c} P_{ij}}{|H|} \text{ where } H \subset V(t) \tag{12}$$

A random walk over the graph G(t) using P is defined as expanding if there exists a constant $\gamma > 0$ and independent of n(the number of nodes in the graph) such that $\Phi(P) \ge \gamma$. This means that the mixing time of P effectively scales as O(log n). That is the random walk using P mixes very fast [11].

Substituting the value of P_{ij} from Eqs. (11) in (12)

$$\Phi(P) = \min_{H \subset V(t):|H| \le n/2} \frac{\sum_{i \in H; j \in H^c} \frac{E_x^t}{\sum_{y \in \Gamma_G(x)} E_y^t}}{|H|}$$

$$\Phi(P) = \left(\frac{E_x^t}{\sum_{y \in \Gamma_G(x)} E_y^t} \right) \min_{H \subset V(t):|H| \le n/2} \frac{E(H, H^c)}{|H|}$$

From definition (2) we know that

$$\Phi(P) \ge \min_{H \subset V(t):|H| \le n/2} \frac{b|H|}{|H|} \ge b \left(\frac{E_x^t}{\sum_{y \in \Gamma_G(x)} E_y^t} \right)$$

Thus the P defined in Eq. (11) has a conductance $b \left(\frac{E_x^t}{\sum_{y \in \Gamma_G(x)} E_y^t} \right) > 0$ and is independent of n. Thus such a graph formed from the probability matrix P is an expander as proposed.

Theorem 2: Let us denote as **u** an arbitrary initiator node of the graph constructed using Algorithm EXCONST. The gossip based information diffusion process initiating from node **u** on the graph requires at most O (log n) rounds to inform all nodes in the network.

Proof: Let us denote by I(t) and H(t) as the set of nodes which are informed at the beginning of round t and the set of nodes which have received the information at time t but have not forwarded the information yet. $H(t) \subseteq I_t$. We divide our analysis into three phases with each phase consisting of several steps. During each phase we consider information dissemination from the node in H(t − 1) that were informed in the previous phase.

Phase 1: The initiator node u is the only informed node at time step t = 0. We know from vertex expansion property of expander graph that the neighborhood $\Gamma(u)$ of u exponentially grows at each time step t. If $d = \omega(\log n)$, then we need to only inform the vertices in neighborhood $\Gamma(u)$. Otherwise if $\Gamma_{t-1}(u)$ is informed at time t − 1 then $\Gamma_t(u)$ can be informed in at most Δ steps. Thus, for some constant Z, after t = log n steps

$$H(t) \geq Z \log n$$
$$I(t) = O(H(t)) \tag{13}$$

Phase 2: From Phase 1 we obtained a set $H(t)$ which comprises of at least log n nodes. Phase 2 aims to inform at least (n/d) vertices. From Eq. (13) and Property 1 of expander graph it can be written that

$$Z_2 \log n \leq I(t) \leq Z_\alpha n/d \tag{14}$$

Thus property 1 guarantees that if there is a set consisting of informed vertices satisfying the vertex expansion property then after k steps, the set of informed vertices strictly rises by a factor greater than 1. So for some constant $Z > 0$,

$$|I_{t+k}| \geq C|I_t| \tag{15}$$

Thus $\log_Z(|I_t|) = \log_Z(Z_\alpha(n/d)) = O(\log n)$ steps are needed to inform $O(n/d)$ vertices.

Phase 3: The aim of Phase 3 is to inform a linear number of vertices. From Property 2 of expander graph we know that a large number of nodes that are uninformed have their neighbors that are members of $I(t)$. Thus from Phases 1 and 2 it can be concluded that by applying Phase 3 log n times we can ensure that a linear number of vertices will be informed.

If we continue in this way and merge all the phases we can write that

$$\left| I_{t+O(\log n)} \right| \approx n. \tag{16}$$

Theorem 3: The global phase of GBTS-F protocol generates $O(n \log \log n)$ messages where n is the number of nodes in the network.

Proof: From Property 3, we consider $d = \log n - 1$. During each gossip interval the initiator node chooses the neighboring nodes having energy greater than the initiator's own energy and sends the message to those nodes. The expected number of nodes that the initiator node contacts (we consider that in the worst case the initiator contacts all the d nodes in its neighborhood):

The probability of i nodes receiving the gossip message is $\theta\left(\frac{1}{i+1} \cdot i\right)$. Thus the expected number of nodes that the initiator contacts for sending the gossip message is $\sum_{i=1}^{d} \theta\left(\frac{1}{i+1} \cdot i\right) = O(\log d)$ and so the expected number of messages are $O(\log d)$. By linearity of expectation it can be written that the expected number of messages sent by all the nodes in the network is $O(n \log d) = O(n \log \log n)$.

Corollary: The GBTS protocol performs better than RTGL in terms of message complexity by a factor of $\frac{\log \log n}{\log n}$.

5   Simulation Results

In this section we present the results of our simulation studies to evaluate GBTS-F, including the gossip protocol on expander graph having low message overhead, the local phase of GBTS-F protocol, and the performance of GBTS-F in the global phase under various task and network conditions. We developed a discrete event simulator in C language to perform the simulation of our proposed method. The simulations are performed by varying various system parameters such as number of nodes and number of tasks. The nodes in our model are connected by communication channels assuming an arbitrary topology. Some comparison studies are performed between GBTS-F, RTGL and RTG-DS in terms of reliability and message overhead. The environment for our simulation is as follows: Intel(R) Core TM i5-3210M CPU (2.5-GHz clock) and 4 GB RAM, Ubuntu 14.04 and C compiler: GCC version 4.8.2

In our simulation, each node performs peer sampling to select its neighbors. We made some assumptions to simplify the simulation.

- Nodes and links can fail at any time during the execution of the algorithm.
- Peer samplings are performed successfully so the nodes know about their neighbors.

Nodes and link failures can occur at any time during the execution of the algorithm. However since each node performs peer sampling periodically it gets updated information about the states of the nodes and the links in the system.

We used uniform distribution for describing arrival times, execution times and termination times of distributable tasks. All tasks are made to make invocations through the same set of nodes in the system. However the order of invocation of the tasks on each node may vary due to different order of scheduling of each section on the nodes. Thus a task may miss its deadline because it arrives at its destination node late. Our simulation study uses a fixed number of shared resources. The resources are assigned to each node dynamically. Each task section dynamically determines the resources that it requires. Each time a resource is required by a task section, a part of execution time of the task section is assigned to the resource.

5.1   Performance of Gossip Protocol in Large Scale Systems

The simplest form of the epidemic (gossip) protocol consists of two states: Susceptible and Infected. This form of the gossip protocol is known as the SI model [11]. To apply the SI model to our algorithm we let $v \in V$ be the initiator node of the load balancing algorithm. Let $I(r) \subset V$ be the set of nodes that are recipient to the ongoing request message at round r. These nodes are considered as infected. Initially let $r = 0$ and $i_0 = v$. Mathematically, the logistic growth function of SI model is defined as follows:

$$I(r) = \frac{i_0 e^{kr}}{1 - i_0 + i_0 e^{kr}} \tag{17}$$

Where $I(r)$ is a function that gives the number of nodes infected at round r and k is the size of the peer list also called the fan out.

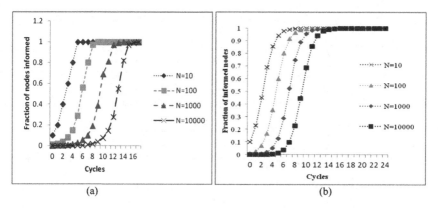

Fig. 1. Fraction of informed nodes for each round (a) theoretical results (b) simulation results

The theoretical results of Eq. (17) are given in Fig. 1(a). The simulation results are given in Fig. 1(b). When the theoretical results and the simulation results are compared, they are not matched exactly due to the random uncertainty nature of gossip algorithm. As we can see larger the network, the more number of rounds it take to inform all the nodes. However we can observe that in both the results as the number of nodes grows exponentially, the required number of rounds increases linearly.

From the graphs we can observe that in large scale systems (each system containing 10–10000 nodes), each theoretical values of I(r) conforms well with the simulation results with a relatively small standard deviation. The gossip protocol used in our algorithm only allows the most recently informed nodes to gossip once in the subsequent round instead of allowing all the informed nodes to gossip repeatedly. In gossip protocols messages are sent at the beginning of a round and they arrive at their destination before the round ends. These messages are not counted in the subsequent round. Consequently the number of messages existing in the network at any time is much less than the total number of messages.

5.2 Performance of Proposed Expander Routing Method Compared to Other Methods of Routing

This experiments focus on the performance comparison of our proposed expander routing protocol to other stochastic methods of routing. The algorithms considered are the *random walking method* and the *receiver-degree routing method*. The gossiping protocols on these algorithms should also perform peer sampling so that nodes have the knowledge of their neighbors. The random walking method is an unbiased routing method where each node uniformly assigns transition probabilities to its neighbors. The receiver-degree routing method for any connected node pair assigns transition probability to a node according to the degree of the receiver node. The transition probability of a neighboring node is higher if it has a higher degree than all other neighbors.

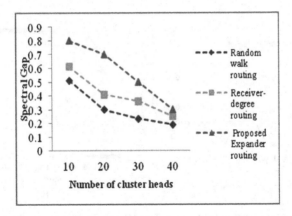

Fig. 2. Variation in spectral gap with respect to variation in the number of nodes

The spectral gap of a matrix is defined as the difference between the largest eigen value and the second largest eigen value of the matrix. We know from spectral graph theory [12], that a graph with large spectral gap has a higher expansion parameter than a graph with low spectral gap. In a large network it is vital to achieve faster convergence rate for routing. The convergence rate of routing depends on spectral gap of the transition probability matrix of the routing method.

The simulations are carried out on random geometric network and a random node is selected as the destination node. We assume that only the nodes in the neighborhood of the destination know about the destination node and all the other nodes in the network are unaware of the destination node.

Figure 2 shows the variations in spectral gaps with respect to the number of cluster heads when the destination node is chosen at random. The spectral gap decreases gradually with the increase in the number of nodes for all methods of routing. The decrease in the spectral gap occurs as convergence time increases with increasing number of nodes. When we compare the spectral gap of our proposed expander routing method with other methods we can observe the considerable improvement of spectral gap of our proposed method with the other routing methods. In the next figures we illustrate the communication overhead of proposed expander routing method with random routing and receiver initiated routing for three different types of network topologies. The final results are obtained from 100 experiments for each size of the network for each network topology and taking the average of the results. From the graphs we can see that for networks with good expansion property like random geometric graph and scale free graph, using our proposed expander routing method requires less number of rounds compared to other routing methods (Fig. 3).

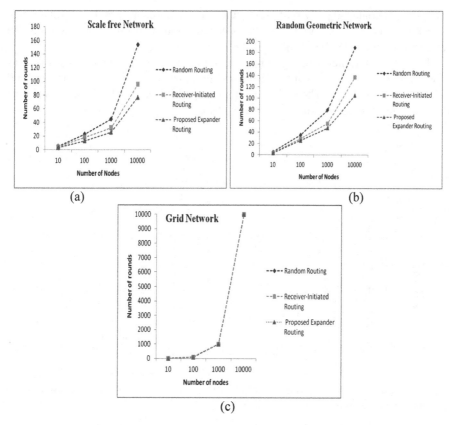

Fig. 3. Number of rounds required for each routing methods for different number of nodes in (a) scale free network (b) random geometric network (c) grid network

5.3   Comparison of GBTS with RTG-L and RTDG-DS

In this section we aim to evaluate the effectiveness of our proposed method GBTS by comparing with RTG-L and RTG-DS, in terms of reliability and message overhead. For simulation we consider a network of 1000 nodes and a fan out of 4. Figure 4 illustrates the task failure probability in a network without node failures (Fig. 4a) and with node failures (Fig. 4b). From the figure we can see that when the number of subtasks increases the probability of success of all the three methods decreases because failure of a single subtask results in the failure of an entire task. However GBTS can still sustain a considerable success probability with a minimum reduction while RTG-L is very susceptible to the change in the number of subtasks. For example the success proba-bility of RTG-L is 0.5 for 5 subtasks while for RTG-DS and GBTS the success probability is 0.8 and 0.99 respectively in simulation for a network without node failures. Beyond that the success probability of RTG-L nearly approaches zero while GBTS keeps the success probability at 0.9. We can notice from the figures that when only a task has only one subtask which means that the present head node can do all the work without contacting any other node, the success probability is 1 for all the three algorithms. Under failure-prone network also the simulation results follow the same pattern as in failure free network.

Under the same network condition, we now consider the message overhead as shown in Fig. 5. For this simulation we consider network of size between 300 to 1000 nodes and a fan out of 4. The message complexity of RTGL and RTG-DS is O (log n) while the message complexity of GBTS is O(log log n) [13].

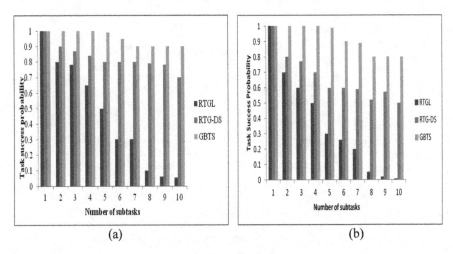

(a)   (b)

Fig. 4. Comparison between RTGL, RTG-DS, GBTS-F on the basis of task success probability (a) without failures (b) in presence of failures.

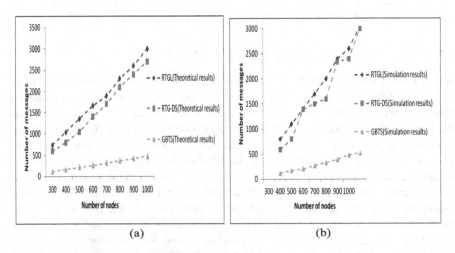

(a)   (b)

Fig. 5. Comparison of RTG, RTG-DS and GBTS-F with respect to message complexity (a) theoretical results (b) simulations

From the figures we can observe that the simulation results also follow the theoretical results in all the three cases. So GBTS achieves much better message complexity than RTGL and RTG-DS.

6   Conclusion

In this paper we present GBTS, a gossip based distributable real time task scheduling algorithm for large scale unreliable networks. The main aim of our proposed algorithm is to use gossip based communication to propagate task scheduling parameters and to determine the next node for execution of task sections. GBTS constructs schedules for local task sections by using slack time for gossiping. Since gossip protocols usually incur high message complexity, we use expander graph to manage the message complexity of our algorithm. We show that by using expander graph we can significantly control the message complexity as well as ensure network connectivity in presence of failures. Analysis of our proposed approach proves that it performs better than other previous protocols. We perform simulation studies validates our theoretical results.

References

1. Han, K., Ravindran, B., Jensen, E.D.: RTG-L: dependably scheduling real-time distributable threads in large-scale, unreliable networks. In: Proceedings of IEEE Pacific Rim International Symposium on Dependable Computing (PRDC) (2007)
2. Zhang, B., Han, K., Ravindran, B., Jensen, E.D.: RTQG: real-time quorum- based gossip protocol for unreliable networks. In: Proceedings of the Third International Conference on Availability, Reliability and Security (2008)
3. Han, K., et al.: Exploiting slack for scheduling dependent, distributable real-time threads in mobile ad hoc networks. In: RTNS, March 2007
4. Huang, F., Han, K., Ravindran, B., Jensen, E.D.: Integrated real-time scheduling and communication with probabilistic timing assurances in unreliable distributed systems. In: 13th IEEE International Conference on Engineering of Complex Computer Systems (ICECCS 2008), Belfast, pp. 79–88 (2008). https://doi.org/10.1109/iceccs.2008.15
5. Wu, H., Ravindran, B., Jensen, E.D.: Energy-efficient, utility accrual real-time scheduling under the unimodal arbitrary arrival model. In: Proceedings of Design, Automation and Test in Europe (2005)
6. Jensen, E.D., et al.: A time-driven scheduling model for real-time systems. In: RTSS, pp. 112–122, December 1985
7. Bettati, R.: End-to-end scheduling to meet deadlines in distributed systems. Ph.D. thesis, UIUC (1994)
8. Reiter, M.K., Samar, A., Wang, C.: Distributed construction of a fault-tolerant network from a tree. In: Proceedings of the 2005 24th IEEE Symposium on Reliable Distributed Systems (SRDS 2005) (2005)
9. Kasprzyk, R.: Diffusion in networks. J, Telecommun. Inf. Technol. **2012**(2), 99–106 (2012)

10. Doerr, B., Friedrich, T., Sauerwald, T.: Quasirandom rumor spreading: expanders, push vs. pull, and robustness. In: Albers, S., Marchetti-Spaccamela, A., Matias, Y., Nikoletseas, S., Thomas, W. (eds.) ICALP 2009. LNCS, vol. 5555, pp. 366–377. Springer, Heidelberg (2009). https://doi.org/10.1007/978-3-642-02927-1_31

11. Shah, D.: Gossip algorithms. Found. Trends Netw. **3**(1), 1–125 (2008). https://doi.org/10.1561/1300000014

12. Wijetunge, U., Perreau, S., Pollok, A.: Distributed stochastic routing optimization using expander graph theory. In: IEEE Australian Communication Theory Workshop AusCTW (2011)

13. Chatterjee, M., Setua, S.K.: Gossip-based real-time task scheduling using expander graph. In: Chaki, R., Cortesi, A., Saeed, K., Chaki, N. (eds.) Advanced Computing and Systems for Security. AISC, vol. 897, pp. 3–14. Springer, Singapore (2019). https://doi.org/10.1007/978-981-13-3250-0_1

14. Chatterjee, M., Mitra, A., Roy, S., et al.: Gossip based fault tolerant protocol in distributed transactional memory using quorum based replication system. Cluster Comput. (2019)

Data-Driven Optimization of Public Transit Schedule

Sanchita Basak, Fangzhou Sun, Saptarshi Sengupta, and Abhishek Dubey[✉]

Department of EECS, Vanderbilt University, Nashville, TN, USA
{sanchita.basak,saptarshi.sengupta,abhishek.dubey}@vanderbilt.edu,
fzsun316@gmail.com

Abstract. Bus transit systems are the backbone of public transportation in the United States. An important indicator of the quality of service in such infrastructures is on-time performance at stops, with published transit schedules playing an integral role governing the level of success of the service. However there are relatively few optimization architectures leveraging stochastic search that focus on optimizing bus timetables with the objective of maximizing probability of bus arrivals at timepoints with delays within desired on-time ranges. In addition to this, there is a lack of substantial research considering monthly and seasonal variations of delay patterns integrated with such optimization strategies. To address these, this paper makes the following contributions to the corpus of studies on transit on-time performance optimization: (a) an unsupervised clustering mechanism is presented which groups months with similar seasonal delay patterns, (b) the problem is formulated as a single-objective optimization task and a greedy algorithm, a genetic algorithm (GA) as well as a particle swarm optimization (PSO) algorithm are employed to solve it, (c) a detailed discussion on empirical results comparing the algorithms are provided and sensitivity analysis on hyper-parameters of the heuristics are presented along with execution times, which will help practitioners looking at similar problems. The analyses conducted are insightful in the local context of improving public transit scheduling in the Nashville metro region as well as informative from a global perspective as an elaborate case study which builds upon the growing corpus of empirical studies using nature-inspired approaches to transit schedule optimization.

Keywords: Timetable optimization · Genetic algorithm · Particle swarm optimization · Sensitivity analysis · Scheduling

1 Introduction

Bus systems are the backbone of public transportation in the US, carrying over 47% of all public passenger trips and 19,380 million passenger miles in the US [18]. For the majority of cities in the US which do not have enough urban forms or budget to build expensive transit infrastructures like subways, the reliance is on buses as the most important transit system since bus systems have advantages

© Springer Nature Switzerland AG 2019
S. Madria et al. (Eds.): BDA 2019, LNCS 11932, pp. 265–284, 2019.
https://doi.org/10.1007/978-3-030-37188-3_16

of relatively low cost and large capacity. Nonetheless, the bus system is also one of the most unpredictable transit modes. Our study found that the average on-time performance across all routes of Nashville bus system was only 57.79% (see Sect. 6.1). The unpredictability of delay has been selected as the top reason why people avoid bus systems in many cities [2].

Providing reliable transit service is a critical but difficult task for all metropolis in the world. To evaluate service reliability, transit agencies have developed various indicators to quantify public transit systems through several key performance measurements from different perspectives [4]. In the past, a number of technological and sociological solutions have helped to evaluate and reduce bus delay. Common indicators of public transit system evaluation include schedule adherence, on-time performance, total trip travel time, etc. In order to track the transit service status, transit agencies have installed AVL on buses to track their real-time locations. However, the accuracy of AVL in urban areas is quite limited due to the low sampling rate (every minute) and the impact of high buildings on GPS devices. To have some basic controls during bus operation, public transit agencies often use time point strategies, where special timing bus stops (time points are special public transit stops where transit vehicles try to reach at scheduled times) are deployed in the middle of bus routes to provide better arrival and departure time synchronizations.

An effective approach for improving bus on-time performance is creating timetables that maximize the probability of on-time arrivals by examining the actual delay patterns. When designing schedules for real-world transport systems (e.g. buses, trains, container ships or airlines), transport planners typically adopt a tactical-planning approach [10]. Conventionally, metro transit engineers analyze the historical data and adjust the scheduled time from past experience, which is time consuming and error prone. A number of studies have been conducted to improve bus on-time performance by reliable and automatic timetabling. Since the timetable scheduling problem is recognized to be an NP-hard problem [28], many researchers have employed heuristic algorithms to solve the problem. The most popular solutions include ad-hoc heuristic searching algorithms (e.g. greedy algorithms), neighborhood search (e.g. simulated annealing (SA) and tabu search (TS)), evolutionary search (e.g. genetic algorithm) and hybrid search [24].

However, there are few stochastic optimization models that focus on optimizing bus timetables with the objective of maximizing the probability of bus arrivals at timepoint with delay within a desired on-time range (e.g. one minute early and five minutes late), which is widely used as a key indicator of bus service quality in the US [1]. A timepoint is a bus stop that is designed to accurately record the timestamps when buses arrive and leave the stop. Bus drivers use timepoints to synchronize with the scheduled time. For example, to quantify bus on-time arrival performance, many regional transit agencies use the range of $[-1, +5]$ minutes compared to the scheduled bus stop time as the on-time standard to evaluate bus performance using historical data [1]. The actual operation of bus systems is vulnerable to many internal and external factors. The

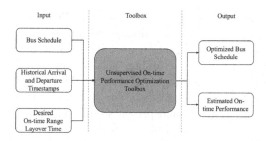

Fig. 1. The proposed toolbox for bus on-time performance optimization. City planners use bus schedule, historical trip information and desired on-time range and layover time, and get outputs of optimized timetable as well as estimated on-time performance.

external factors include urban events (e.g., concerts, sporting events, etc.), severe weather conditions, road construction, passenger and bicycle loading/offloading, etc. One of the most common internal factors is the delay between two consecutive bus trips, where the arrival delay of previous trips causes departure delay of the next trip. Furthermore, there are monthly and seasonal variation in the actual delay patterns, but most transit agencies publish a uniform timetable for the next several months despite the variations. How to cluster the patterns and optimize timetables separately remains an open problem. Furthermore, heuristic optimization techniques have attracted considerable attention, but finding the optimal values of hyper-parameters are difficult, since they depend on nature of problem and the specific implementation of the heuristic algorithms, and are generally problem specific.

Research Contributions: In this paper, the monthly and seasonal delay patterns are studied and outlier analysis and clustering analysis on bus travel times to group months with similar patterns together are carried out. The feature vectors are aggregated by routes, trips, directions, timepoint segments and months encompassing mean, median and standard deviation of the historical travel times. This work significantly extends prior work on the problem as in [23]. Along with a greedy algorithm and a Genetic Algorithm (GA), swarm based optimization algorithm has been introduced in this work, where the semi-autonomous agents in the swarm can update their status guided by the full knowledge of the entire population state. Thus, Particle Swarm Optimization (PSO) [14] algorithm is employed in this work to generate new timetables for month clusters that share similar delay patterns with the goal of testing both evolutionary computing and swarm intelligence approaches. It is observed that the optimized on-time performance averaged across all bus routes has increased by employing PSO as compared to that by GA. Also the execution times of PSO are much less than GA and are more stable indicating lesser variability of results over different runs. Sensitivity analysis on choosing the optimal hyper-parameters for the proposed heuristic optimization algorithms are also presented. A stability analysis of the respective algorithms have been put forward by studying the on-time performance and execution time over several runs. The overall workflow of the proposed optimization mechanisms is illustrated in Fig. 1.

The rest of the paper is organized as follows: Sect. 2 compares our work with related work on transit timetabling; Sect. 3 presents the problem formulation; Sect. 4 presents the details of the transit data stores; Sect. 5 discusses the timetable optimization mechanisms used; Sect. 6 evaluates the performance of the optimization mechanisms and presents sensitivity analysis results; Sect. 7 presents concluding remarks and future work.

2   Related Work and Challenges

This section compares our system with related work on transit timetable scheduling. A number of studies have been conducted to provide timetabling strategies for various objectives: (1) minimizing average waiting time [27] (2) minimizing transfer time and cost [7,12,24], (3) minimizing total travel time [17], (4) maximizing number of simultaneous bus arrivals [9,13], (5) minimizing the cost of transit operation [26], (6) minimizing a mix of cost (both the user's and the operator's) [6].

The design of timetable with maximal synchronizations of bus routes without bus bunching has been researched by Ibarra-Rojas et al. [13]. The bus synchronization strategy has been discussed from the perspective of taking waiting time into account in the transfer stops in the work of Eranki et al. [9]. An improved GA in minimizing passenger transfer time considering traffic demands has been explored by Yang et al. [12]. Traffic and commuter demand has also been considered in the work by Wang et al. [27]. Other than employing optimization algorithms several deep learning techniques [22] have been applied in bus scheduling problems [15].

Nayeem et al. [17] set up the optimization problem over several criteria, such as minimizing travel time and number of transfers and maximizing passenger satisfaction. A route design and neighborhood search through genetic algorithm minimizing number of transfers has been discussed by Szeto et al. [24]. Zhong et al. [29] used improved Particle Swarm Optimization for recognizing bus rapid transit routes optimized in order to serve maximum number of passengers.

2.1   Research Challenges

(a) Clustering Monthly and Seasonal Variations in Historical Arrival Data: Studying the historical travel time at segments can be an effective way to set bus timetables. However, existing work doesn't consider the monthly and seasonal variation in historical monthly data, and the variation can be utilized for better scheduling. Generating one timetable for all months may not be the best solution. As traffic and delay patterns are prone to changes over seasonal variations and various times, we generate clusters grouping months with unsupervised algorithm and develop optimization strategies for the generated clusters.

Table 1. The scheduled time and recorded actual arrival and departure time of two sequential trips that use the same bus of route 4 on Aug. 8, 2016. The arrival delay at the last timepoint of the first trip accumulates at the first timepoint of the second trip.

		Timepoints			
		MCC4_14	SY19	PRGD	GRFSTATO
Trip 1	Scheduled time	10:50 AM	11:02 AM	11:09 AM	11:18 AM
	Actual arrival time	10:36 AM	11:10 AM	11:18 AM	11:27 AM
	Actual departure time	10:50 AM	11:10 AM	11:18 AM	11:30 AM
Trip 2	Scheduled time	11:57 AM	11:40 AM	11:25 AM	11:20 AM
	Actual arrival time	12:11 PM	11:51 AM	11:34 AM	11:27 AM
	Actual departure time	12:11 PM	11:51 AM	11:34 AM	11:30 AM

We evaluate the proposed mechanism via simulation. The cluster-specific schedule is shown to further increase the on-time performance compared to generating one uniform timetable.

(b) Computing Efficiently and Accurately in the Solution Space: Transit performance optimization techniques rely on historical delay data to set up new timetables. However, the large amount of historical data makes it a challenge to compute efficiently. For example, Nashville MTA updates the bus schedule every 6 months but each time there are about 160,000 historical record entries to use. Moreover, the solution space has typically very large under constraints (e.g., sufficient dwelling time at bus stops, adequate layover time between trips, etc.). A suitable optimization algorithm is necessary for efficient and accurate computation. Since this is a discrete-variable optimization problem, gradient-based methods cannot be used and gradient-free methods need to be considered. A naive algorithm for discrete optimization is exhaustive search, i.e., every feasible time is evaluated and the optimum is chosen. Exhaustive search works for a small finite number of choices, and cannot be used for high-dimensional problems. Genetic algorithm [6,7], as well as particle swarm optimization [14] are used commonly in solving heuristic problems . Thus we consider applying genetic algorithm and particle swarm optimization (PSO) in the context. Section 5 describes the key steps of how we apply greedy, genetic and PSO algorithms to solve the timetable optimization problem.

3   Problem Formulation

Typically, transit delay are not only affected by external factors (such as traffic, weather, travel demand, etc.), but also by some internal factors. For example, the accumulated delay occurred on previous trips may cause a delay in consecutive trips by affecting the initial departure time of the next trip. In order to

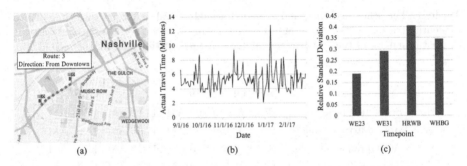

Fig. 2. (a) A route segment on bus route 3 leaving downtown; (b) The variance of actual travel time and (c) the relative standard deviation of actual travel times on a bus route segment in time period between Sept. 1, 2016 and Feb. 28, 2017.

illustrate the problem context with simplicity and without generality, we take two sequential bus trips of route 4 in Nashville as an example (the scheduled time and the actual arrival and departure time recorded on Aug. 8, 2016 are shown in Table 1) to describe the optimization problem. On each service day, after a vehicle of the first trip (121359) arrives at the last stop (Timepoint GRF-STATO) with scheduled time of 11:18 AM, the second trip is scheduled to depart using the same vehicle from the same stop at 11:20 AM. On Aug. 8, 2016, the arrival time at the last stop (Timepoint GRFSTATO) of the first trip (121359) is exceptionally late for 9 min, which contributes to the 10-minute departure delay at the beginning of the second trip. Since the scheduled layover time between the two trips is only 2 min (between 11:18 AM and 11:20 AM), any large delay at the first trip is very likely to transfer to the next trip. Therefore, the optimization problem should involve a process that considers not only the travel delay on segments, but also the improper lay over time between trips.

Figure 2 illustrates the large variation of bus travel time distribution. The example shows travel time data collected from bus trips depart at a specific time of the day on route 3 in Nashville. The coefficient of variation (also known as relative standard deviation) , which is a standardized measure of dispersion of a probability distribution, is very high on all timepoints along the route. The complexity and uncertainty of travel times introduce great challenges to the task of timetable optimization.

3.1   Problem Definition

For a given bus trip schedule b, let $H = \{h_1, h_2, ..., h_m\}$ be a set of m historical trips with each trip passing n timepoints $\{s_1, s_2, ..., s_n\}$. So the *on-time performance* of the bus trip schedule b can be defined as a ratio of an indicator function $I(h_i, s_j)$ summed over all timepoints for all historical trips to the product of the total number of historical trips and total number of timepoints. The indicator function $I(h_i, s_j)$ is 1 if $d_{i,j} \in [t_{early}, t_{late}]$, otherwise 0, where $d_{i,j} = t^{arrival}_{h_i, s_j} - T^{arrival}_{h_i, s_j}$

The objective is to design a schedule optimization problem to generate new $T_{h,s}^{departure}$, ensuring on-time performance maximization. t_{early} and t_{late} are two time parameters pre-defined by the transit authority as a measure of schedule maintenance and $d_{i,j}$ is the actual delay that arriving in timepoint s_j.

4 Data Store

4.1 Data Sources

We established a cloud data store and reliable transmission mechanisms to feed our Nashville Metropolitan Transit Authority (MTA) updates the bus schedule information every six months and provides the schedule to the public via GTFS files. In order to coordinate and track the actual bus operations along routes, MTA has deployed sensor devices at specially bus stops (called timepoints) to accurately record the arrival and departure times. In Nashville, there are over 2,700 bus stops all over the city and 573 of them are timepoint stops. City planners and MTA engineers analyze the arrival and departure records regularly to update the transit schedule. The details of the datasets are as follows:

– *Static GTFS.* This dataset defines the static information of bus schedule and associated geographic information, such as routes, trips, stops, departure times, service days, etc. The dataset is provided in a standard transit schedule format called General Transit Feed Specification (GTFS).
– *GTFS-realtime.* This dataset is recorded real-time transit information in GTFS-realtime format, which include bus locations, trip updates and service alerts. The GTFS-realtime feed is collected and stored in one-minute interval.
– *Timepoints.* This dataset provides accurate and detailed historical arrival and departure records at timepoint stops. The information include route, trip, timepoint, direction, vehicle ID, operator, actual arrival and departure time, etc. The dataset is not available in real-time but collected manually by Nashville MTA at the end of each month.

Even though the same timepoint datasets are utilized in the study, the proposed method is not limited to the timepoint datasets and can use some surrogate data sources: (1) automatic passenger counters (APC) data: APC datasets records both passenger counts and departure/arrival times at stops (2) GTFS-realtime feed: the real-time bus locations reported by automatic vehicle locator (AVL) installed on buses. Compared with timepoint datasets, APC data also provides accurate times at normal stops thus it is the most suitable alternative dataset. However, GTFS-realtime suffers from low sampling rate and low accuracy in the city and may reduce the performance of the proposed mechanism.

4.2 Data Cleaning

Since raw transit dataset often contains missing, duplicate and erroneous samples, preprocessing is a necessary step to prepare a clean and high-quality dataset.

Missing data issue occurs due to hardware or network problems. Generally, there are samples with missing data can be dropped or filled with a specific or average values. Duplicated data (e.g., a bus trip is recorded more than one time) will oversample certain delay values and make the delay dataset biased. We drop the trips with no historical records and remove duplicated records.

Outliers are values that are distant from most of the observed data in presence of which clustering can be inappropriate. K-means clustering algorithm is also sensitive to outliers present in the data. The approach taken here is to calculate Median Absolute Deviation (MAD), a robust measure of statistical dispersion. The MAD of a data set $[X = (x_1, x_2, ..., x_n)]$ can be calculated as: $MAD = median(|x_i - median(X)|)$. For normal distribution the scaled MAD is defined as $(MAD/0.6745)$, approximately equal to the standard deviation. x_i is considered an outlier if the difference between x_i and median is larger than three times of standard deviation (i.e. scaled MAD).

5   Timetable Optimization Mechanisms

5.1   Month Grouping by Clustering Analysis

This section introduces a clustering analysis mechanism that groups months with similar transit delay patterns together and the results will later be used to generate separate timetables for each group.

Feature Engineering. We assume the monthly delay patterns can be represented by the mean, median and standard deviation that derived from historical delay data. Considering a bus trip consists of n timepoints, there are $n - 1$ segments between the timepoints. The mean value μ, the median value m, and the standard deviation σ of the historical travel times for each timepoint segment in each month are integrated to generate feature vectors to represent the historical delay data distribution:

$$[\mu_1, m_1, \sigma_1, \mu_2, m_2, \sigma_2, ..., \mu_{n-1}, m_{n-1}, \sigma_{n-1}] \tag{1}$$

Month Clustering. Clustering is an unsupervised/supervised learning technique for grouping similar data. We employ k-means algorithms to identify the homogeneous groups where months share similar patterns. The trip data per month is first normalized and then clustered using feature vectors (in Eq. 1) by K-Means algorithm:

$$\arg\min_S \sum_{i=1}^{k} \sum_{x \in S_i} \|x - \mu_i\|^2 \tag{2}$$

where μ_i is the mean of all datapoints in cluster S_i. Determining the optimal number of clusters in a data set is a fundamental issue in partitioning clustering. For k-means algorithms, the number of clusters is a hyper-parameter that needs

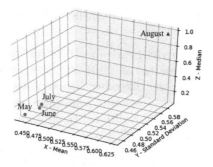

Fig. 3. The feature vectors [mean, standard deviation, median] of the travel time in 4 months of 2016 for a segment (WE23-MCC5_5) on a bus trip of route 5.

to be set manually. An upper bound is set in advance. Elbow [16], Silhouette [19] and gap statistic [25] methods are popular direct and statistical methods to find the optimal number of clusters. Particularly, Silhouette analysis is employed in this study to measure how close each point is to others within one cluster. The silhouette score $s(i)$ is defined as:

$$s(i) = \frac{b(i) - a(i)}{max\{a(i), b(i)\}} \tag{3}$$

where for each data point with index i in the cluster, a_i is the average distance between $data_i$ and the rest of data points in the same cluster, b_i is the smallest average distance between $data_i$ and every other cluster. Some other clustering techniques that can be applied to these kind of problem can be found in [20].

Example. Figure 3 plots the [mean, standard deviation, median] vectors of the monthly travel time for a segment (WE23-MCC5_5) on a bus trip of route 5 (Fig. 6(a)). From Fig. 3 the monthly variation of the data is evident and hence two clusters ([May, June, July] and [August]) can be formed from these four months of data to prepare distinct schedules for the clusters.

5.2 Estimating On-Time Performance of Transit Schedules

Historical Dwell Time Estimation. Travel demand at bus stops is important statistics for setting up proper schedule times. However, for bus systems without automatic passenger counters (APCs), historical travel demand (represented by number of commuters boarding) is not available in original datasets. To get demand patterns, we utilize historical arrival and departure times to estimate the dwell time caused by passengers. Particularly, we consider the following two scenarios in historical records: (1) when a bus turns up at a stop, earlier than scheduled time, the waiting time between the scheduled time and actual departure time is used, (2) on the other hand, when it turns up later than scheduled time, the waiting time between the actual arrival time and departure time is used. As shown in Table 1, for the timepoint SY19 on trip 1 with scheduled time of 11:02 AM:

- For the case when a bus arrived earlier at 10:58 AM instead of the scheduled time at 11:02 AM and departed at 11:04 AM, as the bus would always wait there at least for 4 min (the difference between the actual and scheduled arrival time) irrespective of presence of passengers, the dwell time caused by passengers is calculated as the additional time taken for departure after the scheduled time (11:04 AM − 11:02 AM = 2 min).
- On the other hand, if the bus arrived later at 11:05 AM and departed at 11:06 AM, then the dwell time caused by passengers is calculated as the additional time spent after the actual arrival time (11:06 AM − 11:05 AM = 1 min).

Arrival Time Estimation. The arrival time of a bus at a stop is impacted by two factors: (1) travel times at segments before the stop, and (2) dwell times at the previous stops. We assume that a bus will wait until the scheduled time if it arrives earlier than the scheduled time, and the historical travel time between two timepoints will remain the same in the simulation. In order to obtain an estimate of the arrival time, the historical dwell time caused by commuters (which in turn is representative of the historical travel demand), is factored into account by adding it to the arrival time at any timepoint. The simulation will stall for an additional time till the new scheduled time is reached in the event that the previous sum is earlier than the new scheduled time. By taking into consideration the simulated departure time st_{h,s_j}^{depart} at previous timepoint s_j, the actual travel time $t_{s_{j+1}}^{arrive} - t_{s_j}^{depart}$ between s_j and s_{j+1}, the dwell time $t_{s_{j+1}}^{dwell}$, the simulated departure time $st_{h,s_{j+1}}^{depart}$ at a timepoint s_{j+1} can be found out. The new schedule time $T_{h,ss_{j+1}}^{depart}$ at s_{j+1} is expressed as:

$$st_{h,s_{j+1}}^{depart} = \max(T_{h,ss_{j+1}}^{depart}, st_{h,s_j}^{depart} + (t_{s_{j+1}}^{arrive} - t_{s_j}^{depart}) + t_{s_{j+1}}^{dwell}) \qquad (4)$$

5.3 Timetable Optimization Using a Greedy Algorithm

We employed a greedy algorithm that adjusts the scheduled arrival time greedily and sequentially for the succeeding segments between timepoints. The main objective is to optimize the bus arrival time for succeeding timepoints such that new optimized schedule is guaranteed to maximize the probability of bus arrivals between any two consecutive stops with delay bounded in the desired range of $[t_{early}, t_{late}]$.

We utilized the empirical cumulative distribution function (CDF) to evaluate the percentage of historical delay in desired range instead of assuming that the data is drawn from any specific distribution (e.g. Gaussian distribution).

An empirical CDF is a non-parametric estimator of the CDF of a random variable. The empirical CDF of variable x is defined as:

$$\hat{F}_n(x) = \hat{P}_n(X \leq x) = n^{-1} \sum_{n=1}^{n} I(x_i \leq x) \qquad (5)$$

where $I()$ is an indicator function:

$$I(x_i \leq x) = \begin{cases} 1, & \text{if } x_i \leq x \\ 0, & \text{otherwise} \end{cases} \tag{6}$$

Then the CDF of x in range $[x + t_{early}, x + t_{late}]$ can be calculated using the following equation:

$$\hat{F}_n(x + t_{late}) - \hat{F}_n(x + t_{early})$$
$$= n^{-1} \sum_{n=1}^{n} I(x + t_{early} \leq x_i \leq x + t_{late}) \tag{7}$$

5.4 Timetable Optimization Using Heuristic Algorithms

The performance optimization for scheduling transit vehicles is a multidimensional problem and as such the objective function is nonconvex in nature consisting of several troughs and ridges. Hence, to compute the optimally scheduled routing strategy with acceptable time constraints, an approach powered by high quality of solution estimation techniques such as evolutionary algorithms and metaheuristics can be considered.

Genetic Algorithm. Genetic algorithm [11] is a heuristic optimization algorithm that derives from biology. The basic steps involved in genetic algorithms include initialization, selection, crossover, mutation, and termination. The timetable for each trip is decided by the scheduled departure time at the first stop as well as the scheduled travel time between any two subsequent timepoints along the trip. Since our goal is to update timetables to make the bus arrivals more on time, we assign the scheduled travel times between timepoints as chromosomes in populations, and use the on-time performance estimation mechanism proposed in Sect. 5.2 as objective functions. The chromosome of the individual solutions in the genetic algorithm is a vector of integers representing travel time between subsequent timepoints. In order to reduce the search space and match the real-world scenarios, the travel time in each individual is re-sampled to a multiple of 60 seconds and restricted to the unit of minutes. The performance of this algorithm is governed by different hyperparameters such as population size, crossover and mutation rate controlling the algorithm's exploitation and exploration capability. The choice of such hyperparameters are explained in detail in Sect. 6.

Particle Swarm Optimization. Eberhert and Kennedy [14] proposed particle swarm optimization (PSO) as a stochastic population based optimization algorithm which can work with non-differentiable objective function without explicitly assuming its underlying gradient disparate from gradient descent techniques. The interested reader is directed to [21] by Sengupta et al. for a detailed understanding of the algorithm. PSO has been shown to satisfactorily provide solutions to a wide array of complex real-life engineering problems, usually out of scope of deterministic algorithms [3,5,8]. PSO exploits the collective intelligence arising out of grouping behavior of flocks of birds or schools of fish.This manifestation of grouping is termed as 'emergence', a phenomenon in which a cohort of individuals from a social network is aimed to accomplish a task beyond their individual capability. Likewise, each particle in the swarm, represents a potential solution to the multi-dimensional problem to be optimized.

Initialization. Each particle has certain position which can be thought of as a collection of co-ordinates representing the particle's existence in a specific region in the multidimensional hyperspace. As a particle is a potential solution to the problem, the particle's position vector has the same dimensionality as the problem. The velocity associated with each particle is the measure of the step size and the direction it should move in the next iteration.

Each particle in the swarm maintains an n-dimensional vector of travel times. At first, the position for each particle in the population is initialized with the set of travel time between the timepoints randomly selected between the minimum and maximum of the aggregated actual historical data. With swarm size as p, every particle i ($1 < i < p$) maintains a position vector $x_i = (x_{i1}, x_{i2}, x_{i3}, ..., x_{in})$ and a velocity vector $v_i = (v_{i1}, v_{i2}, v_{i3}, ..., v_{in})$ and a set of personal bests $p_i = (p_{i1}, p_{i2}, p_{i3}, ..., p_{in})$.

Optimization. At each iteration, the position of a particle is updated, and compared with the personal best (*pbest*) obtained so far. If the fitness due to the position obained at current iteration is more (as it is a fitness maximization problem) than the *pbest* obtained upto the previous iteration, then the current position becomes the personal best or *pbest*, otherwise *pbest* remains unchanged. Thus the best position of a particle obtained so far is stored as *pbest*. The global best or *gbest* is updated when the population's overall current best, i.e., the best of the *pbsest*s is better than that found in the previous iteration.

After initializing positions and velocities, each particle updates its velocity based on previous velocity component weighted by an inertial factor, along with a component proportional to the difference between its current position and *pbest* weighted by a cognition acceleration coefficient, and another component proportional to the difference between its current position and (*gbest*), weighted by a social acceleration coefficient. This is socio-cognitive model of PSO and facilitates information exchange between members of the swarm. Since all members are free to interact with each other, the flow of information is unrestricted and the PSO algorithm is said to have a 'fully-connected' topology. While updating the velocity, a particle's reliance on its own personal best is dictated by its cognitive ability, and the reliance on the entire swarm's best solution is dictated by its social interactive nature. Hence those factors in the velocity component are weighted by the cognition acceleration coefficient *c1* and social acceleration coefficient $c2$. The new positions of the particles are updated as the vector sum of the previous positions and the current velocities. Thus the positions of the particles, are updated aiming towards intelligent exploration of the search space, and subsequent exploitation of the promising regions in order to find the optimal solution based on fitness optimization of the stated problem.

After each iteration is completed, the velocity and position of a particle are updated as follows:

$$v_{i,j}(t+1) = w.v_{i,j}(t) + c_1.r_1(t).(p_{i,j}(t) - x_{i,j}(t)) + c_2.r_2(t).(p_{g,j}(t) - x_{i,j}(t)) \quad (8)$$

$$x_{i,j}(t+1) = x_{i,j}(t) + v_{i,j}(t+1) \quad (9)$$

$v_{i,j}$ and $x_{i,j}$ represent the velocity and position of the i-th particle in the j-th dimension. Cognition and social acceleration coefficients are indicated by c_1 and c_2, whereas r_1 and r_2 are random numbers uniformly distributed between 0 to 1. $p_{i,j}$ represents a particle's personal best and $p_{g,j}$ represents the global best of the population. w acts as an inertial weight factor controlling the exploration and exploitation of new positions in the search space and t denotes the number of iterations.

The problem is formulated as fitness maximization problem in order to bring out optimal travel times to improve on-time performance. Hence the personal best of a particle is updated as follows at the end of each iteration.

$$p_{i,j}(t+1) = \begin{cases} p_{i,j}(t), & \text{if } fitness(x_{i,j}(t+1)) \leq fitness(p_{i,j}(t)) \\ x_{i,j}(t+1), & \text{if } fitness(x_{i,j}(t+1)) > fitness(p_{i,j}(t)) \end{cases} \quad (10)$$

Termination. The termination condition set for PSO is the predefined maximum number of iterations. Since the optimized on time performance is different for each trip, the termination condition is not set as any predefined upper limit of the fitness value. With other hyperparameters fixed PSO can produce the optimal solution approximately in 30 iterations for this problem.

The pseudo code for PSO is discussed in Algorithm 1. Historical timepoint datasets are used to conduct the particle swarm optimization algorithm for this problem. The input includes on-time range, maximum number of iterations, number of particles in the population size, inertia factor, cognition and social acceleration coefficient, bus trip and upper limit of number of month clusters.

6 Evaluation of the Results

6.1 Evaluating the Clustering Analysis

To evaluate the effectiveness of the clustering analysis, we compared the optimized on-time performance with and without a clustering analysis step: (1)

ALGORITHM 1. Particle Swarm Optimization algorithm for bus on-time performance optimization

---

Data: $D \leftarrow$ Historical timepoint datasets

Input : (1) $[t_{early}, t_{late}] \leftarrow$ on-time range , (2) $maxIter \leftarrow$ maximum number of iterations $maxIter$, (3) $npop \leftarrow$ number of particles in the population size $npop$, (4) $w \leftarrow$ inertia weight, (5) $c1 \leftarrow$ cognition acceleration coefficient, (6) $c2 \leftarrow$ social acceleration coefficient, (7) $h \leftarrow$ bus trip for optimization, (8) $upperLimit \leftarrow$ upper limit of the number of clusters

Output: Optimized schedule b at timepoints for bus trip h

GetAllTimepoints(D, h);

GetHistoricalData(D, h);

$monthClusters \leftarrow$ ClusterMonthData($upperLimit$);

for $monthCluster \in monthClusters$ **do**

    $P \leftarrow []$;

    for *population size npop* **do**

        Initialize each particle with random position and velocity

        $P \leftarrow P \cup InitialIndividual()$;

    end

    while *maxIter is not reached* **do**

        Evaluate the fitness function (J) for each particle's position (x)

        if $J(x) ¿ J(pbest)$, then $pbest = x$

        $gbest \leftarrow$ Update if the population's overall current best is better than that in previous iteration

        Update the velocity of each particle according to equation *(10)*

        Update the position of each particle according to equation *(11)*

    end

    Give *gbest* as the optimal schedule b at timepoints for bus trip h

end

---

months are not clustered and a single timetable is generated for all months, (2) month clustering is conducted at first and the optimization algorithms is applied on different month groups to generate separate timetables.

Table 2 shows the original and optimized on-time performance on average across all bus routes. Using the genetic algorithm without clustering step improved the original performance from 57.79% to 66.24%. By adding the clustering step which groups months with similar patterns the performance was improved to 68.34%.

Table 2. Comparison of original and optimized on-time performance averaged across all bus routes for GA without and with clustering and PSO with clustering respectively.

	Original	GA w/o. clustering	GA w. clustering	PSO w. clustering
On-time perf.	57.79%	66.24%	68.34%	68.93%

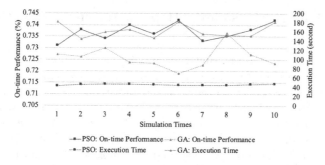

Fig. 4. The chart shows the simulation results of on-time performance and execution times for GA and PSO to run 10 times.

6.2 Comparing Optimization Performance of Greedy, Genetic and PSO Algorithms

The original on-time performance, optimized on-time performance using greedy algorithm, genetic algorithm and PSO are illustrated in Fig. 5. It is observed that while all the algorithms can improve the on-time performance, the genetic algorithm and PSO outperforms the greedy algorithm because they optimize the schedule for all timepoint segments on each trip all together. The original on-time performance of all bus routes is 57.79%. The greedy algorithm improved it to 61.42% and the genetic algorithm improved it further to 68.34%. The PSO algorithm has a slightly better optimized on-time performance of 68.93%. Figure 4 shows the simulation results of the stability analysis for GA and PSO. Even though GA and PSO got similar on-time performance, with PSO surpassing the performance of GA by a small extent, the execution times of PSO are much less than GA and are more stable.

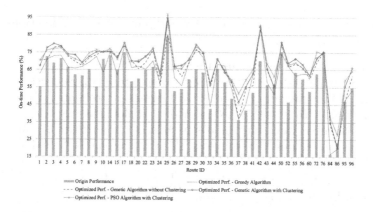

Fig. 5. The original on-time performance and the optimized on-time performance using greedy algorithm, genetic algorithm with/without clustering analysis and PSO algorithm.

6.3 Sensitivity Analysis on the Hyper-parameters of the Genetic Algorithm

We designed three simulations that choose different hyper-parameters: (1) population sizes that range from 10 to 110, (2) crossover rates that range from 0.1 to 1.0, (3) mutation rates that range from 0.1 to 1.0. Real-world data is collected from Route 5, which is one of major bus routes that connects downtown Nashville and the southwest communities in Nashville. The route contains 6 timepoint stops and 5 segments between the 6 timepoint stops. The bus trips with direction from Downtown are selected. The goal is to maximize the on-time performance for these trips by optimizing the schedule time at the 6 timepoint stops.

Figure 6(b) shows the simulation results of choosing different population sizes. Increasing the population size from 10 to 90 results a better on-time performance, however, increasing the size ever further doesn't help making the on-time performance any better. On the other hand, the total time increases linearly as the population size grows. So a population size around 90 is the optimal size to use.

Figure 6(c) illustrates results of using different crossover rates. The optimized on-time performance remains almost the same for the crossover range, but there is a significant difference in terms of the total execution time. The crossover rate impacts the exploitation ability. A proper crossover rate in the middle of the range can faster the process to concentrate on an optimal point.

Figure 6(d) show the simulation results when using different mutation rates. The total execution time is small when the mutation rate is either very small or very large. Mutation rates controls the exploration ability. During the optimization, a small mutation rate will make sure the best individuals in a population do not vary too much in the next iteration and thus is faster to get stable around the optimal points. So we suggest setting a very small mutation rates when running the proposed algorithm.

6.4 Sensitivity Analysis on the Hyper-parameters of Particle Swarm Optimization

We designed four simulation setups that choose different hyper-parameters: (1) The inertial weight factor, w that range from 1 to 8, (2) Social acceleration coefficient $c1$ that range from 1 to 8, (3) Cognition acceleration coefficient $c2$ that range from 1 to 8, and (4) Number of particles that range from 2 to 36. Real-world data regarding bus timings is collected from Route 8, which is one of the major bus routes that connects Music City Central Nashville and the Lipscomb University in Nashville. The route contains 5 timepoint stops and 4 segments between the 5 timepoint stops. The goal is to maximize the on-time performance for these trips by optimizing the schedule time at the 5 timepoint stops.

Figure 6(e) shows the simulation result while optimizing for the inertial weight w by varying it. It is observed that the optimized on-time performance is at its peak when w is nearly equal to 5 with less execution time. Performance

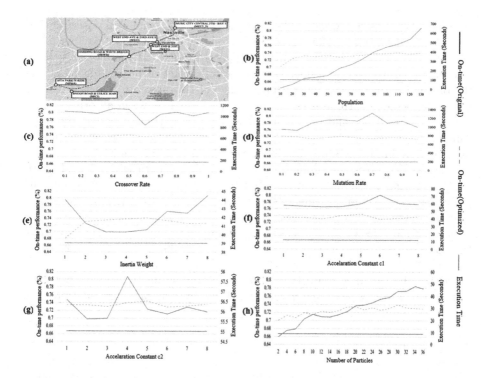

Fig. 6. (a) Timepoints on bus route 5 in Nashville [23], (b) The on-time performance and overall execution time for different population sizes for GA, (c) The on-time performance and overall execution time for different crossover rates, which controls the exploitation ability of the GA, (d) The on-time performance and overall execution time for different mutation rates, which controls the exploration ability of the GA, (e) The on-time performance and overall execution time for different inertia weights, exploring new regions of search space in PSO, (f) The on-time performance and overall execution time for different cognition acceleration coefficients c1, in PSO, (g) The on-time performance and overall execution time for different social acceleration coefficients c2, in PSO, (h) The on-time performance and overall execution time for various population size, in PSO

deteriorates along with an increase in execution time as the selection is moved away from 5. So, an optimal value to choose for w, will be somewhere around 5.

Figure 6(f) shows the simulation result for optimizing the cognition acceleration coefficient $c1$ by varying it. The particle has a velocity component towards its own best position weighted by $c1$, hence the term 'cognitive'. It is observed that the optimized on-time performance is improved when $c1$ increases from 3 to 5 with less execution time. After that the performance deteriorates along with increase in execution time. So, an optimal value to choose for $c1$, will be within the range specified.

Figure 6(g) shows the simulation result for optimizing for the social acceleration coefficient $c2$ by varying it. The particle has a velocity component towards the global best position weighted by c_2, hence the term social. It is observed that the optimized on-time performance is improved when c_2 is equal to 5 with less execution time. Also, c_2 being 4 produces good results, but there is an increase in execution time at that value. But the overall effect of parameter c_2 affects the on-time performance only within a range of two percent. Sometimes, PSO is able to produce optimal or near optimal performance, when all other hyperparameters are fixed, and thus is not sensitive to a particular hyperparameter, which is the case considered here. So an optimal value to choose for c_2, may be close to 5, maintaining approximately a ratio near to 1:1:1 among w, c_1 and c_2.

Figure 6(h) shows the simulation result for optimizing the number of particles by varying the population size. It is observed that the optimized on-time performance is maximized when the number of particles reaches 30. The execution time increases with the number of particles, so it is better to choose such number of particles that produces the best pair in the accuracy-execution time tradeoff. So, the population size can be chosen as 30 in this case as it yields equally efficient results with a relatively small execution time.

Although a good insight about choice of hyperparameters can be obtained from this sensitivity analysis, variations of the hyperparameters may produce better results in specific routes.

7   Conclusion

In this paper, we presented research findings within a bus on-time performance optimization framework that significantly extends our prior work [23] by proposing a stochastic optimization toolchain and presenting sensitivity analyses on choosing optimal hyper-parameters. Particularly, we describe an unsupervised analysis mechanism to find out how months with similar delay patterns can be clustered to generate new timetables. A classical, fully-connected PSO is benchmarked against a greedy algorithm as well as a genetic algorithm in order to optimize the schedule time to maximize the probability of bus trips that reach the desired on-time range. It is observed that the PSO implementation improves the bus on-time performance compared to other heuristics while requiring lesser execution time across all routes. Simulations of optimization performance as well as sensitivity analyses on the hyper-parameters of the GA and PSO algorithms are conducted. The results indicate different strategies for choosing between the genetic algorithm and PSO, and selecting optimal hyper-parameters guided by the problem specificity and resource availability. With the knowledge of this extensive study on applying guided random search techniques for bus on-time performance optimization and the selection of hyperparameters that generate promising results, a possible extension of this generalizable architecture to other real-world optimization problems is worth looking at as future work.

Acknowledgments. This work is supported by The National Science Foundation under the award numbers CNS-1528799 and CNS-1647015 and 1818901 and a TIPS grant from Vanderbilt University. We acknowledge the support provided by our partners from Nashville Metropolitan Transport Authority.

References

1. Arhin, S.A., Noel, E.C., Dairo, O.: Bus stop on-time arrival performance and criteria in a dense urban area. Int. J. Traffic Transp. Eng. **3**(6), 233–238 (2014)
2. American Public Transportation Association: Ridership report archives (2017)
3. Banks, A., Vincent, J., Anyakoha, C.: A review of particle swarm optimization. Part ii: hybridisation, combinatorial, multicriteria and constrained optimization, and indicative applications. Natural Comput. **7**(1), 109–124 (2008). https://doi.org/10.1007/s11047-007-9050-z
4. Benn, H.: Bus route evaluation standards, transit cooperative research program, synthesis of transit practice 10. Transportation Research Board, Washington, DC (1995)
5. Bouyer, A., Hatamlou, A.: An efficient hybrid clustering method based on improved cuckoo optimization and modified particle swarm optimization algorithms. Appl. Soft Comput. **67**, 172–182 (2018). https://doi.org/10.1016/j.asoc.2018.03.011. http://www.sciencedirect.com/science/article/pii/S1568494618301273
6. Chakroborty, P.: Genetic algorithms for optimal urban transit network design. Comput.-Aided Civ. Infrastr. Eng. **18**(3), 184–200 (2003)
7. Chakroborty, P., Deb, K., Subrahmanyam, P.: Optimal scheduling of urban transit systems using genetic algorithms. J. Transp. Eng. **121**(6), 544–553 (1995)
8. Dhabal, S., Sengupta, S.: Efficient design of high pass fir filter using quantum-behaved particle swarm optimization with weighted mean best position. In: Proceedings of the 2015 Third International Conference on Computer, Communication, Control and Information Technology (C3IT), pp. 1–6, February 2015. https://doi.org/10.1109/C3IT.2015.7060145
9. Eranki, A.: A model to create bus timetables to attain maximum synchronization considering waiting times at transfer stops (2004)
10. Fan, W., Machemehl, R.B.: Optimal transit route network design problem with variable transit demand: genetic algorithm approach. J. Transp. Eng. **132**(1), 40–51 (2006)
11. Goldberg, D.E.: Genetic Algorithms in Search, Optimization and Machine Learning, 1st edn. Addison-Wesley Longman Publishing Co., Inc., Boston (1989)
12. Hairong, Y., Dayong, L.: Optimal regional bus timetables using improved genetic algorithm. In: Second International Conference on Intelligent Computation Technology and Automation, ICICTA 2009, vol. 3, pp. 213–216. IEEE (2009)
13. Ibarra-Rojas, O.J., Rios-Solis, Y.A.: Synchronization of bus timetabling. Transp. Res. Part B: Methodol. **46**(5), 599–614 (2012)
14. Kennedy, J., Eberhart, R.: Particle swarm optimization. In: Proceedings of the IEEE International Conference on Neural Networks, vol. 4, pp. 1942–1948, November 1995. https://doi.org/10.1109/ICNN.1995.488968
15. Khiari, J., Moreira-Matias, L., Cerqueira, V., Cats, O.: Automated setting of bus schedule coverage using unsupervised machine learning. In: Bailey, J., Khan, L., Washio, T., Dobbie, G., Huang, J.Z., Wang, R. (eds.) PAKDD 2016. LNCS (LNAI), vol. 9651, pp. 552–564. Springer, Cham (2016). https://doi.org/10.1007/978-3-319-31753-3_44
16. Kodinariya, T.M., Makwana, P.R.: Review on determining number of cluster in k-means clustering. Int. J. **1**(6), 90–95 (2013)
17. Nayeem, M.A., Rahman, M.K., Rahman, M.S.: Transit network design by genetic algorithm with elitism. Transp. Res. Part C: Emerg. Technol. **46**, 30–45 (2014)
18. Neff, J., Dickens, M.: 2016 public transportation fact book (2017)

19. Rousseeuw, P.J.: Silhouettes: a graphical aid to the interpretation and validation of cluster analysis. J. Comput. Appl. Math. **20**, 53–65 (1987)
20. Sengupta, S., Basak, S., Peters, R.A.: Data clustering using a hybrid of fuzzy c-means and quantum-behaved particle swarm optimization. In: 2018 IEEE 8th Annual Computing and Communication Workshop and Conference (CCWC), pp. 137–142, January 2018. https://doi.org/10.1109/CCWC.2018.8301693
21. Sengupta, S., Basak, S., Peters, R.A.: Particle swarm optimization: a survey of historical and recent developments with hybridization perspectives. Mach. Learn. Knowl. Extract. **1**(1), 157–191 (2018). http://www.mdpi.com/2504-4990/1/1/10
22. Sengupta, S., et al.: A review of deep learning with special emphasis on architectures, applications and recent trends. CoRR abs/1905.13294 (2019). http://arxiv.org/abs/1905.13294
23. Sun, F., Samal, C., White, J., Dubey, A.: Unsupervised mechanisms for optimizing on-time performance of fixed schedule transit vehicles. In: 2017 IEEE International Conference on Smart Computing (SMARTCOMP), pp. 1–8. IEEE (2017)
24. Szeto, W.Y., Wu, Y.: A simultaneous bus route design and frequency setting problem for Tin Shui Wai, Hong Kong. Eur. J. Oper. Res. **209**(2), 141–155 (2011)
25. Tibshirani, R., Walther, G., Hastie, T.: Estimating the number of clusters in a data set via the gap statistic. J. Roy. Stat. Soc.: Ser. B (Stat. Methodol.) **63**(2), 411–423 (2001)
26. Ting, C.J., Schonfeld, P.: Schedule coordination in a multiple hub transit network. J. Urban Plann. Dev. **131**(2), 112–124 (2005)
27. Wang, Y., Zhang, D., Hu, L., Yang, Y., Lee, L.H.: A data-driven and optimal bus scheduling model with time-dependent traffic and demand. IEEE Trans. Intell. Transp. Syst. **18**(9), 2443–2452 (2017)
28. Wu, Y., Yang, H., Tang, J., Yu, Y.: Multi-objective re-synchronizing of bus timetable: model, complexity and solution. Transp. Res. Part C: Emerg. Technol. **67**, 149–168 (2016)
29. Zhong, S., Zhou, L., Ma, S., Jia, N., Zhang, L., Yao, B.: The optimization of bus rapid transit route based on an improved particle swarm optimization. Transp. Lett. **10**(5), 257–268 (2018). https://doi.org/10.1080/19427867.2016.1258972

Pattern Mining

Discovering Spatial High Utility Frequent Itemsets in Spatiotemporal Databases

P. P. C. Reddy[1], R. Uday Kiran[2,3(✉)], Koji Zettsu[2], Masashi Toyoda[3],
P. Krishna Reddy[1], and Masaru Kitsuregawa[3,4]

[1] International Institute of Information Technology-Hyderabad, Hyderabad, India
pradeepchandra.p@research.iiit.ac.in, pkreddy@iiit.ac.in
[2] National Institute of Information and Communications Technology, Tokyo, Japan
zettsu@nict.co.jp
[3] The University of Tokyo, Tokyo, Japan
{uday_rage,toyoda,kitsure}@tkl.iis.u-tokyo.ac.jp
[4] National Institute of Informatics, Tokyo, Japan

Abstract. Spatial High Utility Itemset Mining (SHUIM) aims to discover all itemsets in a spatiotemporal database that satisfy the user-specified *minimum utility* (*minUtil*) and *maximum distance* (*maxDist*) constraints. The popular adoption and successful industrial application of SHUIM suffers from the following two limitations: (*i*) Since SHUIM determines the interestingness of an itemset without taking into account its *support* within the data, SHUIM facilitates sporadic itemsets with high utility to be generated as SHUIs. In particular, items in long transactions can combine with each other and be generated as SHUIs. (*ii*) SHUIM is a computationally expensive process because the generated itemsets do not satisfy the downward closure property. This paper introduces Spatial High Utility Frequent Itemset Mining (SHUFIM) to address these two issues. A SHUI in a spatiotemporal database is said to be a SHUFI if its *support* is no less than the user-specified *minimum support* (*minSup*) constraint. The usage of *minSup* not only facilitates the proposed model to be tolerant to the long transactions within the data but also facilitates us to employ additional pruning techniques to reduce the computational cost. A single scan fast algorithm has also been proposed to discover all SHUFIs in a spatiotemporal database. Experimental results demonstrate that the proposed algorithm is efficient. We also demonstrate the usefulness of the proposed model with a real-world application.

Keywords: Data mining · Pattern mining · Utility pattern mining · Spatiotemporal patterns

1 Introduction

Spatial High Utility Itemset Mining (SHUIM) is an important data mining model, which generalizes the renowned high utility itemset mining [12] by taking

© Springer Nature Switzerland AG 2019
S. Madria et al. (Eds.): BDA 2019, LNCS 11932, pp. 287–306, 2019.
https://doi.org/10.1007/978-3-030-37188-3_17

into account the spatiotemporal characteristics of the items within the data. It aims to discover all itemsets in a spatiotemporal database that have *utility* no less than the user-specified *minimum utility* (*minUtil*) and the distance between its items is no more than the user-specified *maximum distance* (*maxDist*). A classic application of SHUIM is traffic congestion data analytics. It consists of analyzing which sets of neighboring roads have observed high levels of congestion over a time period. An example of an SHUI generated in the traffic congestion data is [5]:

$$\{132, 135, 137, 134, 136, 138, 140, 133\} \; [utility = 419, 245].$$

The above spatial high utility itemset (SHUI) provides useful information that a total of 419, 245 m of congestion was observed on the road segments whose identifiers are 132, 135, 137, 134, 136, 138, 140, and 133. At the time of disasters, the above information can be found very useful to the users for various purposes such as diverting the traffic, alerting the pedestrians, and suggesting the patrol path for the police.

The basic model of SHUIs is as follows [5]: Let $I = \{i_1, i_2, \cdots, i_m\}$, $m \geq 1$, be a set of items. Let $p(i_j, ts)$ denote the **external utility** of an item $i_j \in I$ at a timestamp $1 \leq ts \leq n$. Let $P(i_j) = p(i_j, 1) \cup \cdots \cup p(i_j, n)$ denote the set of all external utility values of i_j in the data. The (external) **utility database**, UD, refers to the set of external utility values of all items in I, i.e., $UD = \bigcup_{i_j \in I} P(i_j)$.

A **temporal database**, $D = \{T_1, T_2, \cdots, T_n\}$, where $T_{ts} \subseteq I$ denotes a transaction. Each transaction T_{ts} is associated with a timestamp $ts \in (1, n)$. Every item $i_j \in T_{ts}$ has a positive number $q(i_j, T_{ts})$, called its **internal utility**. Generally, the internal utility of an item represents its *frequency* in a transaction. A spatial database, $SD = \bigcup_{i_j \in I} (i_j, (lat_{i_j}, long_{i_j}))$ is a collection of location points of all items in I. The terms lat_{i_j} and $long_{i_j}$ denote the latitude and longitude information of an item i_j, respectively. The utility of an item i_j in a transaction T_{ts}, denoted as $u(i_j, T_{ts})$, represents the product of its internal and external utility values, i.e., $u(i_j, T_{ts}) = p(i_j, T_{ts}) \times q(i_j, T_{ts})$. Let $X \subseteq I$ be an itemset. An itemset is a k-itemset if it contains k items. The utility of an itemset X in a transaction T_{ts}, denoted as $u(X, T_{ts}) = \Sigma_{i_j \in X} u(i_j, T_{ts})$ if $X \subseteq T_{ts}$; otherwise, $u(X, T_{ts}) = 0$. The utility of an itemset X in a database D is defined as $u(X) = \Sigma_{T_{ts} \in g(X)} u(X, T_{ts})$, where $g(X)$ denotes the set of all transactions containing X. An itemset X is said to be a spatial high utility itemset if its *utility* is no less than the user-specified *minimum utility* (*minUtil*) and the distance between its items is no more than the user-specified *maximum distance* (*maxDist*). That is, X is a SHUI if $u(X) \geq minUtil$ and $\forall i_a, i_b \in X, a \neq b, Dist(i_a, i_b) \leq maxDist$, where $Dist(.)$ is a distance function (e.g. Euclidean distance and Geodesic distance).

Example 1. Let $I = \{a, b, c, d, e, f, g, h\}$ be a set of all items (or road identifiers). A temporal database generated from I is shown in Table 1. Table 2 presents the external utilities (or congestion lengths) of all items at various timestamps.

Table 1. Temporal database

ts	Items
1	$(a,2),(b,3),(c,1),(d,1)(e,4)$
2	$(a,1),(b,2),(f,2),(g,1)$
3	$(a,1),(e,2),(f,3),(g,1)$
4	$(b,3),(c,2),(d,1)$
5	$(b,2),(c,1),(d,3),(e,1),(g,2)$
6	$(a,1),(f,2),(g,1),(h,4)$

Table 2. External utility database

ts/Item	a	b	c	d	e	f	g	h
1	100	200	100	150	200	0	0	0
2	50	100	0	0	0	50	100	0
3	50	0	0	0	150	100	100	0
4	0	50	50	100	0	0	0	0
5	0	50	100	50	150	0	50	0
6	50	0	0	0	0	150	100	300

Note that an item doesn't appear in a transaction if its external utility is zero in that transaction. Let the unit for measuring congestion length be *meters* (or *m*). The external utility of item a at timestamp 1 is $p(a,1) = 100$, and the internal utility of item a at timestamp 1 is $q(a,1) = 2$. The spatial database shown in Table 3 provides the location information of all items in Table 1. The *utility* of an item a (i.e., congestion length at road a) in the first transaction is $u(a,T_1) = p(a,T_1) \times q(a,T_1) = 2 \times 100 = 200$ m. The set of items a and b, i.e., $\{a,b\}$ (or ab in short) is an itemset, which contains two items; therefore, it is a 2-itemset. The utility of itemset ab (or the total length of congestion observed on the roads a and b at the timestamp 1) in the transaction T_1 is $u(ab,T_1) = u(a,T_1) + u(b,T_1) = 200 + 600 = 800$ m. In Table 1, the itemset ab appears in the transactions T_1 and T_2; therefore, $g(x) = \{T_1,T_2\}$. The *utility* of itemset ab (or the total length of congestion observed on the roads a and b) in the entire database is $u(ab) = u(ab,T_1) + u(ab,T_2) = 800 + 250 = 1050$ m. Similarly, the *utility* of itemset cd in the entire database is $u(cd) = u(cd,T_1) + u(cd,T_4) + u(cd,T_5) = 250 + 200 + 250 = 700$ m. If the user-specified $minUtil = 1050$ m and $maxDist = 5$ km, then cd is not a SHUI because $u(cd) \not\geq minUtil$. However, ab is a SHUI because $u(ab) \geq minUtil$ and $Dist(a,b) \leq maxDist$. This SHUI can be expressed as ab [*utility* = 1050 m]. The set of all SHUIs generated from Table 1 are shown in the fifth column of Table 4.

Table 3. Spatial database

Items	location
a	$(0,0)$
b	$(3,4)$
c	$(3,-4)$
d	$(6,0)$
e	$(3,0)$
f	$(9,0)$
g	$(12,0)$
h	$(18,0)$

Table 4. Itemsets generated by different models

S.No.	Itemset	utility	support	SHUIs	SHUFIs
1	ace	1100	1	✓	✗
2	cde	1450	2	✓	✓
3	ce	1150	2	✓	✓
4	ab	1050	2	✓	✓
5	h	1200	1	✓	✗
6	ae	1350	2	✓	✓
7	bd	1250	3	✓	✓
8	bde	1950	2	✓	✓
9	de	1250	2	✓	✓
10	b	1050	4	✓	✓
11	be	1650	2	✓	✓
12	e	1250	3	✓	✓
13	abe	1600	1	✓	✗

Table 5. Neighbors

Item	Neighbors
a	bce
b	ade
c	ade
d	bcef
e	abcd
f	dg
g	f
h	−

The popular adoption and successful industrial application of SHUIM has been hindered by the following two obstacles:

- Since the *utility* of an itemset represents the *sum* of utilities of its individual items, this measure is sensitive to the length of a transaction within the data. In particular, *utility* facilitates items in long transactions to combine with one another and be generated as (spatial) high utility itemsets. Consequently, sporadic itemsets with high utility can be generated as SHUIs.

Example 2. Consider the first transaction, which is a long transaction in Table 1. This transaction facilitates the items *a*, *c* and *e* to combine with each other and be generated as a SHUI (see first row in Table 4). Similarly, this transaction also facilitates the items *a*, *b* and *e* to combine with each other and be generated as a SHUI (see last row in Table 4). Unfortunately, both of these itemsets may be uninteresting to the user as they have appeared only once in the entire data. An itemset may be interesting to the user if it has appeared at least twice in the entire data.

- The itemsets generated by SHUIM do not satisfy the downward closure property [1]. That is, all non-empty subsets of SHUI may not be SHUIs. This increases the search space, which in turn increases the computational cost of finding the SHUIs.

This paper makes an effort to address these two limitations. We propose a generic Spatial High Utility Frequent Itemset Model (SHUFIM) to find all itemsets in the data that satisfy a user-specified *minimum support* (*minSup*), *minUtil*, and *maxDist* constraints. The *minSup* controls the minimum number of transactions in which an itemset must appear within the data. Consequently, this constraint not only facilitates us to prune the sporadic itemsets with high utility in the data but also facilitates us to introduce new pruning techniques which can reduce the computational cost of finding the desired itemsets efficiently. We theoretically show that the search space of SHUFIM is no more than the frequent itemset mining in the worst case.

Finding SHUFIs in spatiotemporal databases is a challenging task because of two main reasons: (*i*) Current algorithms to find SHUIs cannot be directly used to find SHUFIs. It is because these algorithms do not capture the *support* information of an itemset within the data. (*ii*) Most previous studies on (spatial) high utility itemset mining [5] try to reduce the computational cost by exploiting pruning techniques based on *utility* measure alone. On the contrary, the proposed SHU-FIM tries to effectively reduce the computational cost by exploring pruning techniques based on both *utility* and *support* measures.

The contributions of this paper are as follows: (*i*) This paper proposes a generic model to discover SHUFIs in spatiotemporal databases. (*ii*) A novel pruning measure based on *minSup*, *minUtil* and *maxDist* constraints has been introduced to reduce the search space and the computational cost of finding the desired itemsets. We call this pruning measure as **neighborhood suffix utility**. A new pruning technique based on neighborhood suffix utility has been described to identify itemsets whose supersets may generate SHUFIs at higher order. These itemsets are known as *secondary itemsets* (SIs). We theoretically show that the search space of finding SHUFIs using SIs is no more than the search space of finding frequent itemsets within the data. (*iii*) A fast single scan algorithm, called Spatial High Utility-Frequent Itemset Miner (SHU-FIM), is presented to find all SHUFIs in a spatiotemporal database. (*iv*) Experimental results show that SHU-FIM is efficient. (*v*) We also present a case study in which we have applied the proposed model to identify useful information in congestion data.

The remainder of this study is organized as follows. In Sect. 2, we discuss the literature on high utility itemset mining, and spatial high utility itemset mining. In Sect. 3, we introduce the proposed model of SHUFI. In Sect. 4, we introduce the SHU-FIM algorithm. In Sect. 5, we present the experimental results. Finally, in Sect. 6, we provide conclusion of this study and discuss future research directions.

2   Related Work

Yao et al. [12] described high utility itemset mining by taking into account the importance of items and their occurrence *frequency* in every transaction. To circumvent the fact that *utility* is not anti-monotonic and to identify all high utility itemsets (HUIs), algorithms such as IHUP [2], PB [6], Two-Phase [8], BAHUI [9], UP-Growth and UP-Growth+ [10,11], and MU-Growth [14] utilize the *transaction weighted utility* (TWU) to prune the search space. TWU is an upper bound on the utility of itemsets. Note that the above mentioned algorithms operate in two phases. In the first phase, they identify candidates of HUIs by calculating their TWUs to prune the search space. If an itemset has a TWU greater than $minUtil$, then it is considered as a candidate itemset because the itemset or its supersets may be HUIs. Otherwise, if an itemset has a TWU less than $minUtil$, it is discarded. In the second phase, these algorithms scan the database to calculate the exact utility of all candidates to filter those that are low-utility itemsets; unfortunately, all of the above mentioned algorithms suffer from the problem of generating too many candidates. Among the two-phase algorithms, UP-Growth is one of the fastest. It uses a tree-based algorithm inspired by the FP-Growth algorithm for FIM. It was shown to be up to 1000 times faster than Two-Phase and IHUP. More recent two-phase algorithms such as PB, BAHUI, and MU-Growth have introduced various optimizations and different design but only provide a small speed improvement over Two-Phase or UP-Growth (MU-Growth is only up to 15 times faster than UP-Growth).

Recently, to avoid the problem of candidate generation, single phase HUI mining algorithms (such as d2HUP [7], EFIM [13], and HU-FIMi [4]) were developed. These algorithms use upper bounds that are tighter than the TWU to prune the search space and can immediately obtain the exact utility of any itemset to determine if it should be outputted. Unfortunately, all HUIM algorithms disregard the spatiotemporal characteristics of items within the data. Uday et al. [5] described SHUIM to discover interesting high utility itemsets in spatiotemporal databases. An algorithm, called Spatial High Utility Itemset-Miner (SHUI-Miner), was discussed to find all SHUIs within the data. Since SHUIs do not satisfy the downward closure property, SHUI-Miner employs two utility upper-bound measures, namely *Probable Maximum Utility (PMU)* and *Neighborhood Sub-tree Utility (NSTU)*, to reduce the search space. Unfortunately, SHUIM is sensitive to the noise and length of a transaction within the data. As a result, SHUIM may generate sporadic itemsets with high utility as SHUI. The proposed SHUFIM is aimed to address this problem. More important, the pruning measures, *PMU* and *NSTU*, alone are inadequate to address the search space in finding SHUFIs. It is because these measures take into account only the *utility* of an itemset and completely ignore the *support* information of an itemset. On the contrary, the proposed pruning techniques for finding SHUFIs will take into account both *utility* and *support* information of an itemset to reduce the search space and/or computational cost.

3   Extended Model: Spatial High Utility Frequent Itemset

The proposed model of Spatial High Utility Frequent Itemset (SHUFI) extends the model of SHUI by taking into account the *support* information of its itemsets. We now introduce the proposed model of SHUFI.

Definition 1 (*Support* of an itemset in a database). *Continuing with the SHUI model described in Sect. 1, the support of an itemset X in a database D is defined as the number of transactions containing X in D. That is, $sup(X) = |g(X)|$, where $sup(X)$ represents the support of X and $g(X)$ represents the transactions containing X in D.*

Example 3. In Table 1, the itemset ab has occurred in the transactions T_1 and T_2. Therefore, $g(X) = \{T_1, T_2\}$ and $sup(ab) = |g(X)| = 2$. Similarly, the *support* of itemset abe, i.e., $sup(abe) = 1$.

Definition 2 (Frequent itemset X). *An itemset X is a frequent itemset if $sup(X) \geq minSup$, where $minSup$ represents a user-specified minimum support value.*

Example 4. If the user-specified $minSup = 2$, then ab is a frequent itemset because $sup(ab) \geq minSup$. On the contrary, abe is not a frequent itemset because $sup(abe) \ngeq minSup$.

Definition 3 (Spatial high utility frequent itemset X). *An itemset X is a spatial high utility frequent itemset if $sup(X) \geq minSup$, $utility(X) \geq minUtil$, and $\forall i_a, i_b \in X, a \neq b, Dist(i_a, i_b) \leq maxDist$.*

Example 5. If the user-specified $minSup = 2$, $minUtil = 1050$ and $maxDist = 5$, then ab is a SHUFI because $sup(ab) \geq minSup$, $util(ab) \geq minUtil$ and $Dist(ab) \leq Dist(a, b)$. On the contrary, the SHUIs cae and abe are not SHUFIs because $sup(cae) \ngeq minSup$ and $sup(abe) \ngeq minSup$, respectively. Thus, the proposed model is able to prune sporadically occurring high utility itemsets in the data. The complete set of SHUFIs generated from Table 1 are shown in the sixth column of Table 4.

Definition 4 (Problem definition). *Given a non-binary temporal database (D), external utility database (UD), a spatial database (SD), and the user-specified minimum support (minSup), minimum utility (minUtil) and maximum distance (maxDist) constraints, the problem of mining SHUFIs involve discovering all itemsets in D that have support no less than minSup, utility no less than minUtil, and the distance between all items in an itemset is no more than maxDist.*

In the next section, we describe an algorithm to find all SHUFIs in a spatiotemporal database.

4 SHU-FIM

In this section, we first describe the basic idea of the SHU-FIM and then we explain the algorithm in detail. Next, we explain the algorithm using the database shown in Table 1.

4.1 Basic Idea

Since *utility* is not an anti-monotonic function, the generated SHUFIs do not satisfy the downward closure property. That is, all non-empty subsets of a SHUFI may not be SHUFIs. This increases the search space, which in turn increases the computational cost of finding the SHUFIs. The proposed algorithm performs the following steps to reduce the search space and the computational cost of finding the SHUFIs:

1. In the first step, we employ *minSup* and *probable maximum utility* [5] constraints to identify items whose supersets may be SHUFIs. We call these items as ternary items.
2. In the next step, we generate secondary items by sorting the ternary items in *utility* descending order. A novel pruning measure, called *Neighborhood Suffix utility*, has been introduced to identify the secondary items. Secondary items represent items for which the depth-first search (or projection databases) has to be performed to find all SHUFIs within the data. It has to be noted that the depth-first search will not be carried out by the ternary items, which are not the secondary items.
3. In most cases, the secondary items can be directly projected to mine all SHUFIs within the data. However, in a few cases, especially when mining itemsets at low *minUtil* values, neighborhood suffix utility is not sufficient. It is because most ternary items will be generated as secondary items. To handle such cases, we generate primary items by calculating the *neighborhood sub-tree utility* for secondary items. It has to be noted that calculation of *neighborhood sub-tree utility* is a computationally expensive process, and therefore, needs to be avoided when possible.
4. Finally, we recursively mine all primary items to find all SHUFIs.

We now discuss each of these steps in detail.

Step 1: Generation of Ternary Items. Uday et al. [5] described *probable maximum utility (PMU)*, to prune items whose supersets in a database (or projected database) cannot be SHUIs. In this paper, we utilize both *PMU* and *minSup* to prune items whose supersets cannot be SHUFIs. The *pmu* of an item in a database is defined in Definition 5.

Definition 5 *(PMU of an item in a database).* *Let N_{i_j} denote the set of neighbors for an item $i_j \in I$. That is, $\forall i_k \in N_{i_j}, dist(i_j, i_k) \leq maxDist$. The probable maximum utility (PMU) of an item in a transaction T_{ts}, denoted as $pmu(i_j, T_{ts})$, represents the sum of utilities of i_j and its neighboring items in T_{ts}. That is, $pmu(i_j, T_{ts}) = u(i_j, T_{ts}) + \sum_{i_k \in T_{ts} \cap i_k \in N_{i_j}} u(i_k, T_{ts})$. Let $g(i_j)$ denote the set of all transactions containing i_j. The PMU of an item in a database, denoted as $pmu(i_j)$, represents the sum of probable maximum utility of i_j in all transactions. That is, $pmu(i_j) = \sum_{T_{ts} \in g(i_j)} pmu(i_j, T_{ts})$.*

Example 6. In Table 1, the item a appears in the transactions T_1, T_2 , T_3 and T_6. Therefore, $g(a) = \{T_1, T_2, T_3, T_6\}$. The set of all neighbors of item a, i.e., $N_a = \{bce\}$ (see Table 5). The *pmu* of a in T_1, i.e., $pmu(a, T_1) = u(a, T_1) + u(b, T_1) + u(c, T_1) + u(e, T_1) = 200 + 600 + 100 + 800 = 1700$ m. Similarly, $pmu(a, T_2) = 250$ m, $pmu(a, T_3) = 350$ m and $pmu(a, T_6) = 50$ m. The *PMU* of a in the entire database, i.e., $pmu(a) = pmu(a, T_1) + pmu(a, T_2) + pmu(a, T_3) + pmu(a, T_6) = 1700 + 250 + 350 + 50 = 2350$ m. In other words, a with all its neighbors can at most result in a *utility* (or congestion length) of 2350 m.

The above definition captures the *utility* upper bound of an item with respect to its neighbors. We now define *ternary items* using *pmu* and *support* measures.

Definition 6 *(Ternary items).* *An item $i_j \in I$ is a ternary item if $PMU(i_j) \geq minUtil$ and $sup(i_j) \geq minSup$. The term $PMU(i_j)$ represents probable maximum utility of an item $i_j \in I$ and the term $sup(i_j)$ represents the support of an item $i_j \in I$.*

Example 7. The *pmu* values for the items a, b, c, d, e, f, g and h in Table 1 are $2350, 2650, 1850, 2500, 2700, 1000, 1100$ and 1200, respectively. Similarly, the *support* values for the items a, b, c, d, e, f, g and h are $4, 4, 3, 3, 3, 3, 4$ and 1, respectively. If the user-specified $minUtil = 1050$ and $minSup = 2$, then a, b, c, d, e and g are ternary items. On the contrary, items f and h are not the ternary items because $PMU(f) \ngeq minUtil$ and $sup(h) \ngeq minSup$.

It has to be noted that only ternary items and their supersets can be generated as SHUFIs. The correctness is shown in Theorem 1.

Theorem 1. *Let $X \supseteq i_j$. If $pmu(i_j) < minUtil$ or $sup(i_j) < minSup$, then i_j cannot be a SHUFI. In addition, X is also not a SHUFI.*

Proof. The correctness is straight forward to prove from Properties 1 and 2.

Property 1. For an item $i_j \in I$, $pmu(i_j) \geq u(i_j)$. Therefore, if $pmu(i_j) < minUtil$, then $u(i_j) < minUtil$.

Property 2 (Apriori property). If $X \subset Y$, then $sup(X) \geq sup(Y)$. Thus, if $sup(X) < minSup$, then $sup(Y) < minSup$.

We now generalize the Definitions 5 (i.e., *pmu* of an item) and 6 (i.e., ternary item) by taking into account the notion of itemset. This generalization facilitates uses to push the above pruning technique to the lower levels of itemset lattice.

Definition 7 *(Neighboring items that can extend an itemset). Let TI denote the set of ternary items. Let $\succ$ denote any particular order on ternary items. In our approach, we are considering $\succ$ as utility decreasing order on ternary items as this particular ordering prunes more number of ternary items on applying neighborhood suffix utility pruning measure. Let α be an itemset. Let $E(\alpha)$ denote the set of all ternary items that can be used to extend α according to the depth-first search, i.e., $E(\alpha) = \{z | z \in TI \wedge z \succ x, \forall x \in \alpha\}$. Let $N(\alpha) = (\cap_{i_k \in \alpha} N_{i_k}) \cap E(\alpha)$ denote the set of all neighboring items that can be used to extend α according to the depth-first search to find SHUFIs.*

Example 8. The *utility* descending order of ternary items in Table 1, i.e., $TI = \{e, b, d, g, a, c\}$. If $\alpha = ea$, then $E(ea) = \{c\}$. The neighboring items that can extend ea, i.e., $N(ea) = N(e) \cap N(a) \cap E(ea) = \{abcd\} \cap \{bce\} \cap \{c\} = \{c\}$. In other words, the itemset ea will be extended by taking into account only the item c.

Definition 8 *(Probable Maximum Utility (PMU) of an itemset in a transaction). The PMU of an itemset α in a transaction T_{ts}, denoted as $PMU(\alpha, T_{ts}) = u(\alpha, T_{ts}) + \sum_{i \in T_{ts} \wedge i \in N(\alpha) \wedge i \succ x \forall x \in \alpha} u(i, T_{ts})$.*

Example 9. In Table 1, ea has appeared in the transactions T_1. This transaction in *utility* descending order of items is $\{(e, 4), (b, 3), (d, 1), (a, 2), (c, 1)\}$. The PMU of ea in T_1 is the sum of utilities of ea and the neighboring items that are appearing in the sorted T_1 after a. That is, $pmu(ea, T_1) = u(ea, T_1) + u(c, T_1) = 1000 (= u(e, T_1) + u(a, T_1)) + 100 = 1100$.

Definition 9 *(PMU of an itemset in a database). The PMU of an itemset α in a database, denoted as $PMU(\alpha) = \sum_{T_{ts} \in g(\alpha)} pmu(\alpha, T_{ts})$.*

Example 10. The itemset ea has appeared in the transactions T_1 and T_3. Therefore, $g(ea) = \{T_1, T_3\}$. The *pmu* of ea in the entire database, i.e., $pmu(ea) = pmu(ea, T_1) + pmu(ea, T_3) = 1100 + 350 = 1450$. In other words, the maximum utility any superset of ea (containing only the neighboring items of e and a) can have in Table 1 is 1450.

The above definition represents the maximum utility achievable by any superset of an itemset (or an item) in the data. Thus, if the *pmu* of an itemset fails to satisfy *minUtil*, then we stop exploring its supersets. The correctness is based on Property 3. We now define ternary itemset and show that the search space of finding SHUFIs using ternary itemsets is no more than the search space of finding frequent itemsets.

Definition 10 *(Ternary itemset)*. An itemset $X \in I$ is a ternary itemset if $PMU(X) \geq minUtil$ and $sup(X) \geq minSup$.

Relation between ternary itemsets and frequent itemsets. Let TI and FI denote the set of ternary itemsets and frequent itemsets in a database. Since every ternary itemset is a frequent itemset (see Definition 10), the relationship between these two itemsets is $TI \subseteq FI$. Henceforth, finding SHUFIs using ternary itemsets is much less than finding frequent itemsets (though SHUFIs do not satisfy the downward closure property).

Property 3. **(Overestimation using the PMU of an itemset)** Let α be an itemset and z be an item in $N(\alpha)$. Let β be an extension of α such that $z \in \beta$. The relationship $PMU(\alpha \cup z) \geq u(\beta)$ holds (in the projected database of α). In other words, PMU is an overestimating function [5].

Step 2: Generation of Secondary Items Using Neighborhood Suffix Utility. The ternary items generated in previous step constitute of both high utility-frequent items and uninteresting items. To reduce the computational cost of finding the desired itemsets, we need to identify a subset of those ternary items whose depth-first search in the itemset lattice (or projected databases) will result in finding all SHUFIs. To find such items, we introduce a new pruning measure, called *neighborhood suffix utility* (see Definition 11). The proposed measure enhances the *suffix utility* measure [4] by taking into account both the spatial information and support information of items. The neighborhood suffix utility measure facilitates us to identify items (or itemsets) whose depth-first search in the itemset lattice (or set enumeration tree) will result in SHUFIs. The time complexity for measuring neighborhood suffix utility measure is $O(N)$, where N represents the number of ternary items.

Definition 11 *(Neighborhood Suffix utility)*. Let $S = \{i_1, i_2, \cdots, i_k\} \subseteq I$ be an ordered list of ternary items such that $u(i_1) \geq u(i_2) \geq \cdots \geq u(i_k)$. The neighborhood suffix utility of an item $i_j \in S$, denoted as $nsu(i_j)$, is the sum of utilities of remaining items in the list which are neighbors of i_j. That is, $nsu(i_j) = \sum_{p=j+1}^{|S|} u(i_p)$ if $i_p \in N(i_p)$. For the last item in S, $nsu(i_k) = 0$.

Example 11. Continuing with the Example 7, let us order the ternary items in descending order of their *utility* values. Let $\succ$ denote this utility descending order. The ternary items in $\succ$ order are e, b, f, d, g, a and c. Let us consider item e, which is the first item in $\succ$ order. The neighbors of this item are a, b, c and d (see Table 5). Thus, the item e will generate SHUFIs by combining with the items a, b, c and d. Thus, the *neighboring suffix utility* of e, i.e., $NSU(e) = util(a) + util(b) + util(c) + util(d) = 2100$.

We now define secondary items based on the above definition of *neighborhood suffix utility*.

Definition 12 *(Secondary item)*. A ternary item i_j is a secondary item if and only if $utility(i_j) + NSU(i_j) \geq minUtil$.

Example 12. The sum of *utility* and *NSU* values for the ternary items a, b, c, d, e and g in Table 1 are $650, 1800, 300, 700, 3350$ and 400, respectively. If the user-specified $minUtil = 1050$, then e and b be considered as the secondary items. That is, depth-first search on these two items will result in finding all SHUFIs. The depth-first search on remaining ternary items (i.e., 'a, c, d' and 'g') can be avoided as further search on these items will not result in any SHUFI.

The correctness of secondary items is based on Property 4, and shown in Property 5.

Property 4 (**Additive property**). For an itemset X, $util(X) \leq \sum_{i_j \in X} util(i_j)$.

Property 5. For an item $i_j \in ternaryitems$, if $utility(i_j) + NSU(i_j) < minUtil$, then depth-first search in the itemset lattice (or projected databases) of i_j will not yield any SHUFIs.

Step 3: Generation of Primary Items Using Neighborhood Sub-tree Utility. In most cases, the secondary items generated in the previous step form the primary items. However, in a few cases, especially when mining itemsets at low $minUtil$ values, neighborhood suffix utility is not sufficient. It is because most ternery items will be generated as secondary items. To handle such cases, we generate primary items by calculating the *neighborhood sub-tree utility* for secondary items. The calculation of *neighbourhood sub-tree utility* is a computationally expensive step because it needs a scan on the data. At high $minUtil$ values and in large databases, eliminating this step would be beneficial as this step takes time to compute. In this paper, we are providing this step for completeness.

A subset of secondary items whose projections will only generate SHUFIs are known as *primary items*. The measure, *Neighborhood sub-tree utility*, is used to find primary items from secondary items. The $NSTU$ is defined in Definition 13 and illustrated in Example 13. The primary items are defined in Definition 14 and illustrated in Example 14.

Definition 13 (*Neighborhood sub-tree utility*). *Let α be an itemset and $z \in N(\alpha)$ be an item that can extend α according to the depth-first search in the itemset lattice (or set enumeration tree). The neighborhood sub-tree utility of z with respect to α, denoted as $NSTU(\alpha, z) = \sum_{T_{ts} \in g(\alpha \cup z)} [u(\alpha, T_{ts}) + u(z, T_{ts}) + \sum_{i \in T_{ts} \wedge i \in N(\alpha \cup z)} u(i, T_{ts})]$.*

Example 13. The *utility* descending order of ternary items in Table 1 is e, b, d, g, a and c. Let us consider item b, which has the *pmu* value of 2650. Let $\alpha = \emptyset$. The *neighborhood sub-tree utility* of $\alpha \cup b$, $NSTU(\alpha, b) = [600 + 150 + 200] + [200 + 50] + [150 + 100] + [100 + 150] = 950 + 250 + 250 + 250 = 1700$. Since $NSTU(b) \geq minUtil$, it turns out b will generate SHUFIs by projecting it.

Definition 14 (*Primary items*). *Given the utility descending order of secondary items, a secondary item $i_j \in SI$ is said to be a primary item if $NSTU(i_j) \geq minUtil$.*

Example 14. Consider the secondary items e and b. The $NSTU$ values for these two secondary items, i.e., $NSTU(e) = 2700$ and $NSTU(b) = 1700$. Since *neighbourhood subtree utility* values of both items is greater than $minUtil$, e and b are considered as primary items.

Step 4: Recursive Mining of Primary Items. For each primary item, we perform depth first search to find the complete set of SHUFIs. The depth first search is achieved by recursively mining the primary items.

Algorithm 1. SHU-FIM

1: **input** : D: a temporal database, SD: spatial database, UD: external utility database, $minUtil$: a user-specified minimum utility constraint, $minSup$: a user-specified minimum support constraint,$maxDist$: a user-specified maximum distance constraint

2: **output** : A set of spatial high utility frequent itemsets

3: $N = Neighbors(SD, maxDist)$. Let $N(i_j)$ denote the set of all neighboring items for $i_j \in I$. As $Neighbors()$ is a simple and straight forward function, we are not describing it in this paper.

4: Let α denote an itemset that needs to be extended. Initially, set $\alpha = \emptyset$;

5: Scan the temporal database to determine the PMU,*support*,*utility* for every item $i_j \in I$.Bin-arrays can be used to efficiently calculate the PMU,*support*,*utility* of items .

6: $ternary(\alpha) = \{i_j | i_j \in I \wedge PMU(i_j) \geq minUtil \wedge s(i_j) \geq minSup\}$

7: Let $\succ$ be the total order of *utility* descending values on $ternary(\alpha)$;

8: Scan D to remove each item $i \notin ternary(\alpha)$ from the transactions, sort items in each transaction according to $\succ$, and delete empty transactions;

9: Sort transactions in D according to $\succ_T$;[13]

10: Calculate *neighborhood suffix utility* for each item $i \in ternary(\alpha)$.

11: Let $secondary(\alpha)$ denote the set of all items in $ternary(\alpha)$ that have $u(i)+nsu(i) \geq minUtil$;

12: Calculate *neighborhood sub-tree utility* for all items in $secondary(\alpha)$ by scanning the database D once using utility-bin array;

13: $Primary(\alpha) = \{z \in secondary(\alpha) | NSTU(\alpha, z) \geq minUtil\}$;

14: $RecursiveSearch(\alpha, D, Primary(\alpha), ternary(\alpha), minUtil, minSup)$;

4.2 Proposed Algorithm

The SHU-FIM is presented in Algorithms 1 and 2. Since the calculation of PMU, NSU and $NSTU$ for a projected database of α is straightforward and can be calculated efficiently using bin-arrays [3], we have not presented the related algorithms in this paper. The algorithm has the following steps: (*i*) finding ternary items, i.e., items whose supersets can be spatial high utility-frequent itemsets, (*ii*) finding secondary items from ternary items and (*iii*) finding primary items from secondary items, and (*iv*) finding all spatial high utility-frequent itemsets by recursively mining all primary items. We now briefly explain each of these

---

Algorithm 2. RecursiveSearch

1: **input** : α: an itemset, $\alpha - D$: the α projected database, $Primary(\alpha)$: the primary
   items of α, $ternary(\alpha)$: the ternary items of α, $minUtil$: minimum utility, $minSup$:
   minimum support
2: **output**: the set of all SHUFIs that are extensions of α
3: **for each** *item* $i \in Primary(\alpha)$ **do**
4:    $\beta = \alpha \cup \{i\}$;
5:    Scan α-D to calculate $u(\beta)$, $sup(\beta)$. Create a projected database, $\beta - D$.
6:    **if** $u(\beta) \geq minutil$ && $sup(\beta) \geq minSup$ **then**
7:       output β as a spatial high utility frequent itemset
8:    **end if**
9:    **if** $sup(\beta) < minSup$ **then**
10:      *continue.*
11:   **end if**
12:   Calculate $nstu(\beta, z)$, $pmu(\beta, z)$ and $sup(z)$ for all item $z \in ternary(\alpha)$ by scanning β-D once, using two utility-bin array;
13:   $Primary(\beta) = \{z \in ternary(\alpha)|NSTU(\beta, z) \geq minutil$ and $z \in N_\beta$ and $sup(z) \geq minSup$ in β-D$\}$;
14:   $ternary(\beta) = \{z \in ternary(\alpha)|PMU(\beta, z) \geq minutil$ and $z \in N_\beta$ and $sup(z) \geq minSup$ in β-D$\}$;
15:   Search(β, β-D, $Primary(\beta)$, $ternary(\beta)$, $minutil$, $minSup$);
16: **end for**

---

steps by using the databases shown in the Tables 1, 2 and 5. Let $minSup = 2$, $minUtil = 1050$ and $maxDist = 5$.

Definition 15 *(Projected transaction).* *The projection of a transaction* $T \in D$ *using an itemset* α *is denoted as* $\alpha - T$ *and defined as* $\alpha - T = \cup_{\forall i \in T \land i \in N(\alpha)} i$.

Example 15. Let $\alpha = a$. In Table 1, the item a appears in T_3. Therefore, the projected transaction of item a, i.e., $a - T = \{(e, 2)\}$. Other items in T_3 are not included in $a - T$ because they are not the neighbors of a when $maxDist$ allowed is 5.

Definition 16 *(Projected database).* *The projection of a database* D *using an itemset* α *is denoted as* $\alpha - D$ *and defined as the multiset* $\alpha - D = \cup_{\forall T \in D \land \alpha - T \neq \emptyset} \{\alpha - T\}$.

Example 16. In Table 1, item a appears in T_1, T_2, T_3 and T_6. Therefore, $a - D = \{\{(b, 3), (c, 1), (e, 4)\}, \{(b, 2)\}, \{(e, 2)\}\}$. Note that T_6 becomes null so it is not included in the projected database of a.

Finding Ternary Items. The list of items and their corresponding neighbors is shown in Table 5. Initially, α is set $\emptyset$. Next, we calculate the *pmu, utility* and *support* values of all the items in Table 1 using the Bin-arrays. The list of PMU values is $\{\{a : 2350\}, \{b : 2650\}, \{c : 1850\}, \{d : 2500\}, \{e : 2700\}, \{f : 1000\}, \{g : 1100\}, \{h : 1200\}\}$. The list of *support* values is $\{\{a : 4\}, \{b : 4\}, \{c :$

$3\}, \{d : 3\}, \{e : 3\}, \{f : 3\}, \{g : 4\}, \{h : 1\}\}$. The list of *utility* values is $\{\{a : 350\}, \{b : 1050\}, \{c : 300\}, \{d : 400\}, \{e : 1250\}, \{f : 700\}, \{g : 400\}, \{h : 1200\}\}$. As described in the Algorithm 1, ternary items with respect to α are generated by comparing PMU value of each item with $minUtil$ and *support* with $minSup$ (line 6 in Algorithm 1). The set of ternary items generated from Table 1 are $\{a, b, c, d, e, g\}$. Note that the items f and h are not the ternary items because $pmu(f) < 1050(= minUtil)$ and $sup(h) < 2(= minSup)$.

The ternary items are sorted in *utility* descending order $\succ$. Thus, $ternary(\alpha) = \{e, b, d, g, a, c\}$ (line 7 in Algorithm 1). Next, all ternary items in every transaction are sorted in $\succ$ order by pruning non-ternary items (line 8, 9 in Algorithm 1).

Finding Secondary Items. Next, we calculate neighborhood suffix utility (NSU) for each item in $ternary(\alpha)$ (line 10 in Algorithm 1). In our running example the values of $u(i_j) + nsu(i_j)$ for all ternary items are $\{\{e : 3050\}, \{b : 1800\}, \{d : 700\}, \{g : 400\}, \{a : 650\}, \{c : 300\}\}$. The secondary items are e and b since their sum of *utility* and their *neighbourhood suffix utility* $\geq minutility$. Thus, $secondary(\alpha) = \{e, b\}$.

Finding Primary Items. The $NSTU$ values for the secondary items are $\{\{b : 1700\}, \{e : 2700\}$ (line 12 in Algorithm 1). As both items have $NSTU$ greater than $minUtil$, they are considered as *primary items*. That is, $primary(\alpha) = \{e, b\}$ (line 13 in Algorithm 1). In our running example, no secondary item got pruned with $NSTU$ because many ternary items were already pruned by the proposed *neighborhood suffix utility* measure. Next, RecursiveSearch function on e and b will find all SHUFIs from the database (line 14 in Algorithm 1).

(a) Projected database of e (i.e., e-D from table 1)		(b) PMU, NSTU and sup values of items in e-D				(c) Projected database of ea (i.e., ea-D)		(d) PMU, NSTU and sup values of the items in ea-D			
ts	Items	Items	PMU	NSTU	Sup	ts	Items	Items	PMU	NSTU	Sup
T1	(b, 3),(d,1),(a,2),(c,1)	a	2050	1450	2	T1	(c,1)	c	1100	1100	1
T3	(a,1)	b	2150	2150	2						
T5	(b,2),(d,3),(c,1)	c	1650	1150	2						
		d	2150	1450	2						

Fig. 1. Recursive mining of primary items

Recursive Mining of Primary Items. The RecursiveSearch (i.e., Algorithm 2) works as follows. For each primary item, construct its projected database and recursively mine the projected database until the projected database is empty. For our working example it starts with the first item e in the $Primary(\alpha)$. Since $\alpha = \emptyset$, $\beta = e$, we scan D (i.e., $\alpha - D$) and calculate $u(\beta) = u(e) = 1250$ and

$s(\beta) = s(e) = 3$. Next, we construct a projected database $\beta - D$ (i.e., $e - D$) with only neighboring items of e. Figure 1(a) shows the projected database of e. β will be output as SHUFI only if both its *support* and *utility* are greater than their required thresholds. As $u(e) \geq minUtil$ and $s(e) \geq minSup$, e will be generated as a SHUFI. If $s(\beta) < minSup$ then neither β or its supersets will be SHUFIs because of anti monotonic property of frequency, so the algorithm jumps to next item in $primary(\alpha)$ thus pruning all the itemsets which may extend β in set enumeration tree. Next, for each item z in $\beta - D$, we calculate PMU, $NSTU$ and sup values, if $s(\beta) \geq minSup$. Since $sup(e) \geq 2$ the PMU, $NSTU$ and sup values are calculated for all items in $e - D$ and are shown in Fig. 1(b). The items which are neighbors of β and having $NSTU$ greater than or equal to $minUtil$ and sup greater than or equal to $minSup$ are considered as primary items (i.e., $primary(\beta)$). The primary items are $primary(\beta) = \{a, b, c, d\}$. Next, construct projected database for ea i.e., $(ea - D)$ as shown in Fig. 1(c). The utility and support of ea are 1350 and 2 respectively and are caluclated while constructing the projectetd database of ea (line 5 in Algorithm 2). Since $u(ea) \geq minUtil$ and $sup(ea) \geq minSup$ itemset ea is a SHUFI. Since $sup(ea) \geq minSup$, we calculate $PMU, NSTU$ and sup for all neighbouring items in $ea - D$ as shown in Fig. 1(d). As $NSTU(c) \geq minUtil$ and $sup(c) < minSup$ in $ea - D$, we wont consider c as the primary item of ea. As there are no primary items for ea, we wont explore ea any further, rather we move to the next primary item of e. The above process is repeated for all primary items. The complete set of SHUFIs generated by SHU-FIM for the database shown in Table 1 are presented in Table 4.

5   Experimental Results

In this section, we compare both SHUIM and SHU-FIM algorithms at different $minUtil$ and $minSup$ values in various databases, and shown that SHU-FIM performs better than SHUIM.

The algorithms, SHUIM and SHU-FIM, were written in java and executed on a machine with 1.5 GHz processor and 8 GB RAM. Due to page limitations, the experiments have been conducted on only two real-world databases, namely, *Congestion database* and *Pollution database*. The source Congestion database contained 39,873 data points generated from a total of 2,412 items (or RIDs). After transforming the source database into temporal database, the resultant database had 287 rows. The minimum, average and maximum length of each transaction in the resultant Congestion database are 41, 140, and 355, respectively.

The source air pollution database contained 97,651 data points of 1,026 items (or sensor ids). Each data point represents the $PM_{2.5}$ value recorded by a sensor. The unit for measuring PM2.5 is microgram per cubicmeter ($\mu g/m^3$). After transforming the source database into temporal database, the resultant database had 95 rows. The minimum, average and maximum length of each transaction in the resultant Pollution database are 985, 1018, and 1025, respectively. More details on each of these databases is provided in [5].

Figure 2(a) and (b) show the number of SHUFIs generated in Congestion and Pollution databases at different *minUtil* and *minSup* values, respectively. **The two thick straight lines in each of the figures represent the corresponding values generated by SHUIM at different minUtil values.** The following observations can be drawn from these two figures: (i) the SHUIs generated by SHUIM is not influenced by *minSup*. Henceforth, the number of SHUIs generated by SHUIM remains constant. (ii) Increase in *minUtil* results in the decrease of SHUIs and SHUFIs generated by SHUIM and SHU-FIM, respectively. It is because many itemsets have failed to satisfy the increased *minUtil* constraint. (iii) Increase in *minSup* results in decrease of SHUFIs. It is because many itemsets have failed to satisfy the increased *minSup* value. More important, it can be observed that many sporadic itemsets which were generated as SHUIs by SHUIM algorithm have failed to be SHUFIs. Thus, the proposed model of SHUFIM has facilitated the user to prune uninteresting high utility itemsets occurring very few times in the data. (This topic is further discussed in latter parts of this paper.).

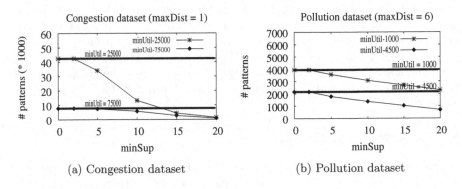

(a) Congestion dataset

(b) Pollution dataset

Fig. 2. Patterns generated at different minUtil and minSup

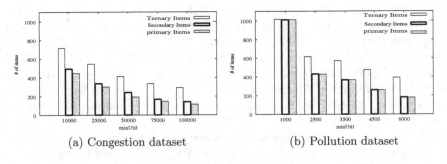

(a) Congestion dataset

(b) Pollution dataset

Fig. 3. Number of ternary and secondary items generated at different minUtil values

Figure 3(a) and (b) show the number of ternary, secondary, and primary items generated in Congestion and Pollution databases at different *minUtil* values, respectively. The following observations can be drawn from these two figures: (i) increase in $minUtil$ decreases number of ternary, secondary and primary items. It is because more items have failed to satisfy the increased $minUtil$ value. (ii) increase in $minUtil$ also increases the gap between the number of ternary and secondary items. On the contrary, increase in $minUtil$ decreases the gap between the secondary and primary items. It is because at high $minUtil$ values, almost all secondary items will be primary items. In pollution dataset, it can be observed that the number of secondary and primary items remains same for different $minUtil$ values. In such cases, we can prevent the computational expensive step of calculating $NSTU$ and mine all SHUFIs by performing depth-first search on the secondary items.

Figure 4(a) and (b) show the number of nodes (or the search space) explored by SHUIM and SHU-FIM algorithms at different $minUtil$ and $minSup$ values in Congestion and Pollution databases, respectively. The following observations can be observed from these figures: (i) The number of nodes explored by both SHUIM and SHU-FIM algorithms decrease with the increase in $minUtil$ value. It is because fewer patterns get generated by both algorithms at higher $minUtil$ values. (ii) The number of nodes explored by SHUIM algorithm is not influenced by the $minSup$. On the contrary, the nodes explored by SHU-FIM algorithm decreases with the increase of $minSup$. It is because fewer SHUFIs get generated with the increase in $minSup$. (iii) Even at $minSup = 0$, the proposed SHU-FIM algorithm has explored fewer nodes as compared to the SHUIM algorithm. It is because the proposed NSU measure facilitated the SHU-FIM algorithm to generate fewer primary items as compared against the SHUIM algorithm.

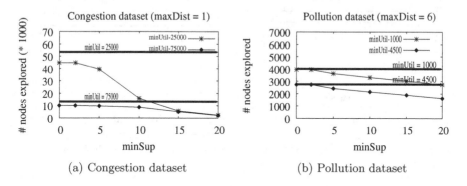

(a) Congestion dataset   (b) Pollution dataset

Fig. 4. Number of nodes explored at different minUtil and minSup

Figure 5(a) and (b) show the runtime comparison of SHUIM and SHU-FIM algorithms at different $minUtil$ and $minSup$ values in Congestion and Pollution databases, respectively. It can be observed that the runtime requirements of both algorithms depends on the number of patterns getting generated at different $minSup$ and $minUtil$ values. Most important, even at $minSup = 0$, the proposed SHU-FIM has taken less runtime as compared against SHUIM. It is because the proposed *NSU* has facilitated SHU-FIM to explore fewer nodes as compared against SHUIM algorithm.

Figure 6(a) and (b) show the memory consumed by both SHUIM and SHU-FIM algorithms at different *minUtil* and *minSup* values in Congestion and Pollution databases, respectively. Similar observations as above (i.e., memory requirements) can be observed in these two figures.

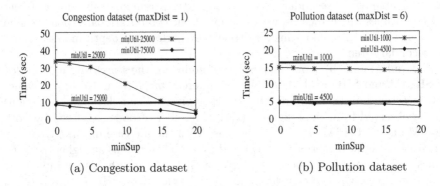

(a) Congestion dataset   (b) Pollution dataset

Fig. 5. Runtime comparison of SHUIM and SHU-FIM algorithms at different *minUtil* and *minSup*

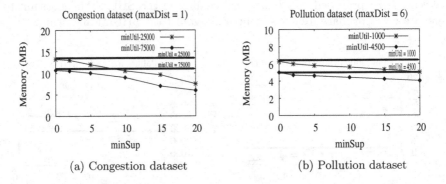

(a) Congestion dataset   (b) Pollution dataset

Fig. 6. Memory comparison of SHUIM and SHU-FIM algorithms at different *minUtil* and *minSup*

5.1 Case Study: Improving Traffic Safety

To demonstrate the usefulness of SHUFIs in traffic congestion data, we have divided congestion data of 17-7-2015 into hourly intervals and applied SHU-FIM on each hourly congestion data to find SHUFIs (i.e., sets of road segments where a lot of congestion has observed within the traffic network over some time). The $maxDist = 1$ km and $minUtil = 25000$ m(=25 km). Figure 7(a)–(c) show the highly congested road segments (or SHUFIs) generated at 01:00 h, 02:00 h and

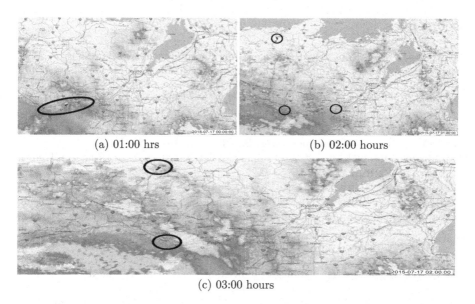

(a) 01:00 hrs   (b) 02:00 hours

(c) 03:00 hours

Fig. 7. SHUFIs generated in congestion data at different hours. Each colored line segments in a dark circle represent a SHUFI

03:00 h, respectively. Road segments with same color represent a spatial high utility Frequent itemset. Hourly XRAIN data of the typhoon Nangka was also overlaid on the road segments. It can be observed roads with heavy congestion (i.e., SHUFIs) vary with each hour and also depends on precipitation. Thus, the knowledge of SHUFIs (or heavy congested roads over some time) can be found very useful to the traffic control room in diverting the traffic.

6   Conclusions and Future Work

This paper proposed a flexible model of spatial high utility frequent itemset that exist in a spatiotemporal database. A novel *utility* upper bound measure, called *Neighborood suffix utility*, has been introduced to reduce the search space of SHUFIM. We have also theoretically shown that the search space of finding all SHUFIs in a spatiotemporal database is no more than the search space of finding frequent itemsets in a spatiotemporal database. A single pass algorithm has also been presented to find all SHUFIs in spatiotemporal databases, effectively. Experimental results demonstrate that the proposed algorithm is efficient.

As a part of future work, we would like to extend the proposed model of spatial high utility frequent itemset by taking into account both positive and negative external utility values.

References

1. Agrawal, R., Srikant, R.: Fast algorithms for mining association rules in large databases. In: Proceedings of 20th International Conference on Very Large Data Bases, VLDB 1994, vol. 1215, pp. 487–499 (1994)
2. Ahmed, C.F., Tanbeer, S.K., Jeong, B., Lee, Y.: Efficient tree structures for high utility pattern mining in incremental databases. IEEE Trans. Knowl. Data Eng. **21**(12), 1708–1721 (2009)
3. Gan, W., Lin, J.C., Fournier-Viger, P., Chao, H., Hong, T., Fujita, H.: A survey of incremental high-tility itemset mining. Wiley Interdisc. Rev.: Data Min. Knowl. Discov. **8**(2), e1242 (2018)
4. Uday Kiran, R., Yashwanth Reddy, T., Fournier-Viger, P., Toyoda, M., Krishna Reddy, P., Kitsuregawa, M.: Efficiently finding high utility-frequent itemsets using cutoff and suffix utility. In: Yang, Q., Zhou, Z.-H., Gong, Z., Zhang, M.-L., Huang, S.-J. (eds.) PAKDD 2019. LNCS (LNAI), vol. 11440, pp. 191–203. Springer, Cham (2019). https://doi.org/10.1007/978-3-030-16145-3_15
5. Kiran, R.U., Zettsu, K., Toyoda, M., Fournier-Viger, P., Reddy, P.K., Kitsuregawa, M.: Discovering spatial high utility itemsets in spatiotemporal databases. In: Proceedings of the 31st International Conference on Scientific and Statistical Database Management, SSDBM 2019, Santa Cruz, CA, USA, 23–25 July 2019, pp. 49–60 (2019)
6. Lan, G., Hong, T., Tseng, V.S.: An efficient projection-based indexing approach for mining high utility itemsets. Knowl. Inf. Syst. **38**(1), 85–107 (2014)
7. Liu, J., Wang, K., Fung, B.C.M.: Direct discovery of high utility itemsets without candidate generation. In: 12th IEEE International Conference on Data Mining, pp. 984–989. IEEE (2012)
8. Liu, Y., Liao, W., Choudhary, A.: A Two-phase algorithm for fast discovery of high utility itemsets. In: Ho, T.B., Cheung, D., Liu, H. (eds.) PAKDD 2005. LNCS (LNAI), vol. 3518, pp. 689–695. Springer, Heidelberg (2005). https://doi.org/10.1007/11430919_79
9. Song, W., Liu, Y., Li, J.: Bahui: fast and memory efficient mining of high utility itemsets based on bitmap. Int. J. Data Warehous. Min. (IJDWM) **10**(1), 1–15 (2014)
10. Tseng, V.S., Shie, B., Wu, C., Yu, P.S.: Efficient algorithms for mining high utility itemsets from transactional databases. IEEE Trans. Knowl. Data Eng. **25**(8), 1772–1786 (2013)
11. Tseng, V.S., Wu, C., Shie, B., Yu, P.S.: Up-growth: an efficient algorithm for high utility itemset mining. In: Proceedings of the 16th ACM SIGKDD International Conference on Knowledge Discovery and Data Mining, Washington, DC, USA, 25–28 July 2010, pp. 253–262 (2010)
12. Yao, H., Hamilton, H.J., Butz, C.J.: A foundational approach to mining itemset utilities from databases. In: Proceedings of the Fourth SIAM International Conference on Data Mining, pp. 482–486 (2004)
13. Zida, S., Fournier-Viger, P., Lin, J.C., Wu, C., Tseng, V.S.: Efim: a fast and memory efficient algorithm for high-utility itemset mining. Knowl. Inf. Syst. **51**(2), 595–625 (2017)
14. Zida, S., Fournier-Viger, P., Wu, C.-W., Lin, J.C.-W., Tseng, V.S.: Efficient mining of high-utility sequential rules. In: Perner, P. (ed.) MLDM 2015. LNCS (LNAI), vol. 9166, pp. 157–171. Springer, Cham (2015). https://doi.org/10.1007/978-3-319-21024-7_11

Efficient Algorithms for Flock Detection in Large Spatio-Temporal Data

Jui Mhatre[1], Harsha Agrawal[1], and Sumit Sen[2(✉)]

[1] Veermata Jijabai Technological Institute, Mumbai, India
{jbmhatre_m17,hagrawal_m17}@ce.vjti.ac.in
[2] Indian Institute of Technology - Bombay, Mumbai, India
sumitssen@gmail.com

Abstract. Increasing availability of location-based applications and sensor devices have necessitated quicker analysis of moving object data streams in order to identify patterns. The efficiency of currently available algorithms used in pattern detection is not adequate to handle large scale data streams that are increasingly available. We focus on the particular problem of flock detection in moving object data and our goal is to detect flocks quickly and using fast algorithms. Firstly, we employ a triangular grid to reduce the search space of clustering algorithms which has a significant effect in case of dense objects. As a second step, we implement a modified flock membership function and pipeline creation that ensures better memory and time performance during cluster detection. We show that this refinement also improves the rate of flock detection. Finally, we parallelize our algorithm to further enhance the handling of massive data streams. Based on an extensive empirical evaluation of these algorithms across a variety of moving object data sets, we show that our method is significantly faster than the existing comparable methods over sliding windows. In particular, it requires lesser time to identify flocks and is 2–4 times faster thus confirming the efficiency and effectiveness of our approach.

Keywords: Moving objects · Clustering · Spatio-temporal groups · Flocks · Parallel execution

1 Introduction

Advances in positioning technologies and sensor availability have together ensured large scale availability of moving object data. Spatio-temporal data stream of human movement, animals, vehicles and even natural objects like icebergs are being captured and analyzed using several techniques but mainly related to GPS logs. Several applications in different domains ranging from animal behavior studies to environmental studies use such data sets both at real time and or in a different mode. It is characteristic of such data streams to have unbounded length and fast arrival rate owing to high frequency of data turnaround [7].

© Springer Nature Switzerland AG 2019
S. Madria et al. (Eds.): BDA 2019, LNCS 11932, pp. 307–323, 2019.
https://doi.org/10.1007/978-3-030-37188-3_18

Detection of patterns in moving object data sets have been the subject of many studies and some of the research on this topic has been surveyed by Tanuja and Govindarajulu [22], and Yuan et al. [24]. Algorithms used to detect such a pattern can be seen to have similar concepts of clustering such as those based on density, hierarchy, partitions, models or grids as surveyed by Yuan et al. [24]. While the application for which the moving object data is being analyzed determines the minimum threshold of performance and efficiency required, it is important to note that with increasing availability of sources as well as rising volume of the data, faster and more efficient methods are urgently required.

In this paper we outline technique to improve both the efficiency and performance of flock detection and is organized as follows. In the next section, we outline previous work in the area of flock detection and outline our motivation in order to provide a background to our proposed approach. In the subsequent section, we provide problem definition along with theoretical proofs of our hypothesis. Section 4 provides implementation and empirical evaluation of these techniques. Finally we discuss implementation of the work and prove conclusions along with discussion and future work.

2   Previous Work

There has been significant interest in querying patterns in moving object data that exemplify 'collaborative' in 'group patterns' [6], Central to such analysis is the assumption that beyond the location information of these 'objects', there is no additional data to identify (or rule out) the membership of an 'object' to a 'group'. The ability to query from a given data set, groups such as 'convoy' [14], flocks [8,21,23] and swarm [17] have been widely reported. Flocks are considered as a group that move together within a disk of a used specified size [23]. While swarms are generally described as patterns where group of objects of significant sizes stay together for sufficient amount of time [10,16]. Convoys are a more general notion of a flock where the rigidity of a disk is replaced by a notion of density connection [13].

While in this paper we are mainly focused on flock pattern detection, we discuss in Sect. 5 how some of approaches generally apply to swarms and convoy queries.

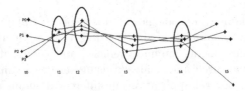

Fig. 1. Flock F (P0, P1, P2, P3) spans for time instances t1 to t4 has membership value, $\mu = 4$, pipeline threshold, $\delta = 4$ and ϵ is the diameter of disks identified. P0 to P3 are trajectory points in time instances t0 to t5

Flocks have been traditionally defined as a large enough subset of objects moving along a path close to each other for a certain predefined time [8]. As shown in Fig. 1, given a set of trajectories $\{T\}_i^n$ and a minimum flocking time interval given by δ, the set of μ entities, in a diameter of ϵ. is representative of flock. While Benkert et al. [8] define this as set (μ, δ, ϵ), Vieira et al. [23] approach the problem of discovery of a flock pattern by defining a disk with a given radius that covers all moving object of the flock long enough. They define such disks and prove that if there is a disk that covers all trajectories in a flock then there exist a another disk at another time instance with the same diameter but different centers as shown in Fig. 2. The implication of their theorem is that it restricts the number of locations in space where flocks can be found and also reduces the problem to polynomial time complexity. This is demonstrated using square grid based structure.

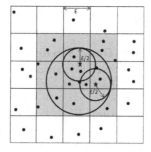

Fig. 2. Square grid for range query in [23]. Side of each cell is equal to ϵ

Cao et al. [11] adapt the flock detection problem in pedestrian movement studies and have reported flock patterns. This extraction algorithm based on a four step approach that relies on spatial neighbourhood computation, membership persistence, degree of freedom filtering and pruning. In a survey of spatio-temporal data mining method [6], trajectory similarity is categorized as one of the many types of analysis that moving object data set present and these are dependent on distance and other similarity metrics. They also highlight the heterogeneity in space and time density and employ clustering algorithm. This is particularly relevant in spatially dense data when looking for cluster.

2.1 Motivation

Our work is motivated in part by the approach and efficiency constraints of the flock detection methods. In particular, we are focused on the time required to identify flocks in a large GPS trace data set. We attempt to efficiently identify flocks such that all possible flocks, under given criteria, are identified. Our flocks identification approach can be applied to any kind of trajectory data. However, we assume that datapoints are available at certain regular intervals.

For empirical evaluation, we have collected human (not necessarily pedestrian) movement data for evaluating our approach. We have collected human movement traces from 197 people. We have processed the traces [4,5] such that location of each person at interval of 30 seconds is fetched for further analysis. We also aimed to provide faster algorithm for analyzing a wide range of data and hence went on to evaluate the performance on other data sets namely,

- A dense data set having information on iceberg activity in the North Atlantic since 1913 taken from International Ice Patrol (IIP) Iceberg sighing database [3].
- A relatively sparse data set of truck movement in Athens city [1]. Trucks dataset consists of 276 trajectories of 50 trucks delivering concrete to several construction places around Athens metropolitan area in Greece for 33 distinct days.

3   Problem Definition

Definition 1. *Spatiotemporal Data Stream:* *Given a set of point objects P whose location is tracked for time T, spatiotemporal datastream S is set of locations (x_{pi}, y_{pi}) of each point object p at each time instant i.*

$$S = \{S_i \mid i \text{ is time at any instant}, 0 \leq i \leq t \text{ and } S_i \text{ is set of points}$$
$$p(x_i, y_i, i) \text{ where } p \in P\}$$

Definition 2. *Database Snapshot:* *For a spatio temporal data set with T time instances, $t_j \in T$ are time instances where location of objects p are captured such that $p_j \in D$. D is given data set and p_j is location of object p at time t_j and $p \in P$. Then database snapshot D_j at time t_j is,*

$$D_j = \{p_j\}$$

Definition 3. *Cluster:* *Given a database snapshot at any time instant i be D_i then cluster C with ϵ radius is set of all points in D_i that are maximum $2 * \epsilon$ distance apart from each other.*

$$C = \{p_j \mid p_k \in C \text{ and } p_k, p_j \in P, \text{ distance}(p_k, p_j) \leq 2 * \epsilon \}$$

where ϵ is radius of cluster. For a given database snapshot, a set of clusters will be formed with count ranging from 1 to n where n is number of points in D_i.

Definition 4. *Trajectory:* *It is temporally ordered set of locations of a given object. For any object p, trajectory of that object is S_p where,*

$$S_p = \{p_j \mid j = 1...n, t_j \in T\}$$

where n is total number of time instances where object locations are intercepted.

Definition 5. *Pipeline: Every cluster is identified using unique points objects in the cluster. For any cluster C_{ij}, where i is cluster identifier and j is database snapshot identifier for D_j in which cluster c_{ij} is identified. Then pipeline P_i is set of all clusters c_{ij} for n consecutive time instances.*

$$P_i = \{c_{ij} \mid c_{ij} \in D_j, \ j = 1...n \text{ and } i \text{ is cluster identifier}\}.$$

where n is length of pipeline.

Definition 6. *Flocks: Flock F_i is a pipeline P_i, such that each cluster in pipe has size with minimum μ (membership threshold) point objects and size of pipe is greater than or equal to δ (pipeline threshold).*

For identification of clusters in a particular database snapshot, we follow a grid based approach. In earlier work, square grid have been used [23]. As shown in Fig. 2 search space is found and disc shaped clusters are identified. We have proposed an approach to reduce the search space using triangular grid instead of square grid in Sect. 3.1.

To identify the flock, we need to identify pipeline. This requires comparison of all the clusters of all the database snapshots. We have proposed an algorithm which uses mid-start approach to reduce the number of comparisons. We have mathematically proved that the search of cluster pipelines is faster when mid-start approach is used where we start comparing clusters in snapshot at middle of pipeline as compared to the TDE approach [23] which starts comparison of clusters from first database snapshot in Sect. 3.2. To harness the multiprocessing capability of CPU, we have implemented an approach to use multiple cores and run the clustering process in parallel and thus improving the performance.

3.1 Triangular Grid for Reducing Search Space in Lieu of Square Grid

Definition 7. *Range Query: It finds all the points $p_k p$ in neighborhood of a given point p at maximum distance ϵ. Say we want to calculate the range of point p then,*

$$Range(p) = \{p_k \mid p_k \in D \text{ and } distance(p_k, p) \leq \epsilon\}$$

When querying for neighbourhood, search space is set of all the points in data set D. It is observed that the search space can be further reduced using grid based approach [23]. In order to reduce the search space, we use a triangular grid.

Theorem 1. *Triangular grid for range query reduces search space more than using square grid.*

Proof : Assume that points are equally distributed over the area D hence area under computation is directly proportional to computational complexity. We

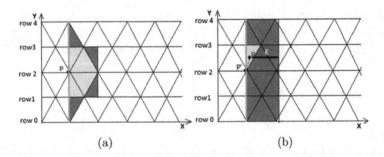

Fig. 3. Figure depicting search space of point p for finding the range. (a) shows the case where p lies on vertex of the triangle, (b) shows the case where p lies within face of the triangle or on any edge. Area in yellow denotes the area which is implicitly under the range of p. Area in blue denotes that area which requires explicit calculation of $\epsilon - neighbourhood$. $E = \epsilon$ is length of triangle side (Color figure online)

compare search space in terms of area of grid to be analyzed. In square and triangle grid, size of each cell is ϵ then,

a_{sq} = area of cell in square grid

a_{tr} = area of cell in triangular grid

$a_{tr} = 0.433 * a_{sq}$

- For range query in square cell, Fig. 2, we need to consider each point for calculating ϵ-neighbourhood in 9 cells (in yellow). Therefore total search space for neighbourhood calculation is,

$= 9 * \epsilon * \epsilon$

$= 9 * a_{sq}$ [23]

- For triangular cells, there are 2 cases based on location of point p,

  1. p lies on any vertex of triangular cell Fig. 3a,

     All points marked in yellow triangles are at a distance $\leq \epsilon$. Therefore all those points can be directly added to range. For the points in the triangles marked blue, we need to check for ϵ-neighbourhood explicitly. Thus search space under computation overhead is reduced to 2 triangles.

     $= 2 * a_{tr}$

     $= 2 * 0.5 * \epsilon * \sqrt{3}/2 * \epsilon$

     $= \epsilon * \sqrt{3}/2 * \epsilon$

     $= 0.866 * a_{sq}$

  2. If p lies on face of triangle or any edge Fig. 3b,

     All points marked yellow are directly taken and ones in blue require explicit ϵ-neighbourhood calculation. During this calculation, we always check the x coordinate, if it exceeds p.x + ϵ then search for that triangle is halted. For any point p on face of triangle, maximum triangles under blue will be 7.5, Therefore search space is,

     $= 7.5 * a_{tr}$

$$= 7.5 * 0.5 * \epsilon * \sqrt{3}/2 * \epsilon$$
$$= 3.25 * \epsilon * \sqrt{3}/2 * \epsilon$$
$$= 2.81 * a_{sq}$$

Both cases of triangular cells have less search space than square cells for range query.

Once we have identified clusters, we have a challenge to compare clusters in consecutive time instances. In Theorem 2, we introduce an approach to reduce the comparison to conclude that flock doesn't exist by proving that the cluster doesn't exist at atleast one time instant in given pipeline line such that one or more points move out of cluster.

3.2   Reducing Database Snapshot Comparisons

Theorem 2. *For a moving object p with location P at start and location P' after time δ and P, P' ϵ C, then the probability of reaching the time instant faster, when object p moves out of cluster, is more when we start comparison from middle snapshot of PP' than the probability when we start comparison from P.*

Proof. Let a be the point when object moves out of cluster, b be the point when object returns from out of the cluster,

x = distance(Pa)
y = distance(bP')
midpoint of PP', m = P + (x + y + ab)/2
  There are 3 cases based on size of x, y:
  Say line segment is pp' having midpoint at m. Let k be mid point of pm. Let a, b be the point that is to be reached.
  There are 3 cases,

- Midpoint m lies outside the cluster (Fig. 4a), which means a, b lie on either sides of m.
  Probability that a lies left to k given (b − m) > (a − p) = 0.25(a is reachable faster from p)
  Probability that a lies left to k given (b − m) < (a − p) = 0.25(b is reachable faster from m)
  Probability that a lies right to k given it is in left of m = 0.5 (a is reachable faster from m)
  Hence the probability that distance(a/b - p), i.e. distance between a and p or distance between b and p, is greater than distance(m−a/b),
  P1(distance(a/b − p) > distance (m − a/b)) = 0.75*0.33
  P1(distance(a/b − p) < distance (m − a/b)) = 0.25*0.33
- Midpoint m lies inside cluster but at point before moving outside cluster (Fig. 4b), which means a and b lie on right side of m.
  Probability of m reaching a faster = 1
  Probability of p reaching a faster = 0, Hence

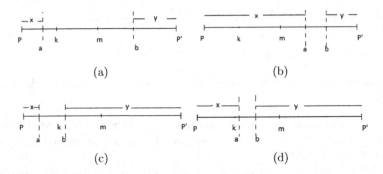

Fig. 4. Figure shows different scenarios when point p travels outside cluster and flock condition fails. Length of segment corresponds to the time spent by point p at different locations. (a) m lies outside the cluster, (b) m lies inside the cluster such that p spends more than half of the time inside cluster before moving outside, (c) m lies inside the cluster such that p spends more than half of the time inside the cluster after returning from outside and x < ab, (d)m lies inside the cluster such that p spends more than half of the time inside the cluster after returning from outside and x > ab

P2(distance(a/b − p) > distance(m − a/b)) = 1*0.33
P2(distance(a/b − p) < distance(m − a/b))= 0*0.33
- Midpoint m lies inside cluster but at a point after returning to cluster from outside (Fig. 4c, d), which means a, b lie on left side of m.
  Probability that a lies to left of k then there are 2 cases,
  • Probability of b lies to left of k = 0.25 (p reaches faster to a)
  • Probability of b lies to right of k = 0.25,then there are equal probabilities that p reaches a faster or m reaches faster to b, each with probability 0.125
  Probability that a lies to right of k then m reaches faster to a/b than m = 0.5, Hence
  P3(distance(a/b − p) > distance(m − a/b)) = 0.625*0.33
  P3(distance(a/b − p) < distance(m − a/b)) = 0.375*0.33

Thus, P(distance(a/b − p) > distance(m − a/b)) = 0.33 (0.75 + 1 + 0.625) = 0.78375 and P(distance(a/b − p) < distance(m − a/b)) = 0.33(0.25 + 0 + 0.375) = 0.20625. Thus we can reach point a or faster if we start from midpoint m as compared to point P. Hence for two points on line segment, and either of the point is to be reached then, there is higher probability of reaching from midpoint than from start of segment. (Given that two points cannot interchange the order).

4   Implementation

Our data set consist of collection of points p such that,

$$P = \{p_i \mid p_i = (x_i, \ y_i, \ t_j) \ and \ 1 \le i \le n \ and \ 0 \le j \le S\}$$

where n is data set size and S is time for which data is captured Every point object in data set has location coordinates (x, y) at any time instant t.

Flock determination is done in 2 stages,

- Clustering of points, in disk shape of diameter ϵ and minimum membership μ using triangular grid for range query.
- Combining clusters with common member points in consecutive δ slots using proposed mid-start approach.

4.1 Triangular Grid for Range Query

Definition 8. *Triangular grid:* *It is set T_I of triangles t such that for any triangle t in a virtual underlying grid T_V that has at least one point of data set is enclosed t.*

$$T_I = \{t \mid t \in T_V \text{ and } p \in P \text{ and } t = \text{ encloses}(p, t) \}$$

The Algorithm 1 to find enclosing triangle takes the point p as coordinates x, y as the input and finds the triangle in which it lies. The y-coordinate gives the row number and x coordinate gives exact triangles to which the point belongs. Triangles 1, 2, 3 in Fig. 5b are the three possibilities of triangles to which the point can belong, when range identified by x is J(x1, y1) and K(x2, y2). Using the angles a, b respectively formed by J, K with a given point p, we can select the final enclosing triangle from the three triangles using following cases.

1. $a \leq 60$ *and* $b > 120$ .... *Triangle* 2
2. $a \leq 60$ *and* $b \leq 120$ .... *Triangle* 3
3. $a > 60$ *and* $b \geq 120$ .... *Triangle* 1

Range Query (Disk Shape with Radius ϵ). We use plane sweep approach [19] and select point p for performing range query. Here range query requires ϵ-neighbourhood points in its right semicircle [21]. When triangular grid is used for range query there are two possibilities,

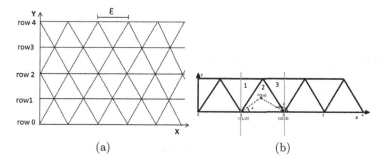

(a)   (b)

Fig. 5. Triangular Grid Construction, (a) Virtual Triangular grid, (b) Triangle Identification from any point p(x, y)

input : point p(x,y,t), time t;
Result: Triangle T such that p lies inside T
initialize T;
height=$\epsilon * \sqrt{3/2}$;
rownum = floor(p.y/height);
if *rownum is even* **then**
| x1=floor(x/ϵ) * height;
else
| **if** x_j $\epsilon/2$ **then**
| | x1=-$\epsilon/2$
| **else**
| | x1=floor(x/ϵ) * ϵ - $\epsilon/2$
| **end**
end
y1=y2=y;
x2=ϵ + x1;
a = angle formed by p(x,y) with (x1,y1);
b = angle formed by p(x,y) with (x2,y2);
T= identify the triangle using a, b as specified in Figure 5b;
return T;

Algorithm 1. Finding enclosing triangle from a given point.

- Range R' of a point p when p lies of vertex of any triangle then,

$$R'(p) = \{p_t \mid p_t \, \epsilon \, t_i, t_o \text{ and } distance(p_t, p) \leq \epsilon\}$$

where t_i, t_o are internal and outer triangles in Fig. 3a.
- Similarly if Range R" of a point p which lies withing the triangle or on edges of triangle then,

$$R''(p) = \{p_t \mid p_t \, \epsilon \, t_i, t_o \text{ and } distance(p_t, p) \leq t\}$$

where t_i, t_o are internal and outer triangles in Fig. 3b.

From the range selected for each point, clusters are created with diameter ϵ using approach discussed by M. Vieira et al. [23].

4.2   Flock Identification Using Mid-start Approach

For a flock to exist, we require clusters in consecutive database snapshots with same members (minimum size is μ) to exist. Say, for a database snapshot at time t_i, set of cluster C_i is given as,

$$C_i = \{c_{ji} \mid i \geq 1 \text{ and } i \leq \delta \text{ and } j \geq 1 \text{ and } j \leq n \text{ and}$$
$$n \text{ is number of clusters in } i^{th} \text{ database snapshot}\}.$$

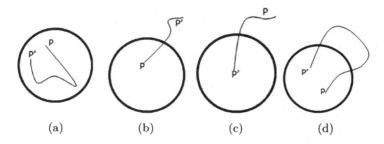

Fig. 6. Figure depicting time spent by any point P in a disk shaped cluster from time instance 1 to $\delta + i - 1$ and finally point comes at P'. (a) point p remains inside the cluster for whole time, (b) point p moves outside the cluster after some time, (c) point p returns inside the cluster from outside, (d) point p starts from inside and moves outside the cluster then returns back to cluster.

In this process, we compare clusters in database snapshots of consecutive time instances and if clusters exist for at least δ time stamps with minimum μ members then that cluster is identified as flock.

In mid-start approach, we check whether clusters with same members exist in first and last snapshot, if they exist then instead of checking sequentially from second snapshot, we check from middle snapshot and then move outwards in both sides, i.e. $s + i$ and $\delta + s - i$ where s is starting snapshot and i ϵ (1, 2, ..$s + \delta/2$).

We use the method of contradiction to identify the flock. To this end we find the time instant where clusters fail to exist for size $\geq \mu$. Once we compare the first and last snapshot and if they have common clusters then we know highest probability that cluster will exist, will be in second and penultimate database snapshot. We can generalize the statement: If, for snapshot $s + i$ and $\delta + s - i$, common clusters exist then highest probability that same clusters would exist is in $s + i + 1$ and $\delta + s - i - 1$. On the contrary, lowest probability of occurrence of same cluster is in $s + \delta/2$. Our approach uses this fact to reduce the total δ comparisons. After comparing the first and last database snapshot, we start comparison from the middle snapshot, $s + \delta/2$, and proceed outwards. Since probability of failure of flock existence is in middle and reduces outwards, we follow the same path and achieve faster decision whether flock exists as discussed in Sect. 3.2.

In Fig. 6a, both Top Down Evaluation(TDE) technique [23] and mid-start approach would compare all δ database snapshots. For the cases in Fig. 6b, c both these approaches make only 2 comparisons. But for case shown in Fig. 6d, TDE approach compares sequentially from start to the time where clusters of size $\geq \mu$ fail to exist. On the contrary, our proposed mid-start approach starts from middle and compares outwards, hence it reaches the point of flock failure faster than TDE approach. Algorithm 2 shows the details of implementation of mid-start approach.

Input : Clusters C of database snapshots from i to $\delta + i$;
Result: Flock F
initialize start = i;
end = $\delta + i$;
if *end-start=0* **then**
| return F
else
| **for** $c_e in C_{end}$ **do**
| | **for** $c_s in C_{start}$ **do**
| | | $F_I = F_I \cup (c_e \cap c_s)$
| | **end**
| **end**
| **if** *end-start=1* **then**
| | return F_I;
| **else**
| | F_{IN}=mid-start(C, i+1);
| | **if** F_{IN} *is empty* **then**
| | | return F_{IN}
| | **end**
| | **for** $c_n in F_{IN}$ **do**
| | | **for** $c_i in F_I$ **do**
| | | | F= F ;
| | | | $\cup (c_n \cap c_i)$
| | | **end**
| | **end**
| | return F;
| **end**
end

Algorithm 2. Mid-start Algorithm

There exist only one case when TDE outperforms mid-start approach which is, say x is length of time traversed by point p from time s till it goes out of cluster. Let y be time spent outside cluster and z be the time in cluster but after return from outside till $s + \delta$ time.

$$midpoint = s + \delta/2 = s + (x + y + z)/2$$
$$if z > x + y \text{ and } 2 * x < midpoint$$

then TDE performs better than mid-start approach. In all other cases, mid-start approach gives better results. We address these cases in our implementation.

4.3 Parallelization of Triangular Grid

We scan the triangular grid using plane sweep approach [19] as shown in Fig. 7. We adopt the approach similar to McKenney et al. [18]. We divide the data set spatially into parts equal to number of Logical processors. Plane sweeping approach is used in each part in parallel, to find clusters. The result is then

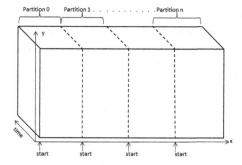

Fig. 7. Plane sweeping approach used to parallelize the triangular grid approach

merged and mid-start approach is applied to identify flocks. We have used 2 cores with 4 Logical Processors on Intel(R) Core(TM) i5-6200U CPU @2.30 GHz.

4.4 Experimental Evaluation

The efficiency of proposed approach is tested against the square grid and TDE approach used by M. Vieira et al. [23]. The evaluation is done on 2 data sets of 2 different types. Dense data set is used for showing the efficiency of triangular grid used in our approach over the square grid for range query. A Comparatively sparse data set is employed to check whether computation cost increases in this case.

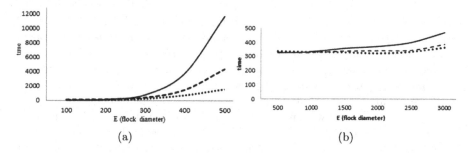

(a)　　　　　　　　　　　　　　　(b)

Fig. 8. Total time to identify flocks as flock diameter ϵ increases in (a) Iceberg dataset (b) Trucks dataset. Results of square grid approach(——), triangular grid(- - - - -), paralleling with triangular grid(·········)

Figure 8a, b show results when varying flock diameters(ϵ) for different data sets. Both plots show time. in seconds, needed to process the data set. It is observed, as ϵ increases, in all 3 methods, time required to process flocks increase. Also, up to some point of ϵ, k, approach discussed in Vieira et al. [23] gives better results but as ϵ increases further, the results using triangular approach outperforms the square grid approach. This is because, in our approach there is some time consumed in building triangular grid which is more when compared to building square grid. This lowers the performance for $\epsilon \leq k$. But once the triangular grid is built, search space for range query decreases and performance increases ($\epsilon > k$). This increase in performance is significant for dense data sets as compared to sparse data sets. Paralleling our approach improves the our results further.

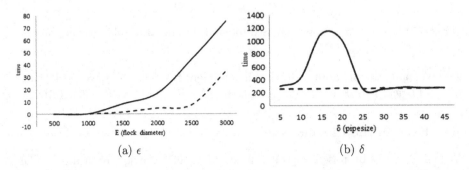

(a) ϵ (b) δ

Fig. 9. Total taken by trucks data to form flocks using TDE approach and mid-start approach. Results of TDE approach(——) and mid-start approach(- - - - -). (a) effect of increasing flock diameter on time taken to identify flocks (b) effect of increasing pipeline size on time to identify flocks

Figure 9a,b shows time required to form flock using TDE approach and mid-start approach when parameters ϵ, δ are increased on trucks data set with 50 database snapshots. In Fig. 9a, it is observed that as ϵ increases, time taken by TDE is more than time taken by mid-start approach. On other hand, when pipeline size (δ) is used, Fig. 9b, it is observed that time taken by TDE increases, but as pipeline spans entire data set then time decreases. In mid-start approach, time taken is much less and it remains somewhat constant throughout.

As discussed earlier, we have collected and used human movement data for our analysis. Figures 10a, b show the results for the same. It is observed that, as flock diameter ϵ increases, time required to process the flocks increases. But this increase is significant as compared to the approach used by Vieira et al. [23], using square grid. Triangular grid with mid-start approach gives efficient results. Paralleling our approach decreases our time requirement further. Similar results are obtained when δ is incremented. Triangular grid with mid-start approach gives better results as compared to square grid and TDE approach. Implementation details and code are available [2].

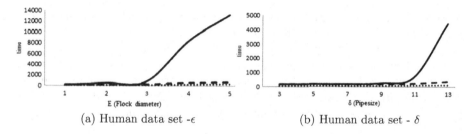

(a) Human data set -ϵ (b) Human data set - δ

Fig. 10. Total time to identify flocks in Human data set as flock diameter ϵ increases. Results of square grid approach(——), triangular grid(- - - - -), paralleling with triangular grid(·········). (a) effect of increasing flock diameter on time taken to identify flocks (b) effect of increasing pipeline size on time to identify flocks

5 Discussion

We have experimented with our approach using different datasets that represent multiple application domains. While clustering with human movement data assists finding activity groups and travel patterns in society, the iceberg dataset analysis is key to our understanding of the combined effect of wind and ocean currents. Flock identification using truck data on the other hand provides an insight into vehicle movement patterns and traffic routines.

Clearly these are very different application domains that have similar spatio-temporal analysis tasks. Availability of spatio-temporal data feeds are increasing in each of these fields and also in others but more importantly, with higher frequency of data turnaround. Whereas the need for faster and better algorithms grows, it is arguable that flock detection in more dense spatial data is required. In this context it is important to revisit the need to consider the heterogeneity of density across the spatial and temporal dimensions [6]. The notion of flocks identified from disks in Vieira et al. [23] clearly focuses on the spatial density (i.e., number of points within the disk). Our approach appears to perform better when the spatial density of points is higher (the iceberg data set has a greater number of points in disks). However, temporal density variation is probably better understood in a convoy detection [14].

Here it is relevant to argue that our approach can be generally applied to other types of patterns including swarms and convoys. For example, reduction of search space using triangular grids can be shown to be advantageous for swarm detection on sliding windows [9] and distributed convoy detection [20]. Similarly, the parallelization of the clustering algorithm can be employed in most approaches. In fact, Tanaka et al. [21] show that the efficiency of their flock detection was improved by employing plane sweep and inverse indexing. In our approach we have combined multiple strategies to improve both the performance and efficiency of flock detection. Trajectories often have different characteristics and the ability to process smaller ones along with long or complicated ones at run-time remains an inherent challenge with large Spatio-Temporal data.

It would be helpful to analyze our multi-strategy approach to evaluate both performance and efficiency with respect to such variability.

Since we wanted to ensure our approach is available for generic cases, we have handled the case where objects do not report locations at the same time. This is done during data pre-processing step. Location for a required time t_b is identified from the two locations at closest times given, t_a and t_c, such that $t_a < t_b < t_c$. It is assumed that point travels in straight line in time from t_a to t_c.

6   Conclusion and Future Work

We have shown that flock detection can be done more efficiently and faster by ensuring a reduced search space. Furthermore, we have shown that by using mid-start approach we can reduce the number of database snapshot comparisons required to identify flocks in a given trajectory dataset.

Our implementation of these improvements to flock detection and their evaluation confirm that our approaches are faster and arguably more efficient in identifying flocks.

The parallelization of the algorithm is seen to drastically reduce the time required to identify the flocks, especially in dense datasets. However, this is only our first effort that employ a parallel plane sweep. We plan to investigate the use of GPU architectures to utilize parallel processing in a more effective manner [12]. We also plan to use further heuristics in our approach to obtain significant flocks in the data set early. In particular, we plan to parallelize the Mid-start algorithm by processing clusters in parallel [15].

References

1. Dataset details. http://www.chorochronos.org/Default.aspx?tabid=71&iditem=31. Accessed 04 Mar 2019
2. Github - onlinereview/flockidentification: Identification of flocks. https://github.com/OnlineReview/FlockIdentification. Accessed 04 Aug 2019
3. Index of /pub/datasets/noaa/g00807. ftp://sidads.colorado.edu/pub/DATASETS/NOAA/G00807. Accessed 04 Mar 2019
4. KML Tutorial—Keyhole Markup Language—Google Developers. https://media.readthedocs.org/pdf/fastkml/latest/fastkml.pdf. Accessed 04 Mar 2019
5. KML Tutorial—Keyhole Markup Language—Google Developers. https://developers.google.com/kml/documentation/kml_tut. Accessed 04 Mar 2019
6. Atluri, G., Karpatne, A., Kumar, V.: Spatio-temporal data mining: a survey of problems and methods. ACM Comput. Surv. (CSUR) **51**(4), 83 (2018)
7. Babcock, B., Babu, S., Datar, M., Motwani, R., Widom, J.: Models and issues in data stream systems. In: Proceedings of the Twenty-First ACM SIGMOD-SIGACT-SIGART Symposium on Principles of Database Systems, pp. 1–16. ACM (2002)
8. Benkert, M., Gudmundsson, J., Hübner, F., Wolle, T.: Reporting flock patterns. In: Azar, Y., Erlebach, T. (eds.) ESA 2006. LNCS, vol. 4168, pp. 660–671. Springer, Heidelberg (2006). https://doi.org/10.1007/11841036_59

9. Bhushan, A., Bellur, U.: Mining swarms from moving object data streams. In: Sarda, N.L., Acharya, P.S., Sen, S. (eds.) Geospatial Infrastructure, Applications and Technologies: India Case Studies, pp. 271–284. Springer, Singapore (2018). https://doi.org/10.1007/978-981-13-2330-0_20

10. Bhushan, A., Bellur, U., Sharma, K., Deshpande, S., Sarda, N.L.: Mining swarm patterns in sliding windows over moving object data streams. In: Proceedings of the 25th ACM SIGSPATIAL International Conference on Advances in Geographic Information Systems. SIGSPATIAL 2017, pp. 60:1–60:4. ACM, New York (2017). https://doi.org/10.1145/3139958.3139988

11. Cao, Y., Zhu, J., Gao, F.: An algorithm for mining moving flock patterns from pedestrian trajectories. In: Morishima, A., et al. (eds.) APWeb 2016. LNCS, vol. 9865, pp. 310–321. Springer, Cham (2016). https://doi.org/10.1007/978-3-319-45835-9_27

12. Diez, Y., Fort, M., Korman, M., Sellarès, J.: Group evolution patterns in running races. Inf. Sci. **479**, 20–39 (2019)

13. Jeung, H., Yiu, M.L., Zhou, X., Jensen, C.S.: Path prediction and predictive range querying in road network databases. VLDB J. **19**(4), 585–602 (2010)

14. Jeung, H., Yiu, M.L., Zhou, X., Jensen, C.S., Shen, H.T.: Discovery of convoys in trajectory databases. Proc. VLDB Endow. **1**(1), 1068–1080 (2008)

15. Khezerlou, A.V., Zhou, X., Li, L., Shafiq, Z., Liu, A.X., Zhang, F.: A traffic flow approach to early detection of gathering events: comprehensive results. ACM Trans. Intell. Syst. Technol. (TIST) **8**(6), 74 (2017)

16. van Kreveld, M., Loffler, M., Staals, F., Wiratma, L.: A refined definition for groups of moving entities and its computation. Int. J. Comput. Geom. Appl. **28**(02), 181–196 (2018)

17. Li, Z., Ding, B., Han, J., Kays, R.: Swarm: mining relaxed temporal moving object clusters. Proc. VLDB Endow. **3**(1–2), 723–734 (2010). https://doi.org/10.14778/1920841.1920934

18. McKenney, M., Frye, R., Dellamano, M., Anderson, K., Harris, J.: Multi-core parallelism for plane sweep algorithms as a foundation for GIS operations. GeoInformatica **21**(1), 151–174 (2017)

19. Nievergelt, J., Preparata, F.P.: Plane-sweep algorithms for intersecting geometric figures. Commun. ACM **25**(10), 739–747 (1982)

20. Orakzai, F., Calders, T., Pedersen, T.B.: Distributed convoy pattern mining. In: 2016 17th IEEE International Conference on Mobile Data Management (MDM), vol. 1, pp. 122–131. IEEE (2016)

21. Tanaka, P.S., Vieira, M.R., Kaster, D.S.: An improved base algorithm for online discovery of flock patterns in trajectories. J. Inf. Data Manag. **7**(1), 52 (2016)

22. Tanuja, V., Govindarajulu, P.: A survey on trajectory data mining. Int. J. Comput. Sci. Secur. (IJCSS) **10**(5), 195 (2016)

23. Vieira, M.R., Bakalov, P., Tsotras, V.J.: On-line discovery of flock patterns in spatio-temporal data. In: Proceedings of the 17th ACM SIGSPATIAL International Conference on Advances in Geographic Information Systems, pp. 286–295. ACM (2009)

24. Yuan, G., Sun, P., Zhao, J., Li, D., Wang, C.: A review of moving object trajectory clustering algorithms. Artif. Intell. Rev. **47**(1), 123–144 (2017)

Local Temporal Compression
for (Globally) Evolving Spatial Surfaces

Xu Teng, Prabin Giri, Matthew Dwyer, Jidong Sun, and Goce Trajcevski[(✉)]

Department of Electrical and Computer Engineering, Iowa State University,
Ames, IA 50014, USA
{xuteng,pgiri,dwyer,jidongs,gocet25}@iastate.edu

Abstract. The advances in the Internet of Things (IoT) paradigm have enabled generation of large volumes of data from multiple domains, capturing the evolution of various physical and social phenomena of interest. One of the consequences of such enormous data generation is that it needs to be stored, processed and queried – along with having the answers presented in an intuitive manner. A number of techniques have been proposed to alleviate the impact of the sheer volume of the data on the storage and processing overheads, along with bandwidth consumption – and, among them, the most dominant is compression. In this paper, we consider a setting in which multiple geographically dispersed data sources are generating data streams – however, the values from the discrete locations are used to construct a representation of continuous (time-evolving) surface. We have used different compression techniques to reduce the size of the raw measurements in each location, and we analyzed the impact of the compression on the quality of approximating the evolution of the shapes corresponding to a particular phenomenon. Specifically, we use the data from discrete locations to construct a TIN (triangulated irregular networks), which evolves over time as the measurements in each locations change. To analyze the global impact of the different compression techniques that are applied locally, we used different surface distance functions between raw-data TINs and compressed data TINs. We provide detailed discussions based on our experimental observations regarding the corresponding *(compression method, distance function)* pairs.

Keywords: Location-aware time series · Triangulated Irregular Network (TIN) · Surface distance · Time series data

1 Introduction and Motivation

The inter-connectivity and collaboration of multiple heterogeneous smart objects enabled by the Internet of Things (IoT) [41] have spurred a plethora of novel

X. Teng—Research supported by NSF grant III 1823267.

P. Giri—Research supported by NSF grant CNS 182367.

J. Sun—Research supported by NSF-REU grant 018522

G. Trajcevski—Research supported by NSF grants III-1823279 and CNS-1823267.

© Springer Nature Switzerland AG 2019
S. Madria et al. (Eds.): BDA 2019, LNCS 11932, pp. 324–340, 2019.
https://doi.org/10.1007/978-3-030-37188-3_19

applications – from smart homes [28], through personalized health care [9] and intelligent transportation system [32], to smart cities [42] and precision agriculture [11]. The multitude of sensors in the devices that define individual smart objects—be it personal (e.g., smart phone and other wearable devices with GPS features) [44] or public (e.g., roadside sensors and traffic cameras) [20] – enable the generation of unprecedented volumes of data which, in turn, provides opportunities for performing variety of analytics, prediction and recommendation tasks integrating variety of sources and contexts [10].

Most, if not all, of the data values are associated with a time-stamp indicating the instant of time at which are particular value was detected. This, in turn, allows for perceiving the data as a time series [40], or even casting it as multidimensional time series [33].

Part of the motivation for this work stems from the traditional and everpresent problem when dealing with Big Data: the Volume. The most common approach to enable storage savings; faster execution time; and saving the bandwidth consumption is to rely on some form of data compression [31]. This topic has been well studied for time-series data [22] and spatio-temporal data [4,37] and many techniques have been proposed in the literature. However, there is another part of the motivation for this work – namely, in many practical applications (e.g., the ones that depend on participatory sensing [17]) – it is often the case that:

- The data is obtained from discrete sources (e.g., measuring carbon footprint or measuring precipitation at given locations/stations [29,43]).
- However, the data is used to "generate" a spatial surface that can be used to represent a continuous distribution of the phenomena of interest over the entire domain (e.g., geo-space).

Hence, the problem that we studied in this work can be succinctly stated as: *How is a spatial shape representing a continuous phenomenon based on values from discrete locations, affected by compressing the corresponding time series with the individually sensed values at each location.*

Towards that, we conducted a series of experiments that were aiming at:

1. Using different compression techniques for time series in order to generate a more compact representation.
2. Using different distance functions to asses the difference between the surfaces obtained from the raw (i.e., uncompressed) time series.
3. Compare the impact of a particular compression technique on a particular distance function.

To our knowledge, this is the first work to systematically address the impact that the compression of location-based time series has on the surface obtained from the discrete set of values from the corresponding locations.

1.1 Organization of the Paper

The rest of this paper is organized as follows: Sect. 2 provided the necessary background and describes the main settings of the problem. Section 3 presents the details of the methodologies that we used Sect. 4 gives a detailed presentation of our experimental observations, along with a discussion of the results, and comparison of advantages and disadvantages of particular involved approach. We conclude the paper in Sect. 5.

2   Preliminaries

We now provide the background for the two main (and complementary) aspects of this work, related to time series comparission and spatial surfaces representation.

2.1   Compressing Time Series Data

A *time series* typically corresponds to a sequence of values $\{t_1, t_2, \ldots, t_n\}$ where each t_i can be perceived as the i-th measurement of a (value of a) particular phenomenon. Often times, the values are assumed to be taken at equi-distant time instants and a time series database is a collection $\{T_1, T_2, \ldots, T_k\}$ where each T_j is a time series – $T_j = \{t_{j1}, t_{j2}, \ldots, t_{jn}\}$.

Time series have attracted a lot of research interest in the past 2–3 decades due to their relevance for a plethora of application domains: from economy and business (stock market, trends detection, economic forecasting), through scientific databases (observations and simulations) databases, to medicine (EEG, gene expressions analysis), environmental data (air quality, hydrology), etc. [15,19]. As a consequence, a large body of works have emerged, targeting problems broadly related to querying and mining (i.e., clustering, classification, motif-discovery) of such data [13,23,26]. One typical feature of the time series databases is that they are very large and, as such, any kind of retrieval may suffer intolerable delays for practical use. Towards that, one would prefer to use the traditional *filter + refine* approach, where the filtering stage uses some kind of an index to prune as much data as possible, without introducing false negatives. However, individual time series also tend to be large – thus, attempting to index them as points in n-dimensional space creates problems in the sense of "dimensionality curse". Hence, one of the first data reduction objectives in time series was to reduce the dimensionality and then use spatial access methods to index the data in the transformed space [30]. The list of desirable properties of an indexing scheme introduced in [15] are:

- It should be much faster than sequential scanning.
- The method should require little space overhead.
- The method should be able to handle queries of various lengths.
- The method should allow insertions and deletions without requiring the index to be rebuilt (from the scratch).
- It should be correct, i.e. there should be no false dismissals.

The list was augmented by two more desiderata in [24]:

- It should be possible to build the index in "reasonable time".
- The index should be able to handle different distance measures, where appropriate.

The notion of data reduction for the purpose of indexing in the context of querying (e.g., similarity search) and mining time series data falls into the category of the, so called, *representation methods*. Essentially, a representation method attempts to reduce the dimensionality of the data, while not loosing the essential characteristics of a given "shape" that it represents – and there are two basic kinds:

- Data Adaptive – where a common representation is chosen for all items in the database, in a manner that minimizes the global reconstruction error.
- Non-Data Adaptive – which exploit local properties of the data, and construct an approximate representation accordingly.

As it turned out, an important property of any representation is the one of being able to have a lower-bound when conducting the search, which would ensure the absence of false negatives/dismissals induced by the pruning [15]).

However, there is another notion brought about in the time series literature which has influenced works in clustering, mining and compressing trajectories' data – namely, the *similarity measure* (equivalently, *distance measure*). Similarity measures aim at formalizing the intuition behind assessing how (diss)similar are two series. More formally, for two time series T_1 and T_2, a similarity function $Dist$ calculates the distance between the two time series, denoted by $Dist(T_1, T_2)$, and the desirable properties that $Dist(T_1, T_2)$ should include (cf. [13]) are:

- Provide a recognition of perceptually similar objects, even though they are not mathematically identical.
- Be consistent with human intuition.
- Emphasize the most salient features on both local and global scales.
- Be universal in the sense that it allows to identify or distinguish arbitrary objects, that is, no restrictions on time series are assumed.
- Abstract from distortions and be invariant to a set of transformations.

While some of the desiderata above may be favored over the others for a particular application domain, another categorization of similarity measures, based more on the way that they treat the matching points of the two series (cf. [40]) can be specified as:

- *Lock-step measures* – the distance measures that compare the i–th point of one time series to the i–th point of another, such as the Euclidean distance and the other Lp norms.
- *Elastic measures* – ones allowing a comparison of one-to-many points (e.g., DTW) and one-to-many/one-to-none points (e.g., LCSS).

To cater to the desirable features of the distance measures, one feature was to enable time warping in the similarity computation. A well known example – DTW (Dynamic Time Warping) distance [40] is used to allow a time series to be "stretched" or "compressed" to provide a better match with another time series (i.e., a "one-to-many" mapping of the data points is allowed for as long as each data point from one series is matched to a data points from another).

In addition to the lock-step and elastic measures, other distance functions have been introduced, motivated by a particular application context. Thus, for example, a group of measures has been developed based on the *edit distance* for strings – e.g., LCSS (*longest common subsequence*) [39] which introduced a *threshold parameter* ε specifying that two points from two time series are considered to match if their distance is less than ε. Other examples include ERP distance [7] which combines the features of DTW and EDR, by introducing the concept of a *constant reference point* for computing the distance between gaps of two time series; SpADe [8], which is a pattern-based similarity measure for time series; etc.

2.2   Triangulated Irregular Networks

The main rationalé behind using TIN is two-fold:

1. They are the most popular method for building a surface from a set of irregularly spaced points [27].
2. They enable focusing on small details in highly variable input feature [14]

TIN is a representation of choice whenever a surface can be constructed from a collection of non overlapping surfaces having triangular facets [27]. In addition, TINs are capable of preserving multiple resolutions [3].

As data structure, they consist of set of vertices (x_i, y_i, z_i) that originate in 2D triangles (the (x_i, y_i) projection), and have a vertical component z_i corresponding to a value measured at (x_i, y_i). An illustration of the 2D collection of triangles and the corresponding TIN is provided in Figs. 1 and 2, respectively.

One can also perceive TINs as a special case of Digital Elevation Model which have the surface of the triangular mesh – i.e., a set of T triangles for the finite set of points S (cf. [16]), that satisfies the following three conditions:

- S are the set of vertices of T.
- Interior angles of any two triangles cannot intersect.
- If boundaries of triangle are intersected, then it should be common edge or vertex.

There are several algorithmic solutions for constructing TIN surfaces[1] and the most popular one is based on Delanuay triangulation. A distinct property of Delaunay triangulation of a given set of planar points S is that for any triplet of points $s_i, s_j, s_k (\in S)$ that are selected to form a triangle, there will

---

[1] The very first implementation dating back to 1973s, due to W. Randolph Franklin.

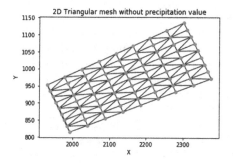

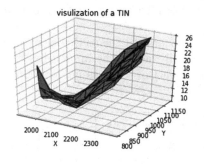

Fig. 1. 2D representation surface in TIN

Fig. 2. 3D TIN with precipitation value

not exist a point $s_m \in S$ that will be in the interior of the circumscribed circle of $\Delta(s_i, s_j, s_k)$ [14]. Equivalently, the Delaunay triangulation maximizes the minimum angle of all the angles of triangles. An illustration of Delaunay triangluation for constructing TIN is shown in Fig. 3 (cf. [27]).

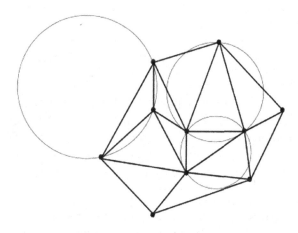

Fig. 3. Delaunay triangulation for constructing TIN

If S is the number of vertices (in our application domain, corresponding to locations of the weather stations) and b is the number of vertices on the boundary of the convex hull of all the points in S, the maximum number of triangles obtained by Delaunay triangulation would be:

$$Number\ of\ Triangles = 2 \times S - b - 2 \tag{1}$$

3   Methodology of Comparative Study

We now present the methodology used for the comparative analysis, and discuss in detail the specific approaches that we used for compression and evaluating the impact of compressing location-based time series on the global TIN.

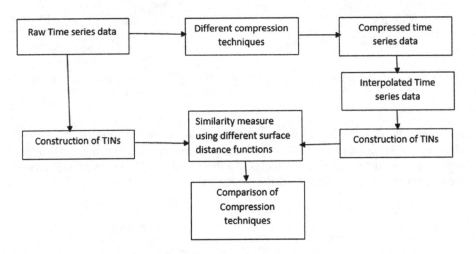

Fig. 4. Flowchart of comparative evaluations

The workflow used is illustrated in Fig. 4. We first collected the data set of locations and measurements in each location from around the globe and categorized it into 50 different clusters based on their latitude and longitude values. The raw data was used to generate Delaunay triangulation and subsequently the TIN – i.e., a collection (or, "time series") of TINs for the time instants of each measurement in each location.

Next, we applied a compression to each of the time series containing the measurement values for the respective location and proceeded with constructing a new collection of TINs. However, due to the compression, certain points in the original time series may not be present. For such points we are using linear interpolation to generate the z-value (i.e., a measurement).

Finally, we compared the TINs at each time instant before the compression and TINs at each time instant after the compression.

3.1   Compression Techniques

In this work we used five different compression approaches which belong to two broad categories of time series compression. They consist of two dimensionality reduction techniques (Piecewise Aggregate Approximation (PAA) and Discrete Fourier Transform (DFT)), and three more native-space compression methods. The details follow

Piecewise Aggregate Approximation: The main idea of Piecewise Aggregate Approximation (PAA) [21] is to divide the original time series into N equal-size frames, where N is the desired dimensionality, and use the average value of all the data in each frame to represent each window. Mathematically, the formula of using PAA over n-dimensional time series to compress it into N dimensionality is shown in Eq. 2:

$$\bar{t}_i = \frac{N}{n} \sum_{j=\frac{n}{N}(i-1)+1}^{\frac{n}{N}i} t_j, i = 1, 2, ..., N \tag{2}$$

There is always a trade-off between compression ratio and information preservation – one extreme case would be selecting $N = n$ and the compressed representation would be identical to the original time series.

Discrete Fourier Transform: Discrete Fourier Transform (DFT) [2] is a widely-used method to find the spectrum of the finite domain signal. In theory, any n-length time series can be transformed into the frequency domain with equal number of sine and cosine waves associated with corresponding amplitudes, which enables us to inverse the transform and reconstruct the original time series. In such sense, to represent the original time series into N dimensionality, only keeping the waves with the largest N amplitudes would be a possible solution without losing too much significant information.

(Adapted) Douglas-Peucker Algorithm: Douglas-Peucker (DP) [12] algorithm is well-known for its capability of reducing the number of points while keeping the outline shape of original data. Given tolerance threshold ε, the steps of DP algorithm are as follows:

1. Construct a line segment by connecting the initiator (first point initially) and terminus (last point initially)
2. Find the "anchor" having the largest distance from the line segment and use it as the new terminus of its left part and new initiator of its right part
3. "Anchor" divides the line segment at step.1 into 2 parts and then repeat step.1 and step.2 until the largest distance in any line segment is less than ε

DP algorithm is a classical polylines compression approach. [36] adapts it into time series compression technique by considering vertical distance in stead of perpendicular. Formally, the vertical distance between a point t_k and line segment (t_i, t_j), $i < k < j$, is calculated by $|t'_k - t_k|$, where t'_k is the intersection of the line segment and the line passing through t_k and perpendicular to time-axis.

Visvalingam-Whyatt Algorithm: "Effective area" is the key concept behind Visvalingam-Whyatt (VW) [38] algorithm, which represents the area of the triangle constructed by a point with its two neighbors. Given a sequence of time series and a error tolerance ε, the algorithm would iteratively drop the middle point of the triangle with the smallest "effective area" and updating the triangles related to that removed point until the "effective area" of any triangle is larger than ε.

(Adapted) Optimal Algorithm: Optimal (OPT) [5] algorithm considers both direction of every time series points; forward and backward. For a time series point $t_i, t_{i+1}, t_{i+2}, ..., t_n$ can be forward points and $t_{i-1}, t_{i-2}, ..., t_1$ is the backward. Here, ith pass of the algorithm draws the circle centered in every forward and backward points with the radius of ϵ. When a new point t_k in forward is being touched, such that k is $i < k \leq n$. Let t_i generates U_k and L_k as the upper and lower spectrum which defines the wedge that is related to point t_k and apex at t_i, while passing through the top and bottom of formed circle centered at point t_k. Highest and lower boundary will be maintained until the intersection of wedges is not empty and if the intersection is empty, denote the point t_k which makes the intersection empty. And then store t_i and t_{k-1} in result and repeat the steps from event point t_{k-1} to forwards and do similar for the backward part.

3.2   Distance Function

We now discuss the distance functions that we used to assess the $|TIN_{raw} - TIN_{compressed}|$, for each of the compression methods.

Hausdorff Distance: Haudsorff distance is a min-max distance measure which defines the property of similarity between two surfaces based on the positions. Hence, it is widely used as a measure of the degree of resemblance between two objects [18].

Mathematically [34], for given two set of finite points A and B, such that $A = a_1, a_2....a_n$ and $B = b_1, b_2....b_n$, Hausdorff's distance is defined as

$$H(A, B) = \max(h(A, B), h(B, A)), \tag{3}$$

where

$$h(A, B) = \max_{a \in A}(\min_{b \in B}(d(a, b))) \tag{4}$$

and

$$h(B, A) = \max_{b \in B}(\min_{a \in A}(d(b, a))) \tag{5}$$

We are going to compare the interpolated surface with the raw/original surface by finding the Hausdorff distance between them.

Volume Based Distance: The second distance function that we used is based on comparing the volumes of the TINs obtained from the original/raw time series and the TINs constructed for each time instant after compression. Volume similarity measure has been widely used as one of the standard techniques to measure the similarity between segments [35].

Each triangle of the TINs are considered as truncated triangular prism having unequal heights at a particular time. Each height is the precipitation value of the corresponding vertex in the base (i.e., coordinates of the weather stations). Figure 5 illustrates a truncated prism as a component of the TIN.

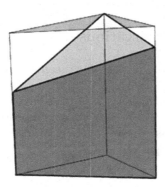

Fig. 5. Truncated triangular prism [25]

Mathematically, for a given four vertices of tetrahedron a, b, c, d, the volume is defined as:

For a given prism with a base consisting of a triplet of vertices a, b, c, and a height C, the volume is:

$$V = \frac{1}{2}|(\boldsymbol{ab} \times \boldsymbol{ac})| \cdot C \tag{6}$$

In our settings, C corresponds to the average value of the precipitation recorded in the three weather stations (i.e., $C = (height(a) + height(b) + height(c))/3$), the locations of which constitute the vertices of the triangle. Volume based distance function will show how strongly the original volume differs from interpolated volume after using compression techniques.

Angular Distance: Angular Distance is a metric which corresponds to an inverse of Cosine Similarity. Cosine similarity is used in 3D surfaces to measure the similarity between the perpendicular vectors of two corresponding triangles For a given triangle with vertices a, b, c having coordinates $(x_a, y_a), (x_b, y_b), (x_c, y_c)$, and each with corresponding height (i.e., measurement value) of z_a, z_b and z_c, the normal vector is calculated as:

$$N = (b - a) \times (c - a) \tag{7}$$

The cosine similarity for the normal vector N_1 and N_2 of two triangles is defined as [6]:

$$CS(N_1, N_2) = \frac{N_1 \cdot N_2}{||N_1|| \cdot ||N_2||} \tag{8}$$

Based on this, their Angular distance can be measured by cos-inverse of cosine similarity [6].

$$Angular\,Distance(N_1, N_2) = \frac{cos^{-1}(CS(N_1, N_2))}{\pi}. \tag{9}$$

4   Experimental Observations

We now present the details of our experimental observations, based on the methodology described in Sect. 3.

We note that, for reproducibility, both the source code of all the compression methods and distance functions used in the experiments, along with the datasets (before and after compression), are publicly available at https://github.com/ XTRunner/Compression_Spatial_Surface.

4.1   Dataset Description

For the study, we took precipitation measurements of different weather stations across the globe [1]. Due to the geographic dispersion – i.e., having subsets of spatially co-located input points that were significantly far from other subsets of such points, we grouped the input location data into fifty clusters. The number of weather stations with precipitation measurements range from 40 to 81 across the clusters. Each location contains a time series corresponding to 50 years of monthly precipitation recordings. During the construction of the Delaunay triangulation, we converted the (*latitude, longitude*) values of the weather stations' locations into (x, y) (i.e., Cartesian) ones using ECEF (Earth Centered, Earth Fixed) methodology.

4.2   Setting of Parameters

We used multiple values for the parameters to ensure reliability and validity of our observations.

For a given dataset D represented by β_D bits, let $\mathcal{C}(D)$ denote its compressed version obtained by applying a particular compression function $\mathcal{C}$. Assume that the size of $\mathcal{C}(D)$ is $\beta_{\mathcal{C}(D)}$ bits. Then the compression ratio of $\mathcal{C}$ on D is calculated as $\mathcal{R}_{\mathcal{C}}(D) = \frac{\beta_D}{\beta_{\mathcal{C}(D)}}$.

In the experiments, the compression ratios for both DFT and PAA were set to $[10, 20/3, 5, 4, 10/3, 2]$. The rationalé is that: (a) if the compression ratio is too low, data will not be compressed much; (b) if the compression ratio is very high, it might lead to an increased loss of fine details (i.e., information).

The native-domain compression techniques (DP, VW and OPT) were implemented in such a way that they could meet a particular error tolerance. Error tolerance values were defined in a manner that ensures they are most comparable to PAA and DFT in term of compression ratio. The values for the error tolerances used in DP, VW and OPT were $[15, 25, 35, 50, 65, 80]$.

4.3   Observations

We firstly present a high-level observation regarding the impact of the compression. Namely, Fig. 6 shows the TIN corresponding to a particular cluster obtained from the raw data at a randomly chosen time instant. For comparison, in Fig. 7 we show the same cluster and at the same time instant – however, the values of the time series corresponding to the locations (i.e., vertices of Delaunay triangulation) correspond to the ones after interpolation has been applied to the compressed ones, obtained using DFT compression (z-axis indicates the precipitation values).

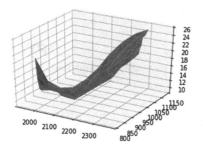

Fig. 6. 3D raw data representation of fifteenth cluster

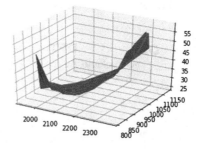

Fig. 7. 3D interpolated data representation of fifteenth cluster after DFT

For each distance measure, we firstly compute the maximum and mean value of the difference between original TINs and compressed ones in each cluster and across all the time instants. To present the results in a more general way, the average of the maximum and mean value among all the geo-clusters are calculated.

Figure 8 illustrates the effectiveness of different compression techniques in terms of Hausdorff distance measurement. Note that the x-axis represents the $1/\mathcal{R}_C(D)$ and the y-axis represents the logarithm (with base 10) value of Hausdorff distance between the original TINs and compressed TINs. From the left side of Fig. 8, which shows the average of all the maximum Hausdorff distance in each cluster, we observe that PAA and DFT algorithms are outperformed by the other three native-domain techniques, especially when the compression ratio is

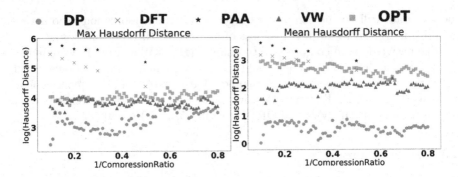

Fig. 8. Results of Hausdorff distance

relatively high. The performances of VW and OPT algorithms are very close to each other. As can be observed, the DP algorithms obtains the best performance when the compression ratio is higher than 2. The right side illustrates another different picture. By using the average of the Hausdorff distance in each cluster as measurement, DP algorithm always achieves the greatest accuracy. Different as the left side, VW algorithm has a distinguishable edge than OPT algorithm. But still, PAA and DFT has the worst performances.

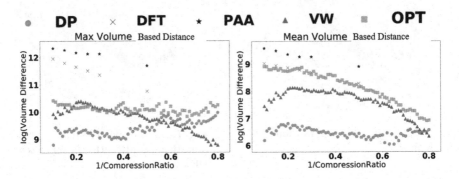

Fig. 9. Results of Volume based distance

In addition to the Hausdorff distance, we also introduce the volume based distance in Sect. 3 to evaluate the performances of different compression techniques. Compared with Figs. 8, 9 presents a different story. From the left figure, we can observe that DP algorithm is not always the best one anymore. When the compression ratio is high, DP algorithm is still able to guarantee the closest result. However, we note that the VW algorithm outperforms the DP algorithm after $1/\mathcal{R}_\mathcal{C}(D)$ is higher than 0.6. Thus, when the required compression ratio is higher than 2, DP algorithm should still be the first choice. But if the accuracy has a higher priority than the compression ratio, VW algorithm can be considered as a compression method of choice. Moreover, we observe that on the right side, the trend is similar to the left, i.e., the performances of VW algorithm is getting closer to DP algorithm and eventually exceeds DP algorithms for compression ratios smaller than 1.25.

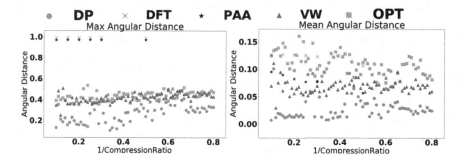

Fig. 10. Results of Angular distance

The third measurement is angular distance of the norm vectors of original and compressed TINs. As shown in the left side of Fig. 10, DP, VW and OPT algorithms have relatively similar results compared with the other two distance functions, especially when the compression ratio is between 1.5 and 2.5. Besides, PAA and DFT algorithms still have the worst performance, while the right side illustrates totally different scenarios. When the compression ratio is higher than 2.5, both PAA and DFT get much better results, which are comparable with VW algorithm and even better than OPT algorithm. And for those three native-domain compression techniques, although the performances are very close for some values of the compression ratio.

The last observation that we report is a single instance of the experiments – with a sole purpose to provide an intuitive illustration for the errors induced by the compression. Specifically, for values of the compression ratio in [0.2, 0.5], we picked the absolute worst-case scenario in terms of the Hausdorff and Volume-based distance between TINs obtained from the raw data and the TINs after the compression has been applied. The results are shown in Fig. 11.

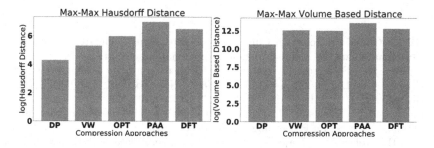

Fig. 11. Worst-case scenarios

5   Concluding Remarks and Future Work

We provided a detailed experimental comparison of the impact that compressing time series data can have when interpreted in a broader context. Specifically, we considered the settings in which each time series data is associated with a discrete location *and* the instantaneous values in each location were used to generate a TIN-based representation of the continuous surface representing a particular phenomenon of interest. When compressing the original/raw data, some of the original measurement values will not be present in the compressed version – and those are obtained by interpolation. However, using the interpolated data in the corresponding locations to generate the TIN, may yield a surface which differs from the one constructed from the raw time series. In this work, we investigated the impact that a particular compression method may have on the "distortion" of the surface, with respect to a particular distance function.

We used five different compression approaches (Discrete Fourier Transform, Piece-wise Aggregrate Approximation, Douglas-Peucker Algorithm, Visvalingam-Whyat Algorithm and Adaptive optimal Algorithm) and two different distance functions (Hausdorff and Volume-based). From among all the combinations of pairs (*compression method, distance function*) we made observations regarding the similarity/difference between the original TIN surfaces and post-compression ones. Our observations indicate, for example, that when it comes to volume-based distances, Douglas Peucker yielded the highest similarity and PAA showed the least similarity. Similarly, when Hausdorff distance was used to calculate the (dis)similarity, we also observed that DP was performing better than the rest. OPT and VW were showing more similarity to raw TINs inc comparison to PAA and DFT, PAA being worst.

As part of our future work, we are expanding the types of compression methods and distance functions used, for the purpose of a more complete classification of the impacts. Another one of our goals is to include applying spatial compression and combining it with the time series data compression, and evaluating the impacts on more heterogeneous datasets (e.g., time series of traffic data). Lastly, we would like to investigate the impact that the compression has on the quality of prediction in time series.

References

1. GPCC: Global Precipitation Climatology Centre. https://climatedataguide.ucar.edu/climate-data/gpcc-global-precipitation-climatology-centre
2. Agrawal, R., Faloutsos, C., Swami, A.: Efficient similarity search in sequence databases. In: Lomet, D.B. (ed.) FODO 1993. LNCS, vol. 730, pp. 69–84. Springer, Heidelberg (1993). https://doi.org/10.1007/3-540-57301-1_5
3. Bertilsson, E., Goswami, P.: Dynamic creation of multi-resolution triangulated irregular network. In: Proceedings of SIGRAD (2016)
4. Cao, H., Wolfson, O., Trajcevski, G.: Spatio-temporal data reduction with deterministic error bounds. VLDB J. **15**(3), 211–228 (2006)
5. Chan, W.S., Chin, F.: Approximation of polygonal curves with minimum number of line segments. Int. J. Comput. Geom. Appl. **6**, 59–77 (1992)

6. Chanwimalueang, T., Mandic, D.: Cosine similarity entropy: self-correlation-based complexity analysis of dynamical systems. Entropy **19**, 652 (2017). https://doi.org/10.3390/e19120652
7. Chen, L., Ng, R.T.: On the marriage of lp-norms and edit distance. In: Proceedings of the Thirtieth International Conference on Very Large Data Bases (VLDB), Toronto, Canada, 31 August– 3 September 2004, pp. 792–803 (2004)
8. Chen, Y., Nascimento, M.A., Ooi, B.C., Tung, A.K.H.: SpADe: on shape-based pattern detection in streaming time series. In: IEEE International Conference on Data Engineering (ICDE) (2007)
9. Cheng, X., Fang, L., Yang, L., Cui, S.: Mobile big data: the fuel for data-driven wireless. IEEE Internet Things J. **4**(5), 1489–1516 (2017)
10. Chudzicki, C., Pritchard, D.E., Chen, Z.: Geosoca: exploiting geographical, social and categorical correlations for point-of-interest recommendations. In: Proceedings of the International Conference On Research and Development in Information Retrieval (SIGIR), pp. 443–452. ACM (2015)
11. Deepika, G., Rajapirian, P.: Wireless sensor network in precision agriculture: a survey. In: 2016 International Conference on Emerging Trends in Engineering, Technology and Science (ICETETS) (2016)
12. Douglas, D.H., Peucker, T.K.: Algorithms for the reduction of the number of points required to represent a digitized line or its caricature. Cartographica: Int. J. Geograph. Inf. Geovisualization **10**, 112–122 (1973)
13. Esling, P., Agon, C.: Time-series data mining. ACM Comput. Surv. **45**, 1 (2012)
14. ESRI: Arcgis desktop help 9.2 - about TIN surfaces (2019)
15. Faloutsos, C., Ranganathan, M., Manolopoulos, Y.: Fast subsequence matching in time-series databases. In: SIGMOD Conference, pp. 419–429 (1994)
16. Floriani, L.D., Magillo, P.: Triangulated irregular network. In: Liu, L., Özsu, M.T. (eds.) Encyclopedia of Database Systems, pp. 3178–3179. Springer, Boston (2009). https://doi.org/10.1007/978-0-387-39940-9_437
17. Gao, H., et al.: A survey of incentive mechanisms for participatory sensing. IEEE Commun. Surv. Tutorials **17**(2), 918–943 (2015)
18. Guo, B., Lam, K.M., Lin, K.H., Siu, W.C.: Human face recognition based on spatially weighted hausdorff distance. Pattern Recogn. Lett. **24**(1), 499–507 (2003)
19. Han, J., Kamber, M.: Data Mining: Concepts and Techniques, 2nd edn. Morgan Kaufmann, Burlington (2012)
20. Jang, J., Kim, H., Cho, H.: Smart roadside server for driver assistance and safety warning: framework and applications. In: Proceedings of the International Conference on Ubiquitous Information Technologies and Applications (2010)
21. Keogh, E., Chakrabarti, K., Pazzani, M., Mehrotra, S.: Dimensionality reduction for fast similarity search in large time series databases. Knowl. Inf. Syst. **3**, 263–286 (2001)
22. Keogh, E., Lonardi, S., Ratanamahatana, C.A., Wei, L., Lee, S.H., Handley, J.: Compression-based data mining of sequential data. Data Min. Knowl. Discov. **14**(1), 99–129 (2007)
23. Keogh, E.J.: A decade of progress in indexing and mining large time series databases. In: VLDB (2006)
24. Keogh, E.J., Chakrabarti, K., Mehrotra, S., Pazzani, M.J.: Locally adaptive dimensionality reduction for indexing large time series databases. In: SIGMOD Conference, pp. 151–162 (2001)
25. Kern, W.F., Bland, J.R.: Solid Mensuration. Wiley/Chapman & Hall, Limited, New York/London (1934)

26. Kotsakos, D., Trajcevski, G., Gunopulos, D., Aggarwal, C.C.: Time-series data clustering. In: Data Clustering: Algorithms and Applications, pp. 357–380 (2013)
27. Liang, S.: Geometric processing and positioning techniques. In: Liang, S., Li, X., Wang, J. (eds.) Advanced Remote Sensing, pp. 33–74. Academic Press, Boston (2012). Chapter 2
28. Maselli, G., Piva, M., Stankovic, J.A.: Adaptive communication for battery-free devices in smart homes. IEEE Internet Things J. **6**, 6977–6988 (2019)
29. Mekis, E., Hogg, W.D.: Rehabilitation and analysis of Canadian daily precipitation time series. Atmos. Ocean **37**(1), 53–85 (2010)
30. Rafiei, D., Mendelzon, A.O.: Similarity-based queries for time series data. In: Proceedings ACM SIGMOD International Conference on Management of Data, SIGMOD 1997, Tucson, Arizona, USA, 13–15 May 1997, pp. 13–25 (1997)
31. ur Rehman, M.H., Liew, C.S., Abbas, A., Jayaraman, P.P., Wah, T.Y., Khan, S.U.: Big data reduction methods: a survey. Data Sci. Eng. **1**(4), 265–284 (2016)
32. Shi, D., et al.: Deep Q-network based route scheduling for TNC vehicles with passengers' location differential privacy. IEEE Internet Things J. **6**, 7681–7692 (2019)
33. Shokoohi-Yekta, M., Wang, J., Keogh, E.J.: On the non-trivial generalization of dynamic time warping to the multi-dimensional case. In: Proceedings of the 2015 SIAM International Conference on Data Mining, pp. 289–297 (2015)
34. Sim, K., Nia, M., Tso, C., Kho, T.: Chapter 34 - brain ventricle detection using hausdorff distance. In: Tran, Q.N., Arabnia, H.R. (eds.) Emerging Trends in Applications and Infrastructures for Computational Biology, Bioinformatics, and Systems Biology. Emerging Trends in Computer Science and Applied Computing, pp. 523–531. Morgan Kaufmann, Boston (2016)
35. Taha, A.A., Hanbury, A.: Metrics for evaluating 3D medical image segmentation: analysis, selection, and tool. BMC Med. Imaging **15**, 29–29 (2015)
36. Teng, X., Züfle, A., Trajcevski, G., Klabjan, D.: Location-awareness in time series compression. In: Benczúr, A., Thalheim, B., Horváth, T. (eds.) ADBIS 2018. LNCS, vol. 11019, pp. 82–95. Springer, Cham (2018). https://doi.org/10.1007/978-3-319-98398-1_6
37. Trajcevski, G.: Compression of spatio-temporal data. In: IEEE 17th International Conference on Mobile Data Management, MDM 2016, 2016 - Workshops, Porto, Portugal, 13–16 June, pp. 4–7 (2016)
38. Visvalingam, M., Whyatt, J.D.: Line generalisation by repeated elimination of points. Cartographic J. **30**, 46–51 (1993)
39. Vlachos, M., Kollios, G., Gunopulos, D.: Elastic translation invariant matching of trajectories. Mach. Learn. **58**(2–3), 301–334 (2005)
40. Wang, X., Mueen, A., Ding, H., Trajcevski, G., Scheuermann, P., Keogh, E.J.: Experimental comparison of representation methods and distance measures for time series data. Data Min. Knowl. Discov. **26**(2), 275–309 (2013)
41. Whitmore, A., Agarwal, A., Xu, L.D.: The Internet of Things: a survey of topics and trends. Inf. Syst. Front. **17**(2), 261–274 (2015)
42. Yao, H., Gao, P., Wang, J., Zhang, P., Jiang, C., Han, Z.: Capsule network assisted IoT traffic classification mechanism for smart cities. IEEE Internet Things J. **6**, 7515–7525 (2019)
43. Yi, W.Y., Lo, K.M., Mak, T., Leung, K.S., Leung, Y., Meng, M.L.: A survey of wireless sensor network based air pollution monitoring systems. Sensors **15**, 31392–31427 (2015)
44. Zhuang, C., Yuan, N.J., Song, R., Xie, X., Ma, Q.: Understanding people lifestyles: construction of urban movement knowledge graph from GPS trajectory. In: International Joint Conference on Artificial Intelligence (IJCAI), pp. 3616–3623 (2017)

An Explicit Relationship Between Sequential Patterns and Their Concise Representations

Hai Duong[1,2], Tin Truong[2(✉)], Bac Le[1], and Philippe Fournier-Viger[3]

[1] VNU-HCMC, University of Science, Ho Chi Minh, Vietnam
lhbac@fit.hcmus.edu.vn
[2] Department of Mathematics and Computer Science,
University of Dalat, Dalat, Vietnam
{haidv, tintc}@dlu.edu.vn
[3] School of Humanities and Social Sciences,
Harbin Institute of Technology (Shenzhen), Shenzhen, China
philfv@hit.edu.cn

Abstract. Mining sequential patterns in a sequence database (SDB) is an important and useful data mining task. Most existing algorithms for performing this task directly mine the set $\mathcal{FS}$ of all frequent sequences in an SDB. However, these algorithms often exhibit poor performance on large SDBs due to the enormous search space and cardinality of $\mathcal{FS}$. In addition, constraint-based mining algorithms relying on this approach must read an SDB again when a constraint is changed by the user. To address this issue, this paper proposes a novel approach for generating $\mathcal{FS}$ from the two sets of frequent closed sequences ($\mathcal{FCS}$) and frequent generator sequences ($\mathcal{FGS}$), which are concise representations of $\mathcal{FS}$. The proposed approach is based on a novel explicit relationship between $\mathcal{FS}$ and these two sets. This relationship is the theoretical basis for a novel efficient algorithm named GFS-CR that directly enumerates $\mathcal{FS}$ from $\mathcal{FCS}$ and $\mathcal{FGS}$ rather than mining them from an SDB. Experimental results show that GFS-CR outperforms state-of-the-art algorithms in terms of runtime and scalability.

Keywords: Sequential pattern mining · Frequent sequence · Generator and closed sequence · Equivalence relation · Partition

1 Introduction

The problem of mining frequent sequences (FS) [1, 13] or sequential patterns plays an important role in data mining and has many real-life applications such as the analysis of web click streams, customer-transaction sequences, medical data, and e-learning data. Mining FS is the task of finding all subsequences satisfying a user-specified minimum support (ms) threshold. Many algorithms have been proposed for discovering the set $\mathcal{FS}$ of all frequent sequences in a sequence database (SDB), such as PrefixSpan [12], SPADE [17], SPAM [4], CM-SPADE [6] and CM-SPAM [6]. Along with mining $\mathcal{FS}$, many researchers have focused on discovering two concise representations of $\mathcal{FS}$, which are the $\mathcal{FCS}$ set of all frequent closed sequences, and the $\mathcal{FGS}$ set of all frequent generator sequences. Some algorithms for discovering $\mathcal{FCS}$ are Clospan [16], BIDE

© Springer Nature Switzerland AG 2019
S. Madria et al. (Eds.): BDA 2019, LNCS 11932, pp. 341–361, 2019.
https://doi.org/10.1007/978-3-030-37188-3_20

[15], ClaSP [9], CM-ClaSP [6] and FCloSM [5], and algorithms for mining $\mathcal{FGS}$ include VGEN [8] and FGenSM [5]. An algorithm for simultaneously mining $\mathcal{FCS}$ and $\mathcal{FGS}$ is FGenCloSM [10]. The main advantage of the sets $\mathcal{FCS}$ and $\mathcal{FGS}$ is that their cardinalities are generally much smaller than that of $\mathcal{FS}$. Thus, they can be mined more efficiently in terms of runtime and memory usage, in big data. Besides, they are lossless and concise representations of $\mathcal{FS}$, i.e., they preserve information about the frequency of all frequent sequences without requiring additional database scans.

Current FS mining algorithms mine the $\mathcal{FS}$ set directly from an SDB. Therefore, for large SDBs and/or low *ms* values, these algorithms often exhibit poor performance because of the extremely large search space and cardinality of $\mathcal{FS}$. In addition, for the problem of mining FSs with constraints, constraint-based mining algorithms (often modified from the above ones) must read the SDB again when the user changes a constraint. This is a drawback when users frequently change constraints.

To address these issues, the current study proposes a novel approach that generates FSs without or with constraints from the sets $\mathcal{FCS}$ and $\mathcal{FGS}$, instead of from an SDB. When an SDB is very large or if a constraint is changed, algorithms based on the proposed approach only need to scan these two sets to generate the result sets. In fact, this approach has been shown to be very effective for the traditional problem of mining frequent itemsets [2, 3, 11, 14]. The authors of these studies proposed efficient algorithms to generate frequent itemsets without and with user-defined constraints from frequent closed itemsets and their generators. Through experiments, the algorithms demonstrated their outstanding advantages compared with state-of-the-art methods for mining frequent itemsets directly from transaction databases. However, to our knowledge, no theoretical results and algorithms have been proposed for the more general problem of mining frequent sequences based on this approach. Note that, this problem is much more challenging than the traditional problem because each sequential pattern can appear multiple times in each input sequence of an SDB. Thus, the following important research question arises.

(*Q*) What is the explicit relationship between $\mathcal{FS}$ and its two concise representation sets $\mathcal{FCS}$ and $\mathcal{FGS}$?

The above question is answered in the current study by proposing novel theoretical results and an algorithm to quickly generate $\mathcal{FS}$ from $\mathcal{FCS}$ and $\mathcal{FGS}$.

The major contributions of this paper are the following. First, we propose an explicit relationship between $\mathcal{FS}$ and the pair $(\mathcal{FCS}, \mathcal{FGS})$ by partitioning $\mathcal{FS}$ into disjoint classes based on an equivalence relation on sequences. This relationship is the theoretical basis not only for designing a method to quickly generate $\mathcal{FS}$ from that pair, but it will also be used for developing novel methods to efficiently mine FSs with constraints. Second, a novel algorithm named GFS-CR is proposed to quickly generate $\mathcal{FS}$ from the sets $\mathcal{FCS}$ and $\mathcal{FGS}$. Third, we show empirical evidence about the correctness of the proposed theoretical results that have been proved mathematically. Besides, the experiments also show that GFS-CR is generally faster than state-of-the-art algorithms that mine $\mathcal{FS}$ directly from an SDB.

The rest of this paper is organized as follows. Section 2 introduces basic concepts related to sequential pattern mining. Sections 3 and 4 present the novel theoretical results and the proposed algorithm, respectively. Section 5 reports experimental results. Finally, Sect. 6 draws a conclusion and discusses future work.

2 Preliminary Concepts and Notations

This section introduces fundamental concepts and notations related to the problem of mining sequential patterns, including the concepts of sequence database, order and equivalence relations on sequences, frequent sequence, generator and closed sequences.

Definition 1 (*Sequence database*). Let $\mathcal{A} = \{a_1, a_2, \ldots, a_M\}$ be a set of items. A subset A of $\mathcal{A}$, $A \subseteq \mathcal{A}$, is called an *itemset*. A *list of itemsets* $\alpha = E_1 \rightarrow E_2 \rightarrow \ldots \rightarrow E_p$ (or $\alpha = <E_1, E_2, \ldots, E_p>$) is called a *sequence*, where $\emptyset \neq E_k \subseteq \mathcal{A}, \forall k = 1, \ldots, p$ and as a convention the *null* sequence is denoted by $\alpha = < >$, when $E_k = \emptyset$, $\forall k = 1, \ldots, p$. A *sequence database* (*SDB*) $\mathcal{D}$ consists of a finite number of input sequences, $\mathcal{D} = \{\Psi_i, i = 1, \ldots, N\}$. Moreover, let $size(\alpha) \overset{\text{def}}{=} p$ and $length(\alpha) \overset{\text{def}}{=} \sum_{k=1}^{p} |E_k|$, where $|A|$ is the number of items in an itemset A.

Definition 2 (*Order relation $\sqsubseteq$ and equivalence relation $\sim$ on sequences*). Let $\alpha = E_1 \rightarrow E_2 \rightarrow \ldots \rightarrow E_p$ and $\beta = E_1' \rightarrow E_2' \rightarrow \ldots \rightarrow E_q'$ be two arbitrary sequences, consider the following *partial order* relation $\sqsubseteq$ and *equivalence* relation $\sim$ on sequences:

a. $\alpha \sqsubseteq \beta$ (or $\beta \sqsupseteq \alpha$) iff[1] $p \leqslant q$ and there exist p natural numbers, $1 \leqslant j_1 < j_2 < \ldots j_p \leqslant q$ such that $E_k \subseteq E_{j_k}', \forall k = 1, \ldots, p$. Then, we say that α is contained in β (or α is a *sub-sequence* of β, β is a *super-sequence* of α), and $\alpha \sqsubset \beta$ iff ($\alpha \sqsubseteq \beta \wedge \alpha \neq \beta$).

b. $\alpha \sim \beta \Leftrightarrow \rho(\alpha) = \rho(\beta)$, where $\rho(\alpha) \overset{\text{def}}{=} \{\Psi \in \mathcal{D} | \sqsupseteq \alpha\}$ – the set of input sequences (of $\mathcal{D}$) containing α. The corresponding *equivalence class* (or briefly *class*) containing α is denoted as $[\alpha]$, $[\alpha] \overset{\text{def}}{=} \{\beta | \rho(\beta) = \rho(\alpha)\}$.

The equivalence relation $\sim$ is used in Sect. 3 to partition $\mathcal{FS}$ into (disjoint) classes.

Definition 3 (*Frequent sequence, generator and closed sequences*). The *support* of α (*supp*(α)) is defined as the number of super-sequences (in $\mathcal{D}$) of α, i.e. $supp(\alpha) = |\rho(\alpha)|$. Then, α is called a *frequent sequence* or *sequential pattern* if its support is no less than a positive user-specified minimum support threshold *ms*, $supp(\alpha) \geqslant ms$. The problem of *frequent sequence mining* is to find the set $\mathcal{FS} = \{\alpha | supp(\alpha) \geqslant ms\}$. Thus, hereafter, we only consider a sequence α if $p(\alpha) \neq \emptyset$ or $supp(\alpha) \geqslant 1$. A sequence α is said to be *closed* (or a *generator*) if there does not exist any super-sequence (or sub-sequence) having the same support as $supp(\alpha)$. Let $\mathcal{FCS}$ and $\mathcal{FGS}$ be two sets of all *frequent closed* sequences and *frequent generators*, respectively. More generally, a sequence σ is called a *closed sequence of* α, if σ is a

---

[1] iff means "if and only if".

closed sequence containing α and σ has the same support as $supp(\alpha)$. A sequence γ is called a *generator sequence of* α, if γ is a generator contained in α and has the same support as $supp(\alpha)$.

Note that each sequence α may have many closed and generator sequences. Thus, let us denote $CloSet(\alpha) \stackrel{\text{def}}{=} \{\sigma \in \mathcal{FCS} \mid \sigma \sqsupseteq \alpha\}$ and $supp(\sigma) = supp(\alpha)\}$ and $GenSet(\alpha) \stackrel{\text{def}}{=} \{\gamma \in \mathcal{FGS} \mid \gamma \sqsubseteq \alpha$ and $supp(\gamma) = supp(\alpha)\}$. It is clear that $CloSet(\alpha) = \{\sigma \in \mathcal{FCS} \mid \sigma \sqsupseteq \alpha\}$ and $\rho(\sigma) = \rho(\alpha)\}$ and $GenSet(\alpha) = \{\gamma \in \mathcal{FGS} \mid \gamma \sqsubseteq \alpha\}$ and $\rho(\gamma) = \rho(\alpha)\}$, because $\forall \beta \sqsupseteq \alpha$, $\rho(\beta) \subseteq \rho(\alpha)$ and $(supp(\beta) = supp(\alpha) \Leftrightarrow \rho(\sigma) = \rho(\alpha))$.

Example 1 (*Running example*). Table 1 shows an example *SDB* $\mathcal{D}$ with $ms = 2$. For brevity, we write the itemset $\{a_1, \ldots, a_k\}$ as $a_1 \ldots a_k$. The sequences $\alpha = d \to a$ and $\beta = d \to ac \to a$ are *frequent* and $\alpha \sim \beta$ since $\rho(\alpha) = \rho(\beta) = \{\Psi_i \mid i = 1, 2, 4\}$ or $supp(\alpha) = supp(\beta) = |\rho(\beta)| = 3 \geqslant ms$. The sequence β is a *frequent closed* sequence of α since $\nexists \gamma : (\gamma \sqsupseteq \beta$ and $supp(\gamma) = 3)$ and the sequence $\delta = d$ is a *frequent generator* of α because $\rho(\delta) = \rho(\alpha)$ (so $supp(\delta) = supp(\alpha) = 3 \geqslant ms$) and $\nexists \delta' : (\delta' \sqsubset \delta$ and $supp(\delta') = supp(\delta))$. Moreover, the cardinalities of $CloSet(\beta) = \{\beta\}$ and $GenSet(\delta) = \{\delta\}$ are equal to 1. Meanwhile, the sets $CloSet(\delta) = \{\beta, d \to af\}$ and $GenSet(d \to ac \to af) = \{d \to a \to f, d \to c \to f\}$ consist of two sequences.

Table 1. A sample *SDB* $\mathcal{D}$.

Ψ_1	$ace \to ab \to ad \to abc \to acdf$
Ψ_2	$b \to ace \to ad \to acdf \to abc$
Ψ_3	$c \to ace \to af$
Ψ_4	$d \to ace \to ag \to af$

The cardinality of $\mathcal{FS}$ is often exponentially larger than the cardinalities of $\mathcal{FCS}$ and $\mathcal{FGS}$, especially for small *ms* values. For example, for $ms = 1$, there are 20,054 frequent sequences in $\mathcal{FS}$, only 20 (0.09% of $|\mathcal{FS}|$) frequent closed sequences in $\mathcal{FCS}$ and 154 (0.77% of $|\mathcal{FS}|$) generator sequences in $\mathcal{FGS}$. Although having much smaller cardinalities, $\mathcal{FCS}$ and $\mathcal{FGS}$ contain essential information about the *SDB* $\mathcal{D}$. Therefore, frequent sequences in $\mathcal{FS}$ can be represented through those in $\mathcal{FCS}$ and $\mathcal{FGS}$, as it will be shown in the next section.

3 An Explicit Relationship Between $\mathcal{FS}$ and Its Concise Representation Sets

To answer the question (Q), this section proposes an explicit relationship between $\mathcal{FS}$ and the two sets $\mathcal{FCS}$ and $\mathcal{FGS}$ using a partitioning method based on the equivalence relation $\sim$, presented in Definition 2.

3.1 Partitioning $\mathcal{FS}$

The $\mathcal{FS}$ set of all frequent sequences can be partitioned into smaller (disjoint) classes $FS(ro)$ as follows. Let $RO \overset{\text{def}}{=} \{\rho(\alpha)|\alpha \in \mathcal{FS}\}$ and $\forall ro \in RO$, $CS(ro) \overset{\text{def}}{=} \{\sigma \in \mathcal{FCS}|\rho(\sigma) = ro\}$, then $RO \overset{\text{def}}{=} \{\rho(\sigma)|\sigma \in \mathcal{FCS}\}$. Since $\sim$ is an equivalence relation, we obtain a rough partitioning of $\mathcal{FS}^2$:

$$\mathcal{FS} = \sum\nolimits_{ro \in RO} FS(ro), \tag{1}$$

where $FS(ro) \overset{\text{def}}{=} \{\alpha \in \mathcal{FS}|\rho(\alpha) = ro\}$. It is obvious that $FS(ro) = \bigcup_{\sigma \in CS(ro)}$ $\bigcup_{\gamma \in GenSet(\sigma)} FS(\sigma, \gamma))$, or $FS(ro)$ is an union of subsets $\mathcal{FS}(\sigma, \gamma) \overset{\text{def}}{=} \{\alpha|\gamma \sqsubseteq \alpha \sqsubseteq \sigma\}$, where $\sigma \in CS(ro)$ and $\gamma \in GenSet(\sigma)$.

Obviously, $\forall ro \in RO$, $\alpha \in FS(ro)$, $supp(\alpha) = |ro| \geqslant ms$, i.e., all frequent sequences in an equivalence class $FS(ro)$ have the same support as $|ro|$. Thus, this partitioning preserves the support and it is not necessary to perform support calculation for all frequent sequences in $FS(ro)$, which is very costly, as in the traditional method of mining $\mathcal{FS}$ directly from an SDB. Besides, the cardinality of $\mathcal{FS}$ is often much larger than that of RO. For example, for the database of the running example and $ms = 1$, $|\mathcal{FS}| = 20{,}054$ but $|RO| = 6$. Then, we store only six support values, that is one for each class $FS(ro)$ (instead of 20,054 support values) and those of all sequences in $\mathcal{FS}$ can be directly deduced from these six classes without scanning the original database.

3.2 The Explicit Representation of $\mathcal{FS}(\sigma, \gamma)$

Based on the partitioning of the $\mathcal{FS}$ set defined by Eq. (1), this section presents an explicit representation of frequent sequences in each subset $\mathcal{FS}(\sigma, \gamma)$.

For any two sequences $\gamma = E_1 \to E_2 \to \ldots \to E_p$ and $\sigma = E'_1 \to E'_2 \to \ldots \to E'_q$ such that $\gamma \sqsubseteq \sigma$, i.e., there exists a list of positions where γ appears in σ, denoted and defined as $lp \overset{\text{def}}{=} [j_1, j_2, \ldots, j_p]$, where $1 \leqslant j_1 < j_2 < \ldots < j_p \leqslant q$ and $E_k \subseteq E'_{j_k}$, $\forall k = 1 \ldots p$. For example, consider $\sigma = ace \to ad \to ac \to ac \in CS(\{\Psi_1, \Psi_2\})$ and a generator γ of σ, $\gamma = a \to c \in GenSet(\sigma)$. It is observed that two itemsets a and c of γ can be subsets of the itemsets ace and ac of σ at positions of 1 and 3, respectively. Then, we say that γ appears in σ according to $lp_1 = [1, 3]$. In addition, a and c of γ also have multiple occurences in σ at other positions such as $lp_2 = [1, 4]$, $lp_3 = [2, 3]$, $lp_4 = [2, 4]$ and $lp_5 = [3, 4]$.

The proposed method for explicitly generating all frequent sequences in $\mathcal{FS}(\sigma, \gamma)$ consists of the following five main steps.

(1) Determine the set $LP(\sigma, \gamma) \overset{\text{def}}{=} \{lp_1, lp_2, \ldots, lp_n\}$ that includes position lists of the different occurrences of γ in σ. For example, for $\sigma = ace \to ad \to ac \to ac$ and $\gamma = a \to c \in GenSet(\sigma)$, then $LP(\sigma, \gamma) = \{lp_i, i = 1 \ldots 5\}$.

---

[2] $\sum_{1 \leqslant i \leqslant n} A_i$ denotes the union of disjoint sets $A_1, A_2, \ldots, A_n$, i.e. $A_i \cap A_j = \emptyset$, $\forall i \neq j$, $1 \leqslant i \leqslant, j \leqslant n$.

(2) Expand γ with the size q of σ according to each $lp \in LP(\sigma, \gamma)$ to obtain $Ex(\gamma, lp) \overset{\text{def}}{=} E_1^{\sim} \to E_2^{\sim} \to \ldots \to E_q^{\sim}$ such that

$$E_i^{\sim} = \begin{cases} \emptyset, & \text{if } i \notin lp \\ E_k, & \text{if } i = j_k \in lp \end{cases}, \forall i = 1, \ldots, q.$$

For instance, consider $lp_2 = [1, 4] \in LP(\sigma, \gamma)$, then $Ex(\gamma, lp_2) = a \to \emptyset \to \emptyset \to c$.

(3) For each $lp = \{j_1, j_2, \ldots, j_p\} \in LP(\sigma, \gamma)$, find the difference sequence of σ and $Ex(\gamma, lp)$ that is denoted and defined as $\Delta(lp) \overset{\text{def}}{=} D_1 \to D_2 \to \ldots D_q$, where $D_i \overset{\text{def}}{=} E_i' \backslash E_i^{\sim}, \forall i = 1, \ldots, q$.

For example, for $lp_2 = [1, 4]$, we have $\Delta(lp_2) = ce \to ad \to ac \to a$, because $D_1 = ace \backslash a = ce$, $D_2 = ad \backslash \emptyset = ad$, $D_3 = ac \backslash \emptyset = ac$ and $D_4 = ac \backslash c = a$.

(4) For each $lp \in LP(\sigma, \gamma)$, generate each sequence $\alpha = F_1 \to F_2 \to \ldots \to F_q$ in $\mathcal{FS}(\sigma, \gamma)$ as follows, $\alpha = Ex(\gamma, lp) \oplus \delta(lp)$, for all $\delta(lp) = d_1 \to d_2 \to \ldots \to d_q \sqsubseteq \Delta(lp)$, where the operator $\oplus$ is the direct sum of two sequences on itemsets appearing at the same positions in $Ex(\gamma, lp)$ and $\delta(lp)$, i.e., $F_i = E_i^{\sim} + d_i$ and $d_i \subseteq D_i, \forall i = 1, \ldots, q$. Then, reduce α by removing all its empty itemsets. For example, with $\Delta(lp_2) = ce \to ad \to ac \to a$, consider a sub-sequence $\delta(lp_2) = e \to a \to \emptyset \to \emptyset$ of $\Delta(lp_2)$, then $\alpha = Ex(\gamma, lp_2) \oplus \delta(lp_2) = a \to \emptyset \to \emptyset \to c \oplus e \to a \to \emptyset \to \emptyset = ae \to a \to \emptyset \to c$. By removing its empty itemsets, we obtain the reduced sequence $\alpha = ae \to a \to c$. Note that the number of sub-sequences $\delta(lp_2)$ of $\Delta(lp_2)$ is $2^{length(\Delta(lp_2))} = 2^7 = 128$. Thus, we have 128 sequences α generated according to $lp_2 = [1, 4]$.

(5) Finally, we obtain the set $\mathcal{FS}'(\sigma, \gamma) \overset{\text{def}}{=} \{\alpha = Ex(\gamma, lp) \oplus \delta(lp) | lp \in LP(\sigma, \gamma), \delta(lp) \sqsubseteq \Delta(lp)\}\}$ that contains all sequences generated from the pair (σ, γ).

From the explicit representation of frequent sequences α in $\mathcal{FS}'(\sigma, \gamma)$ based on the closed sequence σ and the generator γ, the following proposition is derived whose proof is presented in Appendix 1.

Proposition 1 ($\mathcal{FS}'(\sigma, \gamma)$ - an explicit representation of $\mathcal{FS}(\sigma, \gamma)$). For $\forall ro \in RO$, $\forall \sigma \in CS(ro)$ and $\forall \gamma \in GenSet(\sigma)$, we have

$$\mathcal{FS}(\sigma, \gamma) = \mathcal{FS}'(\sigma, \gamma). \tag{2}$$

Note that the partition presented in Eq. (1) is proposed to avoid duplication between different classes $FS(ro)$ for all $ro \in RO$. However, the proposed method may still lead to many duplicate patterns among different subsets $\mathcal{FS}'(\sigma_i, \gamma_j)$ of $FS(ro)$, for each $\sigma_i \in CS(ro)$ and $\gamma_j \in GenSet(\sigma_i)$ as shown in Example 2 below.

Example 2 (Duplication between subsets $\mathcal{FS}'(\sigma, \gamma)$, $\sigma \in CS(ro)$, $\gamma \in GenSet(\sigma_i)$). For the database of the running example, consider $ro = \{\Psi_1, \Psi_2, \Psi_4\}$ and three *closed* sequences, $\sigma_1 = d \rightarrow ac \rightarrow a$, $\sigma_2 = d \rightarrow af$ and $\sigma_3 = ace \rightarrow a \rightarrow af$. Then, $\rho(\sigma_1) = \rho(\sigma_2) = \rho(\sigma_3) = ro$, so $CS(ro) = \{\sigma_1, \sigma_2, \sigma_3\}$ and $|CS(ro)| > 1$. Besides, $GenSet(\sigma_1) = GenSet(\sigma_2) = \{\gamma = d\}$ and $GenSet(\sigma_3) = \{\gamma_1 = a \rightarrow a \rightarrow a, \gamma_2 = a \rightarrow a \rightarrow f, \gamma_3 = e \rightarrow a \rightarrow a, \gamma_4 = e \rightarrow a \rightarrow f\}$.

(i) (*Duplication between two different subsets* $\mathcal{FS}'(\sigma_i, \gamma)$ *and* $\mathcal{FS}'(\sigma_k, \gamma)$, $\forall \sigma_i$, $\sigma_k \in CS(ro)$, $\forall \gamma \in GenSet(\sigma_i) \cap GenSet(\sigma_k)$ *and* $|CS(ro)| > 1$). For the generator $\gamma = d$ of σ_1 and σ_2, the set of all sequences α generated from the pair (σ_1, γ) is $\mathcal{FS}'(\sigma_1, \gamma) = \{d, d \rightarrow a, d \rightarrow c, d \rightarrow ac, d \rightarrow a \rightarrow a, d \rightarrow c \rightarrow a, d \rightarrow ac \rightarrow a\}$, and $\mathcal{FS}'(\sigma_2, \gamma) = \{d, d \rightarrow a, d \rightarrow f, d \rightarrow af\}$. Hence, the sub-sequences d and $d \rightarrow a$ in $\mathcal{FS}'(\sigma_2, \gamma)$ are duplicated in $\mathcal{FS}'(\sigma_1, \gamma)$. Note that both patterns are contained in σ_1. To avoid such duplication, any sequence α generated in $\mathcal{FS}'(\sigma_2, \gamma)$ satisfying the following duplicate condition, $DCondC(\alpha, \sigma_2, CS(ro))$ $\overset{\text{def}}{=} (\exists \sigma_k \in CS(ro) : k < 2, \alpha \sqsubseteq \sigma_k)$ should be eliminated.

(ii) (*Duplication between two different sub-classes* $\mathcal{FS}'(\sigma_i, \gamma)$ *and* $\mathcal{FS}'(\sigma_k, \gamma)$, $\forall \gamma_j$, $\gamma_k \in GenSet(\sigma)$ *and* $|GenSet(\sigma)| > 1$). For two generators $\gamma_1 = a \rightarrow a \rightarrow a$ and $\gamma_2 = a \rightarrow a \rightarrow f$ of σ_3, then $\mathcal{FS}'(\sigma_3, \gamma_1) = \{a \rightarrow a \rightarrow a, ac \rightarrow a \rightarrow a, ae \rightarrow a \rightarrow a, ace \rightarrow a \rightarrow a, a \rightarrow a \rightarrow af, ac \rightarrow a \rightarrow af, ae \rightarrow a \rightarrow af, ace \rightarrow a \rightarrow af\}$ and $\mathcal{FS}'(\sigma_3, \gamma_2) = \{a \rightarrow a \rightarrow f, ac \rightarrow a \rightarrow f, ae \rightarrow a \rightarrow f, ace \rightarrow a \rightarrow f, a \rightarrow a \rightarrow af, ac \rightarrow a \rightarrow af, ae \rightarrow a \rightarrow af, ace \rightarrow a \rightarrow af\}$. Hence, four sub-sequences $a \rightarrow a \rightarrow af$, $ac \rightarrow a \rightarrow af$, $ae \rightarrow a \rightarrow af$ and $ace \rightarrow a \rightarrow af$ generated in $\mathcal{FS}'(\sigma_3, \gamma_2)$ are duplicated with those in $\mathcal{FS}'(\sigma_3, \gamma_1)$. Note that all these duplicate patterns contain γ_1. To deal with this, we should not create any sequence α satisfying the following duplicate condition, $DCondG(\alpha, \sigma_3, \gamma_2, GenSet(\sigma_3)) \overset{\text{def}}{=} (\exists \gamma_k \in GenSet(\sigma_3) : k < 2, \alpha \sqsupseteq \gamma_k)$.

3.3 Fast Generation of All Frequent Sequences in $\mathcal{FS}$ from $\mathcal{FCS}$ and $\mathcal{FGS}$

Based on the explicit representation of $\mathcal{FS}(\sigma, \gamma)$ in Proposition 1, a novel method to generate all frequent sequences in $\mathcal{FS}(\sigma, \gamma)$ and $\mathcal{FS}$ can be developed. Note that in the traditional problem of mining frequent itemsets, each itemset has only a unique closed itemset. But in the current problem of generating $\mathcal{FS}$ from the pair of $\mathcal{FCS}$ and $\mathcal{FGS}$, each frequent sequence α can have many closed and generator sequences and α can appear in each closed sequence at many different positions. This can result in generating a huge number of duplicate sequences during the mining process, leading to very long runtimes and high memory usage. To overcome this problem, this subsection proposes duplicate conditions and corresponding pruning strategies to detect and eliminate all duplicate sequences early without generating them.

As shown in Example 2, the two cases can lead to the generation of duplicate frequent sequences in *different* subsets $\mathcal{FS}'(\sigma, \gamma)$, $\sigma \in CS(ro)$, $\gamma \in GenSet(\sigma)$. Thus, many duplicate patterns may be discovered in each equivalence class $FS(ro)$. Besides, there are other cases that can lead to duplication in each subset $\mathcal{FS}(\sigma, \gamma)$. Therefore, we propose three strategies to overcome this challenge, defined as follows. For each $ro \in$

RO, $CS(ro) = \{\sigma_1, \sigma_2, \ldots, \sigma_n\}$ and $GenSet(\sigma_i) = \{\gamma_1, \gamma_2, \ldots, \gamma_m\}$, consider any two closed sequences σ_k, σ_i in $CS(ro)$ with $k < i$ and a generator γ_j in $GenSet(\sigma_i)$.

Pruning Strategy 1 (*Eliminating duplicate sub-sequences generated according to two different closed sequences σ_k and σ_i in $CS(ro)$*). Any sequence α generated in $\mathcal{FS}'(\sigma_i, \gamma_j)$ satisfying the duplicate condition related to the *closed* sequence σ_i,

$$DCondC(\alpha, \sigma_i, , CS(ro)) \stackrel{\text{def}}{=} (\exists \sigma_k \in CS(ro) : k < i, \alpha \sqsubseteq \sigma_k),$$ is eliminated.

Pruning Strategy 2 (*Eliminating duplicate sub-sequences generated according to two different generator sequences γ_j and γ_l in $GenSet(\sigma_i)$*). Any sequence α generated in $\mathcal{FS}'(\sigma_i, \gamma_j)$ satisfying the duplicate condition related to the *generator* $\gamma_j \in GenSet(\sigma_i)$,

$$DCondG(\alpha, \sigma_i, \gamma_j, GenSet(\sigma_i)) \stackrel{\text{def}}{=} (\exists \in GenSet(\sigma_i) : l < j, \alpha \sqsupseteq \gamma_l)$$ is also eliminated.

Note that for the traditional problem of mining frequent itemsets, each equivalence class has only one closed pattern, so it is unnecessary to use Pruning Strategy 1. However, for the current problem, each class may have many closed sequences and frequent sequences generated in the subset $\mathcal{FS}'(\sigma_i, \gamma_j)$ may be duplicated in $\mathcal{FS}'(\sigma_k, \gamma_j)$, which was produced in the previous steps. Thus, Pruning Strategy 1 can help eliminating a considerable number of duplicate sequences.

Example 3 (*Illustration of Strategies 1-2*). Consider the closed sequences $\sigma_1 = d \rightarrow ac \rightarrow a$, and $\sigma_2 = d \rightarrow af$ in $CS(ro)$ and their generator $\gamma = d$ as shown in Example 2. (*i*) Assume that all sequences in $\mathcal{FS}'(\sigma_1, \gamma)$ have been generated already. When producing two sequences $\alpha_1 = d$ and $\alpha_2 = d \rightarrow a$ in $\mathcal{FS}'(\sigma_2, \gamma)$, we find that the conditions $DCondC(\alpha_1, \sigma_2, CS(ro))$ and $DCondC(\alpha_2, \sigma_2, CS(ro))$ in Pruning Strategy 1 hold. Thus, α_1 and α_2 are eliminated from $\mathcal{FS}'(\sigma_2, \gamma)$ to avoid duplication. Meanwhile, for the two generators $\gamma_1 = a \rightarrow a \rightarrow a$ and $\gamma_2 = a \rightarrow a \rightarrow f$ of σ_3 in Example 2. (*ii*) consider the sub-classes $\mathcal{FS}'(\sigma_3, \gamma_2)$ containing 8 sequences. Obviously, four sequences $\alpha_1 = a \rightarrow a \rightarrow af$, $\alpha_2 = ac \rightarrow a \rightarrow af$, $\alpha_3 = ae \rightarrow a \rightarrow af$ and $\alpha_4 = ace \rightarrow a \rightarrow af\}$ contain γ_1, i.e., the conditions $DCondG(\alpha_k, \sigma_3, \gamma_2, GenSet(\sigma_3))$ for $k = 1..4$ are satisfied. These sequences are then eliminated from $\mathcal{FS}'(\sigma_3, \gamma_2)$, so $\mathcal{FS}'(\sigma_3, \gamma_2) = \{a \rightarrow a \rightarrow f, ac \rightarrow a \rightarrow f, ae \rightarrow a \rightarrow f, ace \rightarrow a \rightarrow f\}$.

Let us denote $\mathcal{FS}^*(\sigma_i, \gamma_j) \stackrel{\text{def}}{=} \{\alpha \in \mathcal{FS}'(\sigma_i, \gamma_j) \mid not(DCondC(\alpha, \sigma_i, CS(\rho(\sigma_i))))$ and $not(DCondG(\alpha, \sigma_i, \gamma_j, GenSet(\sigma_i)))\}$. Then, there are no duplicate sequences between different subsets $\mathcal{FS}^*(\sigma_i, \gamma_j)$ in each class $FS(ro)$, for $\sigma_i \in CS(ro)$, $\gamma_j \in GenSet(\sigma_i)$. Based on (1), (2) and the above strategies, we obtain the following Theorem 1 whose proof is given in Appendix 2.

Theorem 1 (*Partitions of $\mathcal{FS}$*).

a. $\mathcal{FS} = \sum_{ro \in RO} FS(ro)$ (*rough partition*).
b. $FS(ro) = \sum_{\sigma_i \in CS(ro)} \sum_{\gamma_j \in GenSet(\sigma_i)} \mathcal{FS}^*(\sigma_i, \gamma_j)$, $\forall ro \in RO$.

Hence, $\mathcal{FS} = \sum_{ro \in RO} \sum_{\sigma_i \in CS(ro)} \sum_{\gamma_j \in GenSet(\sigma_i)} \mathcal{FS}^*(\sigma_i, \gamma_j)$ (*smooth partition*).

In the rest of the paper, for the database of the running example, let us consider $ro_1 = \{\Psi_1, \Psi_2\} \in RO$, $\sigma = ace \to ad \to ac \to ac \in CS(ro_1)$ and $\gamma = a \to c$ $\in GenSet(\sigma)$. Hence, $LP(\sigma, \gamma) = \{lp_1 = [1, 3], lp_2 = [1, 4], lp_3 = [2, 3], lp_4 = [2, 4]$ and $lp_5 = [3, 4]\}$. Then, $Ex(\gamma, lp_1) = a \to \emptyset \to c \to \emptyset$, $\Delta(lp_1) = ce \to ad \to a \to ac$, $Ex(\gamma, lp_2) = a \to \emptyset \to \emptyset \to c$ and $\Delta(lp_2) = ce \to ad \to ac \to a$.

Note that applying Pruning Strategies 1–2 only ensure non-duplication between different subsets $\mathcal{FS}^*(\sigma_i, \gamma_j)$. However, the biggest challenge that makes the current problem much more difficult than the traditional one of generating frequent itemsets is that each frequent sequence can appear at many different position sets of its closed sequences. For example, the generator $\gamma = a \to c$ appears in the closed sequence $\sigma = ace \to ad \to ac \to ac$ at five position sets lp_{1-5}. This results in a very large number of duplicate sequences in each subset $\mathcal{FS}'(\sigma, \gamma)$. Thus, the following Pruning Strategy 3 is proposed to prune all duplicate sequences early in each subset $\mathcal{FS}'(\sigma, \gamma)$ without generating them.

Assume that X is an arbitrary set with many elements, let us denote $X[k]$ as the kth element of X, e.g. $lp_1[2] = 3$ or $\Delta(lp_1)[3] = a$. The set lp_m is said to be smaller than lp_n, denoted as $lp_m \prec lp_n$, if $(lp_m[1] < lp_n[1])$ or $(lp_m[1] = lp_n[1]$ and $lp_m[2] < lp_n[2])$, or $(lp_m[1] = lp_n[1]$ and $lp_m[2] = lp_n[2]$ and $lp_m[3] < lp_n[3])$, etc. Note that elements in $LP(\sigma, \gamma)$ are sorted in ascending order w.r.t $\prec$. Given $ro \in RO$, $\sigma \in CS(ro)$, $\gamma \in GenSet(\sigma)$, $q = size(\sigma)$, lp_m and $lp_n \in LP(\sigma, \gamma)$ such that $lp_m \prec lp_n$, let $k_0 = FI(m, n)$ be the first index at which lp_m differs from lp_n, e.g. $k_0 = FI(1, 2) = 2$ because $lp_1[1] = lp_2[1] = 1$ and $lp_1[2] = 3 \neq lp_2[2] = 4$. Consider any sequence $\alpha = Ex(\gamma, lp) \oplus \delta(lp)$ generated in $\mathcal{FS}'(\sigma, \gamma)$.

Pruning Strategy 3 (*Eliminating duplicate sequences generated in each subset* $\mathcal{FS}'(\sigma, \gamma)$). For each $lp \in LP(\sigma, \gamma)$ and $\delta(lp) \sqsubseteq \Delta(lp)$, any sequence $\alpha = Ex(\gamma, lp) \oplus \delta(lp)$ in $\mathcal{FS}'(\sigma, \gamma)$ that satisfies one of the following three duplicate conditions, $DCond_{1-3}(\alpha, lp) \overset{def}{=} \{DCond_i(\alpha, lp), i = 1 \ldots 3\}$ and all its appropriate super-sequences will be eliminated. Details of the three duplicate conditions are presented next.

(i) **Duplicate Condition 1** ($DCond_1$ – *used for each lp in* $LP(\sigma, \gamma)$). For any $\delta(lp) = d_1 \to d_2 \to \ldots \to d_q \sqsubseteq \Delta(lp) = D_1 \to D_2 \to \ldots \to D_q$ and $\alpha = Ex(\gamma, lp) \oplus \delta(lp) = F_1 \to F_2 \to \ldots \to F_q$, consider any $i \notin lp$ such that $1 < i \leqslant q$. If there exist a *maximum successive series* of *empty events* in α at positions from $(i - 1)$ down to $n_1 \geqslant 1$ (i.e., $F_k = \emptyset$, $\forall k = n_1, n_1 + 1, \ldots, i - 1$) and an index k such that the non-empty itemset d_i is contained in $D_i \cap D_k$, then α is surely a duplicate of a sequence previously generated according to the index k. Moreover, generating all other frequent sequences, which are formed from α by growing the itemsets d_j in $\delta(lp)$ such that $j \, j \notin \{n_1, \ldots, i - 1\}$ and $d_i \subseteq D_i \cap D_k$, will also result in duplication. Therefore, they can be pruned early.

For example, with $lp_2 = [1, 4]$, $Ex(\gamma, lp_2) = a \to \emptyset \to \emptyset \to c$ and $\Delta(lp_2) = ce \to ad \to ac \to a$, consider $\delta(lp_2) = \emptyset \to \emptyset \to a \to \emptyset \sqsubseteq \Delta(lp_2)$. Then, $\alpha = a \to \emptyset \to a \to c = a \to a \to c$ (after reduction). Since $i = 3 \notin lp_2$ and there exists $k = 2 < i$ such that $F_2 = \emptyset$ and $d_3 = a \subseteq D_3 \cap D_2 = ad \cap ac = a$, i.e., the condition $DCond_1$ holds, and the duplication certainly happens. *Indeed*, for other $\delta'(lp_2) = \emptyset \to$

$a \to \emptyset \to \emptyset \sqsubseteq \Delta(lp_2)$ generated previously according to $It_{k=2} = a$, the sequence $\alpha' = Ex(\gamma, lp_2) \oplus \delta'(lp_2) = a \to \emptyset \to \emptyset \to c \oplus \emptyset \to a \to \emptyset \to \emptyset = a \to a \to \emptyset \to c = a \to a \to c$ is identical with α. Thus, α is eliminated from $\mathcal{FS}'(\sigma, \gamma)$. Moreover, we must not continue producing all super-sequences of α by generating all subsets of $D_1 = ce$ and $D_4 = a$, e.g. $ae \to a \to ac$, because this also leads to duplication. Therefore, by considering a sequence $\alpha = a \to \emptyset \to a \to c$ according to $\delta(lp_2)$, we can prune $2^{length(D_1) + length(D_4)} = 2^3 = 8$ duplicate sequences.

(ii) **Duplicate Condition 2** (*DCond$_2$ – used for two different lp_n and lp_m in $LP(\sigma, \gamma)$*).
For any $lp_n \in LP(\sigma, \gamma)$, $\delta(lp_n) = d_1 \to d_2 \to \ldots \to d_q \sqsubseteq \Delta(lp_n)$ and $\alpha = Ex(\gamma, lp_n) \oplus \delta(lp_n) = F_1 \to F_2 \to \ldots \to F_q$, if there exists a $lp_m \in LP(\sigma, \gamma)$ and an index $i = lp_m[k_0]$ such that $lp_m \prec lp_n$, where $k_0 = FI(m, n)$, and d_i contains $\gamma[k_0]$, then α is duplicated with a sequence α' previously produced from lp_m. Moreover, generating all other frequent sequences that are formed from α by growing all itemsets in $\delta(lp_n)$ will also lead to duplication. Thus, we can stop producing them.

For example, for $lp_2 = [1, 4] \in LP(\sigma, \gamma)$, consider $\delta(lp_2) = \emptyset \to \emptyset \to c \to \emptyset \sqsubseteq \Delta(lp_2) = ce \to ad \to ac \to a$ and $\alpha = Ex(\gamma, lp_2) \oplus \delta(lp_2) = a \to \emptyset \to c \to c$. Then, there exist $lp_1 = [1, 3] \in LP(\sigma, \gamma)$ and the index $i = lp_1[k_0] = 3$ (with $k_0 = 2$) such that $d_3 = c$ in $\delta(lp_2)$ contains $\gamma[2] = c$, i.e., the condition $DCond_2$ is satisfied. Thus, α is a duplicate of a sequence generated from lp_1. Indeed, for $Ex(\gamma, lp_1) = a \to \emptyset \to c \to \emptyset$, $\Delta(lp_1) = ce \to ad \to a \to ac$, consider $\delta(lp_1) = \emptyset \to \emptyset \to \emptyset \to c \sqsubseteq \Delta(lp_1)$, then we have sequence $\alpha' = Ex(\gamma, lp_1) \oplus \delta(lp_1) = a \to \emptyset \to c \to c \equiv \alpha$. In addition, for $\delta_0(lp_2) = ce \to d \to c \to \emptyset \sqsubseteq \Delta(lp_2))$ formed from $\delta(lp_2)$ by adding ce and d to its first and second itemsets, respectively, another sequence $\alpha = Ex(\gamma, lp_2) \oplus \delta_0(lp_2) = ace \to d \to c \to c$ is also duplicated with $\alpha' = Ex(\gamma, lp_1) \oplus \delta_0(lp_1) = ace \to d \to c \to c$, where $\delta_0(lp_1) = ce \to d \to \emptyset \to c \sqsubseteq \Delta(lp_1)$. Similarly, duplication also happens for all sequences $\alpha = Ex(\gamma, lp_2) \oplus \delta'(lp_2)$, where $\delta'(lp_2) \sqsubseteq \Phi \overset{def}{=} \Delta(lp_2) \backslash \delta(lp_2) = ce \to ad \to a \to a$. Hence, by only considering a sequence $\alpha = a \to \emptyset \to c \to c$ according to $\delta(lp_2)$, we can additionally eliminate $2^{length(\Phi)} = 2^6 = 64$ duplicate sequences early, without generating them.

(iii) **Duplicate Condition 3** (*DCond$_3$ – used for two different lp_n and $lp_m \in LP(\sigma, \gamma)$*). For any $lp_n \in LP(\sigma, \gamma)$, $\delta(lp_n) = d_1 \to d_2 \to \ldots \to d_q \sqsubseteq \Delta(lp_n)$ and $\alpha = Ex(\gamma, lp_n) \oplus \delta(lp_n) = F_1 \to F_2 \to \ldots F_q$, if there exists a $lp_m \in LP(\sigma, \gamma)$ such that $lp_m \prec lp_n$ and an index $i = lp_n[k_0]$ such that all itemsets of α from positions $j = lp_m[k_0]$ to $i - 1$ are empty (i.e. $F_k = \emptyset, \forall k = j, j + 1, \ldots, i - 1$) and the itemset d_i of $\delta(lp_n)$ is contained in $\Delta(lp_n)[i] \cap \Delta(lp_m)[j]$, then α is duplicated with a sequence α' produced from lp_m. In addition, the duplication will also occur when generating all other frequent sequences that are formed from α by growing all itemsets d_k in $\delta(lp_n)$ such that $k \notin \{j, \ldots, i - 1\}$ and $d_i \subseteq \Delta(lp_n)[i] \cap \Delta(lp_m)[j]$. Thus, we don't need to generate them.

For example, for $lp_2 \in LP(\sigma, \gamma)$, consider the sequence $\delta(lp_2) = \emptyset \to \emptyset \to \emptyset \to \emptyset \sqsubseteq \Delta(lp_2)$ and $\alpha = Ex(\gamma, lp_2) \oplus \delta(lp_2) = a \to \emptyset \to \emptyset \to c = a \to c$. Since $lp_1 \prec lp_2$,

$i = lp_2[2] = 4$, $j = lp_1[2] = 3$, $F_3 = \emptyset$ and $d_4 = \emptyset$ (of $\delta(lp_2)) \subseteq \Delta(lp_2)[4] \cap \Delta(lp_1)[3] = a$, i.e., the condition $DCond_3$ holds, then α is a duplicate of a sequence produced from lp_1. Indeed, for $lp_1 = [1, 3]$, consider $\delta(lp_1) = \emptyset \to \emptyset \to \emptyset \to \emptyset \sqsubseteq \Delta(lp_1)$. We obtain $\alpha' = a \to \emptyset \to c \to \emptyset \oplus \emptyset \to \emptyset \to \emptyset \to \emptyset = a \to \emptyset \to c \to \emptyset = a \to c \equiv \alpha$. Moreover, for $\delta_0(lp_2) = ce \to \emptyset \to \emptyset \to a(\sqsubseteq \Delta(lp_2))$ formed from $\delta(lp_2)$ by adding ce and a to its first and fourth itemsets, respectively. Then, the new generated sequence $\alpha = a \to \emptyset \to \emptyset \to c \oplus ce \to \emptyset \to \emptyset \to a = ace \to ac$ is also identical with $\alpha' = Ex(\gamma, lp_1) \oplus \delta_0(lp_1) = a \to \emptyset \to c \to \emptyset \oplus ce \to \emptyset \to a \to \emptyset = ace \to ac \equiv \alpha$, where $\delta_0(lp_1) = ce \to \emptyset \to a \to \emptyset \sqsubseteq \Delta(lp_1)$. Similarly, the duplication also occurs for all sequences $\alpha = Ex(\gamma, lp_2) \oplus \delta'(lp_2)$, where $\delta'(lp_2) \sqsubseteq \Phi = ce \to ad \to \emptyset \to a$. Thus, by only checking a sequence $\alpha = a \to \emptyset \to \emptyset \to c$ according to $\delta(lp_2)$, we can prune $2^{length(\Phi)} = 2^5 = 32$ duplicate sequences.

Note that the number of duplicate sequences that are pruned by Pruning Strategy 3 based on the three duplicate conditions $DCond_{1-3}$ is exponential.

To avoid duplication when explicitly producing all sequences of $\mathcal{FS}'(\sigma, \gamma)$, we propose a duplicate-free representation of $\mathcal{FS}'(\sigma, \gamma)$, defined as follows.

$$\mathcal{FS}''(\sigma, \gamma) \overset{\text{def}}{=} \{\alpha = Ex(\gamma, lp) \oplus \delta(lp) \in \mathcal{FS}'(\sigma, \gamma) | not(DCond_{1-3}(\alpha, lp))\}.$$

However, a much more difficult question arises. Besides the three situations presented in Pruning Strategy 3, is there other situations that result in duplicates during the process of generating sequences in $\mathcal{FS}''(\sigma, \gamma)$? Based on Theorem 1 and the three duplicate conditions $DCond_{1-3}$, the following Theorem 2 answers this question negatively. The proof is presented in Appendix 3.

Theorem 2 (*Completely and distinctly generating all frequent sequences in $\mathcal{FS}'(\sigma, \gamma)$*).

a. $\mathcal{FS}'(\sigma, \gamma) = \mathcal{FS}''(\sigma, \gamma)$.
b. All frequent sequences in $\mathcal{FS}''(\sigma, \gamma)$ are generated distinctly. Thus, all frequent sequences in $\mathcal{FS} = \sum_{ro \in RO} \sum_{\sigma_i \in CS(ro)} \sum_{\gamma_j \in GenSet(\sigma_i)} \mathcal{FS}^*(\sigma_i, \gamma_j)$ are also generated completely and distinctly, where $\mathcal{FS}^*(\sigma_i, \gamma_j) \overset{\text{def}}{=} \{\alpha \in \mathcal{FS}''(\sigma_i, \gamma_j) \mid not(DCondC(\alpha, \sigma_i, CS(\rho(\sigma_i))))$ and $not(DCondG(\alpha, \sigma_i, \gamma_j, GenSet(\sigma_i)))\}$.

Based on the above theoretical results, we next design a novel algorithm named GFS-CR (Generating Frequent Sequences from Concise Representations) to efficiently generate $\mathcal{FS}$ from the two sets $\mathcal{FCS}$ and $\mathcal{FGS}$.

4 The GFS-CR Algorithm

From the sets $\mathcal{FCS}^* \overset{\text{def}}{=} \{(\sigma, \rho(\sigma), supp(\sigma)) | \sigma \in \mathcal{FCS}\}$ and $\mathcal{FGS}^* \overset{\text{def}}{=} \{(\gamma, supp(\gamma)) | \gamma \in \mathcal{FGS}\}$ for each $ro \in RO \overset{\text{def}}{=} \{\rho(\sigma) | \sigma \in \mathcal{FCS}\}$, the corresponding sets $\mathcal{FCGS}(ro) \overset{\text{def}}{=} \{(\sigma, GenSet(\sigma), supp(\sigma)) | \sigma \in \mathcal{FCS}, \rho(\sigma) = ro\}$ and $\mathcal{FCGS} = \overset{\text{def}}{=} \{\mathcal{FCGS}(ro) | ro \in RO\}$ are created as the input of the GFS-CR algorithm (Fig. 1). For each pair of the

```
FS GFS-CR(FCGS)
1. FS = ∅;
2. for each FCGS(ro) ∈ FCGS do
3.     for each (σ_i, GenSet(σ_i), supp(σ_i)) ∈ FCGS(ro) do
4.         for each γ_k ∈ GenSet(σ_i) do {
5.             FS*(σ_i, γ_k) = GFS-OneSubSet(σ_i, γ_k);
6.             FS = FS + FS*(σ_i, γ_k);
7.         }
8. return FS;
```

Fig. 1. The GFS-CR algorithm for generating FS from FCS and FGS.

```
FS*(σ_i, γ_k) GFS-OneSubSet(σ_i, γ_k)
1. Find the position list LP(σ_i, γ_k);
2. q = size(σ_i); FS*(σ_i, γ_k) = ∅;
3. for each lp_n ∈ LP(σ_i, γ_k) do {
4.     γ' = Ex(γ_k, q, lp_n);        // expand γ_k to the size q based on lp_n
5.     D = Δ(lp_n) = σ_i ⊖ γ' = D_1→D_2→...→D_q;
6.     duplicate = false; α = <>;
7.     SearchFS(1, q, α, D, γ', lp_n, LP(σ_i, γ_k), false, false, FS*(σ_i, γ_k));
8. } // for each lp_n
9. return FS*(σ_i, γ_k);
```

Fig. 2. The GFS-OneSubSet procedure.

closed sequence σ_i and its generator γ_k, the procedure GFS-OneSubSet is called to generate frequent sequences in each subset $FS^*(\sigma_i, \gamma_k)$ (line 5). Then, the $FS^*(\sigma_i, \gamma_k)$ set is added to FS (line 6). Finally, GFS-CR returns the FS set (line 8).

In the GFS-OneSubSet procedure shown in Fig. 2, for each $lp_n \in LP(\sigma_i, \gamma_k)$ after calculating $Ex(\gamma_k, q, lp_n)$ and $\Delta(lp_n)$ (lines 4–5), the SearchFS procedure (Fig. 3) is called to find all frequent sequences according to lp_n (line 7). It returns as result all frequent sequences in $FS^*(\sigma_i, \gamma_k)$. The SearchFS procedure applies Pruning Strategy 3 by checking the three pruning conditions $DCond_{1-3}$ to detect and eliminate all duplicate sequences in each subset $FS^*(\sigma, \gamma)$ (lines 7, 12 and 18). In addition, Pruning Strategies 1-2 are also utilized by testing the conditions $DCondC$ (line 2) and $DCondG$ (line 24). Finally, SearchFS is recursively called to generate larger sequences (line 28).

The correctness of GFS-CR is guaranteed by Proposition 1 and Theorems 1–2.

SearchFS$(i, q, \alpha, D, \gamma', k, n, LP(\sigma, \gamma_k), \mathcal{FS}^*(\sigma, \gamma_k))$
. Input: $lp \in LP(\sigma, \gamma_k)$, position i, size q, sequence α, $D = \Delta(lp)$, its_i and $\mathcal{FS}^*(\sigma, \gamma_k)$
. Output: the set $\mathcal{FS}^*(\sigma, \gamma_k)$ has been updated.
1. **if** $(j > q)$ **then** {
2.   **if** $(not(DCondC(\alpha, \sigma_i, CS(\rho(\sigma_i)))))$ //Check the pruning condition DCondC
3. $\mathcal{FS}^*(\sigma_i, \gamma_k) = \mathcal{FS}^*(\sigma_i, \gamma_k) \cup \{\alpha\};$
4.   **return**;
5. }
6. **for each** $its_i \subseteq D_i$ **do** {
7.   **if** $(DCond_1(lp_n, i, \alpha, D, its_i)))$ **then** //Duplication condition 1 in Strategy 3
8. $duplicate = true;$
9.   **else** { // $i \in lp_n$
10.   **for each** $lp_m \in LP(\sigma, \gamma_k): m < n$ **do** {
11. $k_0 = FI(m, n);$
12.   **if** $(DCond_2(lp_m, i, k_0, \gamma_k, its_i))$ **then** { // Duplication condition 2 in Strategy 3
13. $duplicate = true;$
14.   **break**; // for each lp_m
15.   }
16. $k = lp_m[k_0];$
17. $intersection = D_i \cap \Delta(lp_m) [k];$
18.   **if** $(DCond_3(lp_m, lp_n, i, k_0, its_i, \alpha, intersection))$ **then** { //DCond₃ in
19. $duplicate = true;$ //Strategy 3
20.   **break**; // for each lp_m
21.   }
22.   } // for each lp_m
23.   }
24.   **if** $(\exists \gamma_{k'} \in GenSet(\sigma): k' < k, \alpha \rightarrow (\gamma'[i] \cup its_i) \sqsupseteq \gamma_{k'})$ **then** //DCondG in Strategy 2
25. $duplicate = true;$
26.   **if** $(duplicate)$ **then continue**; // for each its_i
27. $\alpha = \alpha \rightarrow (\gamma'[i] \cup its_i);$
28. $SearchFS(i{+}1, \alpha, D, \gamma', k, n, LP(\sigma, \gamma_k), \mathcal{FS}^*(\sigma, \gamma_k));$
29. $\alpha.RemoveItemset(i);$
30. } // for each its_i
31. **return**;

Fig. 3. The SearchFS procedure.

5   Experimental Results

Experiments were performed on an Intel(R) Core(TM) i5-2320, 3.0 GHz CPU with 12 GB of memory, running Windows 8.1. All the tested algorithms were implemented in Java 7 SE. The source code of the state-of-the-art algorithms CM-SPADE, CM-SPAM [6] and PrefixSpan [12] were obtained from [7].

Experiments were carried on four databases, two real-life SDBs named Kosarak and BMS [7], which are frequently used to evaluate state-of-the-art algorithms for mining frequent sequences, and two synthetic SDBs D2C7T10N2.5S6I4 and D5C10T6N5S6I4 generated using the IBM Quest data generator (obtained from [7]) with parameters described in Table 2. The characteristics of these databases are shown in Table 3.

Table 2. Parameters of the IBM quest synthetic data generator.

Parameters	Meaning
D	Number of sequences (in thousands) in the database
C	Average number of itemsets per sequence
T	Average number of items per itemset
N	Number of different items (in thousands) in the database

Table 3. Characteristics of datasets.

Dataset	#Sequence	#Item	Avg. seq. length (items)	Type of data
BMS	59,601	497	2.51 (std = 4.85)	Web click stream
Kosarak	10,000	10,094	8.14 (std = 22)	Web click stream
D2C7T10N2.5S6I4	2,000	2,500	53.5	Synthetic
D5C10T6N5S6I4	5,000	5,000	51.0	Synthetic

5.1 Correctness and Runtime of GFS-CR

The first goal of the experiment is to check the correctness of the proposed GFS-CR algorithm, which is ensured by the theoretical results of Sect. 3. In addition, the performance of GFS-CR is also compared with that of CM-SPADE, CM-SPAM and PrefixSpan, which discover sequential patterns directly from databases. The algorithms were run on each SDB while decreasing the ms threshold. The result sets (sequential patterns) discovered by the algorithms were compared, as well as their runtimes.

In terms of result sets, it is found that the sequential pattern set generated by GFS-CR is always identical to those discovered by CM-SPADE, CM-SPAM and PrefixSpan on all the tested SDBs for various ms values. This empirical evidence confirms the correctness of GFS-CR. Because the GFS-CR algorithm generates the $\mathcal{FS}$ set from the $\mathcal{FCS}$ and $\mathcal{FGS}$ sets, for each SDB we recorded the cardinalities of these sets ($\mathcal{FCS}$ and $\mathcal{FGS}$) and calculated the ratio of $|\mathcal{FCS}|$ and $|\mathcal{FGS}|$ to $|\mathcal{FS}|$. In Table 4, the ratios are shown for each tested SDB and various ms values. It is observed in this table that the cardinalities of $\mathcal{FCS}$ and $\mathcal{FGS}$ are often much smaller than that of $\mathcal{FS}$.

Table 4. Comparison of the cardinalities of $\mathcal{FCS}$, $\mathcal{FGS}$ and $\mathcal{FS}$.

Dataset	minsupp (%)	$|\mathcal{FS}|$ (1)	$|\mathcal{FCS}|$ (2)	$|\mathcal{FGS}|$ (3)	(2)/(1) (%)	(3)/(1) (%)
BMS	0.055	69,417,073	99,696	212,348	0.14	0.3
Kosarak	0.14	270,131,364	16,918	20,769	0.006	0.008
D2C7T10N2.5S6I4	0.6	26,339,016	25,671	44,422	0.09	0.017
D5C10T6N5S6I4	0.42	31,557,297	26,514	97,257	0.08	0.031

In terms of runtime, Fig. 4. shows the runtimes of the four algorithms on the tested SDBs for various *ms* values, where the runtime of GFS-CR includes the runtime for mining the sets $\mathcal{FCS}$ and $\mathcal{FGS}$ from the SDBs using the FGenCloSM [10] algorithm and the runtime for generating $\mathcal{FS}$ from the sets. As shown in this figure, the runtimes of GFS-CR are often much less than those of CM-SPAM, CM-SPADE and PrefixSpan, especially for low *ms* values. For example, for *ms* = 0.4% on the SDB D5C10T6N5S6I4, GFS-CR is from 12 to 20 times faster than the remaining algorithms. The reason is that the number of sequential patterns discovered is too large compared to the number of closed sequential patterns and generators (see Table 4). It is notable that for *ms* < 0.059% on BMS and *ms* = 0.5% on D2C7T10N2.5S6I4, the algorithms CM-SPADE and PrefixSpan did not terminate within 1,000 s, while GFS-CR and CM-SPAM completed their work, and the runtime gap between GFS-CR and CM-SPAM considerably increased when *ms* was decreased to lower values.

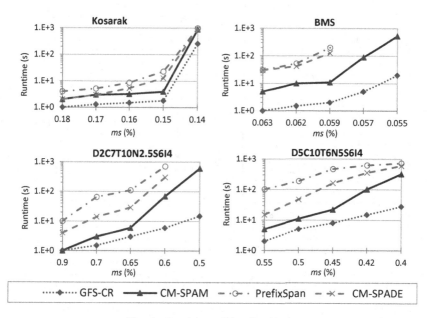

Fig. 4. Runtimes of the algorithms.

5.2 Scalability

To evaluate the scalability of the GFS-CR algorithm, a second experiment was performed where parameters of synthetic datasets (see Table 2) were varied. Figure 5 shows the runtimes of the algorithms on synthetic datasets generated using the IBM Quest Synthetic Data generator. In this figure, a parameter letter followed by the wildcard (*) symbol indicates that the parameter is varied for that dataset.

Results depicted in Fig. 5 show that runtimes increase along with dataset parameters. However, GFS-CR has linear scalability for all dataset parameters, while the runtimes of other algorithms significantly increase for the parameters D, T and N.

Overall, experiments demonstrate that the theoretical results proposed in Sect. 3 not only provide an explicit relationship between $\mathcal{FS}$ and the sets $\mathcal{FCS}$ and $\mathcal{FGS}$, but also that the GFS-CR algorithm based on these results can quickly generate $\mathcal{FS}$ from $\mathcal{FCS}$ and $\mathcal{FGS}$. Experiments shows that this approach outperform the traditional approach of directly mining $\mathcal{FS}$ from an SDB, used by the three other algorithms, and that the performance difference is especially large on big databases.

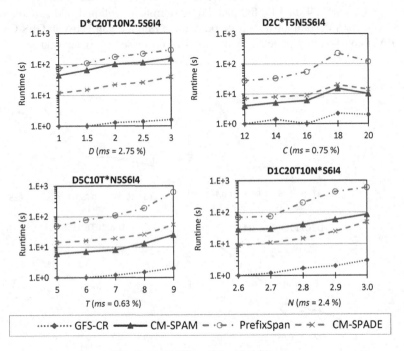

Fig. 5. Scalability of the algorithms.

6   Conclusion and Future Work

This paper has defined an explicit relationship between the set of frequent sequences and the sets of frequent closed and generator sequences. That relationship is the theoretical basis for the novel GFS-CR algorithm, designed to quickly generate $\mathcal{FS}$ from its concise representations, rather than using the traditional approach of directly discovering $\mathcal{FS}$ from an SDB. Results from an experimental evaluation have shown that GFS-CR not only generates the correct result set, but that it also outperforms the state-of-the-art algorithms in terms of execution time and scalability.

The proposed explicit relationship is a theoretical basis that can be used to develop a novel model and efficient algorithms to quickly generate frequent sequences satisfying user-defined constraints from the concise representations, a problem that has many real-life applications. These interesting results, considering different types of constraints, will be presented in our future papers.

Acknowledgment. This work was supported by Vietnam's National Foundation for Science and Technology Development (NAFOSTED) under Grant Number 102.05-2017.300.

Appendices

Appendix 1: Proof of Proposition 1

(i) "$\mathcal{FS}(\sigma, \gamma) \subseteq \mathcal{FS}'(\sigma, \gamma)$": Consider any $\alpha \in \mathcal{FS}(\sigma, \gamma) : \gamma \sqsubseteq \alpha \sqsubseteq \sigma$. Without loss of generality, we can assume that $\gamma = E_1 \rightarrow E_2 \rightarrow \ldots \rightarrow E_p$, $\alpha = F_1 \rightarrow F_2 \rightarrow \ldots \rightarrow F_q$, $\sigma = E_1' \rightarrow E_2' \rightarrow \ldots \rightarrow E_q'$, $F_i \subseteq E_i'$, $\forall i = 1, \ldots, q$ and $\exists lp = \{j_1, j_2, \ldots, j_p\} \in LP(\alpha, \gamma) \subseteq LP(\sigma, \gamma)$, with $1 \leqslant j_1 < j_2 < \ldots j_p \leqslant q$: $E_k \subseteq F_{jk} \subseteq E_{jk}'$, $\forall k = 1, \ldots, p$. Set $d_i \overset{\text{def}}{=} \begin{cases} F_i, & \text{if } i \notin lp \\ F_i \backslash E_k, & \text{if } i = j_k \in lp \end{cases}$, $D_i \overset{\text{def}}{=} \begin{cases} E_i', & \text{if } i \notin lp \\ E_i' \backslash E_k, & \text{if } i = j_k \in lp \end{cases}$, $\delta(lp) \overset{\text{def}}{=} d_1 \rightarrow d_2 \rightarrow \ldots \rightarrow d_q$, $E_i^{\sim} \overset{\text{def}}{=} \begin{cases} \emptyset, & \text{if } i \notin lp \\ E_k, & \text{if } i = j_k \in lp \end{cases}$, $\forall i = 1, \ldots, q$, $Ex(\gamma, lp) \overset{\text{def}}{=} E_1^{\sim} \rightarrow E_2^{\sim} \rightarrow \ldots \rightarrow E_q^{\sim}$, then $d_i \subseteq D_i$ and $F_i = E_i^{\sim} + d_i$, $\forall i = 1, \ldots, q$, so $\alpha = Ex(\gamma, lp) \oplus \delta(lp) \in \mathcal{FS}'(\sigma, \gamma)$.

(ii) "$\mathcal{FS}(\sigma, \gamma) \supseteq \mathcal{FS}'(\sigma, \gamma)$": $\forall \alpha \in \mathcal{FS}'(\sigma, \gamma)$, $\alpha = Ex(\gamma, lp) \oplus \delta(lp) = F_1 \rightarrow F_2 \rightarrow \ldots \rightarrow F_q$, then $F_i = E_i^{\sim} + d_i$, $d_i \subseteq D_i$, $E_i^{\sim} \subseteq F_i \subseteq E_i^{\sim} + D_i \subseteq E_i'$, $\forall i = 1, \ldots, q$. Thus, $\gamma \sqsubseteq \alpha \sqsubseteq \sigma$, $\alpha \in \mathcal{FS}(\sigma, \gamma)$. □

Appendix 2: Proof of Theorem 1

a. *First*, we prove that $\forall \alpha \in \mathcal{FS}$, $\exists \sigma \in CloSet(\alpha)$ such that $ro \overset{\text{def}}{=} \rho(\alpha) = \rho(\sigma)$, so $\sigma \in \mathcal{FCS}$ and $ro \in RO$, then $FS(ro) = [\sigma]$. Indeed, $\beta \in FS(ro) \Leftrightarrow [\beta \in \mathcal{FS} \wedge \rho(\beta) = ro] \Leftrightarrow [\beta \in \mathcal{FS} \wedge \rho(\beta) = \rho(\sigma) = ro] \Leftrightarrow \beta \in [\sigma]$.

(i) "$\mathcal{FS} \subseteq \bigcup_{ro \in RO} FS(ro)$": $\forall \alpha \in \mathcal{FS}$, $\exists \sigma \in CloSet(\alpha)$ and $ro \overset{\text{def}}{=} \rho(\sigma) = \rho(\alpha)$. Then $\alpha \in [\sigma]$, $supp(\sigma) = supp(\alpha) \geqslant ms$. Thus, $\sigma \in \mathcal{FCS}$, $ro \in RO$ and $\alpha \in FS(ro)$.

(ii) "$\bigcup_{ro \in RO} FS(ro) \subseteq \mathcal{FS}$": $\forall ro \in RO$ and $\alpha \in FS(ro)$, then $\exists \sigma \in \mathcal{FCS} : \alpha \sqsubseteq \sigma$ and $ro = \rho(\sigma) = \rho(\alpha)$, so $supp(\alpha) = supp(\sigma) \geqslant ms$ and $\alpha \in \mathcal{FS}$. Hence, $\mathcal{FS} = \bigcup_{ro \in RO} FS(ro)$.

Since $\sim$ is an equivalence relation, then different equivalence classes $[\sigma]$ or $FS(ro)$ are disjoint. Thus, $\mathcal{FS} = \bigcup_{ro \in RO} FS(ro)$.

b. *First*, we prove that all different subsets $\mathcal{FS}^*(\sigma_i, \gamma_j)$, $\forall \sigma_i \in CS(ro)$, $\forall \gamma_j \in GenSet(\sigma_i)$, are disjoint. *Indeed*, assume conversely that $\exists \sigma_m$, $\sigma_i \in CS(ro)$, $\exists \gamma_k \in GenSet(\sigma_m)$, $\exists \gamma_j \in GenSet(\sigma_i)$: $(\sigma_m \neq \sigma_i$ or $\gamma_k \neq \gamma_j)$ and $\exists \alpha \in \mathcal{FS}^*(\sigma_m, \gamma_k) \cap \mathcal{FS}^*(\sigma_i, \gamma_j)(\neq \emptyset)$. Consider the following two cases.

(i) If $\sigma_m \neq \sigma_i$, then without loss of generality, we can assume that $m < i$ and $\exists \gamma \in GenSet(\alpha)$ (because $GenSet(\alpha) \neq \emptyset$). We have $\gamma \in \mathcal{GS}$, $\gamma \sqsubseteq \alpha \sqsubseteq \sigma_i$, $\rho(\gamma) = \rho(\alpha) = \rho(\sigma_i)$, because $\alpha \in \mathcal{FS}^*(\sigma_i, \gamma_j)$. Then $\gamma \in GenSet(\sigma_i)$, $\alpha \in \mathcal{FS}(\sigma_i, \gamma)$ and $\alpha \in \mathcal{FS}^*(\sigma_i, \gamma)$. Moreover, since $\alpha \in \mathcal{FS}^*(\sigma_m, \gamma_k) \subseteq \mathcal{FS}(\sigma_m, \gamma_k)$, then $\alpha \sqsubseteq \sigma_m$ with $m < i$. This contradicts the condition *not* $(DCondC(\alpha, \sigma_i, CS(\rho(\sigma_i))))$ in $\mathcal{FS}^*(\sigma_i, \gamma)$.

(ii) Otherwise, if $\sigma_m \equiv \sigma_i$ and $\gamma_k \neq \gamma_j$, then without loss of generality, we can assume that $k < j$. Since $\alpha \in \mathcal{FS}^*(\sigma_i, \gamma_k) \subseteq \mathcal{FS}(\sigma_i, \gamma_k)$, then $\alpha \sqsupseteq \gamma_k$, with $k < j$. This also contradicts the condition *not* $(DCondG(\alpha, \sigma_i, \gamma_j))$ in $\mathcal{FS}^*(\sigma_i, \gamma_j)$. *Finally*, we prove that $FS(ro) = \sum_{\sigma_i \in CS(ro)} \sum_{\gamma_j \in GenSet(\sigma_i)} \mathcal{FS}^*(\sigma_i, \gamma_j)$, $\forall ro \in RO$.

(iii) "$\subseteq$": $\forall \alpha \in FS(ro)$, then $\alpha \in \mathcal{FS}$ and $\rho(\alpha) = ro$. Since $CloSet(\alpha) \neq \emptyset$ and $GenSet(\alpha) \neq \emptyset$, $[\exists \sigma_i \in CloSet(\alpha), \exists \gamma_j \in GenSet(\alpha) : \sigma_i \in CS, \gamma_j \in \mathcal{GS}, \gamma_j \sqsubseteq \alpha \sqsubseteq \sigma_i, \rho(\sigma_i) = \rho(\gamma_j) = \rho(\alpha)]^{(*)}$, so $\gamma_j \in GenSet(\sigma_i)$, $supp(\sigma_i) = supp(\gamma_j) = supp(\alpha) \geqslant ms$, $\sigma_i \in \mathcal{FCS}$ and $\sigma_i \in CS(ro)$. Hence, $\alpha \in \mathcal{FS}(\sigma_i, \gamma_j)$. Without loss of generality, we can select σ_i and γ_j that are, respectively, the first closed and generator sequences satisfying the condition$^{(*)}$. Thus, $FS(ro) \subseteq \sum_{\sigma_i \in CS(ro)} \sum_{\gamma_j \in GenSet(\sigma_i)} \mathcal{FS}^*(\sigma_i, \gamma_j)$.

(iv) "$\supseteq$": $\forall \sigma_i \in CS(ro)$, $\forall \gamma_j \in GenSet(\sigma_i)$, $\forall \alpha \in \mathcal{FS}^*(\sigma_i, \gamma_j)$, we have $\sigma_i \in \mathcal{FCS}$, $\alpha \sqsubseteq \sigma_i$, $\rho(\alpha) = \rho(\sigma_i) = ro$, so $supp(\alpha) = supp(\sigma_i) \geqslant ms$, i.e. $\alpha \in FS(ro)$. Thus, $\mathcal{FS}^*(\sigma_i, \gamma_j) \subseteq FS(ro)$ and $FS(ro) \supseteq \sum_{\sigma_i \in CS(ro)} \sum_{\gamma_j \in GenSet(\sigma_i)} \mathcal{FS}^*(\sigma_i, \gamma_j)$. □

Appendix 3: Proof of Theorem 2

(a) It is clear that $\mathcal{FS}^{**}(\sigma, \gamma) \subseteq \mathcal{FS}'(\sigma, \gamma)$. In the proof of the above three pruning cases, $\forall \alpha \in \mathcal{FS}'(\sigma, \gamma)$, if $\exists i = 1, 2, 3: DCond_i$ is true, then α is a duplicate of some previously generated sequences. Thus, $\mathcal{FS}'(\sigma, \gamma) \subseteq \mathcal{FS}^{**}(\sigma, \gamma)$ and $\mathcal{FS}'(\sigma, \gamma) = S^{**}(\sigma, \gamma)$.

(b) We will prove the first assertion by contradiction. Assume conversely that there are two sequences $\alpha = Ex(\gamma, lp_m) \oplus \delta(lp_m)$ and $\beta = Ex(\gamma, lp_n) \oplus, \delta(lp_n)$ (in $\mathcal{FS}^{**}(\sigma, \gamma)$) according to two different position lists in $LP(\sigma, \gamma)$: $lp_m \neq lp_n$, but $\alpha \equiv \beta$, i.e., there exist two $lp_m = \{j_1, j_2, \ldots, j_p\}$, $lp_n = \{i_1, i_2, \ldots, i_p\} \in LP(\sigma, \gamma)$ with $m < n$, $1 \leqslant j_1 < j_2 < \ldots j_p \geqslant q$, $1 \leqslant i_1 < i_2 < \ldots < i_p \leqslant q$, $d(lp_m)_i$

$$\overset{\text{def}}{=} \begin{cases} E'_i, & \text{if } i \notin lp_m \\ E'_i \backslash E_k, & \text{if } i = j_k \in lp_m \end{cases}, \quad d_i \overset{\text{def}}{=} d(lp_n)_i \overset{\text{def}}{=} \begin{cases} E'_i, & \text{if } i \notin lp_n \\ E'_i \backslash E_k, & \text{if } i = i_k \in lp_n \end{cases}, \quad F_i \subseteq d(lp_m)_i,$$

$F'_i \subseteq d_i$, $\forall i = 1,\ldots, q$, and two corresponding sequences $\alpha = F_1 \to F_2 \to \ldots \to F_q$ and $\beta = F'_1 \to F'_2 \to \ldots \to F'_q$ that belong to $\mathcal{FS}^{**}(\sigma, \gamma)$ such that $\alpha \equiv \beta$ and $DCond_k(lp_m)$, $DCond_k(lp_n)$ are false, $\forall k = 1, 2, 3$.

After deleting all empty itemsets in α and β, we obtain $\alpha = F_{u_1} \to F_{u_2} \to \ldots \to F_{u_N}$, $\beta = F'_{v_1} \to F'_{v_2} \to \ldots \to F'_{v_N}$ with $N = size(\alpha) = size(\beta)$, $1 \leqslant u_1 < u_2 < \ldots < u_N \leqslant q$, $1 \leqslant v_1 < v_2 < \ldots < v_N \leqslant q$ and $F_{u_i} = F'_{v_i} \neq \emptyset, \forall i = 1,\ldots, N$ (because $\alpha \equiv \beta$). We set $i_0 = j_0 = u_0 = v_0 = k_0 \equiv 0$, $k_{p+1} = N$ and $1 \leqslant k_1 < k_2 < \ldots < k_p \leqslant N$: $j_r = u_{k_r}$ (so $F_{u_{k_r}} = F_{j_r} \supseteq E_r \subseteq F'_{j_r}$), $\forall r = 1, \ldots, p$.

For $\forall r = 1, \ldots, p + 1$, $\forall k = (k_{r-1} + 1),\ldots, k_r$, then $u_{k_{r-1}} = j_{r-1} < u_k \leqslant j_r \leqslant i_r$ and we prove that if $\{(v_h = u_h, \ \forall h = 0, \ \ldots, \ k-1), \ (v_{k-1} < i_r)$ and $(\forall r' = 0, \ldots, r - 1, j_{r'} = i_{r'} = u_{k_{r'}} = v_{k_{r'}})\}$, $(H(r, k))$, then $\{v_k = u_k$, and if $(r \leqslant p)$ then $[(v_k < i_r$, if $k < k_r)$ and $(j_r = i_r = u_{k_r} = v_{k_r})]\}$, $(C(r, k))$.

Since $i_0 = j_0 = u_0 = v_0 = k_0 = 0$ and $i_0 < i_1$, then the hypothesis $H(1, 1)$ is always true. For $\forall r = 1, \ldots, p + 1$, $\forall k = (k_{r-1} + 1), \ldots, k_r$, assume that the hypothesis $H(r, k)$ is true.

(i) *First*, we consider any k such that $[k_{r-1} + 1 \leqslant k < k_r$, for $r \leqslant p]$ or $[k_p + 1 \leqslant k \leqslant N$, with $r = p + 1]$. If $r \leqslant p$, then $i_{r-1} = j_{r-1} = u_{k_{r-1}} < u_k < j_r = u_{k_r}$ and $v_k \leqslant i_r$, because $F_{u_k} = F'_{v_k} \neq \emptyset$ and $v_{k-1} < i_r$. Consider the following three cases.

If $(r \leqslant p$ and $v_k = i_r)$: then $j_{r-1} < u_k < j_r \leqslant v_k = i_r$, $E'_{u_k} \supseteq F_{u_k} = F'_{v_k} = F'_{i_r} \supseteq E_r$, so $u_k \in lp_{pre} \overset{def}{=} \{j_1,\ldots,j_{r-1}, u_k, j_{r+1},\ldots,j_p\}$ (and $lp_{pre} \overset{def}{=} \{u_k, j_2,\ldots,j_p\}$ if $r = 1$), $lp_{pre} \prec lp_m$, $FI(pre, m) = r$, $lp_{pre}[r] = u_k$ and the u_k^{th} itemset (F_{u_k}) contains E_r, so $DCond_2(lp_{pre}, lp_m, u_k)$ is *true*. Therefore, $v_k < i_r$.

If $(u_k < v_k$ and $(v_k < i_r$, if $r \leqslant p))$: then $i_{r-1} = j_{r-1} = u_{k_{r-1}} \leqslant u_{k-1} = v_{k-1} < u_k < v_k$ and $(v_k < i_r$, with $r \leqslant p)$, $v_k \notin lp_n$, $u_k \notin lp_n$, $d(lp_n)_{v_k} = E'_{v_k}$, $d(lp_n)_{u_k} = E'_{u_k}$, $F'_{v_{k-1}} \neq \emptyset$, $F'_{r'} = \emptyset$, $\forall r' = (v_{k-1} + 1),\ldots, (v_k - 1)$, $v_{k-1} + 1 \leqslant u_k \leqslant v_k - 1$. Moreover, for the non-empty itemset $its_{t_{v_k}}$ of β such that $its_{t_{v_k}} = F'_{v_k} = F_{u_k} \subseteq E'_{v_k} \cap E'_{u_k} = d(lp_n)_{v_k} \cap d(lp_n)_{u_k}$, so $DCond_1(lp_n, v_k, u_k)$ is *true*.

If $v_k < u_k$, then $j_{r-1} = u_{k_{r-1}} \leqslant u_{k-1} = v_{k-1} < v_k < u_k$ and $(u_k < j_r$, with $r \leqslant p)$. Similarly, $u_k \notin lp_m$, $v_k \notin lp_m$, $d(lp_m)_{v_k} = E'_{v_k}$, $d(lp_m)_{u_k} = E'_{u_k}$, $F_{u_{k-1}} \neq \emptyset$, $F_{r'} = \emptyset$, $\forall r' = (u_{k-1} + 1),\ldots, (u_k - 1)$, $u_{k-1} + 1 \leqslant v_k \leqslant u_k - 1$. Moreover, for the non-empty itemset its_{u_k} of α such that $its_{u_k} = F_{u_k} = F'_{v_k} \subseteq E'_{u_k} \cap E'_{v_k} = d(lp_m)_{u_k} \cap d(lp_m)_{v_k}$, so $DCond_1(lp_m, u_k, v_k)$ is *true*.

Hence, $u_k = v_k$ and $(v_k < j_r \leqslant i_r$, with $r \leqslant p)$. $\left(C(r, k)^{(1)}\right)$.

(ii) *Second*, if $r \leqslant p$, we consider $k = k_r$. Then $i_{r-1} = j_{r-1} = u_{k_{r-1}} < u_k = u_{k_r} = j_r \leqslant i_r$. Since $F_{u_k} = F'_{v_k} \neq \emptyset$ and $v_{k-1} = v_{k_r-1} < i_r$, then $v_{k_r} \leqslant i_r$. Consider the following cases.

If $v_{k_r} < u_{k_r} = j_r \leqslant i_r$, then $i_{r-1} = j_{r-1} = u_{k_{r-1}} = v_{k_r-1} < v_{k_r} < u_{k_r} = j_r \leqslant i_r$, $E'_{v_{k_r}} \supseteq F'_{v_{k_r}} = F_{u_{k_r}} = F_{j_r} \supseteq E_r$. Thus, $v_{k_r} \in lp_{pre} \overset{def}{=} \{i_1,\ldots,i_{r-1}, v_{k_r}, i_{r+1},\ldots,i_p\}$ (and $lp_{pre} \overset{def}{=} \{v_{k_1}, i_2,\ldots,i_p\}$, if $r = 1$), $lp_{pre} \prec lp_n$, $FI(pre, n) = r, lp_{pre}[r] = v_{k_r}$ and with the itemset $F'_{v_{k_r}}$ of β such that $F'_{v_{k_r}} \supseteq E_r$, so $DCond_2(lp_{pre}, lp_n, v_{k_r})$ is *true* (or $DCond_3(lp_{pre} \overset{def}{=} \{j_1, \ldots, j_{r-1}, v_{k_r}, j_{r+1}, \ldots, j_p\}, lp_m, u_{k_r})$ is *true*).

Therefore, $j_r = u_{k_r} \leqslant v_{k_r} \leqslant i_r$. But the case $u_{k_r} < v_{k_r} \leqslant i_r$ cannot occur.

Indeed, if $j_r = u_{k_r} < v_{k_r} = i_r$, then $i_{r-1} = j_{r-1} = u_{k_{r-1}} = v_{k_{r-1}} \leqslant u_{k_r-1} < u_{k_r}$ $= j_r < v_{k_r} = i_r$. Thus, $F'_{r'} = \emptyset$, $\forall r' = u_{k_r}, \ldots, (v_{k_r} - 1)$, $E'_{u_{k_r}} \supseteq F_{u_{k_r}} = F_{j_r} \supseteq E_r$, $u_{k_r} \in$ $lp_{pre} \overset{def}{=} \{i_1, \ldots, i_{r-1}, u_{k_r}, i_{r+1}, \ldots, i_p\}$ (and $lp_{pre} \overset{def}{=} \{u_{k_1}, i_2, \ldots, i_p\}$, if $r = 1$), $lp_{pre} \prec lp_n$, $FI(pre, n) = r$, $lp_{pre}[r] = u_{k_r}$, $lp_n[r] = i_r = v_{k_r}$. Since $its_{v_{k_r}} = F'_{v_{k_r}} \setminus E_r \subseteq d(lp_n)_{v_{k_r}} = E'_{v_{k_r}} \setminus E_r$ and $F_{u_{k_r}} \setminus E_r \subseteq d(lp_{pre})_{u_{k_r}} = E'_{u_{k_r}} \setminus E_r$, then for itemset $its_{v_{k_r}}$ of β such that $its_{v_{k_r}} = F'_{v_{k_r}} \setminus E_r = F_{u_{k_r}} \setminus E_r \subseteq d(lp_n)_{v_{k_r}} \cap d(lp_{pre})_{u_{k_r}}$. Therefore, $DCond_3(lp_{pre}, lp_n, v_{k_r})$ is *true*.

If $j_r = u_{k_r} < v_{k_r} < i_r$, then $i_{r-1} = j_{r-1} = u_{k_{r-1}} = v_{k_{r-1}} \leqslant v_{k_r-1} = u_{k_r-1} < u_{k_r} = j_r < v_{k_r} < i_r$. Hence, $F'_{r'} = \emptyset$, $\forall r' = (v_{k_r-1} + 1), \ldots, (v_{k_r} - 1)$, $v_{k_r-1} + 1 \leqslant u_{k_r} \leqslant v_{k_r} - 1, u_{k_r}$ and $v_{k_r} \notin lp_n, d(lp_n)_{v_{k_r}} = E'_{v_{k_r}}, F_{u_{k_r}} \subseteq d(lp_n)_{u_{k_r}} = E'_{u_{k_r}}$. Thus, for the non-empty itemset $its_{v_{k_r}}$ of β such that $its_{v_{k_r}} = F'_{v_{k_r}} = F_{u_{k_r}} \subseteq d(lp_n)_{v_{k_r}} \cap d(lp_n)_{u_{k_r}}$, the condition $DCond_1(lp_n, v_{k_r}, u_{k_r})$ is *true*.

Thus, $v_{k_r} = u_{k_r} = j_r \leqslant i_r$. However, if $v_{k_r} = u_{k_r} = j_r < i_r$, then $i_{r-1} = j_{r-1} = u_{k_{r-1}} = v_{k_{r-1}} < v_{k_r} = u_{k_r} = j_r < i_r$, $E'_{j_r} \supseteq F'_{j_r} = F'_{v_{k_r}} = F_{u_{k_r}} \supseteq E_r$. Thus, $j_r \in lp_{pre}$ $\overset{def}{=} \{i_1, \ldots, i_{r-1}, j_r, i_{r+1}, \ldots, i_p\}$ (and $lp_{pre} \overset{def}{=} \{j_1, i_2, \ldots, i_p\}$, if $r = 1$), $lp_{pre} \prec lp_n$, $FI(pre, n) = r$, $lp_{pre}[r] = j_r$ and the itemset F'_{j_r} of β contains E_r, so $DCond_2(lp_{pre}, lp_n, j_r)$ is *true*.

Hence, $v_{k_r} = u_{k_r} = j_r = i_r$. $(C(r, k)^{(2)})$.

From $(C(r, k)^{(1)}) - (C(r, k)^{(2)})$, we have $(C(r, k))$.

Finally, under the *hypothesis* $\alpha \equiv \beta$, then $\forall r = 1, \ldots, p + 1$, $\forall k = (k_{r-1} + 1), \ldots, k_r$, we always have $(v_k = u_k)$, $(j_r = i_r = u_{k_r} = v_{k_r}$, if $r \leqslant p)$. Therefore,

$$(\forall k' = 1, \ldots, N, v_{k'} = u_{k'})^{(*)} \text{ and } (\forall r = 1, \ldots, p, j_r = i_r)^{(**)}.$$

If $lp_m = lp_n$, then $j_r = i_r$, $\forall r = 1, \ldots, p$. Since α and β are generated from two different position lists in $\mathcal{FS}^{**}(\sigma, \gamma)$, so $\exists i_0 = 1, \ldots, N$ such that $u_{i_0} \neq v_{i_0}$. This contradicts $^{(*)}$.

If $lp_m \neq lp_n$, i.e. $\exists m_0 = 1, \ldots, p$ such that $i_{m_0} \neq j_{m_0}$. This also contradicts $^{(**)}$. In other words, the *hypothesis* $\alpha \equiv \beta$ always leads to a contradiction.

Thus, $\alpha \neq \beta$, i.e., all sequences in $\mathcal{FS}^{**}(\sigma, \gamma)$ are distinctly generated.

From Proposition 1 and Theorem 1.a, we have $\mathcal{FS}(\sigma, \gamma) = \mathcal{FS}'(\sigma, \gamma) = \mathcal{FS}^{**}(\sigma, \gamma)$ and the remaining assertions are deduced from Theorem 1.   □

References

1. Agrawal, R., Srikant, R.: Mining sequential patterns. In: Proceedings of 11th International Conference on Data Engineering, pp. 3–14 (1995)
2. Tran, A., Duong, H., Truong, T., Le, B.: Mining frequent itemsets with dualistic constraints. In: Anthony, P., Ishizuka, M., Lukose, D. (eds.) PRICAI 2012. LNCS (LNAI), vol. 7458, pp. 807–813. Springer, Heidelberg (2012). https://doi.org/10.1007/978-3-642-32695-0_77
3. Anh, T., Tin, T., Bac, L.: Structures of frequent itemsets and classifying structures of association rule set by order relations. Intell. Inf. Database Syst. **8**(4), 295–323 (2014)
4. Ayres, J., Flannick, J., Gehrke, J., Yiu, T.: Sequential pattern mining using a bitmap representation. In: Proceedings of 8th ACM SIGKDD International Conference on Knowledge Discovery and Data Mining, pp. 429–435 (2002)
5. Bac, L., Hai, D., Tin, T., Fournier-Viger, P.: FCloSM, FGenSM: two efficient algorithms for mining frequent closed and generator sequences using the local pruning strategy. Knowl. Inf. Syst. **53**(1), 71–107 (2017)
6. Fournier-Viger, P., Gomariz, A., Campos, M., Thomas, R.: Fast vertical mining of sequential patterns using co-occurrence information. In: Tseng, V.S., Ho, T.B., Zhou, Z.-H., Chen, A.L. P., Kao, H.-Y. (eds.) PAKDD 2014. LNCS (LNAI), vol. 8443, pp. 40–52. Springer, Cham (2014). https://doi.org/10.1007/978-3-319-06608-0_4
7. Fournier-Viger, P., Gomariz, A., Gueniche, T., Soltani, A., Wu, C., Tseng, V.S.: SPMF: a Java open-source pattern mining library. Mach. Learn. Res. **15**(1), 3389–3393 (2014)
8. Fournier-Viger, P., Gomariz, A., Šebek, M., Hlosta, M.: VGEN: fast vertical mining of sequential generator patterns. In: Bellatreche, L., Mohania, M.K. (eds.) DaWaK 2014. LNCS, vol. 8646, pp. 476–488. Springer, Cham (2014). https://doi.org/10.1007/978-3-319-10160-6_42
9. Gomariz, A., Campos, M., Marin, R., Goethals, B.: ClaSP: an efficient algorithm for mining frequent closed sequences. In: Pei, J., Tseng, V.S., Cao, L., Motoda, H., Xu, G. (eds.) PAKDD 2013. LNCS (LNAI), vol. 7818, pp. 50–61. Springer, Heidelberg (2013). https://doi.org/10.1007/978-3-642-37453-1_5
10. Hai, D., Tin, T., Bac, L.: Efficient algorithms for simultaneously mining concise representations of sequential patterns based on extended pruning conditions. Eng. Appl. Artif. Intell. **67**, 197–210 (2018)
11. Hai, D., Tin, T., Bay, V.: An efficient method for mining frequent itemsets with double constraints. Eng. Appl. Artif. Intell. **27**, 148–154 (2014)
12. Pei, J., et al.: Mining sequential patterns by pattern-growth: the PrefixSpan approach. IEEE Trans. Knowl. Data Eng. **16**(11), 1424–1440 (2004)
13. Srikant, R., Agrawal, R.: Mining sequential patterns: generalizations and performance improvements. In: Apers, P., Bouzeghoub, M., Gardarin, G. (eds.) EDBT 1996. LNCS, vol. 1057, pp. 1–17. Springer, Heidelberg (1996). https://doi.org/10.1007/BFb0014140
14. Tin, T., Hai, D., Ngan, H.N.T.: Structure of frequent itemsets with extended double constraints. Vietnam J. Comput. Sci. **3**(2), 119–135 (2016)
15. Wang, J., Han, J., Li, C.: Frequent closed sequence mining without candidate maintenance. IEEE Trans. Knowl. Data Eng. **19**(8), 1042–1056 (2007)
16. Yan, X., Han, J., Afshar, R.: CloSpan: mining closed sequential patterns in large datasets. In: Proceedings of SIAM International Conference on Data Mining, pp. 166–177 (2003)
17. Zaki, M.J.: SPADE: an efficient algorithm for mining frequent sequences. Mach. Learn. **42** (1), 31–60 (2001)

Machine Learning

A Novel Approach to Identify the Determinants of Online Review Helpfulness and Predict the Helpfulness Score Across Product Categories

Debasmita Dey[(⊠)] and Pradeep Kumar

Indian Institute of Management Lucknow, Lucknow, UP, India
{fpml7006, pradeepkumar}@iiml.ac.in

Abstract. The proliferation in the number of available online reviews provides an excellent opportunity to use this accumulated enormous information of any product in a more strategic way to improve the quality of the product and services of the e-commerce company. Due to the non-uniform quality of online reviews, it is crucial to identify those helpful reviews from the pile of a large amount of low quality and low informative other reviews. This system will help the customers to form an unbiased opinion quickly by looking at its level of helpfulness. The e-commerce companies measure the helpfulness of a review using the number of votes it gets from other customers. This situation arises problems to newly-authored potentially helpful reviews due to lack of votes. Thus it is essential to have an automated process to estimate and predict helpfulness of any review. This paper identifies the essential characteristics of online reviews influencing the helpfulness of it. This study categorized all characteristics of reviews collected from previous literature in four main categories and then study the combined effect of the four aspects in predicting the helpfulness of a review. The product type (Search or Experience) acts as a control variable in the factors identification model of helpful prediction of a review. An analysis of total 14782 reviews from Amazon.com across five different product category shows the factors influencing the helpfulness of a review varies across product categories. Then a comparative study of two widely used machine learning, Artificial Neural Network and Multiple Adaptive Regression Spline are presented to predict the helpfulness of online review across five different categories and a better method of predicting helpfulness of online reviews are suggested based on the type of product. This study solves the starvation problem of potential newly-authored or infamous reviews without any manual votes along with high accuracy of helpfulness prediction.

Keywords: Online reviews · Helpfulness prediction · Artificial Neural Network · Multiple Adaptive Regression Splines

© Springer Nature Switzerland AG 2019
S. Madria et al. (Eds.): BDA 2019, LNCS 11932, pp. 365–388, 2019.
https://doi.org/10.1007/978-3-030-37188-3_21

1 Introduction

An increasing number of online reviews, e-commerce websites have offered customers their platforms to give their opinions and reviews of products, services, and the seller of the product. These reviews are given by those customers who have already bought or used the products from that e-commerce website or any other sources and are considered as a proxy for the product quality in offline word-of-mouth (WoM) communication. Online customer reviews are defined as "peer-generated product evaluations posted on company or third party websites" [17]. Ecommerce websites allow the customer to share their views about the products and services provided by them in two ways: (a) by numerical star ratings ranging 1 to 5, (b) by providing an area to write your opinion about the product. The availability of customer reviews on an e-commerce website has proved to make a better perception in the customer about its importance, social presence [1], and "stickiness" (time spent on a particular e-commerce site). Increasing the availability of online reviews also proliferate the opportunity of using it more strategically to improve the product and service of the e-commerce company. Now, the quality of these online reviews is not likely to be uniform, and this could range from an excellently vivid description and evaluation to spam with no value addition to make any decision about the purchase. The reviews which are helpful to customers in decision making generally lies under the heap of a large amount of low quality and low informative or fake (and spam) reviews and hence it is challenging for customers to identify and use those helpful reviews to form an unbiased opinion about any product or service.

To solve this problem, e-commerce websites allow the customer to vote for or against the review to show their support or disagreement with that particular review content. For example, "100 out of 140 people found the review helpful" shows the fact that 100 people apart from the reviewer found the review helpful to make their decision and rest 40 people did not find it helpful. The website also allows sorting all customer reviews as per helpfulness of it. However, there is found no such theoretically grounded specific explanation of what are the factors determining the helpfulness of a review and how they are being calculated to sort it. A massive portion of reviews contains no votes or have few votes, and this makes it more challenging to understand their helpfulness and social validity. As per Yang et al. [29], only 20% of the reviews from Amazon dataset [6] have more than five votes, and rest of them either do not have any votes or lesser than five votes. This situation arises problems to the newly authored potential reviews, which does not get the chance to be read by other customers due to lack of votes or infamous products. Therefore, it is essential to have an automated system to estimate helpfulness of any review instead of some manual process (Fig. 1).

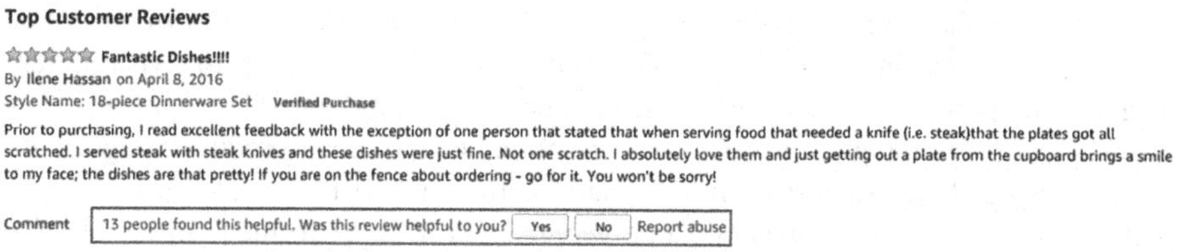

Top Customer Reviews

⭐⭐⭐⭐⭐ **Fantastic Dishes!!!!**
By Ilene Hassan on April 8, 2016
Style Name: 18-piece Dinnerware Set   Verified Purchase

Prior to purchasing, I read excellent feedback with the exception of one person that stated that when serving food that needed a knife (i.e. steak)that the plates got all scratched. I served steak with steak knives and these dishes were just fine. Not one scratch. I absolutely love them and just getting out a plate from the cupboard brings a smile to my face; the dishes are that pretty! If you are on the fence about ordering - go for it. You won't be sorry!

Comment | 13 people found this helpful. Was this review helpful to you?   Yes   No   Report abuse

Fig. 1. An example of the structure of a review on Amazon.com.

The purpose of this study predicts a score denoting the helpfulness of a review automatically by identifying and analyzing linguistic, psychological, and peripheral factors of any review. Previous researches on this issue mostly address either linguistic determinants of psychological features affecting the helpfulness of any review. This study allows focusing on all possible aspect of factors possibly drive the helpfulness of online reviews together. The four identified an aspect of factors that determines helpfulness can be categorized into four types, e.g., Linguistic, Psychological, Text complexity, and peripheral cues. This study identifies the fact that considering only one type of factor to determine review helpfulness might neglect the other aspect of features, which can play a significant part in it. Therefore, apart from the confirmed variables (Rating, Positive emotion, Negative emotion etc.) from previous literature, this study considered not only considered the linguistic features (word count, word per sentence, adjective, etc.) but also the psychological (Analytical thinking, Tone, Authenticity, and Confidence) thought process of the reviewer.

The previous studies [12, 13, 17] also address the fact that product type act as a control variable, and for different types of products (Search good or Experience good), the factors driving helpfulness can differ. Therefore, this study considers this constraint and chooses five different categories of products to address whether and how the determinants of review helpfulness differ as per product type. For example, a highly analytical review may be more useful for cellphone category product customers than the grocery category of products. The five categories are chosen in such a way that it includes not only pure search type (cellphone, clothing) or experience type (beauty, grocery) of good but also products which do not have any physical presence (digital music) and can be considered both search and experience type.

Our research addresses three research questions. First, what are the linguistic and psychological features across different product categories and whether it is needed to study them separately or not? Second, what are the determinants driving the perceived helpfulness of an online review based on different product categories? Third, which method, among the two most used machine learning supervised methods, better predict the helpfulness of an online review?

To address these three research questions, a significantly large number of online reviews from Amazon.com are collected, and these reviews belong to five different categories of products (beauty, grocery, cellphone, clothing, and digital music). The linguistic and psychological variables of online reviews of each product category are extracted using R and Linguistic Inquiry and Word Count (LIWC) software. Then, to address our first research question, one-way ANOVA is performed across all five product categories on all the chosen variables (linguistics and linguistics, etc.) from literature. Next, to explore the factors determining the review helpfulness, the most widely adopted method, Linear regression (LR), is used. Due to the non-linear structure of the data, LR performed poorly and hence a wrapper built around Random Forest classification algorithm, Boruta, is performed, and a subset of variables are selected to predict the helpfulness of reviews of that particular product category reviews. Finally, two widely used machine learning algorithms, Artificial Neural Network (ANN), and Multiple Adaptive Regression Splines (MARS) are used to predict the helpfulness of online reviews using R 3.4.4 software. These two prediction methods are then compared in terms of their Mean Squared Error (MSE) to select the better performing model for each of five product categories.

In short, this study helps to process a large number of online reviews efficiently in an automated way even if reviews do not have any manually entered votes and generate a helpfulness score based on different types of characteristics of a review for each written review. Customers can quickly sort the reviews as per their helpfulness score to make a better purchase decision.

This paper is organized in the following sections. Section 2 presents the previous literature reviews related to these topics. Section 3 presents the method of data collection. Section 4 presents the selection of various variables for this study from the literature. Section 5 gives a clear stepwise idea of the research methodology to solve our research questions. Section 6 presents the result of the experiments and the analysis of the results. Finally, the implication of this study and future scopes are concluded in Sect. 7.

2   Literature Review

Previous researches on the helpfulness of reviews were mainly addressed two typical questions: (i) finding out the critical factors influencing review helpfulness, and (ii) propose a suitable method to predict review helpfulness. The next two sections will address the studies focused on these two issues separately.

2.1   Important Factors Influencing Review Helpfulness

A customer review can have a set of different features, such as Numerical rating, Number of words used in that review, Polarity (Positive, Negative) of the review content, the readability of the text, Style of the writing, etc. Mudambi and Schuff [17] studied the influence of several words in a review, as well as the extremity of a review on the helpfulness of that review. They experimented with their hypothesis using Amazon.com review datasets. The results of this experiment (Tobit Regression) indicate a positive relationship between several words in a review and helpfulness of that review by considering the product type (search or experienced) as a moderator. Korfiatis et al. [10] studied the effect of readability of the review content with the helpfulness of it. In this paper, they measured readability in following four ways: Gunning Fog Index, Flesch reading scale, Coleman-Liau index, and Automated readability index and suggested that readability of review has a more significant impact on helpfulness than several words in it. Ghose and Ipeirotis [3] considered reviewer information, subjectivity, spelling error, and another six type of readability index to study the relationship of these factors with the helpfulness of review. Their study also shows a significant influence of these six readability indexes on review helpfulness. Krishnamoorthy [11], in his research, considered a set of linguistic characteristics (adjective terms, state and action verbs) and compared with other factors (readability, subjectivity). The results from his study showed that rather than considering only readability measure and numerical rating as essential factors, a hybrid model with some linguistic variables could better explain review helpfulness in terms of predictive accuracy. Ghose and Ipeirotis [3] and Forman [2] et al. also supported this explanation given by Krishnamoorthy, showing that a combination of subjective and objective features of review better explains the review helpfulness than considering any of them separately.

In our study, we adopted linear regression primarily to understand the relationship between the dependent (Helpfulness ratio) and an independent variable. Then Boruta algorithm [14] is applied to measure the variable importance as a better explanation of the dependent variable. The methods addressed in the literature did not take into the nonlinear data structure of the Amazon.com dataset. Thus, using linear regression or other methods with linearity assumption will not be appropriate to derive important factors of review helpfulness.

2.2   Prediction Methodology of Review Helpfulness

Mudambi and Schuff [17], Yin et al. [30], Yang et al. [29], Korfiatis et al. [10] and Forman et al. [2] defined helpfulness of a review as the ratio of helpful votes to total number of votes (Helpful votes + Unhelpful votes). They mostly adopted the commonly used method, Linear Regression (LR), to examine the critical factors and prediction of the helpfulness of reviews. Forman et al. [2] transformed the helpfulness ratio in two-class, Helpful, and Unhelpful based on a threshold value and then predicted the helpfulness using linear regression.

Another commonly used method in this research is Support Vector Machine (SVM), as it handles both linear and nonlinear data. Kim et al. [9] applied Support Vector Regression (SVR) for review prediction. Hu and Chen [4] adopted M5P, SVR, and linear regression to measure their prediction performance and did a comparative analysis. Krishnamoorthy [11] used three techniques, e.g., Support vector classification, Random forest, and Naïve Bayes, to predict review helpfulness and then compared them to propose the best prediction model.

Khashei and Bijari [8] proposed a prediction model using Artificial Neural Network (ANN) due to its data-driven and self-adaptive features. Lee et al. [15] adopted a multilayer perceptron neural network (BPN) to predict review helpfulness and compared it with linear regression analysis.

In our study, we adopted three methods (linear regression, Multiple Adaptive Regression Splines, and ANN) and compared their results to find the best suitable method to predict helpfulness. Linear regression was chosen as it is a convenient method addressed in the literature. The other two methods (Multiple Adaptive Regression Splines and ANN) were selected due to their capability of handling nonlinearity in data and better accuracy in prediction.

3   Data Collection

We gathered data for this study from http://jmcauley.ucsd.edu/data/amazon/ [6] since 2005–2014. Amazon product reviews are collected category wise, and the categories are Beauty, Grocery, Cell phone, Clothing, and Digital music. These five product types were chosen in the study based on the following reasons:

- These five categories contain a large number of customer reviews to be analyzed and modeled for training and testing purposes.
- Based on the Nelson [18, 19] study, we included both search (Cell phone) and experience (beauty, grocery) type of products category.
- Our study addresses the fact that a product can exist along a continuum from simple search to pure experience type of product and hence, considers product categories involving mix (Clothing) of search and experience features.
- Digital products are the latest kind of products in the market with no physical presence. Thus, digital music category reviews are included in this study.

For each category mentioned above, all reviews are collected along with their respective numerical rating, Title, review text, the number of helpful and unhelpful votes. After preprocessing and cleansing of data, we excluded those reviews from analysis, which does not have a minimum of ten votes as minimal helpful votes can introduce biases in the model.

A total of 14782 reviews are finally collected after pre-processing to be analyzed after removing all reviews having lesser than ten votes (Table 1).

Table 1. Total number of reviews collected from each of five categories.

	Beauty	Grocery	Cell phone	Clothing	Digital music
Number of reviews	4139	2477	2442	3844	1880

The structure of the data collected is presented below (Fig. 2):

- Product ID: A2ENZ4FESUXXMT
- Reviewer ID: 1400501466
- Review Text: I was looking at expensive tablets that were more like mini notebook computers. I already have a high end notebook. I wanted something that was very portable and not combersome to take on trips, go to coffee shops and etc. I wanted the ability to get email, do limited surfing and read books. The Nook works flawlessly and the display is really nice. I have an N protocol router and the Nook is quick on the Net. I read some negative reviews here. They appear to be written by folks who want to take a $200 unit and turn it into a $500 unit with various apps and other applications. Here's a news flash for the naysayers. Go out and buy the $500 unit and quit complaining. If you want to read books, surf and get email, you'll like this unit.
- Rating: 5.0
- Review Time: 3 December 2012
- The number of votes on helpfulness: 9
- Total number of votes: 10

Fig. 2. An example of the structure of data collected from Amazon.com.

4 Variable Selection

Due to the unstructured form of review text apart from some explicit information (Numerical rating, Number of helpful votes, and several total votes), the review text is transformed into a standard structural format using LIWC (Linguistic Inquiry and Word Count) 2015 software.

LIWC is a text analysis software proposed by Pennebaker [22] to transform the unstructured text into approximately 90 output scores. This 90 output variable evaluates not only the structural and style feature if the text, but also the psychological thought process of reviewer [23–25, 27]. This tool is extensively used and validated by many articles and research papers [7, 20].

The LIWC output is shown below (Fig. 3):

Source (A)	WC	Analytic	Clout	Authentic	Tone	WPS	compare	affect	posemo	negemo	cogproc	insight	cause	percept	see	hear	feel
I did not know that Converse stooped that low to get their products made in Vietnam or ...	28	8.99	7.67	99.00	1.00	28.00	0.00	3.57	0.00	3.57	28.57	3.57	7.14	0.00	0.00	0.00	0.00
this style does not say if it is Sandalfoot or reinforced toe...couldn't buy because of that	17	1.00	1.00	1.00	25.77	17.00	0.00	0.00	0.00	0.00	29.41	0.00	5.88	5.88	0.00	5.88	0.00
** PLEASE NOTE ** I was new to amazon at the time I posted this review, and was angry ...	272	62.14	16.03	64.08	79.41	17.00	0.74	5.88	4.41	1.47	6.46	0.37	2.21	1.84	0.00	0.00	1.84
It is a great watch. Unfortunately I bought two and I received just oneWhere is the other...	26	53.63	22.08	17.46	25.77	8.87	0.00	7.69	3.85	3.85	3.85	0.00	0.00	7.69	7.69	0.00	0.00
The title says it all. I placed my order days ago yet it sits and sits and Amazon has NO NO ...	135	53.54	22.95	92.47	15.37	19.29	0.74	0.74	0.00	0.74	15.56	3.70	2.96	1.48	0.00	1.48	0.00
I purchase these for my husband and when I opened the packaged, I had an instant visual...	41	16.46	50.00	91.58	71.55	20.50	0.00	2.44	2.44	0.00	2.44	2.44	0.00	4.88	2.44	0.00	2.44
I use the item only in swimming pool environments. Works well and is comfortable. The ...	35	85.46	19.58	95.88	99.00	8.75	5.71	8.57	8.57	0.00	11.43	2.86	5.71	0.00	0.00	0.00	0.00
I HAVE YET TO RECIEVE THIS ITEM IT'S BEEN 4 DAYS PAST THE ESTIMATED ARRIVAL DA...	82	32.57	7.18	98.50	25.77	20.50	2.44	2.44	1.22	1.22	6.10	0.00	0.00	0.00	0.00	0.00	0.00
when clicking on link it did not say they did not have them it sent what it did have	19	1.00	50.00	99.00	25.77	19.00	0.00	0.00	0.00	0.00	15.79	5.26	0.00	10.53	0.00	10.53	0.00
after about two weeks and not even an email in regards to my order I looked into what w...	70	63.36	9.94	99.00	25.77	17.50	4.29	0.00	0.00	0.00	5.71	1.43	0.00	2.86	2.86	0.00	0.00
If I bought this I don't know what happened to it. Didn't wear it.	14	1.00	1.00	89.63	25.77	7.00	0.00	0.00	0.00	0.00	21.43	7.14	0.00	0.00	0.00	0.00	0.00

Fig. 3. An example of the output processed by LIWC software.

All variables are selected from previous literature studies and are categorized into four broad categories: (i) Linguistic, (ii) Psychological, (iii) Text complexity, and (iv) Peripheral cues.

The linguistic category of variables is based on the structure of sentence, punctuations, part of speech, polarity or tone of sentences, etc. Example: pronoun, article, preposition, auxiliary verb, word count, word per sentence, adjective.

The psychological category of variables focuses on feeling and thought processes using semantics. Example: comparative words (bigger, better), Analytic, Clout, Percept, Positive emotion, Tone, Negative emotion.

The text complexity category of variables includes those helping the review to understand or read easily or with difficulty. Example: Flesch reading ease index, Syllable, Dictionary word.

The peripheral ques category contains those variables which are independent of review text. Example: Rating, Time of the review posted.

In our study, the dependent variable (Helpfulness) is defined as the ratio of several helpful votes to the total votes (helpful +Unhelpful). For example, for a review where "100 people out of 150 found this review helpful", the helpfulness of the review will be (100/150) 0.67.

The independent variables are chosen from different kinds of literature listed below:

- Linguistic variables:
  - *Compare:* It is defined as the total number of comparative words (bigger, smaller, greater, etc.) used in the review.
  - *Pronoun:* It is measured by the number of pronouns in the text.
  - *Ppron:* It is defined as the number of personal pronouns (I, you, he, she, etc.) in the review and is calculated as the percentage of several pronouns.
  - *Article:* It is defined as the number of articles (a, an, the, etc.) mentioned in the text.
  - *Preposition:* It is measured by the total number of preposition in the review.
  - *Auxiliary verb:* It is defined as the total number of the auxiliary verb (might, must, could, etc.) used in the review.
  - *Adverb:* It is defined as the total number of adverb verb (very, slowly, quickly, etc.) used in the review.
  - *Adjective:* It is defined as the total number of adjectives (better, bright, dull, thick, etc.) used in the review.
  - *AllPunctuation:* It is defined as the total number of sentences with complete and with grammatically correct punctuations used in the review.
  - *I:* It is defined as the percentage number of occurrences of the word 'I' in the review.
- Psychological variables:
  - *Analytic:* It is known as the categorical dynamic index (CDI) and addresses the level of the formal, logical, and hierarchical thought process of the reviewer. A high score implies more formal, logical, and hierarchical thinking.
  - *Clout:* It is defined as the level of expertise or leadership in some context, or how much one is confident about his or her opinions. A higher clout score indicates a more professional and confident opinion, while a lower score indicates a tentative or humble style.
  - *Tone:* It defines the sentimental and emotional tone of the whole text. A score higher than 50 indicates a positive tone, and lower than 50 scores indicate tone with sadness or anxiety or hostility. The exact score of 50 indicates either a lack of emotion or ambivalence.
  - *Authentic:* It is defined as the level of honesty and disclosing thinking of the reviewer, i.e., expressing more personal, humble, and authentic opinions about something. A higher score indicates more honest and vulnerable thinking.
  - *Cogproc:* It is measured as the ratio words evoking the cognitive thought process (cause, know, etc.) of the thinker. A high cogproc score will indicate a more cognitive opinion rather than thorough normal senses.
  - *Percept:* It is measured as the ratio words evoking perceptual thought process ("look," "feeling," etc.) of the thinker. A high percept score will indicate that the opinion is generated and backed up by using the sensed of the reviewer rather than any cognitive information.
  - *Posemo:* It is measured by the ratio of positive emotion words to the total words in the text. It is identified as one of those confirmed factors determining review helpfulness [5, 11, 16, 19, 21, 24, 26].

- *Negemo:* It is measured by the ratio of negative emotion words to the total words in the text. It is identified as one of those confirmed factors determining review helpfulness [5, 11, 16, 19, 21, 24, 26].
- Text complexity:
  - *WC:* It is defined as the total number of words in the review. It is a certain factor of review helpfulness studied in previous literature. It is used as a proxy for text complexity [15, 19].
  - WPS: It is defined as the number of words per sentence. It is used for sentence complexity [2, 3, 9].
  - *Sixltr:* It is defined as the number of words longer than six letters and is used as a proxy for word complexity.
  - *Dic:* It is defined as the percentage of target words captured by the LIWC dictionary.
  - *Flesch Kincaid Readability:* It is defined as the measure of difficulty in reading and understanding a text in English. A higher score indicates that the text piece is easy to read and understand.
- Peripheral Cues:
  - *Rating:* Rating is a numeric score (1 to 5) given by the customer. It is identified as a confirmatory factor of review helpfulness as studies in the literature [5, 9, 15].

5 Research Methodology

The research methodology of this study is described here stepwise below:

Firstly, reviews collected were collected and cleaned, and the preprocessed to get a basic format as below to proceed further (Fig. 4):

helpful votes	total votes	Helpfulness	reviewText	overall	summary
24	24	1	I haven't been a big fan of Prada's fragrances over the years but absolutely fell in love with the sweetness and candy-like scent of this perfume! This smells like a sweet, decadent caramel with tones of vanilla and I'm not sure what else, but it smells great! Although, I must say that this seller is asking for WAY too high a price for this bottle! You could get the 2.7 oz bottle for around the same price at Neiman Marcus (the 1.7 oz for around $80 if you prefer a smaller bottle)!	5	Love the smell of this!
11	14	0.785714286	I bought a similar type of dispenser back in 1995, when I was outfitting my new apartment. Perhaps that was the Dispenser Classic I. That one lasted 11 years without a problem and I threw it out only because it looked old and I was moving. Flash forward to 2011. I decided to buy one for my condo. A day or two after I put in the bottles, I noticed shampoo from the dispenser dripping onto the faucet. It turns out that one of the bottles had a hairline crack in it. It was tiny but enough to cause a serious leak. I examined the bottles closely and realized how thin and cheap the plastic containers really are. Unbelievable. Now I have to decide whether to use Crazy Glue or go through the hassle of emptying the two other containers and sending the whole thing back to Amazon for a refund. If this is the Classic III, I wonder how bad IV will be. I'd be hesitate to order this again.	2	spensers Made of Cheap, Fragile Plastic - Beware!
17	17	1	This really does cover under eye circles and redness but a little goes a long way. Since it's a thick cream stick you should mix in some moisture cream on your hand and smooth it on sparingly. Too much will look cakey and get into wrinkles. Just a very thin layer is best.	5	Excellent
13	13	1	As long as this eyeliner is applied correctly there should be no peeling, fading or flaking. The only time I get some flaking or peeling is if I apply it too thickly. I put this on early in the morning (7am or so) and don't take it off until 11 or 12 pm. And in all that time it stayed where I put it and the color stayed true. I have horrible allergies and I also have tear producing issues...combine these problems and there is really nothing I could wear to line my eyes-until I found Maybelline Lineworks. My eyes can and will at times ooze tears all day.Sometimes it looks as if I'm having a crying jag. BUT even through all this, my liner never goes anywhere! I use Ponds to take it off at night. Easy peasy!The only thing I'm upset with is that Maybelline seems to be dis-continuing many of the colors and while I can understand it, I don't LIKE it.	5	Inexpensive Perfection

Fig. 4. A glimpse of input data after data cleaning and pre-processing.

In the second step, the reviews written in the English language is transformed into a various numeric score using LIWC dictionary. If any target word is matched with the dictionary word, then the corresponding variable's (out of those 90 variables) score is incremented by one. Figure 5 shows an example of this process.

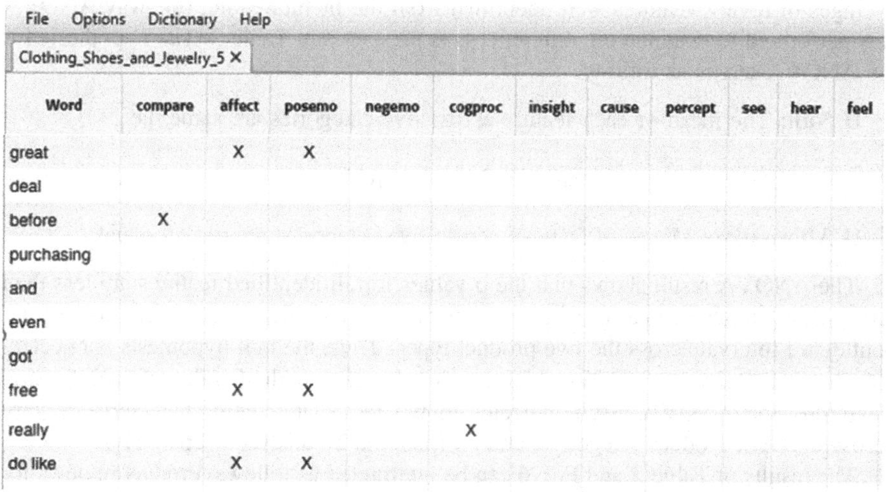

File   Options   Dictionary   Help

Clothing_Shoes_and_Jewelry_5 ×

Word	compare	affect	posemo	negemo	cogproc	insight	cause	percept	see	hear	feel
great		X	X								
deal											
before	X										
purchasing											
and											
even											
got											
free		X	X								
really					X						
do like		X	X								

Fig. 5. Cataloging target words using LIWC 2015 into different linguistic and psychological variables scores.

Next, the exploratory data analysis is performed on our five datasets to calculate the mean and standard deviations of reviews of all categories. To address our first research question, one-way ANOVA is performed on all the variables selected to test whether the product categories are significantly different or not.

In the next step, the determinants of the review helpfulness are explored using linear regression and the Boruta algorithm. Though linear regression is easy to understand and explainable, due to the nonlinear structure of data, it is not suitable to explore the relationship between determinants and the target variable. Hence, Boruta algorithm is performed to explore which independent variables affect the target variable.

In the final step, the helpfulness of online reviews is predicted using the two most widely used techniques for nonlinear data, i.e., Multiple Adaptive Regression Splines (MARS) and Artificial Neural Network (ANN). These two methods to predict online helpfulness of reviews are then compared to determine the suitable method to predict the helpfulness. The helpfulness is predicted considering 70% of training data, and then the mean squared error (MSE) is calculated for both to compare their results for each of five categories.

6 Results and Discussion

6.1 ANOVA Analysis Across Product Categories

Our first research question was whether review characteristics varied across different product categories and if the result is positive, then how they were different. The averages of review features were identified from the literature, and one-way ANOVA was performed to examine the differences, as presented in Table 2. The hypothesis for the ANOVA test is as follow:

H Null: The mean of each feature across five categories are same, i.e.,

$$\mu1 = \mu2 = \mu3 = \mu4 = \mu5$$

H Alternative: Means of features across five categories are not all equal.

The ANOVA result shows that the p values for all identified features are less than 0.01, which indicates that all research variables are significantly different at the 99% confidence interval across the five product types. Thus, the null hypothesis is rejected.

The F critical value is 2.37, which is much lesser than the F value of each feature. Therefore, the ANOVA result indicates to the fact that review characteristics vary across different product categories, and so it should be analyzed separately.

The results of Table 2 and Fig. 6 can be interpreted as follows. Product reviews for the Digital Music category (average 260 words) are found to be the longest among the five product types based on WC, and approximately twice the average length of reviews for Clothing (123 words) and Grocery category (153 words). Moreover, WPS (27.48) and Analytic (72.22) for Digital Music are the highest. This means that the reviews for Digital Music are composed of lengthy and analytical sentences. The reason can be explained as reviewers may require more words to write reviews containing the personal experience and analytical expressions of the music for Digital Music category products. The level of Clout shows the highest score (54.67) for Music, but the lowest score (27.40) for Beauty category.

On the other hand, the Authentic scores showed the opposite results. Reviews for beauty have the highest Authentic scores (58.83), while the video has the lowest Authentic (23.70) scores. In other words, product reviews for Digital Music tend to be written expertly, whereas those of beauty is written authentically in a personal manner. Additionally, Tone (75.72) scores are highly positive in all five categories and are highest in the Clothing category pf products. This can be because reviewers express their personal experience of using and fitting of the product as per their product quality more elaborately than Cell phone category (64.33) products. The score of Percept for beauty Category (5.54) is found to be the highest among five categories as reviewers may use many sensory-based expressions such as "looked," "heard," or "feeling" for beauty, the quality of which is evaluated based on senses. The Flesch Kincaid score of Digital Music category (12.13) is highest as reviewers mostly write their personal feeling of that music is a very informal easy way to express emotion involved with it.

Table 2. Comparison of the average scores for review variables across five product categories.

		Beauty	Grocery	Cellphone	Clothing	Digital Music	F value	p-value
Rating	Mean	4.225	4.188	4.018	4.173	4.182	10.5	0.0
	SD	1.649	1.809	1.845	1.486	1.619		
WC	Mean	174.74	153.067	253.174	123.769	259.439	304.7	0.0
	SD	22070.1	18958.1	93185.413	12296.95	31688.998		
Analytic	Mean	47.811	59.194	63.488	50.829	72.223	491.2	0.0
	SD	518.890	547.656	487.714	613.524	397.078		
Clout	Mean	27.400	36.640	37.199	33.030	54.675	612.8	0.0
	SD	390.315	446.107	394.097	473.148	269.792		
Authentic	Mean	58.834	37.088	46.555	55.935	23.704	738.6	0.0
	SD	789.420	702.204	684.559	846.301	405.750		
WPS	Mean	20.000	20.437	24.439	18.263	27.481	75.0	0.0
	SD	155.647	195.541	1681.128	158.616	566.723		
Compare	Mean	3.172	3.089	2.836	2.989	3.152	12.2	0.0
	SD	4.186	4.834	3.965	5.195	2.592		
Posemo	Mean	4.026	4.671	3.804	5.176	4.547	130.0	0.0
	SD	6.037	7.685	6.859	11.309	3.849		
Negemo	Mean	1.042	1.177	1.134	0.996	1.436	41.8	0.0
	SD	1.422	1.864	1.504	1.772	1.971		
Cogproc	Mean	12.960	11.646	11.296	11.233	10.089	201.4	0.0
	SD	16.960	17.827	13.371	17.069	10.448		
Percept	Mean	5.541	4.513	4.613	3.841	4.654	161.9	0.0
	SD	11.075	9.562	8.270	9.545	3.801		
Tone	Mean	68.487	72.159	64.327	75.724	72.606	73.5	0.0
	SD	787.490	821.067	788.764	772.176	687.242		
Sixltr	Mean	14.635	15.711	15.182	13.213	16.047	162.7	0.0
	SD	22.145	27.198	22.380	23.451	19.888		
pronoun	Mean	16.614	13.778	13.730	15.517	11.546	516.0	0.0
	SD	18.643	21.445	19.703	21.922	15.737		
ppron	Mean	8.975	7.026	6.749	8.895	5.651	511.5	0.0
	SD	9.916	10.591	9.322	15.960	8.494		
article	Mean	6.297	6.762	8.399	7.437	7.799	273.1	0.0
	SD	6.533	7.210	8.209	9.708	5.002		
prep	Mean	11.660	11.983	12.239	11.431	12.361	45.6	0.0
	SD	8.395	10.621	7.944	12.428	6.269		
auxverb	Mean	8.532	8.326	8.285	8.987	7.526	87.1	0.0
	SD	7.644	9.305	7.076	9.929	5.633		
adverb	Mean	5.488	4.914	5.242	5.757	4.693	78.9	0.0
	SD	5.989	6.181	5.624	8.517	4.073		

(*continued*)

Table 2. (*continued*)

		Beauty	Grocery	Cellphone	Clothing	Digital Music	F value	p-value
adj	Mean	6.675	6.836	5.758	7.460	6.188	138.2	0.0
	SD	8.483	9.906	7.513	11.712	4.599		
AllPunc	Mean	15.931	17.457	15.693	17.005	23.219	329.4	0.0
	SD	37.399	69.431	53.109	79.893	72.967		
verb	Mean	16.226	14.001	14.358	15.946	12.895	387.6	0.0
	SD	13.688	16.025	11.414	15.490	10.974		
i	Mean	6.792	4.412	4.561	5.764	2.212	891.1	0.0
	SD	10.935	8.402	7.771	9.390	3.998		
Dic	Mean	85.479	83.038	81.691	85.385	78.655	533.4	0.0
	SD	31.163	48.898	45.262	33.732	43.154		
Flesch	Mean	8.589	9.149	9.885	7.376	12.132	197.9	0.0
	SD	31.562	29.980	54.283	27.541	74.447		

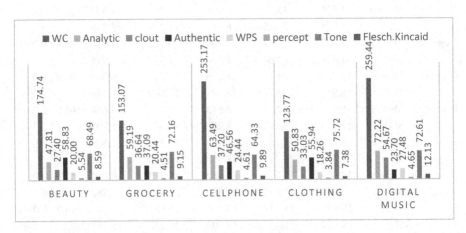

Fig. 6. Comparison of average scores of WC, Analytic, Clout, Authentic, WPS, Percept, Tome, and Flesch Kincaid score across five product categories.

In conclusion, as seen in the previous results, reviews for different product categories have different characteristics. Thus it would be necessary to analyze review helpfulness for each product category separately.

6.2 Factors Determining Review Helpfulness

Our second research question was to identify the determinant factors in the perceived helpfulness of reviews depending on their product category. To do so primarily, Linear regression (LR) is performed across product categories (Table 3).

Table 3. R-square value for each of five categories produced by Linear Regression analysis.

	Beauty	Grocery	Cell phone	Clothing	Digital Music
Multiple R square	0.13	0.21	0.11	0.05	0.47
Degrees of freedom	4113	2451	2416	3818	1854

From the above table, it is seen that the R-square value is very small except for the Digital Music category (0.47). The reason may be due to the nonlinear nature of the data set. To visualize the structure of five data sets, the high dimensional (25 dimensions) data is transferred to a lower dimension (2 dimensions using t-Distributed Stochastic Neighbor Embedding (t-SNE). t-Distributed Stochastic Neighbor Embedding (t-SNE) is a nonlinear method to reduce dimensionality for better visualization of data.

This algorithm works in the following way:

- Calculate the probability of similarity of points in high-dimensional space
- Calculating the probability of similarity of points in the corresponding low-dimensional (2D in this case) space.
- The similarity of data points is calculated as the conditional probability that a point X would select point Y as its neighbor if neighbors were chosen in proportion to their probability density under a Gaussian centered at X.
- The objective function is to minimize the difference between these similarities in high dimensional and low dimensional space to give a suitable representation of data in lower-dimensional space.

However, after this transformation, it is not possible to identify the input features and make any conclusions based on the output (Fig. 7).

Since the data is appeared to be nonlinear in shape, the Pearson correlation method cannot be used here to analyze the relationship between the dependent variable (Helpfulness Ratio) and Independent Variables.

Feature selection is a necessary procedure to reduce the high dimensional data into lower dimensions extracting the important variables among all variables. For this generally, Principal Component Analysis, Singular Value Decomposition, etc. methods are used. However, the primary assumption for the process mentioned above is that the data is linear. Also, these techniques do not consider feature values and target values. Therefore, using these methods shall not be applied in our data set.

The feature selection process can be categorized into three following process:

- Filter Methods: This method does not depend on the machine learning algorithm. Here, features are chosen based on various statistical tests for their correlation with the target variable. Example: Pearson Correlation, Spearman Correlation, Chi-squared test, Fisher's Score, etc.

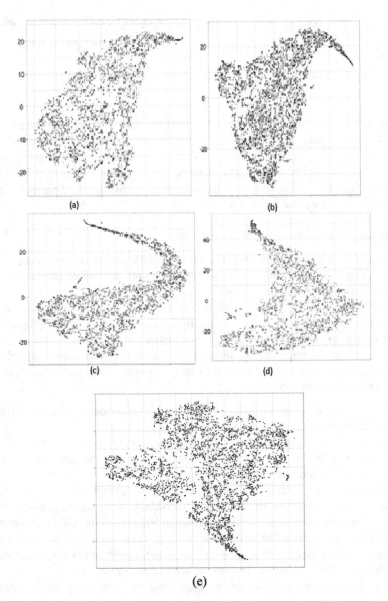

Fig. 7. A two-dimensional transformation of five datasets (a) grocery dataset, (b) beauty dataset, (c) cellphone dataset, (d) music dataset, and (e) clothing dataset.

- Wrapper Methods: This method of feature selection considers subsets of features that allow interaction with other variables by adding or removing features from that subset using a predictive model. Each subset is used to train the model and tested on a hold-out set. This is a computationally extensive algorithm but generally gives the best performing feature set for that model. Example: Recursive feature elimination, Sequential feature selection algorithms, Genetic algorithms, etc.

- Embedded Methods: Embedded feature selection method combines the advantages of both filter and wrapper selection methods and performs feature selection and classification together. This method of selection is not computationally extensive, like wrapper methods. Example: Lasso, Forward selection with Decision trees, Forward selection with Gram Schmidt, etc.

In our study, we used a wrapper built algorithm Boruta [14], which captures essential features with the target variable. Boruta is a wrapper build algorithm implemented in R package. In Boruta, Z score is used as a vital measure to consider the fluctuations of mean. This algorithm decides whether any feature is essential or not to predict the dependent variable. To do so, directly using Z score will not be ideal for measuring the importance of each variable as random fluctuations can mislead in this case. To handle this random fluctuations problem. For each variable, a corresponding shadow variable are defined whose values are assigned by shuffling the actual variables. And then the importance of shadow variables is used to decide the important variables.

Boruta algorithm ensures the randomness in the feature selection procedure and gives a better prediction on the importance of variables. Thus, in this paper, the Boruta algorithm is chosen for the feature selection procedure.

The final features identified important by Boruta algorithm is given below (Fig. 8):

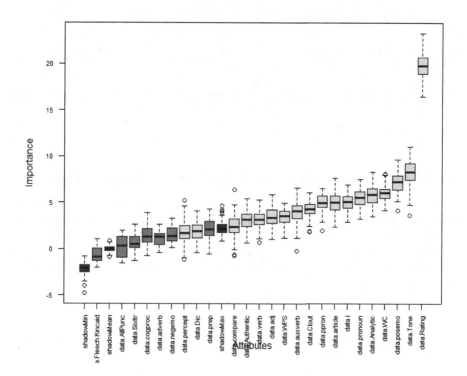

Fig. 8. Graphical plot of confirmed and rejected variables for the product category Clothing generated via Boruta algorithm.

From the above table, it is seen that for all five category product types, Rating, Analytic, WPS, posemo, Tone, Clout, pronoun, ppron, article, aux verb, adj, verb, I and Dic have a significant impact on review helpfulness. This implies these are the significant features to predict helpfulness scores where the rest of the variables only influence the helpfulness score of review specific to some product category. For example, Authentic variable is essential for all five categories of products. This implies a review comprising more honest and personal opinions with high involvement is perceived as more helpful to determine all five types of products. Compare variable is seen critical in Grocery, Cellphone, and Clothing category indicating usage of more comparative words (e.g., bigger, smaller, best, etc.) in the reviews. The negemo variables found vital features to predict the helpfulness score in Grocery and Cellphone category types of products. The percept variable is found necessary for Cellphone and Clothing category, which implies the usage of more perceptual words (e.g., feeling, hearing, etc.) in these product category reviews. The variable sixltr found crucial in only two product categories, namely Grocery and Digital Music, which implies the usage of more complex words (words longer than six letters). Adverb, Flesch Kincaid Readability, and All punc (Punctuation) are found prominent in only Digital Music category reviews. In other words, the usage of adverb words to express a more subjective view of the products in an easily readable manner with punctuation adequately used is an essential feature for Digital Music category products.

This concludes that with the conventional variables, for example, Rating, Word Count (WC), Word per Sentence (WPS), positive emotion (posemo), the other variables used as linguistic features of review, for example, Analytic, Clout, Tone, pronoun, personal pronoun, article, aux verb, adjective, verb, usage of I and Dictionary words are also have a significant impact on review helpfulness of any five categories of product reviews. The other variables (e.g., percept, negemo, compare, Sixltr, Flesch Kincaid Readability score, Allpunc, preposition, and adverb) also influence partially the target variables helpfulness for specific product categories.

The p-value for each of these Boruta results is 0.01, indicating the significance of the process of feature selection.

6.3   Prediction of Review Helpfulness Using Various Datamining Methods

The feature selection process gives us the critical variables affecting the target variable for each of the five category datasets. With the help of these variables, five different prediction models can be developed. In our study, we used Artificial Neural Network and Multiple Adaptive Regression Splines for prediction purposes and presented a comparative analysis of these two prediction models suggesting the best model choose for a specific category.

Table 4. The selected features across five categories using the Boruta algorithm.

	Beauty	Grocery	Cellphone	Clothing	Digital music
Rating	✓	✓	✓	✓	✓
WC	✓	✓	✓	✓	✓
Analytic	✓	✓	✓	✓	✓
Clout	✓	✓	✓	✓	✓
Authentic	✓	✓	✓	✓	✓
WPS	✓	✓	✓	✓	✓
compare		✓	✓	✓	
posemo	✓	✓	✓	✓	✓
negemo		✓	✓		✓
cogproc	✓	✓	✓		✓
percept			✓	✓	
Tone	✓	✓	✓	✓	✓
Sixltr		✓			✓
pronoun	✓	✓	✓	✓	✓
ppron	✓	✓	✓	✓	✓
article	✓	✓	✓	✓	✓
prep	✓	✓			✓
auxverb	✓	✓	✓	✓	✓
adverb					✓
adj	✓	✓	✓	✓	✓
AllPunc					✓
verb	✓	✓	✓	✓	✓
i	✓	✓	✓	✓	✓
Dic	✓	✓	✓	✓	✓
Flesch					✓

Multiple Adaptive Regression Splines (MARS) proposed by Friedman is a non-parametric statistical procedure to determine the relationship between a set of input variables and the dependent variable. This algorithm does not make any prior assumptions about the relationship between independent and dependent variables and works perfectly fine in both linear and nonlinear relationships. This flexibility in determining any relationship gives the idea of using MARS in this case (Fig. 9).

Artificial Neural Network (ANN) tries to model the human brain with the most straightforward definition where the building blocks are neurons. In multilayer artificial Neural Networks, each neuron is connected to others with some coefficients, and learning of the network is done by proper distribution of information through these connections. The capability of processing parallel and nonlinear behavior gives the reason use this algorithm in this paper (Fig. 10).

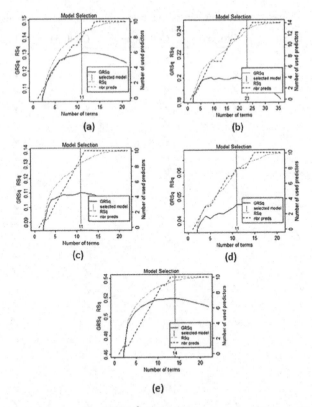

Fig. 9. Model summary capturing GCV R^2 (left-hand y-axis and solid black line) based on the number of terms retained (x-axis), which is based on the number of predictors used to make those terms (right-hand side y-axis) for (a) beauty dataset. (b) grocery dataset, (c) cellphone dataset, (d) Clothing dataset and (e) digital music dataset.

The selected features across five categories in Table 4 used to build five different models using MARS and ANN each to calculate the Mean Squared Error (MSE). The MSE gives the idea of the accuracy of the model across product categories. From the above Fig. 11, it is clear that Multiple Adaptive Regression Splines (MARS) produces lesser MSE than Artificial Neural Network (ANN) except the clothing category. Hence, it can be concluded that MARS gives more accurate results than the Artificial Neural Network for all categories except clothing product category. Therefore, it can be suggested that using the MARS algorithm would be better to predict review helpfulness in Beauty, Grocery, Cellphone and Digital Music category. The ANN produces lower MSE than MARS and hence can be used to predict the review helpfulness in case of clothing category.

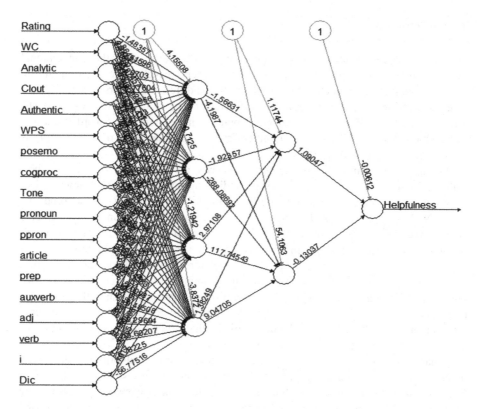

Fig. 10. Model summary for Artificial Neural Network capturing inputs and weights at each layer for grocery dataset.

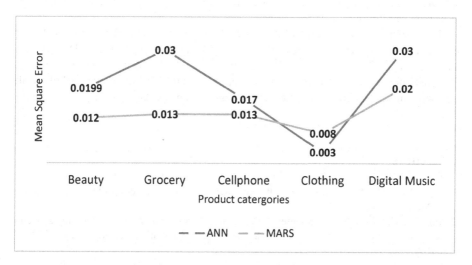

Fig. 11. Comparisons of mean square error: Multiple Adaptive Regression Splines Vs. Artificial Neural Network.

7 Conclusion

This study addresses three research questions. First, along with the conventional explicit variable (Rating), this paper explored several psychological and linguistic features from directly the product reviews across five different categories and examined whether these features are different for a different category using ANOVA analysis. The review of the Digital music category found to have the highest word count and written more analytically with maximum criticism. The authenticity of this category is least among all five categories.

On the other hand, Beauty category reviews are found to have the highest authenticity score but the lowest clout score indicating low expertise of reviewer. Also, it contained the most comparison words to describe the quality of the products. The ANOVA result shows that there are significant differences in review features among five different categories of product reviews at a 99% confidence interval. Secondly, the critical variables influencing the target variable (Helpfulness ratio) are explored using boruta algorithm. It is found that for all five category product types, Rating, Analytic, WPS, posemo, Tone, Clout, pronoun, ppron, article, aux verb, adj, verb, I, and Dic have a significant impact on review helpfulness. The other variables (e.g., percept, negemo, compare, Sixltr, Flesch Kincaid Readability score, Allpunc, preposition, and adverb) also influence partially the target variables helpfulness for specific product categories.

Finally, among two extensively used machine learning algorithms, the better method for review helpfulness prediction is determined. The result shows that both Multiple Adaptive Regression Splines and Artificial Neural Network performs very well as their Mean squared error is less than 5%. However, the MSE of the MARS algorithm is much lower than ANN except for the clothing category. Hence, the MARS algorithm should be used to predict review helpfulness for Beauty, Cellphone, Grocery, and Digital Music category reviews, and ANN should be used to predict the helpfulness score of Clothing category reviews.

This paper also solves the cold start problem, which arises when reviews do not receive any manual votes but have the potential to be a helpful review. Due to starvation, most of the reviews generally do not get the chance to get voted all the time. This prediction method will solve this problem by identifying important psychological, linguistic, and explanatory variables and producing a helpfulness score in the absence of any manual vote.

References

1. Chevalier, J.A., Mayzlin, D.: The effect of word of mouth on sales: online book reviews. J. Mark. Res. **43**, 345–354 (2006)
2. Forman, C., Ghose, A., Wiesenfeld, B.: Examining the relationship between reviews and sales: the role of reviewer identity disclosure in electronic markets. Inf. Syst. Res. **19**, 291–313 (2008)

3. Ghose, A., Ipeirotis, P.G.: Estimating the helpfulness and economic impact of product reviews: mining text and reviewer characteristics. IEEE Trans. Knowl. Data Eng. **23**, 1498–1512 (2011)
4. Hu, Y.H., Chen, K.: Predicting hotel review helpfulness: the impact of review visibility, and interaction between hotel stars and review ratings. Int. J. Inf. Manag. **36**, 929–944 (2016)
5. Hu, N., Koh, N.S., Reddy, S.K.: Ratings lead you to the product, reviews help you clinch it? The mediating role of online review sentiments on product sales. Decis. Support Syst. **57**, 42–53 (2014)
6. McAuley, J., Targett, C., Shi, Q., van den Hengel, A.: Image-based recommendations on styles and substitutes. In: SIGIR (2015)
7. Kacewicz, E., Pennebaker, J.W., Davis, M., Jeon, M., Graesser, A.C.: Pronoun use reflects standings in social hierarchies. J. Lang. Soc. Psychol. **33**, 125–143 (2013)
8. Khashei, M., Bijari, M.: An artificial neural network (p, d, q) model for time series forecasting. Expert Syst. Appl. **37**(1), 479–489 (2010)
9. Kim, S.M., Pantel, P., Chklovski, T., Pennacchiotti, M.: Automatically assessing review helpfulness. In: Proceedings of the 2006 Conference on Empirical Methods in Natural Language Processing, Sydney, Australia, 22–23 July 2006, pp. 423–430 (2006)
10. Korfiatis, N., Garcia-Bariocanal, E., Sanchez-Alonso, S.: Evaluating content quality, and helpfulness of online product reviews: the interplay of review helpfulness vs. review content. Electron. Commer. Res. Appl. **11**, 205–217 (2012)
11. Krishnamoorthy, S.: Linguistic features for review helpfulness prediction. Expert Syst. Appl. **42**, 3751–3759 (2015)
12. Kuan, K.K., Hui, K.L., Prasarnphanich, P., Lai, H.Y.: What makes a review voted? An empirical investigation of review voting in online review systems. J. Assoc. Inf. Syst. **16**, 48–71 (2015)
13. Kumar, N., Benbasat, I.: The influence of recommendations on consumer reviews on evaluations of websites. Inf. Syst. Res. **17**(4), 425–439 (2006)
14. Kursa, M., Rudnicki, W.: Feature selection with the Boruta package. J. Stat. Softw. **36**(11), 1–13. http://dx.doi.org/10.18637/jss.v036.i11
15. Lee, S., Choeh, J.Y.: Predicting the helpfulness of online reviews using multilayer perceptron neural networks. Expert Syst. Appl. **41**(6), 3041–3046 (2014)
16. McAuley, J., Leskovec, J.: Hidden factors and hidden topics: understanding rating dimensions with review text. In: Proceedings of the 7th ACM Conference on Recommender Systems, RecSys', Hong Kong, China, 12–16 October 2013, pp. 165–172 (2013)
17. Mudambi, S.M., Schuff, D.: What makes a helpful online review? A study of customer reviews on Amazon.com. MIS Q. **34**, 185–200 (2010)
18. Nelson, P.: Information and consumer behavior. J. Polit. Econ. **78**(20), 311–329 (1970)
19. Nelson, P.: Advertising as information. J. Polit. Econ. **81**(4), 729–754 (1974)
20. Newman, M.L., Pennebaker, J.W., Berry, D.S., Richards, J.M.: Lying words: predicting deception from linguistic style. Pers. Soc. Psychol. Bull. **29**, 665–675 (2003)
21. Pan, Y., Zhang, J.Q.: Born unequal: a study of the helpfulness of user-generated product reviews. J. Retail. **87**, 598–612 (2011)
22. Pennebaker, J.W., Booth, R.J., Francis, M.E.: Linguistic inquiry and word count (LIWC2007), LIWC, Austin, TX, USA (2007). http://www.liwc.net. Accessed 27 Apr 2018
23. Pennebaker, J.W., Francis, M.E.: Cognitive, emotional, and language processes in disclosure. Cogn. Emot. **10**, 601–626 (1996)
24. Pennebaker, J.W., Boyd, R.L., Jordan, K., Blackburn, K.: The development and psychometric properties of LIWC2015. http://hdl.handle.net/2152/31333. Accessed 27 Apr 2018

25. Pennebaker, J.W., Chung, C.K., Frazee, J., Lavergne, G.M., Beaver, D.I.: When small words foretell academic success: the case of college admissions essays. PLoS ONE **9**, e115844 (2014)
26. Sen, S., Lerman, D.: Why are you telling me this? An examination into negative consumer reviews on the web. J. Interact. Mark. **21**, 76–94 (2007)
27. Tausczik, Y.R., Pennebaker, J.W.: The psychological meaning of words: LIWC and computerized text analysis methods. J. Lang. Soc. Psychol. **29**, 24–54 (2010)
28. Willemsen, L.M., Neijens, P.C., Bronner, F., De Ridder, J.A.: "Highly recommended!" The content characteristics and perceived usefulness of online consumer reviews. J. Comput. Mediat. Commun. **17**, 19–38 (2011)
29. Yang, Y., Yan, Y., Qiu, M., Bao, F.: Semantic analysis and helpfulness prediction of text for online product reviews. In: Proceedings of the 53rd Annual Meeting of the Association for Computational Linguistics and the 7th International Joint Conference on Natural Language Processing, Beijing, China, 26–31 July 2015, pp. 38–44 (2015)
30. Yin, D., Bond, S., Zhang, H.: Anxious or angry? Effects of discrete emotions on the perceived helpfulness of online reviews. MIS Q. **38**, 539–560 (2014)

Analysis and Recognition of Hand-Drawn Images with Effective Data Handling

Mohit Gupta and Pulkit Mehndiratta[(✉)]

Department of Computer Science and Engineering,
Jaypee Institute of Information Technology, Noida, India
mohitatjammu@gmail.com, pulkit.mehndiratta@jiit.ac.in

Abstract. Everyday data is being produced at a mind-boggling rate and at current pace, we are generating more than 2.5 quintillion bytes of data each day. With more data, there comes the chance of generating and understanding more useful information out of it. With this view in mind, we are presenting our work on Quick Draw dataset. It is a repository of approximately more than 50 million hand-drawn drawings of about 345 different objects, contributed by over 15 million users. The work we have presented is a useful approach to perform a country-wise analysis of the styles that people use while making strokes to draw an object or image. Since we are dealing with a huge dataset, we are also presenting a powerful technique to effectively reduce the data and its dimension which results in much more simplified and concise data. Based on the simplified data, we are enhancing the performance of our models to recognize these hand-drawn objects based on the strokes that a user has used for drawing them. For the recognition of objects, we have worked on traditional machine learning models - Nearest Neighbor (K-NN), Random Forest Classifier (RFC), Support Vector Classifier (SVC) and Multi-Layer Perceptron model (MLP) by selecting the best hyper-parameters. To make the recognition results better, we have also presented our work on deep learning models - CNN, LSTM, and WaveNet networks.

Keywords: Machine learning · Deep learning · Image recognition · Data handling · Neural networks · Classification · Hyper-parameter selection

1 Introduction

The penetration of the internet and technology has increased many folds in last two decades, which tries to challenge the existence of pen and paper but, a sometimes and a small notebook and pen can be more effective and convenient compared to mouse and keyboard. As far as human learning is concerned, there are four basic learning styles through which any human being starts to learn. They are auditory skills, visual skills, tactile and kinesthetic skills. Sketches and hand-drawn images are part of tactile and kinesthetic learning making it more learning. On the contrary, sketches and hand-drawn images are very ambiguous and ubiquitous at the same time. We draw images while we think about something, to solve some problems and also at times to communicate with others. The language of signs is older than actual formal languages itself. To

S. Madria et al. (Eds.): BDA 2019, LNCS 11932, pp. 389–407, 2019.
https://doi.org/10.1007/978-3-030-37188-3_22

understand these sketches many systems have been developed from basic CAD systems which offer very less freedom compared to sketch-understanding systems that let users to interact with computers by helping users and to draw images as naturally as possible. It has many industrial as well as educational benefits too, as sketch recognition is widely deployed in-circuit recognition, architectural drawing, and modeling; also it has a very important role to play in mechanical engineering.

The field of recognition of hand-drawn images and sketches has received its due attention and popularity in the past few years. Several types of analysis and interpretation techniques can be associated now to handwritten notes and images. Measures and techniques have been proposed by many researchers to perceive and interpret the human thinking process and converting it into a more useful and deployable in many fields. But, to understand and learn the art of recognition of these images, a lot of training and feature analysis is to be performed in these images resulting in more and more data. Ours is a study that tries to reduce the data by converting it into a more simple and concise format so as to reduce the processing and increasing the performance of the overall system.

2   Related Work

The recognition and identification of the hand-drawn image are not new to the world of computer science and information technology. Shape definition languages have been around since the late '70 s like shape grammar [1]. It was largely used for generation of the shape rather than recognition and identification of it from a sketch. Also, it lacked a basic functionality of providing the stroke order which is very helpful for recognition purposes. More recent work like in [2], authors try to come up with a definition language that helps to parse the diagram. But this also lacks in terms of providing online support as well as very less help to edit and display any shape. With more advancements in the field of recognition of the hand-drawn and sketch, techniques like Bimber [3] tries to give more support for generating more information about the shape but lacks in specifying domain information like the order of strokes, display, and editing.

In a technique [4], authors' tries to use fuzzy-based logic to generate and describe shape but, this technique also lacks the features like shape grouping, display and, editing. Quill [5] a gesture recognition system that tries recognizes a gesture and generate its sketch provides more help and support for how a human will be able to remember this sketch. This system is different as it does not work on a feature rather on the way how the shape is to be drawn which is similar to this work. Apart from this a lot of other languages and projects have been created to recognize and identify any images like The Electronic Cocktail Napkin project [6]. Another dimension of this work is a translation of the images, where the visual language tries to map the difference between the grammar generated and arrangement of the icons given to check for its validity of the syntax.

Many other studies as proposed in [7–9] uses deep learning concepts like convolutional neural networks for images recognition task but no significant work has been done for the freehand-drawn images using mouse and keyboard. In a paper [10], authors have used hand strokes generated by the user to interpret the actual geometrical information of the image using the pixels produces. Finally, it tries to categorize them as ovals, lines, or any other arbitrary shape.

Ours is a technique that tries to recognize the hand-drawn images by making data more simple and concise. Not only this, we have tried to improve the performance and actually performed a comparison between various machine learning and deep learning models to check for the more accurate and correct results. The rest of the paper is as follows: Sect. 3 provides the layout of our technique used. The algorithm we have used and the analysis of the pattern drawn is given in Sect. 4. Section 5 discusses the process of making the data more useful, simple and concise. Section 6 and 7 tries to explain the various machine learning and deep learning models that help us to understand how these techniques improve the process of identification of hand-drawn images. Finally, Sect. 8 provides result and conclusion is provided in Sect. 9 of this article.

3 Methodology

3.1 Dataset

For our research work, we have taken Quick Draw dataset [26] provided by Google. This data is a repository of approximately more than 50 million hand-drawn drawings of about 345 different objects, contributed by over 15 million users from all over the world. The data stores the list of co-ordinates of continuous strokes that users have drawn when asked to draw a sketch of a given object. Each of these strokes is stored as $[(x_1, x_2, x_3, x_4...), (y_1, y_2, y_3, y_4...)]$ where x and y are images co-ordinates different from the normal coordinate system.

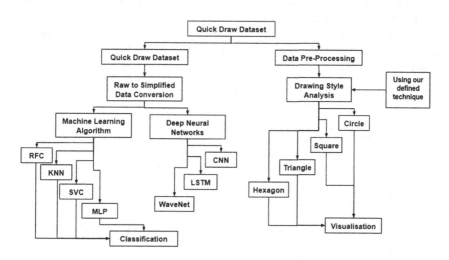

Fig. 1. Depicts the flow of our work

The attributes of the Quick Draw dataset are 1) key_id (uniquely identifies each drawing in the data), country code (It is a two-letter code of the country to which the user (who has drawn the image) belongs), drawing (the co-ordinates of the strokes, which the user has made to draw the drawing), recognized (whether the drawing correctly matches with the word (drawing object class) given), timestamp (the time at which the drawing was made), word (class to which the user drawing belongs) as shown in Fig. 1. For our proposed research work we have taken – 'country code', 'drawing' and 'word' attributes into consideration (Fig. 2).

	countrycode	drawing	key_id	recognized	timestamp	word
0	US	[[93, 38, 0], [8, 63, 96]], [[13, 19], [108, ...	6695476885716992	True	2017-03-17 20:20:25.234330	hexagon
1	US	[[65, 62, 4, 0, 24, 101], [2, 10, 78, 89, 114...	4509550780612608	True	2017-03-04 14:54:41.166690	hexagon
2	BR	[[0, 76, 93, 192, 200, 255, 246, 165, 166, 15...	5401676863242240	True	2017-01-30 13:19:15.170880	hexagon
3	US	[[124, 113, 102, 22, 0, 51, 68, 126, 148, 156...	5388880746381312	True	2017-01-27 17:19:10.248280	hexagon
4	US	[[58, 29, 2, 1, 11, 35, 121, 151, 183, 224, 2...	5724753627185152	True	2017-03-05 02:59:37.331420	hexagon

Fig. 2. Depicting the Quick Draw dataset

3.2   Implementation

We have started our research work with the analysis of the drawing style or pattern in which users around the world draws. For this analysis, we are using various simple shapes that users have drawn. The shapes under analysis are – triangle, square, circle and hexagon. Since we have such a huge dataset, so to analyze the patterns of drawing, we have come up with a very effective technique (mentioned in Sect. 4) in terms of both results and computation. We are focusing to see instinctively, which direction the user's hand go and then finding a pattern in the behavior of a user, belonging to a particular region. Our analysis can help researches to build a model that can automatically keep on suggesting the next stroke move when a user wishes to or starts drawing an object (Fig. 3).

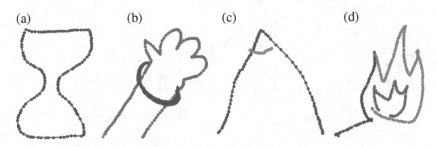

(a)   (b)   (c)   (d)

Fig. 3. (a), (b), (c), (d) Depicts the drawing of the classes – watch glass, hand, mountain, campfire respectively.

After the analysis of the user's drawing pattern, we are moving on to the recognition of the object drawings. But to train models and extracting features from such a huge dataset is a computationally very expensive job to do. So now in this paper, we have presented a method to deal with the dimensions and size of our data. We have used easy mathematics in an effective manner to reduce our stroke co-ordinate data but still maintaining all the relevant information. So, after converting our raw data to simplified forms as mentioned in Sect. 5, we are now ready to propose our recognition task on the simplified dataset.

We have started our recognition task with the machine learning models - Nearest Neighbor (K-NN), Random Forest Classifier (RFC), Support Vector Classifier (SVC) and Multi-Layer Perceptron model (MLP). We have selected the best performing hyper-parameters for these models and then used them to classify the object using the stroke co-ordinate knowledge we are given in the dataset. To improve our work, we have finally used – Deep Learning models CNN and LSTM and to add further advancement to the research work we have worked on WaveNet networks which because of their extremely well-designed structures have increased the classification results of our work.

4 Drawing Pattern Analyses

4.1 Methodology

To analyze how people, draw an object we are using basic shapes from Quick Draw dataset. Our analysis and results are based on the following basic shapes – triangle, square and hexagon. The analysis aims to see in which direction user draw the strokes instinctively when asked to draw an object. The measure we have come up to perform this analysis is the measure of clockwise movements in the strokes for each of the sketches in the dataset (Fig. 4).

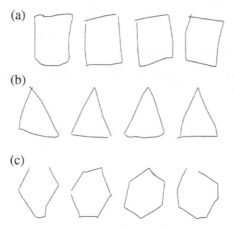

Fig. 4. (a), (b), (c) Drawing classes under consideration – square, triangle, and hexagon respectively.

Each of the strokes in the data is stored as [(x1, x2, x3, x4...), (y1, y2,y3, y4...)] where x and y are images co-ordinates different from the normal coordinate system. In the case of image co-ordinates, the y-axis points downwards that means (1, 1) is at the top left corner. Intuitively, it is very clear that if strokes keep on turning left then that means it is clockwise and counterclockwise if the strokes turn right. So based on this intuition, we have separated the segments from a stroke and then the angle between consecutive segments is compared. But there is a possibility that the curve is a minor jitter or just a tiny turn or any other curve around the diagram. So, considering the angle alone means, just count the number of loops in the curves drawn. To make our approach meaningful, we are also taking the relative length of each line segment into consideration. We are working on how much each turn is turning left or right as compared to the previous line segment. So, finally, we are calculating the sine of the turn angle divided by the length of the previous line segment.

$$\textbf{Score} = \sum_i^{n-1} \frac{\textbf{sine}\left(\textbf{angle between segment}_i \textbf{ and segment}_{i+1}\right)}{\textbf{length of segment}_i} \tag{1}$$

Where n is the total no. of stroke segments in the drawing.

Our score for each drawing has the following properties – or the use of calculating scores

- When negative and positive scores of each stroke turn are summed up to calculate a net score for a drawing, they get canceled out if strokes turn equivalently in opposite directions.
- So, a minor change in direction is not having a great impact on the net score as we are taking the sine parameter normalized by the relative length of the stroke line segment.
- The score for each segment is in the range of −1 and 1, therefore a single turn score can't dominate the score of the drawing as in whole.
- A negative score means counter-clockwise and a positive score means that the user prefers to draw clockwise.

4.2 Algorithm for Score Calculation

In this section we provide the algorithm for calculation score for each stroke to be made

---

Algorithm 1: SCORE CALCULATION ALGORITHM

Input: Drawing stoke coordinates
          for each image

    {

        Extract stroke segment
        //stroke segments are in the form of $[(x_1, x_2, ..) (y_1, y_2,..)]$
        Transposing the stroke segment
        //now we have $[(x_1, y_1) (x_2, y_2) (...)]$
        y=255-y
        //Converting image coordinates to real life coordinates

        foreach stroke segment
        {

            $\mathbf{a}:= [(x_i,y_i),(x_{i+1},\ y_{i+1}),(x_{i+2},y_{i+2})]$
            $\mathbf{b}:= [(x_{i+1},\ y_{i+1}),(x_{i+2},y_{i+2}),(x_{i+3},y_{i+3})]$
            $\mathbf{Vector_a}:= [x_{i+1}-x_i,\ y_{i+1}-y_i]$
            $\mathbf{Vector_b}:= [x_{i+2}-x_{i+1},\ y_{i+2}-y_{i+!}]$
            $\mathbf{Sine}:= \mathbf{Vector_a} - \mathbf{Vector_b}$
            $\mathbf{Score} := \dfrac{\textit{Sine}}{\textit{Eucledian}(\mathbf{Vector(a)},\ \mathbf{Vector(b)})}$
            Total_Score:= Totals_Score+ Score
        }
        Return Total_Score

}

Output: Total Score for each drawing

4.3   Analysis

We have performed a country-wise analysis of the drawing patterns of the user. Based on our analysis, we have observed some very interesting points which are mentioned in Sects. 4.3.1 and 4.3.2.

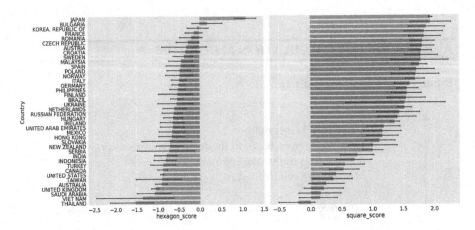

Fig. 5. Country-wise analysis of user pattern for Hexagon and Square drawings.

4.3.1 Hexagon and Square Analysis

Most of the users around the world follow a clockwise (negative score) pattern to draw a hexagon. But the average user in Japan and Bulgaria draws the shape counter-clockwise. Also, the nearby countries – Thailand and Vietnam are sharing the top two clockwise scores and thus having their position on the clockwise scale (Fig. 5).

In the case of a square, Japan is not the odd one, all most all the users from different countries are drawing a square in a counter-clockwise manner. Average users from Thailand are an exception to this analysis.

4.3.2 Circle and Triangle Analysis

To our surprise, except Taiwan and Japan (again), most of the users around the world draw a circle in a counter-clockwise fashion. Users in Japan and Taiwan typically draw a circle in a clockwise manner (Fig. 6).

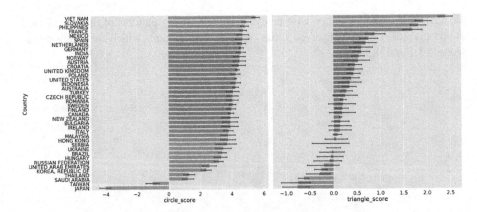

Fig. 6. Country-wise analysis of user pattern for Circle and Triangle drawings

The analysis of triangle shows varying results as countries like – Bulgaria, Spain, Hungary, Croatia, Italy, United Arab Emirates, Brazil, Thailand, and Saudi Arabia follows clockwise pattern i.e. having a negative score. Countries like Vietnam and Sweden have neither positive nor negative score that means there is no particular pattern that the users follow for drawing a triangle.

4.4 Distribution Analysis

Even for simple shape, the type of drawing has a varying impact on users drawing pattern around the world. Surprisingly, even in the case of hexagon and circle, average user pattern differs (Fig. 7).

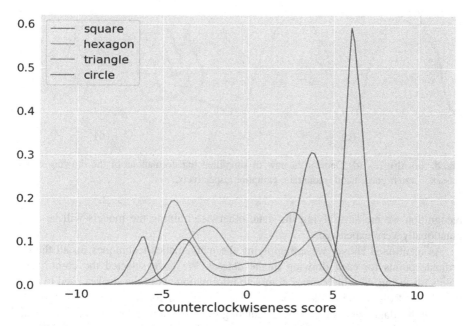

Fig. 7. Counter clockwise score distribution for different drawings under analysis.

We can see a clear bimodal pattern for each drawing distribution for each drawing. This clearly indicates that the drawings show particular direction preferences.

5 Raw to Simplified Form

We are dealing with such huge dataset with more than 50 million hand-drawn drawings having about 345 different object classes. These drawings are contributed by over 15 million users. Further, each instance of this huge dataset contains a large number of stroke coordinates that are used to make a particular drawing. So, we have to deal with huge dimensions and large data size. So, before proceeding to our task of object

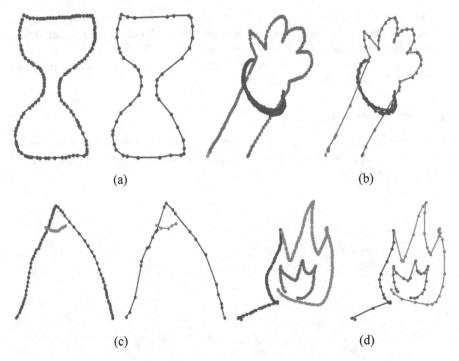

(a)   (b)

(c)   (d)

Fig. 8. (a), (b), (c), (d) Depicts the raw to simplified transformations of the drawing of the classes – watch glass, hand, mountain, campfire respectively.

recognition we are simplifying the data, otherwise training the models will be computationally very expensive.

As mentioned above, we are applying the mathematics techniques on all the co-ordinate points for each drawing in the dataset. We have selected the co-ordinates points (as in Fig. 8) which stores complete information about the strokes used to draw the object. Rest of the stroke c-ordinates are discarded for the recognition task. To perform this data reduction, we are calculating the minimum and maximum coordinate point of the stroke till which its orientation angle is not changing much and then eliminating the co-ordinates in between the min and max points on the considered stroke line segment. Thus, we are simplifying the data yet retaining all the meaningful information. Nowour data is ready for our models to get trained and perform the classification task on the drawing using the informative stroke co-ordinates.

6   Machine Learning Models

6.1   Support Vector Classifier

Support Vector Machine (SVM) [11] is a supervised machine learning model which we are using as classifier [12, 13] for our research purpose. Here, we are classifying the object drawing class using the stroke coordinate information of the drawings. When working as a classifier, SVC finds the optimal parameter to find the best hyper-parameter that can differentiate the classes (drawing object classes) in the best possible manner. It builds up the data points (stroke co-ordinates) in the vector space. Then, it works to find the best possible hyper-parameter. Default kernel for SVC is 'RBF'. We are analyzing the kernel parameters – linear, sigmoid and RBF. After working on different hyper-parameters, we are using SVC with 'RBF' kernel

6.2 Random Forest Classifier

Random forests [14] are the Decision Tree ensembled algorithm. We are using Random Forest Classifier (RFC). Random Forest implements multiple decision trees and uses their performance for classification or regression task. The features are selected from each decision tree and are chosen randomly thus, the splitting is performed randomly. This randomness in splitting is the reason how Random Forest overcomes the problem of over-fitting. Normally up to a threshold, the accuracy results increase with the increase in the number of ensembled trees. By default, the number of ensembled trees in RFC is 10. We have tested RFC on different tree number parameters ranging from 10 to 2000 to get the best fit value. RFC with tree number parameter equals to 10 is performing the best, so we are setting out RFC to this parameter.

6.3 K-Nearest Neighbor (K-NN)

K-Nearest Neighbour (K-NN) [15–17] is a very simple instance-based machine learning algorithm which can be used for both classification and regression tasks. The model looks for the 'k' closest training instances in the given feature vector space i.e. data points which are stroke coordinates for each drawing. The model predicts the class based on the majority to which the instance in the feature space belongs. By default, the 'k' value for K-NN classifier is 5. We have worked on different parameters to see which 'k' value gives us the best classification results and the default parameter is working best for the K-NN model.

6.4 Multi-layer Perceptron Model (MLP)

Multi-Layer Perceptron model (MLP) [18] is a feed-forward artificial neural network-based algorithm. In its working, it mimics the brain. The comprises of following basic components – an input layer, hidden layers and an output layer, where each of these layers comprises of various nodes. These are connected nodes with randomly weighted connections (ranging from 0 to 1) in between with a biased value assigned to these connections. Also, these nodes have activation function assigned to them and these functions add non-linearity to the kernels and thus, try to make the hypothesis much more accurate. The MLP network outputs, calculate errors, back propagates the information and then again update the weights using gradient descent and thus keep on improving its performance with the set number of iterations and keeps on training by gaining the more accurate values of weights and reducing errors. While training the model we have worked on different activation functions and compared the performance on the set parameters. After testing our model on various hyper-parameters, we are finally using MLP with 2 hidden layers, ReLu activation function, and Adam as gradient optimizer.

7 Deep Neural Networks

7.1 Convolutional Neural Network (CNN)

With the advancement in technology and computation, neural networks have gained so much popularity because of their capability to achieve state-of-the-art results. CNN or ConvNets [19–21] are the class of feed-forward deep neural networks. It was developed by Yann LeCun in the 1990s. These neural networks are a stack of the various input layer, multiple hidden layers that finally pass on the information to the output layer. One can consider them to be regularized version of MLP where each of the hidden layers consists of convolutional layer consists of a convolutional layer, pooling layer, and fully connected layers. It mimics animal visual cortex is working. CNN extracts simple and small pattern to identify more complex patterns and thus have gained popularity in image recognition. The biggest reason to use ConvNets id their ability in reducing the images to an easy to process forms without losing features. So, when we are working on such a huge data where we have more than 50 million drawings and each of these drawing data holds a huge list of stroke information, CNN proves to be a very good choice not only in terms of learning features but is also scalable to massive datasets.

Convolutional layers capture the low-level features such as edges and curves in Quick Draw dataset object drawings by striding the filter kernels. With more and more layers, network adapts to capture high-level features. Then the polling layers reduce the spatial size of the features convolved by the convolutional layers which are very important for our work so as to extract features from a large number of coordinate points. After passing through these layers, our model now understands the features and the flattened output is fed to the fully connected neural network which classifies the object drawing based on the extracted information from the stroke co-ordinates (Table 1).

Table 1. CNN architecture used for our proposed work.

Layer (type)	Output shape	Parameters
conv2d (Conv2D)	(None, 32, 32, 32)	320
max_pooling2d (MaxPooling2D)	(None, 16, 16, 32)	0
conv2d_1 (Conv2D)	(None, 16, 16, 64)	18496
max_pooling2d_1 (MaxPooling2D)	(None, 8, 8, 64)	0
dropout (Dropout)	(None, 8, 8, 64)	0
flatten (Flatten)	(None, 4096)	0
dense (Dense)	(None, 680)	27895960
dropout_1 (Dropout)	(None, 680)	0
dense_1 (Dense)	(None, 340)	231540

Total params: 3,036,316
Trainable params: 3,036,316
Non-trainable params: 0

7.2   Long-Short Term Memory (LSTM)

Long-Short Term Memory (LSTM) [22] is Recurrent Neural Networks (RNN) [23] variants architecture and an artificial neural network. They are very powerful as they can process both single data points as well as an entire sequence of data. So, we are treating each line segment stroke-coordinate points as a sequence to feed to the LSTM network. LSTM overcomes the major problem of RNN i.e. vanishing and boosting gradient descent with the help of its well-designed architecture. LSTM architecture consists of cells and gates – input, output and forgets gates. These gates control and protect the information in each cell state and let this information flow through them.

$$\mathbf{f_t} = \sigma\big(\mathbf{x_t U^f} + \mathbf{h_{t-1} W^f}\big) \tag{2}$$

$$\mathbf{i} = \sigma\big(\mathbf{x_t U^i} + \mathbf{h_{t-1} W^i}\big) \tag{3}$$

$$\mathbf{c_t} = \mathbf{c_{t-1}} * \mathbf{f} + \mathbf{g} * \mathbf{i} \tag{4}$$

$$h_t = \tanh(c_t) * o \tag{5}$$

$$o = \sigma(x_t U^o + h_{t-1} W^o) \tag{6}$$

Cells in our LSTM architecture are using sigmoid activation function with point wise multiplicative operations for the gates in the network. Gates are closed when sigmoid gives 0 and are open in case of 1. LSTM starts working with keeping the information in control to feed it to the cell states. This task is performed using forget

Table 2. LSTM Architecture used for our proposed work.

Layer (type)	Output shape	Params
Batch_normalization_1 (Batch)	(None, None, 3)	12
conv2d_1 (Conv2D)	(None, None, 48)	768
dropout_1 (Dropout)	(None, None, 48)	0
conv2d_2 (Conv2D)	(None, None, 64)	15424
dropout_2 (Dropout)	(None, None, 64)	0
conv2d_3 (Conv2D)	(None, None, 96)	18528
dropout_3 (Dropout)	(None, None, 96)	0
Cu_dnnlstm_1 (CuDNNLSTM)	(None, None, 128)	115712
dropout_4 (Dropout)	(None, None, 128)	0
Cu_dnnlstm_2 (CuDNNLSTM)	(None, 128)	132096
dropout_5 (Dropout)	(None, 128)	0
dense_1 (Dense)	(None, 512)	66048
dropout_6 (Dropout)	(None, 512)	0
dense_2 (Dense)	(None, 340)	174420

Total params: 523, 008
Trainable params: 523, 002
Non-trainable params: 6

gates which assigns either 1 or 0 to h$_{t-1}$ and tx output. 0 indicates to ignore the information else retain the information from the previous cell state, C$_{t-1}$ till further processing. Now if the information has to be flown, then which new information has to be stored in the cell state. To perform the task sigmoid layer selects the information to update and tanh function creates new information for the cell state. Now after processing and updating the information, old cell state C$_{t-1}$ is updated to new cell state, C$_t$. All these steps are involved in forwarding the stroke information, updating and retaining the meaningful stroke feature information only (Table 2).

7.3   WaveNet Model

WaveNet [24, 25] model is an autoregressive and fully deep neural network. We are using it is a discriminative model, returning back the promising results for the hand-drawn object recognition. This network architecture is modeling joint probabilities over pixels, achieving state-of-the-art results. To deal with large dimensions and the huge size of Quick-Draw dataset, we are working on this network architecture as it is based on dilated causal convolutions thus, exhibiting large field reception power. The joint probability of the pixels is factorized as conditional probability products which are –

$$p(x) = \prod_{i=1}^{n} p(x_i|x_1, x_2.....x_n) \tag{7}$$

$$p(y) = \prod_{i=1}^{n} p(y_i|y_1, y.....y_n) \tag{8}$$

Where n is the total no. of co-ordinate points in the stroke segments of a drawing.

The calculated conditional probability distribution is modeled as the stack of convolutional layers. But the surprising part is we need no polling layer in this architecture. A categorical distribution is given as output over the next (X$_i$, Y$_i$) value with the softmax layer. This output is optimized using log-likelihood of the given data with respect to the parameters and also helps to check for under fitting or over fitting.

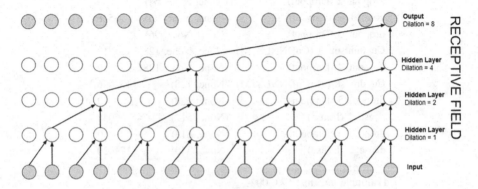

Fig. 9. Visualization of WaveNet model.

Stack of dilated masked convolutions is the main powerful ingredient of this network. At the time of training, conditional probabilities are passed in parallel and the predictions are sequential. After predicting, each simple is passed back to the network to predict the next sample (Fig. 9).

These are faster to train as compared to the RNN network as casual convolutions don't have any recurrent connection. As shown in Fig. 7, the receptive field is growing exponentially on increasing the dilation factor and the network capacity and receptive field size increase with the increase in the dilation factor. The thing is that the WaveNet model needs many layers for increasing its receptive power. So, it is equivalent to a CNN network but is definitely more effective in terms of better and faster training as contrary to CNN, this model doesn't need any stride or pooling convolutions (Table 3).

Table 3. WaveNet architecture used for our proposed work.

Layer (type)	Output shape	Parameters
original_input (Input layer)	(None, 196, 3)	0
dilated_conv_1 (Conv1D)	(None, 196, 192)	1344
dilated_conv_2_tanh (Conv1D)	(None, 196, 192)	73920
dilated_conv_2_sigm (Conv1D)	(None, 196, 192)	73920
gated_activation_1 (Multiply)	(None, 196, 192)	0
skip_1 (Conv1D)	(None, 196, 192)	37056
residual_block_1 (Add)	(None, 196, 192)	0
dilated_conv_4_tanh (Conv1D)	(None, 196, 192)	73920
dilated_conv_4_sigm (Conv1D)	(None, 196, 192)	73920
gated_activation_2 (Multiply)	(None, 196, 192)	0
skip_2 (Conv1D)	(None, 196, 192)	37056
residual_block_2 (Add)	(None, 196, 192)	0
dilated_conv_8_tanh (Conv1D)	(None, 196, 192)	73920
dilated_conv_8_sigm (Conv1D)	(None, 196, 192)	73920
gated_activation_3 (Multiply)	(None, 196, 192)	0
skip_3 (Conv1D)	(None, 196, 192)	37056
residual_block_3 (Add)	(None, 196, 192)	0
dilated_conv_16_tanh (Conv1D)	(None, 196, 192)	73920
dilated_conv_16_sigm (Conv1D)	(None, 196, 192)	73920
gated_activation_4 (Multiply)	(None, 196, 192)	0
skip_4 (Conv1D)	(None, 196, 192)	37056
residual_block_4 (Add)	(None, 196, 192)	0
dilated_conv_32_tanh (Conv1D)	(None, 196, 192)	73920
dilated_conv_32_sigm (Conv1D)	(None, 196, 192)	73920
gated_activation_5 (Multiply)	(None, 196, 192)	0
skip_5 (Conv1D)	(None, 196, 192)	37056
residual_block_5 (Add)	(None, 196, 192)	0
dilated_conv_64_tanh (Conv1D)	(None, 196, 192)	73920

(*continued*)

Table 3. (*continued*)

Layer (type)	Output shape	Parameters
dilated_conv_64_sigm (Conv1D)	(None, 196, 192)	73920
gated_activation_6 (Multiply)	(None, 196, 192)	0
skip_6 (Conv1D)	(None, 196, 192)	37056
residual_block_6 (Add)	(None, 196, 192)	0
dilated_conv_128_tanh (Conv1D)	(None, 196, 192)	73920
dilated_conv_128_sigm (Conv1D)	(None, 196, 192)	73920
gated_activation_7 (Multiply)	(None, 196, 192)	0
skip_7 (Conv1D)	(None, 196, 192)	37056
residual_block_7 (Add)	(None, 196, 192)	0
dilated_conv_256_tanh (Conv1D)	(None, 196, 192)	73920
dilated_conv_256_sigm (Conv1D)	(None, 196, 192)	73920
gated_activation_8 (Multiply)	(None, 196, 192)	0
skip_8 (Conv1D)	(None, 196, 192)	37056
skip_connections (Add)	(None, 196, 192)	0
activation_1 (Activation)	(None, 196, 192)	0
conv_5 ms (Conv1D)	(None, 196, 192)	147648
downsample_to_200 Hz (AveragePooling)	(None, 49, 192)	0
conv_500 ms (Conv1D)	(None, 49, 192)	295104
conv_500 ms_target_shape (Conv1D)	(None, 49, 340)	522580
downsample_to_2 Hz (Average Pooling)	(None, 7, 340)	0
final_conv (Conv1D)	(None, 7, 340)	693940
final_pooling (Global Average Pooling)	(None, 340)	0
final_activation (Activation)	(None, 340)	0

Total params: 3,139,784
Trainable params: 3,139,784
Non-trainable params: 0

8   Results

We have calculated the performance of our proposed approach on the QuickDraw dataset which is a collection of more than 50 million hand-drawn drawings having about 345 different object classes. We have divided the entire data in 80-20 ratio for training and testing respectively. The approach of converting raw data to simplified forms has helped a lot in reducing the computation for training the models as now the models have only refined meaningful data to process and extract features.

Each instance in the data has the co-ordinates of stroke segments that are drawn to make the object (belonging to the class of 345 different objects). We have given trained and tested our models on the basis of these stroke co-ordinates and classes. We have compared the models based on accuracy, precision, and recall (Table 4).

Table 4. Results of the models used in our approach

	Model	Accuracy	Recall	Precision
Machine learning models	RFC	0.71	0.69	0.71
	SVC	0.69	0.70	0.70
	KNN	0.74	0.73	0.73
	MLP	0.72	0.72	0.71
Deep neural networks	CNN	0.78	0.78	0.78
	LSTM	0.79	0.78	0.79
	WaveNet	0.82	0.82	0.81

We have taken 4 machine learning models - Nearest Neighbor (K-NN), Random Forest Classifier (RFC), Support Vector Classifier (SVC) and Multi-Layer Perceptron model (MLP). The final models are compared after selecting the best hyper-parameter for each of these models. We are using KNN with 5 neighbors, SVC having 'RBF' kernel, RFC with 1200 ensemble tree and MLP having 2 hidden layers with ReLu activation function and 'Adam' optimizer. To our surprise, KNN with 5 neighbors as a parameter is giving the best result with 74% accuracy. RFC, MLP, and SVC are classifying the object drawings with the accuracies 71%, 72%, and 69% respectively.

Deep neural networks, on the other hand, are performing very well, the reason being their architecture to extract features and this property has enhanced the performance as we are working on such a huge dataset. Our choice of using WaveNet model has proved to be successful as we are achieving the state-of-the-art results using WaveNet network and too with computational efficiency. CNN and LSTM network are working well and has classified the results with an accuracy of 78% and 79%. Because of the special property of stacked dilated convolutions in WaveNet model. It is extracting features without any stride and polling layer. The flow of sequential data is growing with the growth in the receptive field. These special properties of WaveNet architecture has given it the advantage over CNN and LSTM in our research work and we are achieving the accuracy result of 82% which is very magnificent when we are talking about such huge dataset with so many classes to classify and each of its instances further having the huge amount of co-ordinate points to process.

9 Discussion and Future Scope

Working on such a huge dataset which is further having huge data to process at each instance was a big challenge for our research work. In this paper, we have presented the country-wise analysis of user drawing patterns (as in Sect. 4) and come up with some very interesting findings which one can very easily relate after seeing our work. We have given a clear knowledge about the user's moves. Our analysis can be productive if any researcher is working in the area of stroke suggestion to the user. That means if a user from a country chooses an object e.g. car, then our model using this analysis can suggest strokes to the user automatically and the user will then just be following the suggested strokes to draw a nice car.

Now, our work in the paper also includes object recognition. But data dimensionality and size are the biggest hurdles in our progress. To overcome this problem, we have used the mathematical technique to convert our raw data to simplified forms as mentioned in Sect. 5. This conversion proved to be very effective in refining our information and fastening the training of our model for the recognition task. The machine learning models we used – KNN, MLP, SVC, and RFC. Using machine learning models, we are finally achieving a result with 74% accuracy. So, to improve the performance, we are using neural networks – CNN, LSTM, and WaveNet, as these networks are powerful enough to extract features from the kind of dataset we are using. And as expected, these neural networks enhanced the result accuracy. WaveNet has performed the best of all the models we used. The clear reason behind such exceptional performance is the architecture of WaveNet which enhances the feature extraction and reduces the training time and computation. These models pass on information segment as LSTM but are not relying on gains and extracting features so efficiently but need no stride or polling layer. Thus, making the performance result of our work better.

Our work can be used in many application-like children education [26] hand gesture recognition, sketch-based search [27], signature matching where co-ordinates from the input can be extracted and after plotting them in feature vector space, data can be very easily used in whatever way researchers or developers wish to use.

References

1. Stiny, G., Gips, J.: Shape grammars and the generative specification of painting and sculpture. In: IFIP Congress (2), vol. 2, no. 3, August 1971
2. Futrelle, R.P., Nikolakis, N.: Efficient analysis of complex diagrams using constraint-based parsing. In: Proceedings of 3rd International Conference on Document Analysis and Recognition, vol. 2, pp. 782–790. IEEE, August 1995
3. Bimber, O., Encarnacao, L.M., Stork, A.: A multi-layered architecture for sketch-based interaction within virtual environments. Comput. Graph. **24**(6), 851–867 (2000)
4. Mahoney, J.V., Fromherz, M.P.: Three main concerns in sketch recognition and an approach to addressing them. In: AAAI Spring Symposium on Sketch Understanding, pp. 105–112, March 2002
5. Long, A.C., Landay, J.A., Rowe, L.A.: Quill: a gesture design tool for pen-based user interfaces. University of California, Berkeley (2001)
6. Gross, M.D., Do, E.Y.L.: Demonstrating the electronic cocktail napkin: a paper-like interface for early design. In: Conference on Human Factors in Computing Systems: Conference Companion on Human Factors in Computing Systems: Common Ground, vol. 13, no. 18, pp. 5–6, April 1996
7. Krizhevsky, A., Sutskever, I., Hinton, G.E.: ImageNet classification with deep convolutional neural networks. In Advances in Neural Information Processing Systems, pp. 1097–1105 (2012)
8. Girshick, R., Donahue, J., Darrell, T., Malik, J.: Rich feature hierarchies for accurate object detection and semantic segmentation. In Proceedings of the IEEE Conference on Computer Vision and Pattern Recognition, pp. 580–587 (2014)

9. Donahue, J., et al.: Decaf: a deep convolutional activation feature for generic visual recognition. In International Conference on Machine Learning, pp. 647–655, January 2014
10. Davis, R.: Position statement and overview: sketch recognition at MIT. In: AAAI Sketch Understanding Symposium (2002)
11. Suykens, J.A., Vandewalle, J.: Least squares support vector machine classifiers. Neural Process. Lett. **9**(3), 293–300 (1999)
12. Zhang, M., Zhang, D.X.: Trained SVMs based rules extraction method for text classification. In: 2008 IEEE International Symposium on IT in Medicine and Education, pp. 16–19. IEEE, December 2008
13. Lau, K.W., Wu, Q.H.: Online training of support vector classifier. Pattern Recogn. **36**(8), 1913–1920 (2003)
14. Breiman, L.: Random forests. Mach. Learn. **45**(1), 5–32 (2001)
15. Keller, J.M., Gray, M.R., Givens, J.A.: A fuzzy k-nearest neighbor algorithm. IEEE Trans. Syst. Man Cybern. **4**, 580–585 (1985)
16. Beyer, K., Goldstein, J., Ramakrishnan, R., Shaft, U.: When is "nearest neighbor" meaningful? In: Beeri, C., Buneman, P. (eds.) ICDT 1999. LNCS, vol. 1540, pp. 217–235. Springer, Heidelberg (1999). https://doi.org/10.1007/3-540-49257-7_15
17. Dudani, S.A.: The distance-weighted k-nearest-neighbor rule. IEEE Trans. Syst. Man Cybern. **4**, 325–327 (1976)
18. Jain, A.K., Mao, J., Mohiuddin, K.M.: Artificial neural networks: a tutorial. Computer **29**(3), 31–44 (1996)
19. Kim, Y.: Convolutional neural networks for sentence classification, August 2014. https://arxiv.org/abs/1408.5882
20. Zhang, X., Zhao, J., LeCun, Y.: Character-level convolutional networks for text classification. In: Proceedings of Advances in Neural Information Processing Systems, pp. 649–657 (2015)
21. Xiao, Y., Cho, K.: Efficient character-level document classification by combining convolution and recurrent layers, February 2016. https://arxiv.org/abs/1602.00367
22. Hochreiter, S., Schmidhuber, J.: Long short-term memory. Neural Comput. **9**(8), 1735–1780 (1997)
23. Du, Y., Wang, W., Wang, L.: Hierarchical recurrent neural network for skeleton based action recognition. In: Proceedings of the IEEE Conference on Computer Vision and Pattern Recognition, pp. 1110–1118 (2015)
24. Paine, T.L., et al.: Fast wavenet generation algorithm (2016). arXiv preprint arXiv:1611.09482
25. Oord, A.V.D., et al.: WaveNet: a generative model for raw audio (2016). arXiv preprint arXiv:1609.03499
26. Cao, Y., Wang, H., Wang, C., Li, Z., Zhang, L., Zhang, L.: MindFinder: interactive sketch-based image search on millions of images. In: Proceedings of the 18th ACM International Conference on Multimedia, pp. 1605–1608. ACM, October 2010
27. Paulson, B., Eoff, B., Wolin, A., Johnston, J., Hammond, T.: Sketch-based educational games: drawing kids away from traditional interfaces. In: Proceedings of the 7th International Conference on Interaction Design and Children, pp. 133–136. ACM, June 2008

Real Time Static Gesture Detection Using Deep Learning

Kalpdrum Passi$^{(\boxtimes)}$ (iD) and Sandipgiri Goswami

Laurentian University, Sudbury, ON P3E 2C6, Canada
{kpassi,sgoswami}@laurentian.ca

Abstract. Sign gesture recognition is an important problem in human-computer interaction with significant societal influence. However, it is a very complex task, since sign gestures are naturally deformable objects. Gesture recognition contains unsolved problems for the last two decades, such as low accuracy or low speed, and despite many proposed methods, no perfect result has been found to explain these unsolved problems. In this paper, we propose a deep learning approach to translating sign gesture language into text. In this study, we have introduced a self-generated image data set for American Sign language (ASL). This dataset is a collection of 36 characters containing A to Z alphabets and 0 to 9 number digits. The proposed system can recognize static gestures. This system can learn and classify specific sign gestures of any person. A convolutional neural network (CNN) algorithm is proposed for classifying ASL images to text. An accuracy of 99% on the alphabet gestures and 100% accuracy on digits was achieved. This is the best accuracy compared to existing systems.

Keywords: Sign gestures · Image processing · Deep learning · Convolutional neural networks

1 Introduction

The World Health Organization (WHO) estimated that 250 million people in the world are deaf as well as dumb [1]. These groups of people use symbolic language to communicate with other people. This symbolic language is called sign language. Sign Language is built for communication and is used worldwide among hard of hearing and deaf people. Sign language is not a unique language and is designed differently in different countries. Neither is sign language a recent improvement. There is proof that speaking through gestures has been around since the start of human development [2]. Different counties have their own sign language such as American Sign Language, French Sign Language, Indian Sign Language and Puerto Rican Sign Language, to name a few. Gesture-based communication is dependent on region and has significant differences from other languages. It is very important to understand sign language when we communicate with deaf or young children and their families. Lack of understanding results in significant challenges in understanding this community and may result in miscommunication. Sign Language is a language which is used to convey messages by hand movements, facial expression and body language for

© Springer Nature Switzerland AG 2019
S. Madria et al. (Eds.): BDA 2019, LNCS 11932, pp. 408–426, 2019.
https://doi.org/10.1007/978-3-030-37188-3_23

communication. It is mainly used by deaf and by people who can hear but cannot speak. Sometimes family members and relatives must learn sign language to interpret, which enables deaf to communicate with wider communities.

In this paper, image classification and deep learning have been used for interpreting American Sign Language. For image classification, computer vision algorithms were used to capture images and to process data set for filtering as well as reducing noise from images. Finally, the data set is trained using deep learning algorithm, a convolutional neural network (CNN) for classification of ASL images and measuring the accuracy of the training data set. The abstract view of the derived approach combining the image classification and deep learning for American Sign Language is shown in Fig. 1.

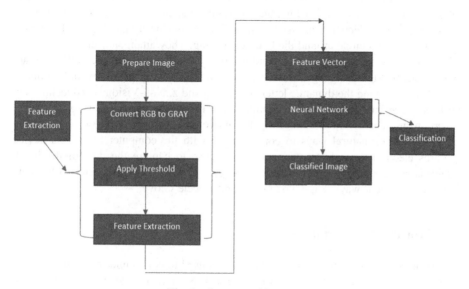

Fig. 1. System architecture

2 Related Work

Machine learning is most commonly used for image recognition. Hidden Markov Model (HMM) and Dynamic Time Warping (DTW), two kinds of machine learning methods, are widely applied to achieve high accuracies [6–8]. These are mostly good at capturing time-based patterns, but they require clearly characterized models that are defined before learning. Sterner and Pentland [6] used a Hidden Markov Model and a 3-Dimensional glove that detects hand movement. Since the glove can attain 3-Dimensional detail from the hand regardless of spatial orientation, they achieved the best accuracy of 99.2% on the test set. Hidden Markov Model uses time series data to track hand gestures and classify them based on the position of the hand in recent frames.

As per research point of view, a linear classifier is easy to work with because linear classifiers are relatively simple models, it requires sophisticated feature extraction and preprocessing methods to get good results [3–5]. Singha and Das [3] achieved an accuracy of 96% on ten classes for images of gestures of one hand using Karhunen-Loeve Transforms. These translate and rotate the axes to build up a new framework based on the variance of the data. This technique is useful after using a skin color detection, hand cropping and edge recognition on the images. They use a linear classifier to recognize number signs including a thumbs up, first and index finger pointing left and right, and numbers only. Sharma [5] has done research using Support Vector Machines (SVM) and k-Nearest Neighbors (KNN) to illustrate each color channel after background noise deletion and noise subtraction. Their research suggests using contours, which is very useful to represent hand contours. They got an accuracy of 62.3% using a Support Vector Machines on the segmented color channel model. Suk [8] suggested a system for detecting hand gestures in a continuous video stream using a dynamic Bayesian Network or DBN model. They try to classify moving hand gestures, such as creating a circle around the body or waving. They attain an accuracy of nearly 99%, but it is worth noting that all hand gestures are different from each other and are not American Sign Language. However, the motion-tracking feature would be applicable for classifying the dynamic letters of ASL: j and z. Non-Vision based technology such as Glove-based hand shape recognition usually has the person wearing gloves and a certain number of wires connecting the glove to a computer. These methods are difficult and non-natural ways to communicate with the computer [21]. This device requires electricity or electromagnetic interference to capture the data of the hand, which is sufficient to describe a handshape gesture [22]. Scientists refer to the data gloves in different ways, e.g. CyberGlove and Accele Glove.

3   Dataset and Variables

American Sign Language (Figs. 2 and 3) [13] is used to communicate among the deaf community and others. A dataset for American Sign Language was created as no standard dataset is available for all countries/subcontinents [18]. In the present scenario, the requirement for large vocabulary dataset is in demand [18]. Moreover, the existing dataset is incomplete and additional hand gestures had to be included. In future, this research can help other researchers to develop their own dataset based on their requirements. This dataset is a collection of 36 characters which contains A to Z alphabets and 0 to 9 numerical digits. Right hand was used to capture 1000 images for specific alphabets and numbers. Code was implemented to flip the images from the right to the left-hand image. The height and width ratios vary significantly but average approximately 50 × 50 pixels. The dataset contains over 72,000 images in grayscale. Additional images can be added to this dataset. Figure 4 shows images of A to Z alphabet. Table 1 describes the dataset properties.

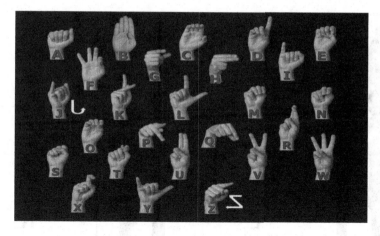

Fig. 2. American Sign Language manual alphabet [13]

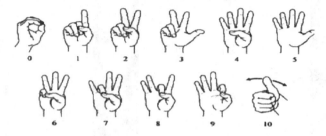

Fig. 3. American Sign Language numbers [13]

3.1 Capturing Images for Dataset

To detect hand gestures using skin color, there are different approaches including skin color-based methods. In this research, after detecting and subtracting the face and other background objects, skin recognition and a contour comparison algorithm were used to search for the hand and discard other background color objects for every frame captured from a webcam or video file. First, we need to extract and store the hand contour and skin color which is used to compare with the frame in order to detect the hand. After detecting the skin area for each captured frame, the contours of the detected areas were compared with the saved hand histogram template contours to remove other skin like objects existing in the image. If the contour comparison of the spotted skin area complies with any one of the saved hand histogram contours, the hand gesture was successfully captured. To capture images of our dataset no special environment setup or a high-resolution camera was required. Images were captured at different times and different lighting conditions. Changes in the lighting conditions might require updating stored histogram template.

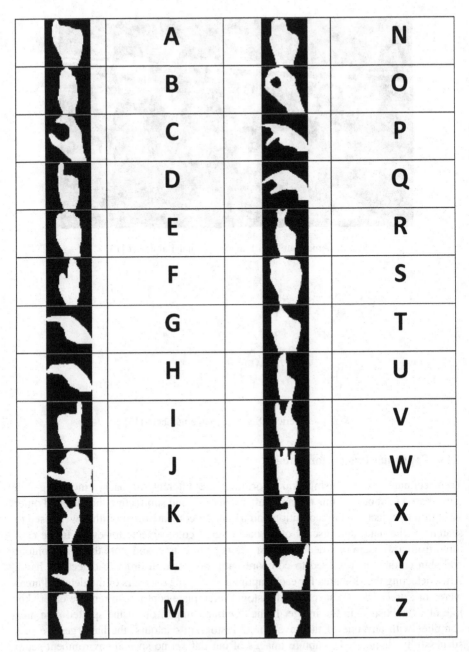

Fig. 4. Dataset images of alphabets A to Z

Table 1. Image properties and their description

Property	Description
Alphabets	A to Z
Numbers	0 to 9
Color	Greyscale
Dimensions	50 × 50
Height	50 pixels
Width	50 pixels
File type	JPEG

4 Hand Gesture Detection

An integrated system is proposed for detection, segmentation, and tracking of the hand in a gesture recognition system using a single webcam. As compared to other methods that use color gloves [10, 11], our method can detect the hand gesture by integrating two useful features: skin color detection and contour matching. The proposed hand gesture detecting algorithm has real-time performance and is strong against rotations, scaling, cluttered background, and lighting conditions. The strength of our proposed hand gesture detection algorithm is based on comparison with other methods mentioned in Sect. 4.1. Detecting the human hand in a cluttered background boosts the performance of hand gesture recognition systems. In this method, the speed and result of recognition is same for any frame size taken from a webcam such as 640 × 480, 320 × 240 or 160 × 120 and the system is robust against a cluttered background as only the hand posture area is detected.

To detect the hand gesture in the image, a four-phase system was designed as shown in Fig. 5. First, load hand contour template which is used to compare and detect hand skin area pattern from webcam using the contours comparison algorithm. Then we open a camera which has a square box to capture hand gesture. The hand must be placed fully within the square box. The skin color locus (captured skin contour template) for the image is removed from the user's skin color after face deletion. In the last step, the hand gesture is spotted by removing false positive skin pixels and identifying hand gesture and other real skin color regions using contour matching with the loaded hand gesture pattern contours.

Fig. 5. Hand posture detection steps

4.1 Skin Detection

Skin detection is a useful approach for many computer vision applications such as face recognition, tracking and facial expression, abstraction, or hand tracking and gesture recognition. There are recognized procedures for skin color modelling and recognition that allow differentiating between the skin and non-skin pixels based on their color. To get a suitable distinction between skin and non-skin areas, a color transformation is needed to separate luminance from chrominance [13].

The input images normally are in the Color format (RBG), which has the drawback of having components dependent on the lighting situations. The misunderstanding between the skin and non-skin pixels can be decreased using color space transformation. There are different approaches to detect skin color components in other color spaces, such as HSV, YCbCr, TSL or YIQ to provide better results in parameter recovery under changes in lighting condition. Research has shown that skin colors of individuals cluster closely in the color space for all people from different societies, for example, color appearances in human faces and hands vary more in intensity than in chrominance [12, 14]. Thus, taking away the intensity V of the original color space and working in the chromatic color space (H, S) provides invariance against illumination situations. In [13], it was established that removal of the Value (V) component and only using the Hue and Saturation components can still allow for the detection of 96.83% of the skin pixels. In this research hue, saturation, value (HSV) color model was used since it has shown to be one of the most adapted to skin-color detection [15]. It is also well-matched with human color perception. In addition, it has real-time execution and it is more robust in cases of rotations, scaling, cluttered background, and changes in lighting condition. The proposed hand gesture detection algorithm is real-time and robust against the mentioned previous changes. The other skin like objects existing in the images are removed from the contour when compared to the loaded hand posture prototype contours.

The HSV color space is gained by a nonlinear transformation of the essential RGB color space. The conversion between RGB and HSV is described in [16]. Hue (H) is a section that characterizes a pure color such as pure yellow, orange or red, whereas saturation (S) provides a measure of the degree to which a pure color is diluted by white light [17]. Value (V) attempts to represent brightness along the gray axis such as white to black, but since brightness is subjective, it is thus difficult to measure [17].

According to [18] and Fig. 6, Hue is estimated in HSV color space by a position with Red starting at 0, Green at 120 and Blue at 240°. The black mark in the diagram at the lower left on the screen determines the hue angle.

Saturation is a ratio that ranges between 0.0 along the middle line of the cone (the V axis) to 1 on the edge of the cone. The value starts from 0.0 (dark) and goes up to 1.0 (bright). According to [12], the HSV model can be resulting from the non-linear transformation from an RGB model according to the calculations in Eqs. (1)–(4).

$$H = \begin{cases} \theta, & G \geq B \\ 2\pi - \theta, & G < B \end{cases} \tag{1}$$

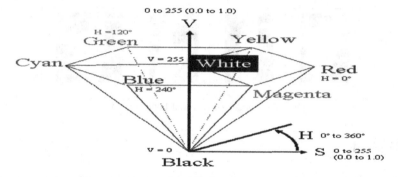

Fig. 6. HSV color space [18] (Color figure online)

$$S = \frac{\max(R, G, B) - \min(R, G, B)}{\max(R, G, B)} \tag{2}$$

$$V = \frac{\max(R, G, B)}{255} \tag{3}$$

$$\theta = arocs\left\{\frac{\lceil(R - G) + [R - B]/2\rceil}{\left[(R - G)^2 + (R - G)(G - B)\right]^{\frac{1}{2}}}\right\} \tag{4}$$

As per an image segmentation point of view, the skin-color detection is divided into two class problem: skin-pixel vs non-skin-pixel classification. Currently, different known image segmentation approaches exist such as thresholding, Gaussian classifier, and multilayer perceptron [19, 20].

In this research, thresholding is used that allows getting a good result for higher computation speed when compared with other techniques, given our real-time requirements. This thresholding is used to find image segmentation and the values between two components H and S in the HSV model where we removed the Value (V) component from the image. Usually, a pixel can be observed as being a skin-pixel when the following threshold values are satisfied: 0° < H < 20° and 75° < S < 190°.

4.2 Contour Comparison

Once the skin color has been detected, the contours of the detected skin color are recovered and then compared with the contours of the hand gesture patterns. Once skin color contours are recognized as belonging to the hand gesture contour patterns, that area is identified as a region of interest (ROI) which is then used for tracking the hand movements and saving the hand posture in JPEG format in small images as shown in Fig. 7. Stored images are further used to extract the features needed to recognize the hand gestures in the testing stage.

Fig. 7. Images of detected hand postures.

5   Method

Our overarching approach was one of the basic supervised learning methods which is most commonly used. This method needs training data with a specific format. Each instance must have assigned label. These labels make available supervision for the learning algorithm. The training process of supervised learning is constructed on the following principle. First, the training data is fed into the model to produce estimates of output. This estimate is compared to the assigned label of the training data in order to evaluate model error. Based on this error the learning algorithm alters model's parameters in order to reduce it.

5.1   Architecture

Convolutional Neural Network (CNN) architecture consists of multiple convolution and dense layers. The CNN architecture includes three types of three convolution layers and each layer has its own max pooling layer and one group of fully connected layer followed by a dropout layer and the output layer as shown in Fig. 8.

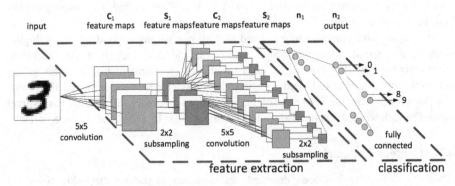

Fig. 8. Convolutional Neural Network (CNN) architecture

The CNN architecture was implemented using the Kera's deep learning framework with Python and Tensor flow backend. The same network was later used for alphabet classification. The proposed CNN contains three convolutional layers and three max-pooling layers. The only fully-connected layer is the final layer and the layer after the last convolutional layer. The input layer takes in batches of images of size

$50 \times 50 \times 1$. ReLU activation function was used between the hidden layers, and a SoftMax activation is used in the output layers multiple class classification and detection. The first convolutional layer has 16 filters of size 3×3. The filter is also called the weight. Filter is used to extract features from the image. It is followed by a pooling layer that uses the max pooling operation with the size of 2×2. When pooling is applied in a forward direction, it is called Forward Propagation. Max pooling is used for sub sampling from an image. The second convolutional layer has 32 filters with a capacity of 3×3. Similar to the first convolutional layer, it is followed by a max-pooling layer with the kernel size of 2×2. The third convolutional layer has 64 filters with the same kernel size as a previous convolutional layer. The 2×2 max pooling is applied yet again. The parameters of the described layers are also illustrated in Fig. 9.

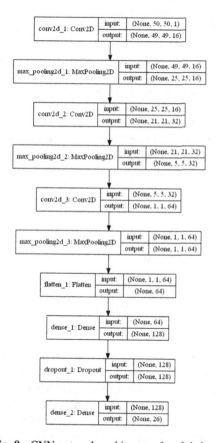

Fig. 9. CNN network architecture for alphabets.

- Convolutional layer with 16 feature maps of size 3×3.
- Pooling layer taking the max over 2 * 2 patches.
- Convolutional layer with 32 feature maps of size 3×3.
- Pooling layer taking the max over 2 * 2 patches.

- Convolutional layer with 64 feature maps of size 3 × 3.
- Pooling layer taking the max over 2 * 2 patches.
- Dropout layer with a probability of 20%.
- Flatten layers.
- Fully connected layer with 128 neurons and rectifier activation.
- Fully connected layer with ten neurons and rectifier activation.
- Output layer.

5.2 Hardware and Software Configuration

Training of Neural Networks is notoriously computationally expensive and requires a lot of resources. From the bottom level perspective, it translates into many multiplications of matrices. Modern Central Processing Units (CPUs) are not made of such computations and therefore are not very efficient. On the other hand, modern GPUs are designed to perform exactly these operations.

At present on the market, there are two main parallel computing platforms CUDA and OpenCL. They both have their own advantages and disadvantages, but the major difference is that CUDA is proprietary, while OpenCL is available free. This divide translates into hardware productions as well. CUDA is mostly supported by NVIDIA and OpenCL is supported by AMD. NVIDIA with its CUDA platform is presently a leader in the domain of deep learning. Therefore, for the training of CNN models, GPUs from NVIDIA were selected. The selected model was GIGABYTE GeForce GTX 1080. Detailed information about hardware configuration is given in Table 2.

Table 2. Hardware configuration

Processor	Configuration
GPU	GeForce GTX 1080 4 GB
CPU	Intel(R) Core(TM) i7-8550 CPU @ 2.00 GHz
Memory	DIMM 1333 MHz 8 GB

From the list of considered software tools, Keras was selected as it is written in python, an easy to program language and as it satisfies all considered factors. Support for efficient GPU implementation in Keras relies on either Tensor flow or Theano backend. From the different user perspectives, it doesn't matter either way, but Tensor flow was selected because it was observed as faster of the two and has GPU-accelerated library package of primitives for deep neural networks. Detailed information about software configuration is given in Table 3.

Table 3. Software configuration

Software	Configuration
Keras	2.04
Tensorflow	1.1.0
CUDA	7.5
Python	3.53
Operating system	Windows 10
Open CV	2.0

5.3   Feature Extraction

The filter is applied across the entire layer and moved one pixel at a time. Each position results in an activation of the neuron and the output is collected in the feature map. Each filter spots a feature at every location on the input. The feature detected from the input is spatially transferred to the next layer, otherwise the undetected input is left unchanged. In other words, starting from top-left corner of the input image, each patch is moved from left to right, one pixel at a time, as shown in Fig. 10. Once it reaches the top-right corner, the filter is moved one pixel in the sliding direction, and again the filter is moved from the left to the right, one pixel at a time. This process repeats until the filter reaches the bottom-right corner of the image.

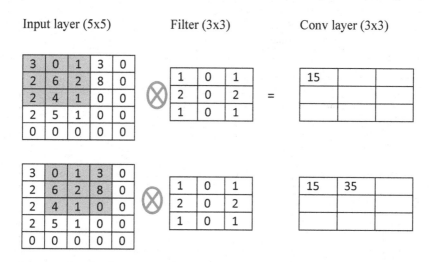

Fig. 10. Example of an input layer being convolved by a kernel.

6   Results

On our self-generated dataset, we achieved an accuracy of 99% to detect hand gestures for alphabets and 100% accuracy was achieved to detect hand gestures for digits. Real-time testing was done with five different students and estimation per student took

20 min for alphabets and approximately 7 to 8 min for digits. Testing was done in non-controlled form, i.e. in different lighting conditions and with different backgrounds. For alphabets 50 epochs were applied, and for digits 20 epochs were applied. Both networks used Adam optimizer and a learning rate of 0.001. Loss function was cross-entropy due to multiple-class classification. The training and testing sets contained 70:30 ratio, respectively in both models. Figure 11 shows the validation accuracy for different epochs for the digits. Figure 12 shows the validation accuracy for different epochs for alphabets. The confusion matrices of both networks are illustrated in Figs. 13 and 14. From both confusion matrices, it is evident that the classification accuracy of both models is almost identical. The only difference is the number of false negatives and true positives. Recall, precision, accuracy and F measure were used as classification evaluation metrics. Table 4 shows the comparison with other researchers results and classification methods.

Table 4. Results from researchers own dataset

Sign Language	Classification Method	Result
Chinese Sign Language	Hypothesis comparison guided cross validation (HC-CV)	88.5
American Sign Language	Pseudo two-dimensional hidden Markov models	98
Chinese Sign Language	Local linear embedding	92.2
American Sign Language	Vector Quantization PCA (VQPCA) Non-Periodic Signs Periodic Signs Total	97.30 97.00 86.80

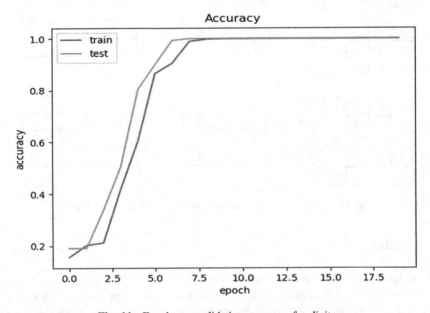

Fig. 11. Epochs vs. validation accuracy for digits.

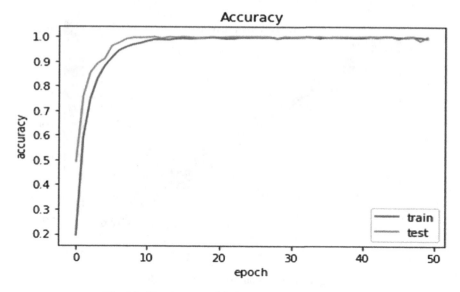

Fig. 12. Epochs vs. validation accuracy for alphabets

Figure 13 shows a 10×10 confusion matrix for digits. There are ten different classes for ten digits on that confusion matrix. The number of correctly classified images is the sum of the diagonal elements in the matrix, all others are incorrectly predicted. Figure 14 shows the confusion matrix for alphabets. In this confusion matrix, there are 26 different classes for the 26 alphabets. For computation purposes,

- TP_0 refers to the positive tuples that are correctly predicted (POSITIVE) by the classifier in the first row-first column, i.e. 353.
- TP_1 refers to the positive tuples that are correctly predicted (POSITIVE) by the classifier in the second row-second column, i.e. 353.
- TP_9 refers to the positive tuples which are correctly predicted (POSITIVE) by the classifier in the ninth row-ninth column, i.e. 375.

The accuracy of the correctly classified images comes out to be 1.

Table 5 shows the precision, recall, F measure, and support for the digits. Table 6 shows the performance metrics for the alphabets.

ROC Area Under the Curve (AUC): The Receiver Operating Characteristics (ROC) charts is a method for organizing classification and visualizing the performance of the trained data using the classifier. AUC for all the digit classes is 1, which shows that the performance evaluation of all classes is excellent in the data set. Also, the weighted value of the AUC is 1 (100%) for the data classified using the CNN network, which correctly classified the data with the best accuracy and best performance. AUC curves for the digits and alphabets are shown in Figs. 15 and 16, respectively.

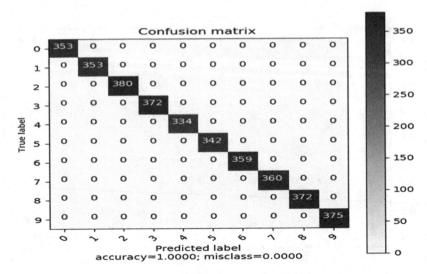

Fig. 13. Confusion matrix for 0 to 9 digits.

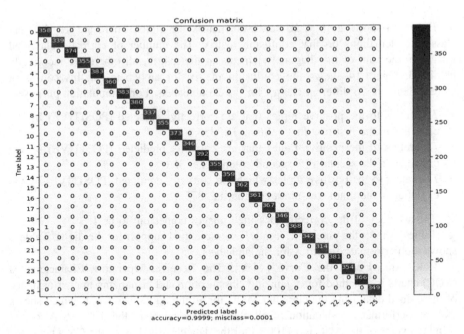

Fig. 14. Confusion matrix for A to Z alphabets.

Table 5. Performance on digits.

	Precision	Recall	F1-score	Support
0	1.00	1.00	1.00	353
1	1.00	1.00	1.00	353
2	1.00	1.00	1.00	380
3	1.00	1.00	1.00	372
4	1.00	1.00	1.00	334
5	1.00	1.00	1.00	342
6	1.00	1.00	1.00	359
7	1.00	1.00	1.00	360
8	1.00	1.00	1.00	372
9	1.00	1.00	1.00	375
Weighted avg	1.00	1.00	1.00	3600

Table 6. Performance on alphabets.

	Precision	Recall	f1-score	Support
0	1.00	1.00	1.00	358
1	1.00	1.00	1.00	339
2	1.00	1.00	1.00	374
3	1.00	1.00	1.00	355
4	1.00	1.00	1.00	383
5	1.00	1.00	1.00	360
6	1.00	1.00	1.00	383
7	1.00	1.00	1.00	380
8	1.00	1.00	1.00	337
9	1.00	1.00	1.00	355
10	1.00	1.00	1.00	373
11	1.00	1.00	1.00	346
12	1.00	1.00	1.00	392
13	1.00	1.00	1.00	355
14	1.00	1.00	1.00	359
15	1.00	1.00	1.00	362
16	1.00	1.00	1.00	361
17	1.00	1.00	1.00	367
18	1.00	1.00	1.00	346
19	1.00	1.00	1.00	369
20	1.00	1.00	1.00	342
21	1.00	1.00	1.00	314
22	1.00	1.00	1.00	381
23	1.00	1.00	1.00	354
24	1.00	1.00	1.00	366
25	1.00	1.00	1.00	349
Weighted avg	1.00	1.00	1.00	9360

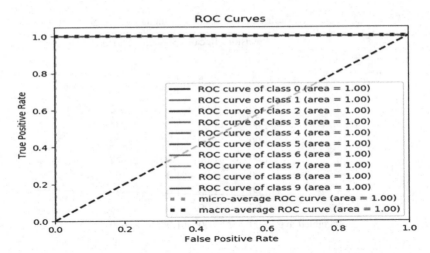

Fig. 15. ROC AUC graph for 0 to 9 digits.

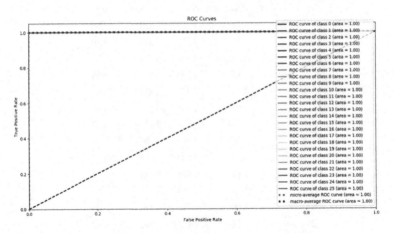

Fig. 16. ROC AUC graph for A to Z alphabets

7 Conclusion and Future Work

In this paper we developed a system to recognize American Sign Language gestures using a skin color model, thresholding and CNN. We have tested with different lighting condition and in different places. The dataset collected in the ideal conditions has proved to be the most efficient dataset in terms of accuracy and gives 99% accuracy on alphabets and 100% accuracy on digits.

Sign gesture recognition still has a long way to go in the research path, especially for 2D systems. This study offers fascinating ideas for future research. Some of these possibilities are defined in this section. As this paper focused only on static sign gesture recognition, one next step forward is to recognize the dynamic sign gesture for the ASL. Even though that study introduces a self-generated new dataset with rather more gestures for American Sign Language, it still does not offer all the possible movements for American Sign Language. Videos with rotation in 3Dimension, words and expressions are examples of how this dataset can be extended.

References

1. WHO calls on private sector to provide affordable hearing aids in developing world [Internet]. Current neurology and neuroscience reports. U.S. National Library of Medicine (2001). https://www.ncbi.nlm.nih.gov/pubmed/11887302. Accessed 22 Jan 2019
2. Emond, A., Ridd, M., Sutherland, H., et al.: The current health of the signing Deaf community in the UK compared with the general population: a cross-sectional study. BMJ Open **5**, e006668 (2015). https://doi.org/10.1136/bmjopen-2014-006668
3. Das, K., Singha, J.: Hand gesture recognition based on Karhunen-Loeve transform, pp. 365–371 (2013)
4. Aryanie, D., Heryadi, Y.: American sign language-based finger-spelling recognition using k-Nearest Neighbors classifier. In: 2015 3rd International Conference on Information and Communication Technology (ICoICT), pp. 533–536 (2015)
5. Sharma, R., Nemani, Y., Kumar, S., Kane, L., Khanna, P.: Recognition of single handed sign language gestures using contour tracing descriptor. In: Lecture Notes in Engineering and Computer Science, vol. 2 (2013)
6. Starner, T., Pentland, A.: Real-time American Sign Language recognition from video using hidden Markov models. In: Proceedings of International Symposium on Computer Vision - ISCV, Coral Gables, FL, USA, pp. 265–270 (1995). https://doi.org/10.1109/iscv.1995.477012
7. Jebali, M., Dalle, P., Jemni, M.: Extension of hidden Markov model for recognizing large vocabulary of sign language. Int. J. Artif. Intell. Appl. **4** (2013). https://doi.org/10.5121/ijaia.2013.4203
8. Suk, H.-I., Sin, B.-K., Lee, S.-W.: Hand gesture recognition based on dynamic Bayesian network framework. Pattern Recognit. **43**, 3059–3072 (2010). https://doi.org/10.1016/j.patcog.2010.03.016
9. Vicars, W.: ASL University. ASL [Internet]. Children of Deaf Adults (CODA). http://www.lifeprint.com/. Accessed 29 Jan 2019
10. Aran, O., Keskin, C., Akarun, L.: Computer applications for disabled people and sign language tutoring. In: Proceedings of the Fifth GAP Engineering Congress, pp. 26–28 (2006)
11. Tokatlı, H., Halıcı, Z.: 3D hand tracking in video sequences. MSc thesis, September 2005
12. He, J., Zhang, H.: A real time face detection method in human-machine interaction. In: 2008 2nd International Conference on Bioinformatics and Biomedical Engineering (2008)
13. Zhu, Q., Wu, C.-T., Cheng, K.-T., Wu, Y.-L.: An adaptive skin model and its application to objectionable image filtering. In: Proceedings of the 12th Annual ACM International Conference on Multimedia - MULTIMEDIA 2004 (2004)
14. Kelly, W., Donnellan, A., Molloy, D.: Screening for objectionable images: a review of skin detection techniques. In: 2008 International Machine Vision and Image Processing Conference, pp. 151–158 (2008)
15. Zarit, B., Super, B., Quek, F.: Comparison of five color models in skin pixel classification. In: Proceedings International Workshop on Recognition, Analysis, and Tracking of Faces and Gestures in Real-Time Systems in Conjunction with ICCV 1999, pp. 58–63 (Cat No PR00378) (1999)

16. Ford, A., Roberts, A.: Color space conversions. Westminster University, London, UK (1998)
17. Gonzalez, R., Woods, R., Eddins, S.: Digital Image Processing Using MATLAB. Englewood Cliffs, NJ (2004)
18. Hughes, J.F.: Computer Graphics: Principles and Practice. Addison-Wesley, Upper Saddle River (2014)
19. Nallaperumal, K., et al.: Skin detection using color pixel classification with application to face detection: a comparative study. In: Proceedings of IEEE International Conference on Computational Intelligence and Multimedia Applications, vol. 3, pp. 436–441 (2007)
20. Greenspan, H., Goldberger, J., Eshet, I.: Mixture model for face-color modeling and segmentation. Pattern Recogn. Lett. **22**(14), 1525–1536 (2001)
21. Mitra, S., Acharya, T.: Gesture recognition: a survey. IEEE Trans. Syst. Man Cybern. Part C (Appl. Rev.) **37**(3), 311–324 (2007)
22. Nagi, J., et al.: Max-pooling convolutional neural networks for vision-based hand gesture recognition. In: 2011 IEEE International Conference on Signal and Image Processing Applications (ICSIPA) (2011)

Interpreting Context of Images Using Scene Graphs

Himangi Mittal[1]([✉]), Ajith Abraham[2], and Anuja Arora[1]

[1] Department of Computer Science Engineering,
Jaypee Institute of Information Technology, Noida, India
himangimittal@gmail.com, anuja.arora@gmail.com
[2] Machine Intelligence Research Labs (MIR Labs), Auburn, WA 98071, USA
ajith.abraham@ieee.org

Abstract. Understanding a visual scene incorporates objects, relationships, and context. Traditional methods working on an image mostly focus on object detection and fail to capture the relationship between the objects. Relationships can give rich semantic information about the objects in a scene. The context can be conducive in comprehending an image since it will help us to perceive the relation between the objects and thus, give us a deeper insight into the image. Through this idea, our project delivers a model which focuses on finding the context present in an image by representing the image as a graph, where the nodes will the objects and edges will be the relation between them. The context is found using the visual and semantic cues which are further concatenated and given to the Support Vector Machines (SVM) to detect the relation between two objects. This presents us with the context of the image which can be further used in applications such as similar image retrieval, image captioning, or story generation.

Keywords: Scene understanding · Context · Word2Vec · Convolution neural network

1 Introduction

Computer Vision has a number of applications which needs special attention of researchers such as semantic segmentation, object detection, classification, localization, and instance segmentation. The work attempted in the paper lies in the category of semantic segmentation. Semantic segmentation has two phases – segmentation, detection of an object and semantic, is the prediction of context.

Understanding a visual scene is one of the primal goals of computer vision. Visual scene understanding includes numerous vision tasks at several semantic levels, including detecting and recognizing objects. In recent years, great progress has been made to build intelligent visual recognition systems. Object detection focuses on detecting all objects. Scene graph generation [1–4] recognizes not only the objects but also their relationships. Such relationships can be represented

© Springer Nature Switzerland AG 2019
S. Madria et al. (Eds.): BDA 2019, LNCS 11932, pp. 427–438, 2019.
https://doi.org/10.1007/978-3-030-37188-3_24

by directed edges, which connect two objects as a combination of the subject – predicate - object. In contrast to the object detection methods, which just result in whether an object exists or not, a scene graph also helps in infusing context in the image. For example, there is a difference between a man feeding a horse and a man standing by a horse.

This rich semantic information has been largely unused by the recent models. In short, a scene graph is a visually grounded graph over the object instances in an image, where the edges depict their pairwise relationships. Once a scene graph is generated, it can be used for many applications. One such is to find an image based on the context by giving a query. Numerous methods for querying a model database are based on properties such as shape and keywords have been proposed, the majority of which are focused on searching for isolated objects. When a scene modeler searches for a new object, an implicit part of that search is a need to find objects that fit well within their scene. Using a scene graph to retrieve the images by finding context has a better performance than comparing the images on a pixel level. An extension to the above application is clustering of similar images. Recent methods cluster the image by calculating the pixel-to-pixel difference. This method does not generalize well and works on images which are highly similar. Also, this method may lead to speed and memory issues. The approach of scene graph infused with context can help to cluster the images even if there"s a vast pixel difference. This method is also translation invariant, meaning, that a girl eating in the image can be anywhere in the image, but the context remains the same. Since this method uses semantic information, it enhances speed and memory.

The paper is structured in the following manner - Sect. 2 discusses the related work done in this direction, highlighting the scope of work to design a better solution. Importance and significance of work are discussed in Sect. 3. Section 4 is about the dataset available and used to perform experiments. Section 5 discusses the solution approach followed by Sect. 6 which covers finding of object and context interpretation using scene graph. Finally, concluding remark and future scope is discussed in Sect. 7.

2   Related Work

The complete work is can be divided into two tasks – Object detection and Context interpretation. Hence, a plethora of papers have been studied to understand the various approaches defined by researchers in order to achieve an efficient and scalable outcome in both directions. Initially, in order to get an idea about deep learning models used in the field of computer vision, paper [6] is studied. This paper [6] covers the various deep learning models in the field of computer vision from about 210 research papers. It gives an overview of the deep learning models by dividing them into four categories - Convolutional Neural Networks, Restricted Boltzmann Machines, Autoencoder, and Sparse Coding. Additionally, their successes on a variety of computer vision tasks and challenges faced have also been discussed.

In 2016, Redmon et al. [5] delivers a new approach, You Look Only Once (YOLO) for object detection by expressing it as a regression problem rather than a classification problem. It utilizes a single neural network which gives the bounding boxes coordinates and the confidence scores. Detection of context in images is an emerging application. Various methods ranging from scene graph to rich feature representation have been employed for the same. In 2018, Yang et al. have developed a model Graph RCNN [4] which understands the context of the image by translating the image as a graph. A pipeline of object detection, pair pruning, GCN for context and SGGen+ has been employed. Similar sort of work is done by Fisher et al. [7], they represent scenes that encode models and their semantic relationships. Then, they define a kernel between these relationship graphs to compare the common substructures of two graphs and capture the similarity between the scenes.

For effective semantic information extraction, Skipgram [8] model has been studies works for learning high quality distributed vector representation. In addition to this, several extensions [9] of Skipgram have been experimented with to improve the quality of vectors and training speed. Two models have been proposed in the work [8] which is an extension to Word2vec to improve the speed and time. Their architecture computes continuous vector representations of words from very large data sets. Large improvements have been observed in the accuracy at a much lower computational cost. The vectors are trained on a large dataset of Google for 1.6 billion words. In 2018, a new method "deep structural ranking" was introduced which described the interactions between objects to predict the relationship between a subject and an object. Liang et al. [10] makes use of rich representations of an image – visual, spatial, and semantic representation. All of these representations are fused together and given to a model of structural ranking loss which predicts the positive and negative relationship between subject and object.

The work [11] aims to capture the interaction between different objects using a context-dependent diffusion network (CCDN). For the input to the model, two types of graphs are used - visual scene graph and semantic graph. The visual scene graph takes into account the visual information of object pair connections and the semantic graphs contain the rich information about the relationship between two objects. Once the features from visual and semantic graphs are taken, they are given as an input to a method called Ranking loss, which is a linear function. Yatskar et al. work [12] is an extension to the predicate and relationship detection. It introduces a method where it focuses on the detection of a participant, the role of the participants and the activity of the participants. The model has coined the term "FrameNet" which works on a dataset containing 125,000 images, 11,000 objects, and 500 activities.

3   Importance and Significance of Work

This work is having its own importance and significance in varying application due to the following:

- An extension to the object detection by finding the underlying relationship between object and subject. Object detection merely works on the presence of the objects giving us partial information about the images. Context can give us the true meaning of the image.
- Classifies the image as similar on the basis of the underlying context. Object detection classifies the images as similar on the basis of the presence of specific objects. However, the images can be quite different than each other based on context. Incorporating the context will give a deeper insight into an image.
- If the context is employed on prepositions as well as verbs (future work), rich semantic information can be used to generate interesting captions and stories related to images.
- No pixel-to-pixel level similarity/clustering calculation. One of the applications of incorporating context is to find similar images. Conventional techniques involve pixel by pixel calculations, thus increasing the overhead. Scene graphs save time by considering the visual and semantic features.
- Useful in query processing, image retrieval, story generation, image captioning. Once the context is detected, it can be used in various applications like query processing in search engines, image retrieval using captioning [13,14], as well as story generation.

4   Datasets

Most famous datasets used for Scene understanding applications are MS-COCO [15], PASCAL VOC, and Visual Genome, and Visual Relationship Detection- VRD.

VRD dataset contains 5000 images, 100 object categories, and 70 predicates. It is most widely used for the relationship detection for an object pair in testing since it contains a decent amount of images. COCO [15] is large scale object detection, segmentation, and captioning dataset. This dataset is used in several applications- Object segmentation, Recognition in context, Super pixel stuff segmentation. It has over 330K images (200K labeled), and 80 object categories. Also, it has 5 captions per image which can be used for image captioning methods.

To perform the experiments, VRD dataset has been taken. Visual Relationship Detection (VRD) with Language Priors is a dataset developed by Stanford aiming to find the visual relationship and context in an image. The dataset contains 5000 images with 37,993 thousand relationships, 100 object categories and 70 predicate categories connecting those objects together. Originally, in the dataset, we are given a dictionary file of training and testing data which we convert into training set with 3030 images, test set of 955 images, and validation set of 750 images.

Statistics of the number of objects and visual relationships in every image is shown in Figs. 1 and 2 respectively. In Fig. 2, the file with an unusual number of 134 relationships in image '3683085307.jpg'.

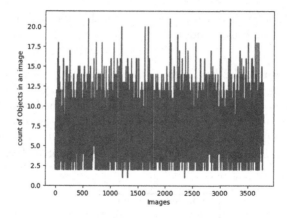

Fig. 1. Statistics of the number of objects in images

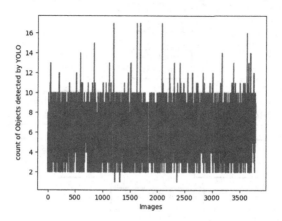

Fig. 2. Statistics of the number of objects detected by YOLO

5   Solution Approach

The purpose of this project is to extend the object detection techniques to find context and hence, understand the underlying meaning of an image. Our idea uses 2D images of various scenes which have objects interacting with each other. The interaction can be in the form of prepositions like (above, on, beside, etc) or activity form like (cooking, eating, talking, etc). The model considers the scenery. The solution approach is basically divided into four modules. These modules are clearly depicted in Fig. 3. The first phase is the object detection for which YOLO object detector has been used. YOLO will provide an image with a bounding box of detected objects. This will be used to identify the semantic and visual features of the image. VGG-16 is used to generate visual features and Word2Vec is used for semantic feature identification. These features are concatenated and given as input to a SVM which provides a probability distribution over all the predicate classes (see Figs. 3 and 4).

Object Detection. The first step of the solution approach is to detect the objects present in an input image. Recent research works have used various deep learning approaches and models. These are developed in order to achieve high efficiency and high accuracy for object detection. Approaches used in literature include YOLO [5], R-CNN [15], Fast-RCNN [16], Faster-RCNN [17], Mask-RCNN [19], and Single-Shot MultiBox Detector (SSD) [20].

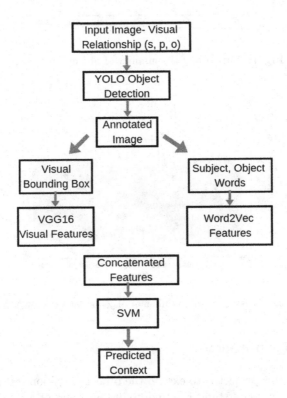

Fig. 3. Flow diagram of solution approach

Here, YOLO (You Only Look Once) has been used for object detection. It has an advantage that instead of using local filters, it looks at an image globally and delivers results. YOLO is very fast since it treats frame detection as a regression problem. The model consists of CNN similar to GoogleNet and instead of using residual block 1*1 convolutional layers are used. With 24 convolutional layers and pre-trained on ImageNet dataset, the model is trained for 135 epochs on PASCAL VOC dataset with a dropout of 0.5. Due to 1*1, the size of the prediction is the same as the feature space. The dimension of the feature space is in the format: 4 box coordinates, 1 objectness score, k class score for every box on the image. Object score represents the probability that an object is contained inside a bounding box. Class confidences represent the probabilities of the detected object belonging to a particular class. YOLO uses softmax for class scores.

5.1 Semantic and Visual Embedding

Once the objects are detected, pairs for every object are created giving us nC2 number of visual relationships. For the visual features, the bounding box of subject and object are taken. The predicate is the intersection over union (IoU) of subject and object bounding boxes. All the three bounding boxes are concatenated and given to a VGG16 network with predicate word as the ground truth label. VGG16 is used for the classification task for the images. It's last layer provides good visual representations for the objects in an image. Hence, it is extracted to get visual relationship features for the concatenated bounding boxes. Further, for the semantic embedding, Word2Vec is used over the subject and object word. It is a powerful two layer neural network that can convert text into a vector representation. Word2Vec converts the subject and object word into a 300 sized feature representation which is concatenated and given to a neural network. The output layer before the application of activation function is extracted to get the semantic embedding for the visual relationship. The generated semantic embedding are stored in a dictionary format. The index is the object id and value is the embedding. We store the object, predicate and their embedding (found by word2vec) in the following formats shown in Table 1.

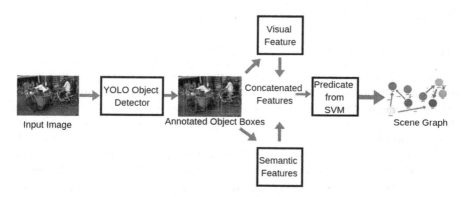

Fig. 4. Phase division for object detection and context interpretation

Considering that the visual and semantic embedding take the rich information about the image which is not limited to only object detection, but also to the semantic information present in an image, other information which can also be taken is the spatial feature representation which considers the location of an object in an image with respect to the other objects.

Table 1. File details of object, predicate, and embedding

File name	Function	Dictionary format
Objects_dict.pkl	Hashing of object	Index-object name
Predicate_dict.pkl	Hashing of predicates	Index-predicate name
Objects_embedding.pkl	Objects Word2Vec embedding	Object name-word2vec embedding
Predicate_embedding.pkl	Predicates Word2Vec embedding	Predicate name-word2vec embedding

5.2 Predicate Detection

The type of predicates in the dataset include the spatial predicates like above, beside, on, etc. Predicates can be of many types depicting the spatial context and activity context like cooking, eating, looking, etc. After the semantic and visual embeddings are extracted, the embedding is concatenated for a visual relationship in an image. Some other methods can also be used when using both the semantic and visual features which include, multiplying both the feature, however, this requires both the representation to be of the same size. The dataset includes around 70 predicates. Since the classes are quite distinct from each other, a decision boundary between the classes would serve as a good strategy to classify between the predicate classes and SVM is a powerful discriminative model to achieve this task. It is used as a classifier to give a class distribution probability over all the 70 classes. The class with the maximum probability is the predicted class. For the scene graph, top 3 predicates are taken. The predicate detected depicts the context shared between the subject and object and thus delivers the meaning of the image.

5.3 Scene Graph Generation

An image contains k number of visual relationships of the format (subject, predicate, object). The predicate was detected in the previous step. Now, the scene graph is generated with nodes as objects/subjects and edges as the predicate. Here, we use a directed scene graph so that there is a differentiation between subject and object.

For example: for a statement, a person eating food, the relationship format of (subject, predicate, object) would be (Person, eating, Food). Here, the person is the subject, food is the object, and eating is a predicate. If an undirected edge is used, this statement loses the distinctive property of the person being subject and food being object. The roles can be reversed due to undirected edges leading to erroneous relationships. Therefore, the use of directed edges is preferred. After the generation of the scene graph, it can be traversed accordingly to generate captions or summary of an image. The scene graph can also be termed as a context graph.

6   Findings

An outcome for a sample VRD dataset image is shown in Fig. 5. The image after YOLO is shown in Fig. 6 which shows the annotated image after YOLO object detection. The objects detected in the shown image in the boundary box are Person, wheel, cart, plant, bike, shirt, basket, and pants. Mean Average Precision (MAP) is taken as a performance measure to test the outcome. It considers the average precision for recall of the detected objects and is a popular metric for the object detectors. For the training set, YOLO had an object detection accuracy of 55 MAP.

Fig. 5. Input image from the VRD-dataset

Fig. 6. Annotated image using YOLO object detector

Finally, a scene graph is generated based on these YOLO detected visual features and semantic features. The scene graph of Fig. 5 is depicted in Fig. 7. Relationships identified for which scene graph is formed are shown in Table 2 showing the scene description using the subject predicate-object relationship.

The loss in the Neural network for a semantic feature and CNN for the visual feature is shown in Figs. 8 and 9 respectively. It is clearly observable in Fig. 8 that the training loss dropped with every epoch. The validation, however, increased after the 50th epoch more than the training. The point where the validation loss increases the training loss depicts the point where the model starts to overfit. Hence, the weights of the network at the 50th epoch were taken for further processing. One of the possible reason of overfitting can be attributed to the dataset being small. The model tries to fit to this small dataset and does not learn the ability to generalize well.

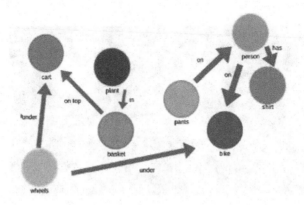

Fig. 7. Scene graph result of the input image

Table 2. Details about scene, objects, and their relationships

Scene	Relationships (s, p, o)
Wheel under cart	(Wheel, under, cart)
Basket on top cart	(Basket, on top, cart)
Plant in basket	(Plant, in, basket)
Wheel under bike	(Wheel, under, bike)
Pants on person	(Pants, on, person)
Person on bike	(Person, on, bike)
Person has shirt	(Person, has, shirt)

The CNN and Neural Network were trained till they reached an accuracy of 95% and 99% respectively. The accuracy for predicate detection from SVM came out to be 60.57%. The SVM was run for a total for 100 epochs. In our previous approach of scene graph generation using Word2vec solely, the accuracy reached till 40% only. However, once we incorporated the visual features also, the accuracy increased to 60%.

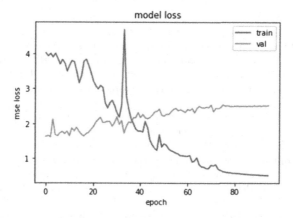

Fig. 8. Training and validation loss for visual features

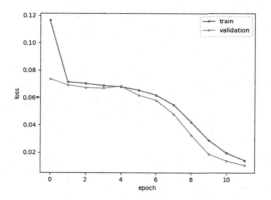

Fig. 9. Training and validation Loss for semantic features

7 Conclusion and Future Scope

Our work leverages the techniques of object detection by finding out the context of the image in addition to the detected object. We are detecting the context from the visual and semantic features of an image. This is achieved by the application of deep learning models YOLO for object detection and Word2Vec for semantic feature representation generation. A neural network is used for the semantic feature of image and VGG16 for the visual feature generation. Context can be used to find out the subtle meaning of the image. Future work includes extending the context to verbs like cooking, eating, looking, etc since our work is covering only the preposition predicates such as on, above, etc. Moreover, in addition to the semantic and visual features, spatial features can be incorporated which will be helpful in determining the location of the objects. Lastly, better object detection models like Faster-RCNN can be employed for more accurate object detection in the first step because if an object is not detected in the first step, it can't be used for processing of visual relationships in the further stages.

References

1. Xu, D., et al.: Scene graph generation by iterative message passing. In: Proceedings of the IEEE Conference on Computer Vision and Pattern Recognition (2017)
2. Li, Y., et al.: Factorizable net: an efficient subgraph-based framework for scene graph generation. In: Proceedings of the European Conference on Computer Vision (ECCV) (2018)
3. Li, Y., et al.: Scene graph generation from objects, phrases and region captions. In: Proceedings of the IEEE International Conference on Computer Vision (2017)
4. Yang, J., et al.: Graph R-CNN for scene graph generation. In: Proceedings of the European Conference on Computer Vision (ECCV) (2018)
5. Redmon, J., et al.: You only look once: unified, real-time object detection. In: Proceedings of the IEEE Conference on Computer Vision and Pattern Recognition (2016)
6. Guo, Y., et al.: Deep learning for visual understanding: a review. Neurocomputing **187**, 27–48 (2016)
7. Fisher, M., Savva, M., Hanrahan, P.: Characterizing structural relationships in scenes using graph kernels. ACM Trans. Graph. (TOG) **30**(4) (2011)
8. Mikolov, T., et al.: Efficient estimation of word representations in vector space. arXiv preprint arXiv:1301.3781 (2013)
9. Mikolov, T., et al.: Distributed representations of words and phrases and their compositionality. In: Advances in Neural Information Processing Systems (2013)
10. Liang, K., et al.: Visual relationship detection with deep structural ranking. In: Thirty-Second AAAI Conference on Artificial Intelligence (2018)
11. Cui, Z., et al.: Context-dependent diffusion network for visual relationship detection. In: 2018 ACM Multimedia Conference on Multimedia Conference. ACM (2018)
12. Yatskar, M., Zettlemoyer, L., Farhadi, A.: Situation recognition: visual semantic role labeling for image understanding. In: Proceedings of the IEEE Conference on Computer Vision and Pattern Recognition (2016)
13. Chen, X., Zitnick, C.L.: Mind's eye: a recurrent visual representation for image caption generation. In: Proceedings of the IEEE Conference on Computer Vision and Pattern Recognition (2015)
14. Gao, L., Wang, B., Wang, W.: Image captioning with scene-graph based semantic concepts. In: Proceedings of the 2018 10th International Conference on Machine Learning and Computing. ACM (2018)
15. Lin, T.-Y., et al.: Microsoft COCO: common objects in context. In: Fleet, D., Pajdla, T., Schiele, B., Tuytelaars, T. (eds.) ECCV 2014. LNCS, vol. 8693, pp. 740–755. Springer, Cham (2014). https://doi.org/10.1007/978-3-319-10602-1_48
16. Girshick, R., et al.: Rich feature hierarchies for accurate object detection and semantic segmentation. In: Proceedings of the IEEE Conference on Computer Vision and Pattern Recognition (2014)
17. Girshick, R.: Fast R-CNN. In: Proceedings of the IEEE International Conference on Computer Vision (2015)
18. Ren, S., et al.: Faster R-CNN: towards real-time object detection with region proposal networks. In: Advances in Neural Information Processing Systems (2015)
19. He, K., et al.: Mask R-CNN. In: Proceedings of the IEEE International Conference on Computer Vision (2017)
20. Liu, W., et al.: SSD: single shot multibox detector. In: Leibe, B., Matas, J., Sebe, N., Welling, M. (eds.) ECCV 2016. LNCS, vol. 9905, pp. 21–37. Springer, Cham (2016). https://doi.org/10.1007/978-3-319-46448-0_2

Deep Learning in the Domain of Near-Duplicate Document Detection

Rajendra Kumar Roul$^{(\boxtimes)}$ (iD)

Department of Computer Science, Thapar Institute of Engineering and Technology,
Patiala 147004, Punjab, India
raj.roul@thapar.edu

Abstract. Increasing of web users due to the popularity of the internet increases the digital documents on the web, and among them many are duplicates and near-duplicates. Identifying duplicate and near-duplicate documents in a huge collection is a significant problem with widespread application and hence, detection and elimination of those documents are the need of the day. This paper proposes a technique to detect the near-duplicate documents on the web which has four main aspects: the first aspect is related to the selection of the important terms from a corpus of documents by developing a new *correlation-based feature selection (CBFS)* mechanism which enhance the performance of the classifier. The second aspect is to compute the similarity scores between each pair of documents of the corpus. The third aspect concerns with combining these similarity scores with the class label of each pair of documents to generate the feature vector for training the Multi-layer ELM (deep learning architecture) and other established classifiers and the fourth and final aspect introduces a heuristics method to rank the near-duplicate documents based on their similarity scores. The empirical results on DUC datasets witness the effectiveness of the proposed approach using Multi-layer ELM as highly appreciable compared to other state-of-the-art classifiers including the deep learning classifiers.

Keywords: Deep learning · LDA · Multilayer ELM · Near-duplicate · Ranking

1 Introduction

Internet is growing at a lightening speed consists of millions of digital documents. These documents can easily replicate and become the roadblock for the search engine to retrieve the relevant results from the web. This replication divides the digital documents into two categories - *Duplicate* and *Near-duplicate*. If two documents have similar content then they are considered as duplicates. In other words, two documents are regarded as duplicate, if they are completely similar. Similarly, documents having minor modifications but are similar to a maximum range are known to be near-duplicates such as timestamps, advertisements, and counters. Near-duplicate documents are not bit-wise similar although they display striking similarities. Duplicate is an inherent problem of every search engines. It becomes interesting problem in late 1990, with the growth of the internet [1]. It has been observed from a study which includes 238,000

© Springer Nature Switzerland AG 2019
S. Madria et al. (Eds.): BDA 2019, LNCS 11932, pp. 439–459, 2019.
https://doi.org/10.1007/978-3-030-37188-3_25

hosts that 10% of these hosts are mirrored [2]. To carry out the analysis of duplicate and near-duplicate documents detection, generally two conditions should meet (i) link-based ranking model should be employed by the collection and (ii) one should disable the collection-security. There are many standard techniques available which can easily identify the duplicate documents but identifying the near-duplicate documents is a tough task [3,4]. Hence detecting and eliminating near-duplicate documents are still remain unanswered [5–7]. In the recent years, near-duplicate document detection has received more attention due to the overwhelm requirements [8–12].

Focusing our work in this direction, the proposed approach detects the near-duplicate documents in four stages as follows:

i. Pre-processing of documents and term clustering: All the documents of a corpus are pre-processed initially and converted into vectors. Similar terms are clustered together by a heuristic method which uses topic modeling to generate k clusters.
ii. Important terms selection and new corpus generation: This stage develops a novel feature selection technique named as *CBFS* which selects top m% terms by using term-term correlation measure and discard the noise and redundant terms from each of the k clusters. Next, a new corpus is generated with those documents which contain these top m% terms, hence reducing the size of the corpus. Altogether eleven similarity measures are used to calculate the similarity scores between every pair of documents in the new corpus.
iii. Training feature vector generation: By combining these similarity scores and the class label ('1' used for 'near-duplicate' and '0' for 'not near-duplicate'), the training feature vector gets generated. To test whether a new input document 'd' is near-duplicate or not to any of the documents in the corpus, the predicted label is compared with the known test label of 'd'.
iv. Ranking the near-duplicate documents: A heuristic technique is proposed to rank the documents by taking a weighted average of the features from the classifier and then computing an aggregate score of each document.

Empirical results of Multilayer-ELM is compared with both shallow and deep learning classifiers to test its efficiency. The novelty and some of the advantages of the proposed approach are as follows:

i. Introducing deep learning to detect near-duplicate documents can shed light in the domain of machine learning. The common restrictions found in traditional classifiers are not found in deep learning [13–15]. To our understanding (as evident from existing literature), no research work has been done before where deep learning has been used for near-duplicate detection. This may be the first work where Multilayer ELM has been tested extensively to detect the near duplicate documents.
ii. Most of the research works have used content-based and less have used semantic-based similarity to detect the near-duplicate documents. However, detection of near-duplicate documents using content-based combined with semantic-based similarity has much importance due to the semantic nature of the web which the present work has considered.
iii. From the literature it is evident that semantic-based similarity measures such as *Normalized Google Distance*, *Word2Vec*, *WordNet-based* techniques are never

used before to detect the near duplicates pages, which are included in the proposed similarity measures to make it more robust.

iv. Researchers have focused only on detection of near-duplicates, but ignore to rank the documents in order to find the percentage of similarities between the documents. The proposed approach not only detects the near-duplicates by combing the content and semantic-based similarity between the documents but also introduces a heuristic approach to rank the near-duplicate documents for finding the percentage of similarities with respect to all the documents of a corpus.

Experimental works on four benchmark categories of DUC datasets show that the results of the proposed approach using Multi-layer ELM are more promising compared to the existing shallow and deep learning classifiers.

2 Preliminaries

2.1 Multilayer ELM

Multilayer ELM (ML-ELM) is an artificial neural network having multiple hidden layers [16,17] and it is shown in Fig. 1. Equations 1, 2 and 3 are used for computing β (output weight vector) in ELM Autoencoder where H represents the hidden layer, X is the input layer, n and L are number of nodes in the input and hidden layer respectively.

i. $n = L$

$$\beta = H^{-1}X \tag{1}$$

ii. $n < L$

$$\beta = H^T \left(\frac{I}{C} + HH^T \right)^{-1} X \tag{2}$$

iii. $n > L$

$$\beta = \left(\frac{I}{C} + H^T H \right)^{-1} H^T X \tag{3}$$

Here, $\frac{I}{C}$ generalize the performance of ELM [18] and is known as regularization parameter.

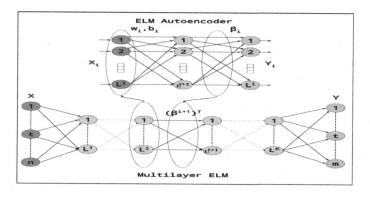

Fig. 1. Architecture of multilayer ELM

Multilayer ELM uses Eq. 4 to map the features to higher dimensional space and thereby makes them linearly separable [18–20].

$$h(\mathbf{x}) = \begin{bmatrix} h_1(\mathbf{x}) \\ h_2(\mathbf{x}) \\ \cdot \\ \cdot \\ \cdot \\ h_L(\mathbf{x}) \end{bmatrix}^T = \begin{bmatrix} g(w_1, b_1, \mathbf{x}) \\ g(w_2, b_2, \mathbf{x}) \\ \cdot \\ \cdot \\ \cdot \\ g(w_L, b_L, \mathbf{x}) \end{bmatrix}^T \tag{4}$$

where, $h_i(\mathbf{x}) = g(w_i.\mathbf{x} + b_i)$. $h(\mathbf{x}) = [h_1(\mathbf{x})...h_i(\mathbf{x})...h_L(\mathbf{x})]^T$ can directly use for feature mapping [19,21,22]. w_i is the weight vector between the input nodes and the hidden nodes and b_i is the bias of the ith hidden node, and $\mathbf{x}$ is the input feature vector with g as the activation function. No kernel technique is required here and that reduces the cost at higher dimensional space because kernel function uses dot product very heavily in the higher dimensional space to find the similarity between the features. ML-ELM takes the benefit of the ELM feature mapping [18,23] extensively as shown in Eq. 5.

$$\lim_{L \to +\infty} ||y(\mathbf{x}) - y_L(\mathbf{x})|| = \lim_{L \to +\infty} ||y(\mathbf{x}) - \sum_{i=1}^{L} \beta_i h_i(\mathbf{x})|| = 0 \tag{5}$$

Using Eq. 6, Multilayer ELM transfers the data between the hidden layers [24].

$$H_i = g((\beta_i)^T H_{i-1}) \tag{6}$$

2.2 Similarity Measures

The following are the similarity measure techniques used in the proposed approach.

1. *Content-based/lexical-based similarity measure*: Different content-based similarity between two documents (d_p, d_q) are as follows:
   (i) *Dice Coefficient* is a statistical measure used for computing the similarity between two documents and is calculated as follows:

   $$dice(d_p, d_q) = \frac{2(d_p.d_q)}{||d_p||^2 + ||d_q||^2}$$

   (ii) *Extended Jaccard Coefficient* and Jaccard's distance are the measurement of asymmetric information on binary attributes and is computed between two documents as follows:

   $$e\text{-}jaccard(d_p, d_q) = \frac{d_p.d_q}{||d_p||^2 + ||d_q||^2 - (d_p.d_q)}$$

   (iii) *Cosine Similarity* is a technique that generates a metric which says how related are two documents by measuring the cosine of the angle between them. It is calculated as follows:

   $$cosine(d_p, d_q) = \frac{d_p.d_q}{||d_p|| * ||d_q||}$$

(iv) *Euclidean Distance* between two documents is represented as follows:

$$euclidean(d_p, d_q) = \sqrt{\sum_{p=q=1}^{n}(a_p - b_q)^2}$$

where $d_p = \{a_1, a_2, \cdots, a_n\}$ and $d_q = \{b_1, b_2, \cdots, b_n\}$ are the co-ordinates of d_p and d_q respectively.

(v) *Simple matching coefficient* founds how many words are matching between two documents is computed as follows:

$$smc(d_p, d_q) = \frac{d_p.d_q}{length(d_p) + length(d_q)}$$

(vi) *Overlap coefficient* determines the overlap of words between two documents and is computed as follows:

$$overlap(d_p, d_q) = \frac{d_p.d_q}{min\{length(d_p), length(d_q)\}}$$

2. *Structure-based similarity measure*: Structure-based similarity is measured by computing how many such terms are there that are common to both d_p and d_q and also they should have the same order in the two documents. The following formula is used to compute such similarity between two documents.

$$struct(d_p, d_q) = \frac{a}{b}$$

where 'a' represents the number of terms that common to both d_p and d_q and maintain the same order in both the documents and 'b' represents total number of terms that common to both d_p and d_q.

3. *Semantic-based similarity measure*: Semantic-based similarity between a pair of documents is computed using different measures and are discussed below:

   i. *Normalized Google Distance* (NGD) [25] compares the number of web hits returned by the Google search engine to find the distance between two terms. We have used nouns extracted from documents using NLTK toolkit[1] to measure the value of NGD between two documents. The value of NGD between two nouns n_p and n_q is calculated as follows:

$$NGD(n_p, n_q) = \frac{max(logT(n_p), logT(n_q)) - logT(n_p, n_q)}{logN - min(logT(n_p), logT(n_q))}$$

   where $T(n_p, n_q)$ and $T(n)$ represent the number of web hits returned by Google for two nouns and one noun respectively and N indicates the number of web documents retrieved by Google. The similarity between each pair of noun is computed as follows:

$$NGD\text{-}similarity(n_p, n_q) = 1 - NGD(n_p, n_q)$$

---

[1] https://www.nltk.org/.

The similarity between two documents d_p and d_q is computed using Eq. 7.

$$NGD\text{-}similarity(d_p, d_q) = \frac{count(NGD\text{-}simialrity(n_p, n_q)) > \alpha}{Z_p \times Z_q} \quad (7)$$

Here, Z_p and Z_q are total number of nouns of d_p and d_q respectively and α is a certain threshold which is set as 0.61 and is decided by experiment.

ii. *word2vec-based similarity* uses architecture of two-layer neural network for capturing the linguistic contexts of the words. It groups similar words together in vector space and makes word embedding. The word2vec-based similarity between d_p and d_q is shown in the Eq. 8.

$$word2vec(d_p, d_q) = \frac{\sum_{w_p \in d_p} \max_{w_q \in d_q} cosine\_similarity(w_p, w_q)}{length\ of\ d_p} \quad (8)$$

iii. *WordNet-based similarity*[2] between the two documents d_p and d_q is shown in the Eq. 9 where *synset* represents a set of cognitive synonyms that are cluster of nouns, verbs, adverbs, and adjectives present in a document.

$$wordnet(d_p, d_q) = \frac{\sum_{w_p \in d_p} \max_{w_q \in d_q} similarity(synset(w_p), synset(w_q))}{length\ of\ d_p}$$
$$(9)$$

4. *Topic-based similarity measure*: It measures how far the topics generated from a corpus are similar between two documents. For this Latent Dirichlet Allocation (LDA) needs to be run for the entire corpus and it generates the topics by combining different words of the corpus. In this similarity measure, a probability distribution of each topic is assigned to every document in the corpus and is computed using the following formula.

$$topic(d_p, d_q) = \frac{topic_p.topic_q}{||topic_p|| * ||topic_q||}$$

3   Proposed Approach

This section discussed the approach to detect the near-duplicate documents in a corpus using the following steps and Algorithm 3 shows the details. Word and term/feature are interchangeably used in the approach which have the same meaning. Only for content and structure similarity between two documents, the TF-IDF vector space is used but for semantic and topic-based similarity, the TF-IDF vector space for the documents are not required where we need only the term-frequency matrix. For example, in semantic similarity, either the words of a document are directly used for computing the similarity or they can be represent in other feature space such as word2vec.

1. *Pre-processing the corpus of documents*:
   Consider a corpus P having a set of classes $C = \{c_1, c_2, \cdots, c_n\}$ and documents

---

[2] http://wordnet.princeton.edu/.

$D = \{d_1, d_2, \cdots, d_b\}$. The documents are pre-processed which includes removal of stop words, stemming[3], lemmatizating[4] and converting the documents to a Bag-Of-Words format where each word is stored along with its word count in a dictionary. Documents use different forms of a word, such as *operate, operating, operation* and *operative*. Stemming and lemmatization are done separately because we require the dictionary of stemmed words for Word2Vec, LDA, and all other content-based techniques while the lemmatized dictionary of words is needed to perform NGD and WordNet techniques. Stemming collapses related words by a heuristic process that chops off the ends of the words resulting in a term which may not be present in the English dictionary. Lemmatization on the other hand, collapses the different inflectional forms of a word, resulting in a base term of a dictionary word. The dimension of P is $b \times l$, where b and l are number of documents and terms (W) respectively, and represented in vector form using their TF-IDF[5] values as shown in Table 1. A new corpus of the documents is generated from the given dataset by pooling the pre-processed lists of documents together. Thus, two corpus's are created - one for the stemmed words and another for the lemmatized words.

Table 1. Document-term matrix

	W_1	W_2	W_3	$\cdots$	W_l
d_1	w_{11}	w_{12}	w_{13}	$\cdots$	w_{1l}
d_2	w_{21}	w_{22}	w_{23}	$\cdots$	w_{2l}
d_3	w_{31}	w_{32}	w_{33}	$\cdots$	w_{3l}
$\vdots$	$\vdots$	$\vdots$	$\vdots$	$\ddots$	$\vdots$
d_b	w_{b1}	w_{b2}	w_{b3}	$\cdots$	w_{bl}

2. *Document-term cluster formation*:

A heuristic method is proposed and it is run on the corpus P which generates k doc-term clusters $dt = \{dt_1, dt_2, dt_3, \cdots, dt_k\}$ where, each dt_i is of dimension $b \times n$ (number of terms get reduced by clustering). The entire process of clustering using the heuristic method is discussed below.

  i. Topic modeling using Latent Dirichlet Allocation (LDA) is run on the corpus P to generate m topics.

  ii. As LDA is a stochastic process due to its probabilistic nature, hence the goodness-of-fit of LDA models is compared with varying numbers of topics to decide the suitable number of m topics that can be generate from the corpus P.

  iii. The goodness-of-fit of an LDA model is evaluated by computing the perplexity (standard topic modeling toolbox)[6] of a set of documents. The perplexity

---

[3] https://tartarus.org/martin/PorterStemmer/.

[4] https://algorithmia.com/algorithms/StanfordNLP/Lemmatizer.

[5] http://www.tfidf.com.

[6] https://nlp.stanford.edu/software/tmt/tmt-0.4/.

represents how well the model describes a set of documents. A lower perplexity indicates a better fit.

iv. It is well known that a collection of similar words constitute a topic, hence we consider each topic as a cluster.

Now, the aim is to select the important terms from each of the k clusters for maintaining the uniformity without excluding any collection.

3. *Correlation-based features selection (CBFS) from a doc-term cluster:*

(i) Frequency-based correlation (FC) calculation:

Equation 10 is used to calculate the frequency-based correlation measure between every pair of terms i and j of the cluster dt_p where, f_{im} and f_{jm} represent the frequency of ith and j^{th} terms in the mth document of the cluster dt_p.

$$FC_{ij} = \sum_{m \in p} f_{im} * f_{jm} \tag{10}$$

(ii) Association matrix construction:

An association matrix is constructed where each row consists of the association values (or correlation values) between term W_i and W_j that generates a semantic component of term vector $\overrightarrow{W_i}$. $FC_{ij} = FC_{ji}$ (i.e., $FC_{ij} = FC_{ij}^T$).

(iii) Normalizing correlation measure (NCM)):

The association score of FC_{ij} is normalized using the Eq. 11 which float the correlation values between 0 and 1. All the diagonal values of NCM are 1, as $i = j$.

$$NCM_{ij} = \frac{FC_{ij}}{FC_{ii} + FC_{jj} - FC_{ij}} \tag{11}$$

(iv) Centroid vector generation:

For each term $\overrightarrow{W_i}$, the mean of all NCM are calculated, which generates a n-dimensional semantic centroid vector $\overrightarrow{sc_p}$ using Eq. 12. Each component of $\overrightarrow{sc_p}$ is shown below.

$$sc_{p_i} = \frac{\sum_{j=1}^{n} NCM_{ij}}{n} \tag{12}$$

(v) Selecting the important terms from each doc-term cluster (dt_p):

a. *Silhouette coefficient calculation:*

The silhouette coefficient ($silhout$) of the term $\overrightarrow{W_i} \in dt_p$ is computed using the Eq. 13.

$$silhout(\overrightarrow{W_i}) = \frac{sep(\overrightarrow{W_i}) - coh(\overrightarrow{W_i})}{\max\left(coh(\overrightarrow{W_i}), sep(\overrightarrow{W_i})\right)} \tag{13}$$

where, cohesion (coh) measures how cohesive is the term, $\overrightarrow{W_i} \in dt_p$ to the centroid, $\overrightarrow{sc_p} \in dt_p$, the Euclidean distance is computed between $\overrightarrow{W_i}$ and $\overrightarrow{sc_p}$ using Eq. 14.

$$coh(\overrightarrow{W_i}) = (||\overrightarrow{sc_p} - \overrightarrow{W_i}||) \tag{14}$$

and separation (*sep*) measures how well separated a term, $\overrightarrow{W_i} \in dt_p$ from the semantic centroid of other clusters, $\overrightarrow{sc_m}$, $\forall m \in [1, k]$ and $m \neq p$ which is shown in Eq. 15.

$$sep(\overrightarrow{W_i}) = \min\left(\|\overrightarrow{sc_m} - \overrightarrow{W_i}\|\right) \qquad (15)$$

where, $\overrightarrow{sc_m}$ is the semantic centroid of the mth cluster.

   b. Finally, all terms are ranked based on their silhouette coefficient scores and among them top 'm%' terms are selected from the cluster dt_p.

(vi) By repeating the step 3 (i–v) for all k doc-term clusters, top m% terms is obtained from each doc-term cluster.

The details of the *CBFS* technique is generalized in Algorithm 1.

Algorithm 1. CBFS Algorithm

1: **Input:** document term matrix with term frequency of a cluster dt_p
2: **Output:** $Top[]$ ← important features of dt_p
3: $FC[][] \leftarrow \phi$ // correlation measure matrix
4: $NCM[][] \leftarrow \phi$ // normalized correlation measure matrix
5: $Silhoutte[] \leftarrow \phi$ // stores the silhoutte coefficient score of all the terms
6: $Top[] \leftarrow \phi$
7: $\overrightarrow{sc_p} \leftarrow \phi$ // semantic centroid of dt_p
8: **for all** terms $(i, j) \in dt_p$ **do**
9:     $sum \leftarrow \phi$
10:     **for all** document $k \in dt_p$ **do**
11:        $sum \leftarrow sum + (f_{ik} * f_{jk})$
12:     **end for**
13:     $FC_{ij} \leftarrow sum$
14: **end for**
15: **for all** terms $(i, j) \in FC$ **do**
16:     $NCM_{ij} \leftarrow \frac{FC_{ij}}{(FC_{ii} + FC_{jj} - FC_{ij})}$
17: **end for**
18: **for** $i \in [1, n]$ **do**
19:     //where n is the total number of terms
20:     $sum \leftarrow \phi$
21:     **for** $j \in [1, n]$ **do**
22:        $sum \leftarrow sum + NCM_{ij}$
23:     **end for**
24:     $sc_i \leftarrow sum/n$ // semantic centroid
25: **end for**
26: **for** $i \in [1, n]$ **do**
27:     $Silhoutte[i] \leftarrow$ silhouette$(\overrightarrow{sc_p}, \overrightarrow{W_i})$
28: **end for**
29: $Top[] \leftarrow$ select top m% terms from $Silhoutte[]$ after ranking them
30: return $Top[]$

---

4. *New corpus generation:*
   After generating top $m\%$ terms from each doc-term cluster, those documents are discarded which does not contain any of these top terms (i.e. irrelevant documents). This way the dimension of each cluster get reduced (both documents and terms). Now, the relevant documents of all the clusters are merged together to generate a new corpus P'.

5. *Training Feature vector generation:*
   For every pair of documents in the new corpus P', the eleven similarity measures (as discussed in Sect. 2.2) are calculated. These eleven similarity measures are appended as a tuple for each pair of documents along with the class label (discussed in step 6) to form twelve dimensional training feature vector. Algorithm 2 illustrates the computation of different similarity measures between every pair of documents of the new corpus P'.

Algorithm 2. Generating training feature vector

1: **Input:** New generated corpus P'
2: **Output:** Feature vector s with eleven similarity measures
3: **for all** document $d_p \in P'$ **do**
4:     **for all** document $d_q \in P'$ **do**
5:         $s \leftarrow s.append(cosine(d_p, d_q))$
6:         $s \leftarrow s.append(euclidean(d_p, d_q))$
7:         $s \leftarrow s.append(e\text{-}jaccard(d_p, d_q))$
8:         $s \leftarrow s.append(dice(d_p, d_q))$
9:         $s \leftarrow s.append(smc(d_p, d_q))$
10:        $s \leftarrow s.append(oc(d_p, d_q))$
11:        $s \leftarrow s.append(struct(d_p, d_q))$
12:        $s \leftarrow s.append(ngd(d_p, d_q))$
13:        $s \leftarrow s.append(word2vec(d_p, d_q))$
14:        $s \leftarrow s.append(wordnet(d_p, d_q))$
15:        $s \leftarrow s.append(topic(d_p, d_q))$
16:    **end for**
17: **end for**
18: return s

---

6. *Finding class label for each pair of documents in the new corpus P':*
   For each pair of documents, a target class of near-duplicate/not near-duplicate is assigned as shown in Table 2. Near-duplicate is assigned the class label '1' while not near-duplicate is assigned the class label '0'. The details about how two documents are decided as near-duplicate or not is explained in Sect. 4.

7. *Training the Classifiers:*
   Multilayer ELM along with other traditional shallow and deep learning classifiers get trained with the training feature vector. The test feature vector (excluding the class label) is passed to the classifier for obtaining the predicted label vector by classifying a given pair of unseen documents as near-duplicate or not. The implementation details are discussed in Algorithm 3.

Table 2. Class label between the documents

	d_1	d_2	d_3	$\cdots$	d_n
d_1	1	0	0	$\cdots$	0
d_2	0	1	1	$\cdots$	1
d_3	0	1	1	$\cdots$	0
$\vdots$	$\vdots$	$\vdots$	$\vdots$	$\ddots$	$\vdots$
d_n	0	1	0	$\cdots$	1

Algorithm 3. Duplicate document detection

---

1. Identify the important terms in each doc-term cluster (i.e., top m terms) using *CBFS* mechanism.
2. Generate a feature vector where each feature is the similarity score between every pair of documents in the corpus.
3. Generate class label or test label vector (as described in step 6 of Sect. 3). Append this class label to the feature vector generated in step 2.
4. Train ML-ELM and other traditional and deep learning classifiers using the training feature vector having 12 features which includes the class label.
5. Pass the test feature vector (excluding the class label) to the classifiers for obtaining a predicted label vector by classifying a given pair of unseen documents as near-duplicates or not.
6. Calculate the precision, recall, F-measure, and accuracy to measure the performance of ML-ELM and other conventional classifiers by comparing the predicted label vector with the known test label vector.

---

8. *Ranking the near-duplicate documents based on the similarity scores:*
   The following steps are used to rank the near-duplicate documents:
   i. After the classifiers get tested, the decision of that classifier is considered which gives the highest F-measure. The output can be interpreted as a feature matrix, which is a list of feature vectors for each near-duplicate document pair (as decided by the classifier).
   ii. This feature matrix is used to find the ranking of near-duplicate documents. Now, the aggregate similarity score between the pair of near-duplicate documents is computed by taking an *average (AVG)* of the scores obtained from the 11 similarity measures used for each document pair and it is shown in Eq. 16.

$$Similarity\text{-}score(d_p, d_q)_{AVG} = \frac{\sum_{i=1}^{11} a[i]}{11} \qquad (16)$$

where, $a[i]$ is the score of each similarity measure between documents d_p and d_q. Alternatively, a *weighted average* of the scores from the 11 measures is decided as the aggregate similarity between each pair of documents. These weights are determined from the parameter thresholds set by the classifier for its input.

iii. Then the similarity score of each document is computed in the corpus individually by averaging the aggregate similarity score of document d_p with all the other documents it is paired with and this gives the composite similarity score ($Composite\text{-}score$) of the document d_p, and is shown in Eq. 17 where N is the total number of documents in the corpus P.

$$Composite\text{-}score(d_p)_{AVG} = \frac{\sum_{q \in P} Similarity\text{-}score(d_p, d_q)_{AVG}}{N} \quad (17)$$

iv. The near-duplicate documents are ranked in a descending order based on their composite similarity scores ($Composite\text{-}score$). Algorithm 4 illustrates the details.

Algorithm 4. Ranking the near-duplicate documents

1: **Input:** Feature matrix (FM) of near-duplicate documents and the corpus P'
2: **Output:** Ranking of near-duplicate documents
3: $Similarity\text{-}score_{avg}[]\ \leftarrow \phi$ //stores the average of similarity between two documents
4: $Composite\text{-}score_{avg}[]\ \leftarrow \phi$ //stores the overall average similarity score of a document
5: $final\_list \leftarrow \phi$
6: **for all** (d_i, d_j) pairs $\in$ FM **do**
7:     $Similarity\text{-}score_{avg}[d_i, d_j] \leftarrow$ avg. of the score of each similarity between d_i and d_j
8: **end for**
9: **for all** $d_i \in$ FM **do**
10:    **for all** $d_j \in P'$ **do**
11:        $Composite\text{-}score_{avg}[d_i] \leftarrow$ overall average similarity score of d_i with respect to P'
12:    **end for**
13: **end for**
14: //Ranking of near-duplicates documents
15: sort all $d_i \in P'$ in descending order based on their $Composite\text{-}score_{avg}[d_i]$ and stored it in $final\_list$.
16: return $final\_list$

---

4   Experimental Framework

For experimental purpose, DUC[7] (Document Understanding Conference) datasets are used. Different sets (or classes) are contained in different folders and the documents in the same folder are *near-duplicates* to each other while those in two different folders are *not near-duplicates*. The 'near-duplicate' and 'not near-duplicate' documents are classified into class '1' and class '0' respectively. To achieve this, during pre-processing, the relative path of the documents is extracted and folder's name (which are basically numbers in the dataset) are added as terms in the dictionary. Term-ids corresponding to the folder numbers are noted and an attribute to the vectors in step 1 of Sect. 3 is added to denote the folder from which the document is coming, while simultaneously making the TF-IDF value against these added term-ids as zero (these added term-ids will have a non-zero value for a document if the document is within the corresponding

---

[7] http://www.duc.nist.gov.

folder) so that the same doesn't reflect while running the feature selection step (step 3 of Sect. 3). Here, documents with complete or partial similar contents are considered as 'near-duplicates' and totally dissimilar contents as 'not near-duplicates'. For the purposes of convenience, the folders of DUC datasets are renamed in incremental order (from 1–30 (in DUC-2000, 2001, and 2002) and 1–50 (in DUC-2005)) and the documents in a similar fashion. The number of topics (i.e., clusters) generated from DUC-2000, DUC-2001, DUC-2002, and DUC-2005 are 52, 46, 34, and 22 respectively.

4.1 DUC-2000 and DUC-2002 Datasets

For these dataset, the test folder is organized into subfolders (30 in this case). Each of those subfolders contains 9–11 documents having summaries pertaining to a specific topic. The documents within those subfolders contain 100–400 word summaries, and each of those documents have similar content and hence are duplicates. Each set having on an average of 10 documents constituting a total of 296 documents and 43956 pairs of documents which are used to detect the near-duplicates. Out of 43956 pairs of documents, 1350 pairs of documents are not near-duplicates as they are come from same clusters. The remaining 42606 pairs of documents are used as near-duplicates.

4.2 DUC-2001 Dataset

This dataset has 30 sets of documents with sets defined by different types of criteria such as event set, opinion set, etc. Each set having around 10 documents constituting a total of 299 documents and 44551 pairs of documents which are used to detect the near-duplicates. Out of 44551 pairs of documents, 1350 pairs of documents are not near-duplicates as they are come from same clusters. The remaining 43201 pairs of documents are used as near-duplicates.

4.3 DUC-2005 Dataset

This dataset has 50 document sets. Each set having 25–50 documents constituting a total of 1503 documents and 1128753 pairs of documents which are used to detect the near-duplicates. Out of 1128753 pairs of documents, 23,250 pairs of documents are not near-duplicates as they are come from same sets. The remaining 1105503 pairs of documents are used as near-duplicates.

4.4 Experimental Setup

Parameter setting of different classifiers are as follows:

  i. *LinearSVM:* Cs = [0.001, 0.01, 0.1, 1, 10], gammas = [0.001, 0.01, 0.1, 1], param_grid = { 'C': Cs, 'gamma': gammas}
 ii. *Multinomial NB:* Prior Probabilities = [0.65, 0.35], alpha = 0.1.
iii. *Gaussian NB:* alpha = 1.0, binarize = 0.0, class_prior = none, fit_prior=True
 iv. *Adaboost:* n_estimators = 12, learning_rate = 1, max_depth = 5, subsample = 0.5, random_state = 0,
  v. *Decision Trees:* min_samples_leaf=1, random_state = none, min_samples_split = 2, min _weight_fraction_leaf = 0.0, presort=False, splitter = 'best'.

vi. *Random Forest:* bootstrap = True, min_samples_leaf = 1, min_samples_split = 2, min_weight_fraction_leaf = 0.0, n_estimators = 10, n_jobs = 1, oob_score = False.

vii. The number of hidden layers for Multilayer ELM is set as three for DUC-2000, 2001 and 2002 datasets whereas it is set as five for DUC-2005 dataset[8]. The number of hidden layer nodes for ELM and Multilayer ELM are set as 250 on all datasets.

viii. The training feature vector length is 11. The size of the length of the feature vectors of all the classifiers for top 1%, 5%, and 10% (the value of m as discussed in step 4 of Sect. 3) are decided accordingly.

4.5   Parameters for Performance Evaluation

i. *Precision (p)* can be defined as follows.

$$p = \frac{|A_d| \cap |I_d|}{|I_d|}$$

where I_d represents the number of documents identified by the proposed approach as near-duplicate and A_d is the actual number of near-duplicate documents present in the given corpus.

ii. *Recall (r)* can be represent as follows where all the symbols have same meaning as defined above.

$$r = \frac{|A_d| \cap |I_d|}{|A_d|}$$

iii. *Accuracy (a)* is measured as follows:

$$a = \frac{tp + tn}{N}$$

where,

tp: number of documents retrieved by the approach and those are near-duplicate,

fp: number of documents retrieved by the approach and those are not near-duplicate,

tn: number of documents not retrieved by the approach and those are not near-duplicate,

fn: number of documents not retrieved by the approach and those are near-duplicate.

$N = tp + fp + tn + fn$.

4.6   Discussion

The performance of proposed *CBFS* technique is compared with the conventional techniques and are shown in Tables 3, 4, 5, 6, 7, 8, 9, 10, 11, 12, 13 and 14 where bold

---

[8] The decision is taken based on the experiment on which the best results are obtained.

Table 3. Top 1% (DUC-2000)

Classifier	Chi-square	BNS	MI	IG	CBFS
LinearSVC	0.89120	0.87162	0.87360	0.88948	0.87514
Linear SVM	0.89234	0.88967	0.89151	0.89444	0.89428
Gaussian NB	0.88227	0.86307	0.86992	0.87514	0.87782
Multinomial NB	0.86303	0.83860	0.83796	0.85794	**0.88663**
Random Forest	0.87322	0.88314	0.88381	0.88472	0.87335
Decision Trees	0.85993	0.85847	0.85902	0.84243	0.85584
Adaboost	0.89672	0.87606	0.88643	0.88235	0.88762
ELM	0.89242	0.87041	0.87542	0.89575	0.88420
Multilayer ELM	0.91703	0.91237	0.90511	0.91668	**0.93802**

Table 4. Top 5% (DUC-2000)

Classifier	Chi-square	BNS	MI	IG	CBFS
LinearSVC	0.93461	0.92818	0.92816	0.93014	0.93453
Linear SVM	0.94378	0.93462	0.93728	0.93374	**0.95508**
Gaussian NB	0.93516	0.88816	0.90252	0.92878	0.93314
Multinomial NB	0.93122	0.90103	0.91607	0.92511	0.92191
Random Forest	0.89075	0.88366	0.86261	0.87112	0.87184
Decision Trees	0.85995	0.86256	0.85811	0.85763	0.85618
Adaboost	0.90426	0.88001	0.90223	0.89423	0.89717
ELM	0.93550	0.92674	0.93011	0.93683	0.92332
Multilayer ELM	0.93873	0.94882	0.94664	0.95584	**0.96746**

Table 5. Top 10% (DUC-2000)

Classifier	Chi-square	BNS	MI	IG	CBFS
LinearSVC	0.94743	0.93733	0.93649	0.94376	**0.94922**
Linear SVM	0.94285	0.94558	0.93646	0.94653	0.94537
Gaussian NB	0.93996	0.91012	0.93353	0.93992	0.91133
Multinomial NB	0.93822	0.92344	0.93272	0.93732	0.91937
Random Forest	0.88266	0.86261	0.86341	0.87253	0.87686
Decision Trees	0.86373	0.84924	0.86607	0.85913	0.85642
Adaboost	0.89294	0.89243	0.89471	0.89274	**0.90573**
ELM	0.93229	0.94052	0.93442	0.94673	0.92887
Multilayer ELM	0.95635	0.96881	0.95566	0.95813	**0.96927**

Table 6. Top 1% (DUC-2001)

Classifier	Chi-square	BNS	MI	IG	CBFS
LinearSVC	0.91492	0.88287	0.89941	0.91785	0.90145
Linear SVM	0.92633	0.90662	0.92184	0.92798	0.91671
Gaussian NB	0.83281	0.79881	0.87911	0.86601	0.83021
Multinomial NB	0.84197	0.76852	0.80705	0.85317	**0.88163**
Random Forest	0.89162	0.88094	0.88437	0.88986	0.88393
Decision Trees	0.86046	0.84001	0.85312	0.86365	0.85874
Adaboost	0.91773	0.89212	0.91534	0.91801	0.88714
Multilayer ELM	0.94651	0.92222	0.91885	0.94563	**0.95708**

Table 7. Top 5% (DUC-2001)

Classifier	Chi-square	BNS	MI	IG	CBFS
LinearSVC)	0.94596	0.92333	0.94105	0.94302	**0.95655**
Linear SVM	0.96511	0.93916	0.94052	0.94794	0.96021
Gaussian NB	0.92632	0.90763	0.91492	0.90218	0.90987
Multinomial NB	0.93814	0.92327	0.93095	0.94525	**0.94667**
Random Forest	0.87345	0.88184	0.84973	0.85482	0.84585
Decision Trees	0.84927	0.84845	0.84887	0.85554	0.84842
Adaboost	0.91893	0.91891	0.91656	0.91921	0.89554
ELM	0.94533	0.92522	0.94677	0.94177	0.94575
Multilayer ELM	0.96668	0.94888	0.96253	0.96548	0.96541

Table 8. Top 10% (DUC-2001)

Classifier	Chi-square	BNS	MI	IG	CBFS
LinearSVC	0.94547	0.92125	0.92338	0.96869	0.96551
Linear SVM	0.94542	0.92652	0.92758	0.97287	0.96768
Gaussian NB	0.92031	0.90688	0.91548	0.91054	0.89722
Multinomial NB	0.95097	0.94561	0.94778	0.95352	**0.96918**
Random Forest	0.85482	0.85482	0.85482	0.84587	0.84587
Decision Trees	0.85071	0.84738	0.85208	0.84574	0.85082
Adaboost	0.91526	0.92682	0.91857	0.91437	0.91672
ELM	0.92334	0.94674	0.95636	0.96944	0.95631
Multilayer ELM	0.97107	0.95876	0.96888	0.97941	**0.98079**

results shows the highest F-measure achieved by a feature selection technique using the corresponding classifier (mentioned in the classifier column). It can be seen from the results that in most of the cases either the proposed *CBFS* technique outperform the traditional feature selection techniques or comparable with them. F-measure comparisons of ML-ELM with other classifiers on *CBFS* technique are shown in Table 15. Similarly, the performances of different classifiers for detecting duplicate pages on DUC datasets are shown in Tables 16, 17, 18 and 19. Figures 2 and 3 show the F-measure and accuracy comparisons of Multilayer ELM with other shallow learning classifiers. Similarly, we compared the performance of Multilayer ELM with the existing deep learning classifiers and the results are shown in Figs. 4 and 5 respectively. From the tables and figures,

Table 9. Top 1% (DUC-2002)

Classifier	Chi-square	BNS	MI	IG	CBFS
LinearSVC	0.93364	0.92336	0.92967	0.93971	0.91525
Linear SVM	0.93242	0.93918	0.93146	0.94952	0.94924
Gaussian NB	0.86441	0.85532	0.85344	0.85343	0.85147
Multinomial NB	0.87202	0.84181	0.85837	0.86014	**0.87526**
Random Forest	0.64006	0.64004	0.64002	0.76296	**0.77929**
Decision Trees	0.89168	0.88852	0.89003	0.89517	0.88447
Adaboost	0.92232	0.92944	0.92361	0.93203	0.90924
ELM	0.94566	0.95022	0.93142	0.93375	0.93326
Multilayer ELM	0.95413	0.95672	0.95679	0.96758	0.94602

Table 10. Top 5% (DUC-2002)

Classifier	Chi-square	BNS	MI	IG	CBFS
LinearSVC	0.95122	0.94781	0.94007	0.95424	0.94074
Linear SVM	0.95228	0.94387	0.94643	0.96554	**0.96889**
Gaussian NB	0.82963	0.87127	0.85762	0.85328	0.84876
Multinomial NB	0.90244	0.89552	0.90328	0.91178	0.90667
Random Forest	0.63822	0.65845	0.62866	0.67425	**0.73312**
Decision Trees	0.90288	0.90603	0.90374	0.90352	0.89667
Adaboost	0.92823	0.92323	0.93327	0.91607	0.91752
ELM	0.94566	0.94896	0.95606	0.93451	0.94453
Multilayer ELM	0.96394	0.96313	0.95836	0.96877	**0.96939**

Table 11. Top 10% (DUC-2002)

Classifier	Chi-square	BNS	MI	IG	CBFS
LinearSVC	0.95732	0.94172	0.94452	0.94694	0.95476
Linear SVM	0.95487	0.94617	0.94684	0.94818	**0.96782**
Gaussian NB	0.79521	0.84727	0.83482	0.81192	0.79144
Multinomial NB	0.90078	0.90556	0.90812	0.90972	0.88692
Random Forest	0.63822	0.63841	0.63427	0.63416	**0.64808**
Decision Trees	0.90556	0.89853	0.90682	0.89941	0.90652
Adaboost	0.91902	0.91692	0.91904	0.91981	0.91075
ELM	0.92333	0.94665	0.94541	0.95501	0.94555
Multilayer ELM	0.96722	0.95763	0.96776	0.96323	**0.96956**

Table 12. Top 1% (DUC-2005)

Classifier	Chi-square	BNS	MI	IG	CBFS
LinearSVC	0.84018	0.81563	0.83201	0.77057	0.81837
Linear SVM	0.85637	0.82417	0.84035	0.81685	**0.87784**
Gaussian NB	0.65441	0.63747	0.67972	0.53565	0.63932
Multinomial NB	0.68786	0.65052	0.66733	0.61532	0.63171
Random Forest	0.83837	0.81692	0.81552	0.79221	**0.84301**
Decision Trees	0.79122	0.77398	0.77502	0.74937	0.76598
Adaboost	0.84677	0.82845	0.83197	0.81567	0.81503
ELM	0.81266	0.80383	0.84785	0.78942	0.80535
Multilayer ELM	0.87503	0.85366	0.86634	0.83654	0.84519

Table 13. Top 5% (DUC-2005)

Classifier	Chi-square	BNS	MI	IG	CBFS
LinearSVC	0.87807	0.86862	0.87753	0.88152	0.86491
Linear SVM	0.87913	0.88137	0.88866	0.88722	0.85949
Gaussian NB	0.70749	0.74333	0.73445	0.70754	0.71887
Multinomial NB	0.77866	0.75184	0.76398	0.75469	0.76444
Random Forest	0.79832	0.81162	0.81848	0.80972	**0.82283**
Decision Trees	0.78482	0.78762	0.79158	0.78827	0.77701
Adaboost	0.83867	0.82968	0.84828	0.84488	0.83075
ELM	0.86722	0.83252	0.84514	0.87525	0.84676
Multilayer ELM	0.89674	0.88569	0.90035	0.90676	**0.90912**

Table 14. Top 10% (DUC-2005)

Classifier	Chi-square	BNS	MI	IG	CBFS
LinearSVC	0.88881	0.88006	0.86353	0.88186	0.87572
Linear SVM	0.87053	0.88834	0.86881	0.88044	0.87601
Gaussian NB	0.70036	0.73239	0.71893	0.70537	0.70294
Multinomial NB	0.78502	0.77861	0.78698	0.77904	**0.79622**
Random Forest	0.81145	0.79832	0.79835	0.80098	0.80889
Decision Trees	0.78081	0.79573	0.79055	0.78478	0.77785
Adaboost	0.83461	0.83171	0.83143	0.82886	0.81984
ELM	0.86623	0.88767	0.88011	0.87562	0.84249
Multilayer ELM	0.90234	0.90431	0.90535	0.90986	**0.91277**

it can be observed that F-measure and accuracy of ML-ELM are better than other established classifiers. For experimental purposes, we have shown similarity scores of 15 documents (DUC-2001) in Table 20 and top 15 documents scores among all document of DUC-2001 in Table 21. Similarly, the semantic similarity scores of 15 documents (DUC-2001) are shown in Table 22. The obtained results of the proposed approach on DUC datasets suggested that Multilayer ELM is well suit for near-duplicate document detection in comparison with other established classifiers. Some of the plausible reasons for ML-ELM outperforming other shallow and deep learning classifiers are as follows:

i. one best quality in Multilayer ELM is that it can easily represents the training feature vector in a high dimensional space and there is no back propagation.

Table 15. Comparing multilayer ELM with other classifiers using *CBFS*

Classifier	DUC-2000			DUC-2001			DUC-2002			DUC-2005		
	1%	5%	10%	1%	5%	10%	1%	5%	10%	1%	5%	10%
LinearSVC	87.514	93.454	94.921	90.146	95.655	96.551	91.521	94.074	95.476	81.837	86.491	87.572
Linear SVM	89.428	95.509	94.536	91.670	96.021	96.768	**94.924**	96.889	96.782	**87.784**	85.949	87.601
Gaussian NB	87.782	93.314	91.133	83.021	90.987	89.722	85.147	84.876	79.144	63.932	71.887	70.294
Multinomial NB	88.662	92.191	91.937	88.163	94.667	96.918	87.526	90.667	88.692	63.171	76.444	79.622
Random Forest	87.335	87.184	87.686	88.394	84.586	84.586	77.928	73.314	64.800	84.302	82.282	80.883
Decision Trees	85.584	85.618	85.642	85.874	84.842	85.082	88.447	89.667	90.652	76.598	77.702	77.785
Adaboost	88.762	89.717	90.573	88.714	89.554	91.672	90.924	91.752	91.075	81.503	83.075	81.984
ELM	88.420	92.332	92.887	90.182	94.575	95.631	93.326	94.453	94.555	80.535	84.676	84.249
Multilayer ELM	**93.802**	**96.746**	**96.927**	**95.707**	**96.541**	**98.079**	94.602	**96.939**	**96.956**	84.519	**90.912**	**91.277**

Table 16. DUC-2000

Classifier	DUC-2000			
	Precision	Recall	F-Measure	Accuracy
ELM	73.65	76.46	75.03	86.67
Multilayer ELM	**83.81**	**81.36**	**82.57**	**95.67**
LinearSVC	77.43	79.64	78.52	85.68
Linear SVM	71.45	69.14	70.28	79.23
*k*NN	54.87	56.49	55.67	67.35
Gaussian NB	60.97	62.38	61.67	73.28
Multinomial NB	66.56	68.01	67.28	76.52
Bernoulli NB	60.29	56.47	58.32	65.35
Decision Trees	56.73	52.47	54.52	67.67
Random Forest	68.30	70.49	69.38	80.35
Gradient Boosting	66.69	64.00	65.32	76.42
Extra Trees	70.56	70.48	70.52	79.37

Table 17. DUC-2001

Classifier	DUC-2001			
	Precision	Recall	F-Measure	Accuracy
ELM	80.53	82.42	81.46	90.23
Multilayer ELM	**84.76**	**86.98**	**85.86**	**95.86**
Linear SVC	80.98	82.51	81.74	92.89
Linear SVM	72.34	70.42	71.37	76.57
*k*NN	50.65	52.11	51.37	63.46
Gaussian NB	61.27	65.61	63.37	72.71
Multinomial NB	64.54	63.11	63.82	74.75
Bernoulli NB	58.34	56.72	57.52	68.32
Decision Trees	48.98	49.58	49.28	61.38
Random Forest	65.51	61.64	63.52	73.43
Gradient Boosting	63.65	64.71	64.18	78.52
Extra Trees	59.45	63.22	61.28	74.38

Table 18. DUC-2002

Classifier	DUC-2002			
	Precision	Recall	F-Measure	Accuracy
ELM	79.43	77.92	78.67	91.46
Multilayer ELM	81.49	**85.13**	**83.27**	**97.32**
LinearSVC	**81.89**	79.86	80.86	93.67
Linear SVM	70.56	68.18	69.35	75.42
*k*NN	51.64	55.21	53.37	70.03
Gaussian NB	59.87	60.77	60.32	73.23
Multinomial NB	61.87	62.67	62.27	77.27
Bernoulli NB	57.93	55.27	56.57	71.18
Decision Trees	55.87	55.25	55.65	72.23
Random Forest	69.40	71.24	70.31	77.46
Gradient Boosting	61.26	63.72	62.47	73.38
Extra Trees	67.89	63.59	65.67	75.08

Table 19. DUC-2005

Classifier	DUC-2005			
	Precision	Recall	F-Measure	Accuracy
ELM	75.45	78.18	76.79	90.35
Multilayer ELM	**80.29**	**83.41**	**81.82**	95.32
LinearSVC	78.43	76.92	77.67	**95.46**
Linear SVM	69.56	67.18	68.35	69.72
*k*NN	54.64	52.15	53.37	64.43
Gaussian NB	65.87	61.05	63.37	68.83
Multinomial NB	63.87	58.81	61.24	77.65
Bernoulli NB	54.92	54.44	54.68	70.38
Decision Trees	56.87	58.67	57.76	67.23
Random Forest	75.34	73.89	74.61	77.46
Gradient Boosting	64.26	62.83	63.54	83.68
Extra Trees	62.39	69.76	65.87	81.48

Table 20. Similarity scores (DUC-2001)

Document name	Similarity score
SJMN91-06184003	0.5808
SJMN91-06184088	0.5909
SJMN91-06187248	0.5566
SJMN91-06254192	0.5626
SJMN91-06290185	0.5783
WSJ910702-0078	0.5385
WSJ910702-0086	0.5582
WSJ910910-0071	0.5457
WSJ911011-0071	0.5757
WSJ911016-0124	0.5466
WSJ911016-0126	0.5747
AP880908-0264	0.5134
AP890616-0192	0.5179
AP901213-0236	0.5133
LA012689-0139	0.5605

Table 21. Top 15 documents (DUC-2001)

Rank	Document name	Similarity score
1	LA031790-0064	0.6070
2	WSJ920302-0142	0.4022
3	AP890901-0154	0.4120
4	AP880509-0036	0.4354
5	SJMN91-06130055	0.4139
6	AP900207-0041	0.4871
7	SJMN91-06324032	0.4682
8	AP880714-0187	0.4079
9	AP890123-0150	0.4492
10	FT931-10514	0.4901
11	AP880909-0135	0.4270
12	AP891005-0230	0.4031
13	FBIS4-23474	0.4375
14	AP881109-0149	0.4609
15	LA110490-0034	0.4555

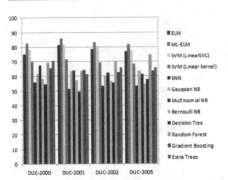

Fig. 2. F-measure (shallow learning)

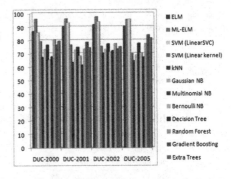

Fig. 3. Accuracy (shallow learning)

ii. able to manage large dataset and extremely fast learning speed.

iii. a better hierarchical or representation learning is generated due to multiple layers and hence a different form of the input is learn by each layer within the network.

iv. using multiple layers, Multilayer ELM obtained higher level abstraction, where as the networks having single layer are fails to achieve it.

v. by considering the number of nodes in the hidden layer more than the input layer, the feature vector is mapped to an extended space. This produces a very good performance of Multilayer ELM on all datasets.

vi. no iteration takes place during the training time and hence unsupervised training takes place very fast.

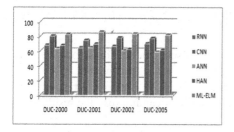

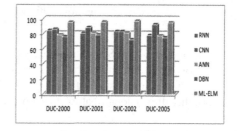

Fig. 4. F-measure (deep learning)   **Fig. 5.** Accuracy (deep learning)

Table 22. Semantic similarity scores (DUC-2001)

First document	Second document	Word2Vec	WordNet	Topic-based	NGD
WSJ911004-0115	WSJ880510-0079	0.6557	0.3192	0.4409	0.5431
AP900816-0166	AP890127-0271	0.7493	0.4763	0.3023	0.4561
SJMN91-06267062	WSJ870804-0097	0.6558	0.4812	0.3023	0.3014
AP890527-0001	AP880316-0061	0.4997	0.5111	0.3023	0.3967
FT922-10715	FT924-12193	0.5309	0.4782	0.3023	0.4385
AP891005-0216	LA113089-0118	0.6245	0.3998	0.7415	0.5118
AP900514-0011	LA112389-0104	0.5308	0.4958	0.3828	0.7103
SJMN91-06041121	SJMN91-06067177	0.7184	0.5008	0.4013	0.3311
SJMN91-06294205	AP890525-0127	0.7183	0.4506	0.4305	0.3431
SJMN91-06254192	AP891125-0090	0.7495	0.4640	0.3023	0.5643
AP880316-0208	AP891125-0090	0.5308	0.4436	0.3023	0.6113
AP891005-0230	WSJ891026-0063	0.2187	0.4237	0.9999	0.4667
SJMN91-06270171	SJMN91-06165057	0.6558	0.4103	0.3009	0.3001
LA102189-0067	AP881022-0011	0.5307	0.4193	0.7087	0.3144
WSJ910620-0075	AP900816-0139	0.6558	0.5466	0.3023	0.6113

5   Conclusion

The paper suggested a near-duplicate document detection approach using Multilayer ELM which is briefed as follows: Initially, the corpus is divided into k clusters. Next, top $m\%$ terms are identified in each cluster using term-term correlation technique and a new corpus of documents are generated. Eleven similarity scores are calculated using different similarity measures. By combining all similarity scores of each pair of documents and their corresponding class label, the training feature vector is generated. The performance of different classifiers are quantified by comparing the predicted class label with the known test label. Finally, all the duplicate documents are ranked using a heuristic approach. Experiment clearly demonstrated that Multilayer ELM provide significant enhancement compared to other deep and shallow classifiers in the domain of duplicate document detection. This work further can be extended on the following lines:

i. by adding Latent Semantic Indexing for checking semantic relationship that exists between the documents before the training feature vector get generate.

ii. by running other state-of-the-art classifiers such as SVM, Random Forest etc., on higher dimensional feature space of Multilayer-ELM can improve the results further.

iii. an alternate approach to rank the documents is to generate a query from the semantic content of the documents of the corpus and this query can be used to determine the similarity among the documents. Different techniques can be developed as the future work to identify and optimize such query.

References

1. Manber, U.: Finding similar files in a large file system. In: Proceedings of the USENIX Winter 1994 Technical Conference on USENIX Winter 1994, vol. 94, pp. 1–10 (1994)
2. Bharat, K., Broder, A.: Mirror, mirror on the web: a study of host pairs with replicated content. Comput. Netw. 31(11), 1579–1590 (1999)
3. Oghbaie, M., Zanjireh, M.M.: Pairwise document similarity measure based on present term set. J. Big Data 5(1), 52 (2018)
4. Asghar, K., Habib, Z., Hussain, M.: Copy-move and splicing image forgery detection and localization techniques: a review. Aust. J. Forensic Sci. 49(3), 281–307 (2017)
5. Roul, R.K., Mittal, S., Joshi, P.: Efficient approach for near duplicate document detection using textual and conceptual based techniques. In: Kumar Kundu, M., Mohapatra, D.P., Konar, A., Chakraborty, A. (eds.) Advanced Computing, Networking and Informatics- Volume 1. SIST, vol. 27, pp. 195–203. Springer, Cham (2014). https://doi.org/10.1007/978-3-319-07353-8_23
6. Shah, D.J., Lei, T., Moschitti, A., Romeo, S., Nakov, P.: Adversarial domain adaptation for duplicate question detection. In: Proceedings of the 2018 Conference on Empirical Methods in Natural Language Processing, Association for Computational Linguistics, pp. 1056–1063 (2018)
7. Hassanian-esfahani, R., Kargar, M.-J.: Sectional MinHash for near-duplicate detection. Expert Syst. Appl. 99, 203–212 (2018)
8. Zhou, Z., Yang, C.-N., Chen, B., Sun, X., Liu, Q., Wu, Q.M.J.: Effective and efficient image copy detection with resistance to arbitrary rotation. IEICE Trans. Inf. Syst. 99(6), 1531–1540 (2016)
9. Zhou, Z., Wu, Q.J., Sun, X.: Encoding multiple contextual clues for partial-duplicate image retrieval. Pattern Recogn. Lett. 109, 18–26 (2018)
10. Zhou, Z., Wu, Q.J., Huang, F., Sun, X.: Fast and accurate near-duplicate image elimination for visual sensor networks. Int. J. Distrib. Sens. Netw. 13(2), 1–12 (2017)
11. Zhou, Z., Mu, Y., Wu, Q.J.: Coverless image steganography using partial-duplicate image retrieval. Soft. Comput. 23, 1–12 (2018)
12. Yang, Y., Tian, Y., Huang, T.: Multiscale video sequence matching for near-duplicate detection and retrieval. Multimed. Tools Appl. 78(1), 311–336 (2019)
13. Ding, S., Zhang, N., Xu, X., Guo, L., Zhang, J.: Deep extreme learning machine and its application in EEG classification. Math. Probl. Eng. 2015, 1–11 (2015)

14. Roul, R.K., Asthana, S.R., Kumar, G.: Study on suitability and importance of multilayer extreme learning machine for classification of text data. Soft. Comput. **21**(15), 4239–4256 (2017)
15. Roul, R.K.: Detecting spam web pages using multilayer extreme learning machine. Int. J. Big Data Intell. **5**(1–2), 49–61 (2018)
16. Kasun, L.L.C., Zhou, H., Huang, G.-B., Vong, C.M.: Representational learning with extreme learning machine for big data. IEEE Intell. Syst. **28**(6), 31–34 (2013)
17. Roul, R.K., Sahoo, J.K., Goel, R.: Deep learning in the domain of multi-document text summarization. In: Shankar, B.U., Ghosh, K., Mandal, D.P., Ray, S.S., Zhang, D., Pal, S.K. (eds.) PReMI 2017. LNCS, vol. 10597, pp. 575–581. Springer, Cham (2017). https://doi.org/10.1007/978-3-319-69900-4_73
18. Huang, G.-B., Zhou, H., Ding, X., Zhang, R.: Extreme learning machine for regression and multiclass classification. IEEE Trans. Syst. Man Cybern. Part B Cybern. **42**(2), 513–529 (2012)
19. Roul, R.K.: Suitability and importance of deep learning feature space in the domain of text categorisation. Int. J. Comput. Intell. Stud. **8**(1–2), 73–102 (2019)
20. Roul, R.K., Agarwal, A.: Feature space of deep learning and its importance: comparison of clustering techniques on the extended space of ML-ELM. In: Proceedings of the 9th Annual Meeting of the Forum for Information Retrieval Evaluation, pp. 25–28. ACM (2017)
21. Huang, G.-B., Chen, L., Siew, C.K., et al.: Universal approximation using incremental constructive feedforward networks with random hidden nodes. IEEE Trans. Neural Netw. **17**(4), 879–892 (2006)
22. Huang, G.-B., Chen, L.: Convex incremental extreme learning machine. Neurocomputing **70**(16), 3056–3062 (2007)
23. Roul, R.K., Rai, P.: A new feature selection technique combined with ELM feature space for text classification. In: Proceedings of the 13th International Conference on Natural Language Processing, pp. 285–292 (2016)
24. Roul, R.K., Bhalla, A., Srivastava, A.: Commonality-rarity score computation: a novel feature selection technique using extended feature space of ELM for text classification. In: Proceedings of the 8th Annual Meeting of the Forum on Information Retrieval Evaluation, pp. 37–41. ACM (2016)
25. Cilibrasi, R.L., Vitanyi, P.M.: The Google similarity distance. IEEE Trans. Knowl. Data Eng. **19**(3), 1–15 (2007)

Author Index

Printed in the United States
By Bookmasters